1977

may

N DA

THE ECOLOGY OF THE SEAS

THE ECOLOGY OF THE SEAS

EDITED BY D.H. CUSHING

AND J.J. WALSH

W.B. SAUNDERS COMPANY

PHILADELPHIA TORONTO

North and South American
distribution rights assigned to

W.B. SAUNDERS COMPANY
West Washington Square, Philadelphia, Pa 19105
833 Oxford Street, Toronto M6Z 5T9, Canada

Library of Congress Catalog Card Number: 76–9641

SBN 0–7216–2812–5

Printed in Great Britain

Contents

Preface ix

1 Introduction *by D.H. Cushing* 1

Part I. The Sea and the Organisms that Live in It

2 Waters of the Sea: The Ocean's Characteristics and Circulation *by R.L. Smith* 23

3 Plants and Animals of the Sea *by T. Wyatt* 59

Part II. The Structure of Life in the Sea

4 The Structure of Life in the Sea *by T.R. Parsons* 81

5 Patchiness *by J.H. Steele* 98

6 Vertical Migration *by A.R. Longhurst* 116

Part III. Functions in the Marine Ecosystem

7 Nutrient Cycles *by R.C. Dugdale* 141

8 Primary Production in the Sea *by C.J. Lorenzen* 173

9 Herbivore Production *by D.J. Tranter* 186

10 Production on the Bottom of the Sea *by K.H. Mann* 225

11 Growth of Fishes *by R. Jones* 251

Part IV. Yield from the Sea

12 Production and Catches of Fish in the Sea *by J.A. Gulland* 283

Part V. Evolutionary Consequences

13 Biology of Fishes in the Pelagic Community *by D.H. Cushing* 317

14 Food Chains in the Sea *by T. Wyatt* 341

Part VI. Theory

15 Sampling the Sea *by J.C. Kelley* 361

16 Models of the Sea *by J.J. Walsh* 388

References 409

Author Index 447

Subject Index 455

Index of Plants and Animals 463

Geographical Index 466

Contributors

D. H. CUSHING, Ministry of Agriculture, Fisheries and Food, Fisheries Laboratory, Lowestoft, Suffolk, U.K.

R. C. DUGDALE, Bigelow Laboratory for Ocean Sciences, McKown Point, West Boothbay Harbor, Maine, 04575, U.S.A.

J. A. GULLAND, Fisheries Department, FAO, Via delle Terme di Caracalla, Rome, Italy.

R. JONES, Department of Agriculture and Fisheries for Scotland, Marine Laboratory, P.O. Box 101, Victoria Road, Aberdeen AB9 8DB, U.K.

J. C. KELLEY, School of Natural Sciences, San Francisco State College, San Francisco, California 94132, U.S.A.

A. R. LONGHURST, Institute of Marine and Environmental Research, 67–9 Citadel Road, Plymouth PL1 3DH, U.K.

C. J. LORENZEN, Department of Oceanography, University of Washington, Seattle, Washington 98105, U.S.A.

K. H. MANN, Marine Ecology Laboratory, New Bedford Institute, Dartmouth, Nova Scotia, Canada.

T. R. PARSONS, Department of Oceanography, University of British Columbia, Vancouver 8, British Columbia, Canada.

R. L. SMITH, Department of Oceanography, Oregon State University, Corvallis, Oregon 97331, U.S.A.

J. H. STEELE, Department of Agriculture and Fisheries for Scotland, Marine Laboratory, P.O. Box 101, Victoria Road, Aberdeen AB9 8DB, U.K.

D. J. TRANTER, C.S.I.R.O., Cronulla, N.S.W. 2006, Australia.

J. J. WALSH, Division of Oceanographic Sciences, Brookhaven National Laboratory, Upton, New York, 11973, U.S.A.

T. WYATT, Ministry of Agriculture, Fisheries and Food, Fisheries Laboratory, Lowestoft, Suffolk, U.K.

Preface

Marine ecology is a diverse subject composed of many facts, many concepts and few testable theories. Any ecological study is an inventory, a budget, or a statement of account. Much natural history expounds an inventory such as Cuvier's *Règne Animal,* White's *Natural History of Selborne* or Fisher's more recent study of fulmars in the North-east Atlantic, and indeed, such accounts reveal much and yield considerable material for speculation. However, we feel that marine ecology has now emerged from its preliminary descriptive phase and that quantification of the interactions of an organism with its environment is slowly leading to refutable hypotheses.

The works of Lohmann and others in the Baltic at the turn of the century and of Birge and Juday in Lake Mendota in the 1920s and 1930s were quantitative, as were the works of the early fisheries biologists, but, quantitative collections are not enough to yield an understanding of underlying ecological processes. Hypotheses are needed for this purpose, the more complex ones being called models; if they are small enough they can be tested and refuted. The notable models created by Riley nearly thirty years ago differ little in basic structure from the plethora being developed today. However, there is now much more information on the nature of the parameters used and these can be tested independently within the general scientific framework. This development was largely due to John Strickland in the decade before he died.

Many authors have contributed to this text and all are quantitative specialists in particular fields of marine ecology. Not all facts and concepts are discussed, but the present state of knowledge of the quantitative nature of marine ecosystems is described. Although a single author might have put a case more effectively than we have done, we believe that our multi-authored text is more broadly based than a single author text could have been, and has a much more extensive survey of the recent or relevant literature. We have solved the problem of cohesion in this text by grouping chapters into sections, each section being preceded by an introductory paragraph summarizing the contents of the section, and linking it to its predecessor.

The first chapter serves as an introduction to the book, expressing in part our attitude towards ecology as a numerate study of organisms, their populations, and their ecosystems, and tracing the development of a somewhat selected history. It consists of those ideas that have led the science towards the discipline which emerges in the structure of the text. Such points of discipline are possibly more important than the detailed structure of models with which we might have filled the text. However, only one is on display and this is important in that the biological processes described are successfully set within a hydrodynamic framework, but, more than that, it is a powerful reminder that the text was written for scientists who are starting to work from ships by others who have spent their working lives at sea.

Lowestoft, 1975 D.H. Cushing
Long Island, 1975 J.J. Walsh

Chapter 1. Introduction

D. H. Cushing

The nature of ecology

The study of populations

The study of ecosystems

The development of marine ecology
The development of description
The development of techniques
The development of concepts

The development of marine ecology *continued*
Nutrient limitation
The production cycle
Grazing in the production cycle
The migration of fish
Fish population dynamics

Conclusion

The nature of ecology

The Greek word *oikos* means 'house' and the derived word 'ecology' ('The study of the house') has connotations of 'home', 'niche', 'territory' and even 'housekeeping'. This term was introduced by Haeckel (1866) who wrote that 'an individual was the product of cooperation between environment and heredity'; such a pervasive definition embraces much of the current experimental and environmental work in the field. Later there was a separation into the ecology of individuals (autecology) and that of populations (synecology), which today is not insisted upon too vigorously. Park (1963) wrote that there were four elements in the study of ecology, the individual, the population, the group of populations and lastly the community (or ecosystem). This very complete definition does not really conflict with Haeckel's, if one admits that individuals are produced as a result of population processes.

Some of the earliest work on aquatic ecology was carried out in lakes, a foretaste of similar work in the sea. For example, Forel studied the deep water fauna in a number of Swiss lakes (1884) in relation to light penetration, temperature, pressure and nutrient salts and showed that inter-community and environmental differences could be correlated. Another considerable study was that of Birge and Juday in Lakes Mendota and Monona (Wisconsin, U.S.A.) from the early decades of this century (a convenient summary is given in Juday 1940); and much of the work of G. A. Riley (e.g. 1946) in the sea is linked historically to that of Lindeman (1942) and of Juday.

Any of Park's four components of ecology can be examined in the field on its own and much of the older study of natural history would now be described as ecology. The work of Philip Gosse (e.g. 1853) on the sea-shore and that of Darwin (1854) on barnacles are ecological studies; indeed many of the older works on the taxonomy of species described as they were taken from the sea, include a great deal of natural history (for example, Day's *Fishes of India*, 1876). There are many modern examples of such work, for instance, studies on luminescence in bathypelagic animals. Such animals were described in Murray and Hjort (1912), but in recent years with the development of photomultipliers, a zone has been discovered near the limit of the penetration of daylight at midday, that is lit entirely by the luminescent organs of the hunting animals that live there (Clarke & Denton 1962, Nicol 1967). The buoyancy of marine animals is another problem of the same character; an intense exploration of the physiology of buoyancy has been made by Denton and his colleagues which has revealed many points of purely ecological interest (e.g., Denton 1964). Perhaps the most remarkable buoyancy mechanism is the spermaceti organ of the sperm whale which allows the animal to dive to a thousand fathoms or more, at which great depths it can pick up squid, its preferred food, near the bottom (Clarke 1970). Ecologists, then, have had much to learn from earlier studies in natural history, the present science's forerunner.

Other examples of marine natural history of considerable interest to the ecologist include the study of the fur seal herd on the Pribilov Islands (Elliott 1884); the seals' mating behaviour on the beaches, by which the large, older males corral harems of females, provides a reserve of bachelors, or 'holluschickie' that supports the fishery. Kipling's story in *The Jungle Book* provides a useful summary of fur seal behaviour that derives directly from Elliott's account (Cushing 1971b). The detailed studies of barnacles in physiological and ecological terms (Crisp & Barnes 1954) yield a considerable body of information that can be used to interpret their ecology. The studies by Fryer and Iles (1972) on the fishes of Lake Tanganyika are the fullest to be based on a link between functional morphology and natural history and provide a model for the type of work that could well be carried out in analogous habitats in the sea. One such example is the Great Barrier Reef, where Savile Kent (1893) studied the exploration of the economic potentialities of the region, a community or ecosystem study in itself.

Ecological studies have since diverged into individual and population studies, an apparent return to Haeckel's definition. Lohmann's (1933) description of the house of *Oikopleura* and his studies of the food organisms found there revealed the existence in the sea of numbers of algal groups that had not always been recorded at all; indeed the search for a method of estimating the numbers of such flagellates in the sea continues today.

Particular behavioural responses are of considerable value in any study of predation. For example, the mechanisms of food location with electric organs have been known amongst the mormyrids for some time (Lissmann 1958), but it has recently been suggested that the same principle might be part of a predator's armament (Kalmijn 1971): certain dogfish are equipped to search for plaice buried in the sand with an electric field location technique. Thus, problems of feeding and predation have been illuminated by studies of functional morphology and of physiology.

Population analysis is one of the main ecological techniques, but sometimes more is revealed than the mere estimates of parameters. Merriman (1941) noted that 492 striped bass were released into San Francisco Bay in 1881 and that within two decades the annual catch had reached 500 tons. Not only was this a simple description of colonization, it also showed that the increment in a population can reach many orders of magnitude given such an opportunity for expansion. Expressed by a logistic curve, the theory of balance requires that the two exponents, net rate of increase and saturation level, should be approximately equal, if the population is to remain in a steady state. In the first generations subsequent to the release of the fish, the net rate of increase would have been very high. Then as it tended to match the saturation level, this rate would have decreased.

The same techniques can be used to show ecological trends, slow departures from a steady state over a period of time and how such trends depend on climatic variation. Ottestad (1942) related cod catches in the Vestfjord in northern Norway to periodicities in climate as indicated by width of rings on the pine trees in the same area. More recently Dickson (1971) has described the climatic changes in the north-east Atlantic in physical terms and such information provides a firm basis for an extension of the type of work started by Ottestad. Results may be expressed as a variation in one or both the exponents of the logistic curve or deduced from a study of the factors that determine recruitment, as indicated in a later chapter on the pelagic community (see Chapter 13).

From the study of the biology of fishes in that chapter, a fish population appears to 'summarize' the annual explorations of the environment made by its recruiting cohorts and to insulate itself to some degree from the harsher variability to which it might be subject. For example, the Arcto-Norwegian cod stock has spawned in the Vestfjord fairly steadily at least since the twelfth century, the period of earliest records. In contrast, the Atlanto-Scandian herring fishery has appeared and disappeared in a periodic manner since the fourteenth century in such a way as to imply stock changes of several orders of magnitude. It is now well known that the climate changes in a periodic manner (Lamb 1972) and it appears that the herring stock can rectify such changes by altering its net rate of increase profoundly at particular stages in the climatic cycle. A similar conclusion on a much broader scale emerges from a study of the Russell cycle (Russell & Demir 1971) in which it was shown that the whole ecosystem in the western English Channel underwent far-reaching changes in 1930/1 which reversed about forty years later; winter phosphorus rose by a third and macroplankton by a factor of four. A famous principle in human physiology was stated by Claude Bernard; the constancy of the internal environment was the condition for free and independent life. To be secure from the vagaries of the environment is a physiological requirement of the individual. Perhaps an analogous homeostatic principle could be applied to populations, such that if a population cannot withstand the environmental changes as does the cod stock, it should respond to them, as does the herring.

Ecology today is an extensive field almost as all-embracing as biology itself. On the one hand, knowledge of the genetics and the physiology of its individuals may be needed in a population study. On the other hand, an ecosystem, or an assemblage of communities, expressed partly in energy and to some extent in numbers, requires an analysis in which populations are lumped and are assumed to remain in a steady state. Any ecological study may limit itself to that of populations or cover the whole ecosystem.

In the sea, as opposed to the land, there are certain characteristics that become imposed on the material collected. The sea is a relatively uniform environment although there are of course considerable differences in irradiance according to depth, time of day and season. Other differences can be measured across considerable distances; although a cod can detect differences of about 1°C in a kilometre or so off Bear Island in June (Lee 1952), most differences in temperature or salinity of this degree are only found across distances of some tens or hundreds of kilometres. It is true that sharper boundaries are found in estuaries, in the Baltic or on the edge of the Kuroshio current (where a siome is found where the boundary rumbles and hisses and can be heard from the deck of a ship; Uda 1937); but these are exceptional.

Systems in the sea are studied on a broad scale, partly because the rather uniform environment requires it, but also because the current systems and the contained fish populations can only be assessed if the scale of search at sea is in hundreds of kilometres. Indeed, because of this scale of study, populations and ecosystems may be examined almost alongside each other. Another important characteristic of the sea is the pattern of turbulence that governs regional differences in productivity and local differences in the onset of production. Again, the maximization of biomass in a cohort or generation may depend in detail upon the variability of food in time and perhaps other environmental variations. Such variability depends upon the differences in rates of death and reproduction, but also upon physical variability; for example, the distribution of algae has been shown to reflect such variation in some detail (Platt 1972), indeed on a very small scale. One might imagine that physical variability could affect any predator–prey system on two scales, that of the food and that of the predator itself and the difference between them in distance might be as much as two orders of magnitude or more.

Although marine ecology as studied at sea is a time-consuming and sometimes arduous discipline it is also a rewarding one, for a number of reasons. The first is that some of the systems are relatively simple and the environment is more or less uniform and therefore easier to study. Thus systems can be examined on a broad scale. Indeed, the cost of ships presupposes this; to master the weather, ships must be large, and thus expensive. The third reason is that it may soon be possible to combine models in physical and biological terms (see Chapter 16) and one can perhaps foresee a time when experiments can be made at sea and initial observations processed in a shipborne computer model to be tested on the spot.

The study of populations

In its broadest sense, a population is a group of individuals, whether stars, motor cars or beetles. Populations of plants and animals are such groups that reproduce themselves. This is not to imply that any population lives in isolation or that a stock of fish has genetic integrity (see Chapter 13), although the work is much easier if measures of emigration or immigration are not needed. Since the study of populations is numerical, such estimates are in themselves a little difficult. The important point is that the study of populations is a numerical one; in contrast, that of an ecosystem may be purely an energetic one.

The actuarial study was originated by Laplace and Pascal and is now used by the population analyst who estimates the chances of birth and of death and produces stochastic models. These are associated with the name of Bartlett (1960), who also estimated the chances of extinction. Such methods are used to predict changes in human populations (although somewhat unsuccessfully at the present time in the developed countries because of contemporary changes in fertility). However, stochastic models, to be accurate, demand many more observations than are available to ecologists; indeed, they are as demanding as those of the physical sciences. Ecologists sometimes make use of the logistic curve (Pearl 1930), which describes the changes in numbers in a population in terms of its net rate of increase and the saturation level or carrying capacity of the environment. The curve is adequate in a descriptive way, but the adaptive mechanisms in the population remain hidden, for example, those responsible for stabilization.

An animal population is designed from an evolutionary point of view to obtain stable, or more or less stable numbers, and a stable structure, that is, a regularity in geographical position and in seasonal migration (see Chapter 13). There are three processes of importance in the maintenance of stable numbers, first, the production of adequate numbers of successfully feeding larvae (the potential recruitment), second, the control of numbers by density-dependent processes and third, the maintenance of competition from generation to generation. There are a number of ways of achieving this, but in general they may be grouped into two categories, the modification of reproduction and the modification of growth. In one sense the two are difficult to distinguish, because in larval and juvenile fishes the growth rates and death rates are linked, but it is possible that in the sea, the potential recruitment is progressively modified between hatching and recruitment. Many of the animals in the sea, including fish, are fecund and the system may well depend upon the maximization of biomass between hatching and maturity (see Chapter 13). It is interesting, however, to note that the top predators, marine mammals and sharks are not fecund at all, relying upon parental care in the one and upon very large yolk stores in the other for a high survival rate. Hence the end control in the ecosystem is independent of the maximization of biomass between hatching and recruitment at lower levels.

In fisheries biology, the study of populations has taken a somewhat different course in isolating the factors in age, that is, in establishing the components of recruitment, growth and mortality independently of the numbers in the population itself. With the use of the logistic curve, the net rate of increase can be established from the matrix of numbers in age and in time, but the carrying capacity of the environment can be estimated from the differences in stock, knowing the net rate of increase. Fisheries biologists can estimate growth and mortality independently of the stock so a model can be synthesized that does not depend on the stock numbers themselves and the logistic curve is tending to disappear in fisheries science. Then the adaptive nature of the population can be studied from the basis of a model, for example, in density-dependent growth and in density-dependent mortality. The responses of fish populations to climatic change, as noticed above, are examples of the immediate and adaptive changes that they can make.

There are two wings of progress in the general study of biology, first, the study of the individual, its physiology, behaviour and the genetic control of development at the molecular level or at the organ level. The second is the study of populations, individually or within ecosystems. The two are linked through genetics and evolutionary studies. Physiological and behavioural constants are widely used in the simulation models that are being developed by ecologists at the present time, but these constants are applied to whole animals, their oxygen consumptions or search fields; the physiology of organs, cells or molecules not being used in the study of populations. Similarly, population genetics is an extensive study important to the understanding of evolutionary events, but can rarely be used in ecology, although some attempts to produce models have been made; this is primarily because the time span of observational data is usually too short to detect any but the most trivial of genetic changes. In ecology the study of populations ignores genetic changes except in theoretical terms and concerns itself with the physiology and behaviour of whole animals, rather than their cells or organs.

The study of ecosystems

Dice (1952) defined an ecosystem as a community within its environment and from this concept developed the idea of energy flowing through the system. Birge and Juday expressed the compartments of the ecosystems in Lake Mendota in units of organic material, the ecosystem being considered as a productive machine, by which inorganic substances and energy are together elaborated into a finished product, measured in gC/m^2. There are obvious practical advantages to such an approach but, if expressed in energy units only, the whole system expresses a flow from the sun's energy through various channels to end in completely degraded forms, and the transfer from one compartment to another can be readily estimated in terms of efficiencies. The engine of production is considered in ideal terms as the death of food animals contributing to the growth of predatory ones, provided that the predator and its food can be properly distinguished within the system being studied or equally well as a flux of material (for example, mass) or an element, usually carbon or nitrogen, as each is readily converted to energy. There is an advantage in using an element like nitrogen because it is readily measured in the sea and so can be used to keep a continuous budgetary-check on modelled processes; dry mass or carbon cannot be used in such a way.

Predator/prey oscillations are described in the Lotka–Volterra equations in which the reactions between the two affect the birth and death rates of each. When an oscillation is established it converges nicely, which means that the two populations can obtain stability in numbers within the system described. However, if there is more than one predator and more than one prey, an oscillation need not converge (Steele 1974), but Steele has suggested that convergence can occur if the sets of reactive constants (i.e. food consumption, etc.) do not vary too much. To put it another way such constants should be adaptive, continuously modifying themselves in time as suggested for the stabilizing mechanisms in a fish population (see Chapter 13). The productive engine obviously has an efficient governor or set of governors, by which variable quantities are stabilized. As indicated in later chapters the nature of the mechanisms is unknown although some attempts are being made to describe it.

Differences in stability are sometimes estimated in a rough and ready way with a diversity index. Such an index describes the sets of choices as energy passes up a food chain or through a food web. In the rough form usually used, one species represents a single choice and so if there are many choices, as for example amongst the large number of copepod species in the tropical ocean, the diversity index is high and, conversely, it is low amongst the few copepod species in the spring outburst in temperate waters. Hence the index is inversely related to food abundance; the food is not only scattered into more channels in the sparser diverse seas of the tropics but there is less of it available. Presumably, the diverse ecosystem is one in which the variance of food supply in each channel, or each unit of information, is low. As a consequence, competition itself might be low and so many species are allowed to survive. If the governor is continuously modified in time by variance in the food supply, it is probably proportional to abundance (or high variance in food generates high competition), which might explain the inverse relation between diversity index and abundance.

There are obviously many answers to the question of what stabilizes an ecosystem. At the level of a single population in the sea one answer is to express the stabilizing capacity of a fish stock as the ratio of density dependent mortality to total mortality between hatching and recruitment to the adult stock (effectively, Ricker's 1954 solution). In more general terms, Lack (1954) suggested that stabilization depended upon food and some of the present models of stabilization in fish populations are little more than elaborations of this theme. The question of stabilizing an ecosystem has only been examined in a general way, although May (1973) has explored the mathematics of the problem with considerable insight.

There is, however, another point of view. Each population in the system must stabilize itself to a given food level or carrying capacity. Then the ecosystem must acquire stability in numbers (or any unit of energy), merely in the stabilization mechanisms of its component populations. As each population maintains or modifies its biomass with respect to its competitors, the problem becomes that of the competitive sharing of available food. It has already been suggested that the essential mechanism in population control is the maximization of biomass within the cohort or generation; or to put it another way, the cumulated excess of growth over death during the reproductive part of the life cycle represents the success of that cohort or generation in passing on to its offspring the best endowment it can. Thus, the question of stability in an ecosystem as distinct from its component populations may not even arise.

Many ecosystems have been studied in energetic terms as an end in itself. The use of common units, the expression of growth increments in calories (rather than weight that includes fat or protein in different proportions of energy) and the derivation of various transfer coefficients are enough justification to validate the analysis of the system in these terms. However, only the productive engine is described in this model and to this extent the rough figures obtained are useful to a fisheries biologist who wishes to assess the quantity of material produced each year in the world ocean (Gulland 1970). Nevertheless, the ecosystem is probably not a superorganism with constraints and boundaries of its own or any shared facilities, although it has characteristics of its own such as the Eltonian pyramid of numbers and the sustenance of the food webs. The only mechanism that links the component populations into a larger unit is competition and in the sea this is perhaps a part of the adaptive maximization of biomass in each cohort or generation. Thus, if competitive advantage is obtained, then the structure of a trophic level is altered and hence the structure and nature of the ecosystem. There may be a sense in which competition between populations maintains what structure there is in an ecosystem and changes its nature from time to time.

The development of marine ecology

Any science develops in descriptions of the world, in techniques to advance such descriptions and in the establishment of relationships that summarize them and synthesize them. In an exact science like physics the most important part is the synthesis of relationships, the interconnecting set of rules into which any new concept is fitted. In a less-developed science like marine ecology such a structure is not so well established.

The development of description

Life in the sea has been described throughout history going back to Aristotle, who systematized the knowledge of fishermen. Much of the main marine exploration took place during the expeditions of the nineteenth century. The earlier ones were geographical or navigational, for example to discover the North West passage, and indeed those in the Antarctic and Southern Ocean in the first decades of the present century also served this purpose. Animals and plants were always collected on such expeditions and formed the basis of much of our present knowledge of systematics. However, the *Challenger* expedition had the particular purpose of exploring the depths of the ocean, to look for animals there, because Forbes' conclusion that below 300 fms the waters were lifeless and empty was doubted with the discovery of animals on the transoceanic telegraph cables. A quite different form of exploration was started by Hensen (1911) on the Plankton expedition, when he found that the quantities of plankton were greater in temperate and high latitudes than in the subtropical seas. Hensen was a remarkable man in a number of ways. Not only did he design and make the first quantitative plankton net (shown to be 95 per cent efficient by Harvey, 1948), but he devised a method of estimating the production of fish from the numbers of their eggs in the sea. Lastly, when faced with the high variability of catches in his plankton net, he derived a contagious distribution to account for it. This form of quantitative exploration has continued with the *Discovery* investigations in the Southern Ocean and in upwelling areas, and in many expeditions all over the world ocean, for example, in the work of Zenkevitch on the epifauna on the sea bed in the seas of the U.S.S.R., in the recent American explorations like NORPAC, and EASTROPAC and the international combinations such as NORWESTLANT (organized by the International Council for the Exploration of the Sea) and the International Indian Ocean expedition.

In recent years, a new phylum has been found in the sea, the Pogonophora (Ivanov 1955). Benthic animals have been seen and caught on the sea bed in the deepest trench (Zenkevitch *et al.* 1954, 1955). New fishes have been discovered in the bathypelagic zone in the deep ocean (Bertelsen 1957). Large midwater trawls have sampled the mesopelagic ocean revealing new species there (Krefft 1968). The rarest fishes till the time of the Kon-Tiki expedition were eaten on that voyage (Heyerdahl 1950) and are now caught readily with large midwater trawls. Lastly, the great expansion of the fishing fleets across the world ocean has extended our knowledge of the distribution of stock densities of commercial fishes enormously. Although new animals will still be found (like the coelacanth), and better estimates of quantity will be made for a long time to come, there is a sense in which much of the exploration has been completed.

One of the results of these explorations was the discovery of biological structures which corresponded with the oceanographic structures being discovered at the same time; Cleve (1895–7) distinguished plankton communities (for example characterized by the diatom *Rhizosolenia* or by *Chaetoceros didymus*) in the North Atlantic, in much the same way as fossils indicate the strata in which they are found. Recently, the plankton recorder network across the North Atlantic has shown that differences between animals occur across enormous distances in that ocean, probably associated with extensive physical structures (Edinburgh Oceanographic Laboratory 1973). In a rather similar way, the upwelling systems of offshore divergences have characteristic communities or ecosystems because the zone of high production extends offshore for hundreds of kilometres (Cushing 1971a). In such ways the populations, communities and ecosystems have been described in terms that correspond to the physical structures in the sea.

The development of techniques

By far the most important developments in technique are those associated with capture. The gear used is often that developed by fishermen over the centuries and, indeed, the scientists still depend upon them to make their gear fully effective. The earlier expeditions used dredges, beam trawls and little midwater nets; today all the highly developed bottom trawls and large midwater trawls are used from stern trawlers to sample considerable depths in the sea, as was the fishermen's original intention.

Plankton nets were first used by J. Vaughan Thompson (see Hardy 1956) and were made into quantitative sampling instruments by Hensen, as indicated above. There are a variety of nets in common use, most of which are as efficient as was Hensen's, which itself was very carefully designed from a hydrodynamic standpoint. The numbers caught can be expressed as numbers per unit volume if the volume filtered is estimated with a flowmeter in the net during the act of capture. Nowadays, high speed nets towed horizontally catch adequate numbers of fish larvae and euphausids. Algae and the smallest animals are trapped in water samples from bottles. Thus, generally, plants and animals can be well sampled, with one or two exceptions (for example, naked flagellates), and they can usually be expressed in numbers or weight per unit volume.

Some of the most useful techniques are chemical in origin. Nutrients are of considerable use in the study of production in the sea particularly as a quasi-conservative property in a dynamic system. Since the 1920's methods have been developed to measure phosphates, nitrates, nitrites and silicates in the sea with simple colorimetric procedures and a high degree of sensitivity. In recent years the rate of sampling has been increased enormously with the autoanalyzer by which such chemical observations are sampled and recorded continuously from a ship at sea. In the same way, chlorophyll can now be estimated continuously with a fluorometer. The radiocarbon method allows us to estimate the increments of production very quickly. All are quick methods that allow rapid sampling at sea and hence the quick accumulation of observations.

Since the Second World War, acoustic techniques have advanced considerably. The Deep Scattering Layer in the deep ocean was revealed by echo sounding and probably comprises much of the zone of mesopelagic fishes. Such echo sounders have been used to sample fish populations in the ocean in numbers of a size group per unit volume. Sonar has been used to examine animals on different scales. Sector scanning sonar can examine fish shoals in detail (Cushing 1973) or be used to observe a single plaice change its behaviour in midwater at slack tide (Greer Walker *et al.* 1973). GLORIA, the long-range side scan sonar on R.R.S. *Discovery* has been used to chart the distribution of fish shoals and their behaviour in an area of about one hundred square kilometres (Rusby *et al.* 1973). Long-range sonars have been used to detect fish out to 50 km, in one example the drift of targets with the tide over a period of weeks was shown (Weston *et al.* 1969). One of the problems facing a marine ecologist is to establish the scale in which the processes studied take place and such equipment may be used to this end in the future.

Techniques have developed in three ways. The first is in sampling the numbers of plants and animals in the sea and such methods are developing continuously to improve the estimates of numbers per unit volume or below unit surface. The chemical or quasi-chemical methods have been developed for many purposes, but today many are being used to estimate rates of nutrient transfer at one level or another. Lastly, acoustic methods can be used to put the processes studied on much broader scales than have been used hitherto.

The development of concepts

Many ideas have played their part in the development of marine ecology, but some have had much greater influence than others and have dominated the science for a period of time. Their development is of great interest historically, but of more importance is the present status of the concept. In this light a few of the more important concepts will be reviewed briefly below.

NUTRIENT LIMITATION

The present status of nutrient limitation is well reviewed by Dugdale in Chapter 7. A notion, developed from the Law of the Minimum propounded by Justus von Liebig, stated that production (on land) was limited by that nutrient in least supply. Brandt (1899) and Raben (1905) measured the quantity of phosphate and nitrate in the sea and found the quantities to be low. From these observations developed the 'agricultural model' of production in the sea; Atkins (1925) had calculated the production of algae in the western Channel to be 1400 tons wet wt/km^2/yr merely by dividing the annual decrement of nutrient in spring and summer by the quantity of nutrient per cell. When nutrient was exhausted at the surface in July and August, production was supposed to stop. I have called this the agricultural model because it supposes that production continues linearly to the limit and that the quantity of nutrient per cell is constant. Today, the quantities estimated by this method are fairly accurate, but they may represent the quantity transferred to animal flesh rather than the quantity of algae produced, and three important factors, the restraint by grazing, the turnover of nutrient in regenerative processes and the varying quantity of nutrient per cell are not taken into account.

The next stage in the development of ideas was Ketchum's notable experiment (1939), in which it was shown that the quantity of algae produced in a short period of time was proportional to the initial amount of nutrient available; it was then concluded that the algal reproductive rate was restrained at a rather high nutrient level. Consequently, in the models of the production cycle developed by Riley (1946), the algal reproductive rate was underestimated and therefore the quantity eaten by the herbivores was also underestimated.

Another important experiment was that of Mackereth (1953), who showed that cells of *Asterionella formosa* Hass. stored phosphorus in high amounts before the spring outburst in Windermere, a lake in northern England, and that the quantity of phosphorus per cell declined in proportion to cell number. It was shown that growth stopped at a low limiting value of nutrient per cell. Before the spring outburst, the cells stored twenty-five times the limiting value of phosphorus per cell, or cell quota. Although Lund (1950) suggested that the production of *Asterionella* in Windermere was eventually restrained by lack of silica, it appears that the algal cells have evolved mechanisms by which the expected loss of phosphorus (in this case) is stored before the spring outburst. The relevance of luxury consumption and cell quota to the theory of nutrient limitation have recently been discussed by Droop (1973).

That phosphorus (and presumably some other nutrients, but not all) might be regenerated directly by the activity of grazing animals was first suggested by Gardiner (1937). Later work by Pomeroy *et al.* (1963) and by Johannes (1964) showed that this certainly occurred and at a relatively high rate, particularly amongst the numerous small animals that are not caught by some of the standard plankton nets. If phosphorus, for example, is held loosely at up to twenty-five times the cell quota for 'luxury consumption', then the destruction of the cells must release the nutrient into the sea; of course, there are also direct releases from the animals themselves in the form of excretion. The problem then is to differentiate between the release from cells destroyed, from excretion and from faecal pellets in such a way that the rate of nutrient turnover can be assessed. The nutrient budget is a useful check in any model of the production cycle in that the quantities in total should always be the same despite the changes in items within the budget.

The most important development in the study of nutrient limitation in recent years is that by Dugdale (1967), described fully in Chapter 7. In the Michaelis–Menten equation, algal reproductive rate is related to the nutrient content in the water. Not only is the nutrient limit rigorously defined, but the nutrient content of the water includes the products of all the regenerative processes. Another way of using the same system is to express all parameters in units of the limiting nutrient and measure the flow through the various compartments; there is no theoretical reason why both the numerical processes of predation and the energetic processes of food conversion should not eventually be expressed in units of nutrient. The real point is that although the problem of nutrient limitation of the algal reproductive rate has been well defined, the nutrient budget in a running ecosystem has not yet been fully calculated.

THE PRODUCTION CYCLE

The production cycle in temperate waters has been described as a spring outburst because of the rapid build-up of algae and animals. The existence of such mechanisms was first indicated by Lohmann (1909) in his study of productive processes throughout the year in Kiel Bay; in order to create the quantity of animals observed he estimated an algal reproductive rate of 0·3/d. Because the food supply was limited, Pütter (1909) calculated that the animals should starve, and put forward his famous suggestion that they might subsist on dissolved organic material, which was later denied by Krogh (1931) on purely physiological grounds. The classic study by Harvey *et al.* (1935) of the spring outburst off Plymouth showed that the copepods were probably eating about 40 per cent of their body weight/day (a figure derived from the decrement of phosphorus divided by the quantity of phosphorus per animal during the decline of the spring outburst). The observations off Plymouth were fitted by a theoretical curve developed by Fleming (1939) on the assumption that the production cycle was a predator–prey relationship driven only by the reproductive rate of the algae and the grazing mortality generated by the herbivores.

It has already been noted that in Riley's models the algal reproductive rate was underestimated and that the animals did not get enough to eat. Because the ratio of reproductive rate to grazing rate was well estimated, the fit to observations in the model was good. Riley's most valuable contribution was to express the modification of the algal reproductive rate in terms of environmental factors, such as light and vertical mixing. Some progress in modelling the production cycle was made during the fifties, but in the following decade the most valuable progress was made on the parameters of which any model is comprised, on nutrient limitation as described above, on the sinking of algae and on grazing mechanisms. Smayda (1970) summarized the experimental measurements made on the sinking rates of algae and it appears that cells of average size sink at the rate of about 0·6 m/d, which is much less than had been thought earlier; larger cells sink much more quickly and senescent populations sink away from the photic layer. The grazing activity of the herbivores was modelled by Cushing (1968) using an encounter theory and by Parsons *et al.* (1967), using a form of the Ivlev curve that relates ration to food density.

With the parameters estimated independently, it is now possible to rewrite models of the production cycle in a fairly reliable way. For example, Steele (1974) has modelled the production cycle in the Northern North Sea for a period of 300 d with a fairly small number of parameters. If each is considered valid (that is, tested independently of the model itself—under experimental conditions) then a test under natural conditions is easily conceivable in statistical terms; of course, large numbers of observations would still be required. Another current model is that of di Toro *et al.* (1973), which simulates the cycle in Lake Erie in 1930 before eutrophication and in 1970, after it; the intermediate steps of nutrient increment are not explained but the important point is that the phenomenon can be accounted for in terms of chlorophyll or carbon, if not in the change of species, by standard methods. In Chapter 16 Walsh describes in some detail the successful use of such techniques at sea in the upwelling areas. This represents the most advanced development yet available.

This short history of the development of production-cycle theory is really one of the application of models to the marine ecosystem. The problems faced by Lohmann, Pütter and Harvey are very like those facing the model makers today even to the details of how much food an animal of a particular size should have. The first point is to decide which are the significant parameters and then to critically determine their influence on methods in independent sets of experiments. The development of nutrient limitation in the history of production cycles is most interesting from this point of view. With Dugdale's introduction of the Michaelis–Menten equation the question was reduced from one of speculation to one of experimentation. The science could then proceed.

GRAZING IN THE PRODUCTION CYCLE

Since the time of Harvey *et al.* (1935) and of Riley (1946), it has been generally agreed that most of the algal production is transferred to animal flesh by grazing, which implies that the death rate of algae by other causes is low. Cushing (1955) tried to show this by counting the number of empty frustules of diatoms. Cushing and Vúcetìc (1963) related the total mortality of algae to the weight of grazers and separated grazing mortality from other sources during the North Sea production cycle; on average, the ratio of grazing mortality to other mortality was 1·65 and at the end of the cycle it was 3·51. Hence, the earlier conclusions were confirmed.

There are two areas of conflict on the role of grazing in the production cycle. The first is the difference between field observations of high grazing mortality (Cushing & Vúcetìc 1963, Tigeguchi 1973) and low observed experimental levels (summarized by Marshall 1973). The second is whether the peak of algal production in the sea is determined by grazing or by a reduction in the algal reproductive rate by low nutrients, or by both. The two are linked because models of the production cycle are constructed on the basis of experimental observations.

By usage, the estimate of algal mortality due to grazing has been called the volume 'filtered' or the volume 'swept clear', but this need bear no relation to the physical volume swept by a copepod, which is very low. If the animal is shown to filter more than the physical volume, then it must to some degree select particles; the work of Harvey (1938), Mullin (1963) and Richman and Rogers (1968) shows that *Calanus*, a large copepod, selects larger diatoms, as might be expected. Recently, Corner *et al.* (1973) have shown that *Calanus* could filter up to 800 mls/d, when fed upon a large diatom, *Biddulphia sinensis*; such a value approaches those in excess of a litre/d observed in the field. An approach to the problem was put forward by Cushing (1968) in a form of encounter theory, that could potentially take into account the different sizes of algae and of grazers. If size of algal cell varies inversely with numbers, the death rate should be attributable to the size range sampled by the animal; further, the death rate of algae might increase from a threshold (Adams & Steele 1964, but see Frost 1972) to a peak in algal density, above which it declines. Perhaps the phrases 'volume swept clear' or 'volume filtered' should be replaced by a term such as 'mortality'.

The history of the second conflict started with Fleming (1939) who described the whole course of the production cycle in the balance of two exponents, the algal reproductive rate and the algal mortality rate due to grazing. Later, Riley (1946) allowed the algal reproductive rate to decline with decreasing phosphorus content, with the consequence noted in the previous section that the modelled animals did not get enough to eat. Lund (1950) observed that the peak of the outburst of *Asterionella formosa* Hass. in Windermere occurred very frequently at a constant low level of silica in the water and concluded that production was restrained by silica lack. Cushing and Vúcetìc (1963) observed that during production cycle in the central North Sea there were high levels of nutrient (phosphate, nitrate and silicate) throughout, and that the algal reproductive rate increased during this period. What is now needed to resolve this point is a reliable method of measuring algal reproductive rates in lakes or in the sea. The total mortality of algal numbers can be estimated quite readily if a reliable measure of the algal reproductive rate is available, and assuming that grazing is the major source of mortality, the second problem can be solved.

Elsewhere in this volume reference is made to the use of particle-size distributions (Chapters 4, 14) in the description of ecological processes. Eppley (1973) has shown that smaller algal cells divide more quickly than larger ones and bigger grazers select larger cells, leaving the smaller ones to be taken by the smaller herbivores. The rates should be properly averaged across the size distributions. Then the transfer of energy from algae to herbivores can be properly described.

THE MIGRATION OF FISH

When Meek (1917) wrote his book *Migration of Fish*, the real problems were the apparent offshore movements which he described and the contranatant migrations of prespawning adults and the denatant ones of spent fish and larvae. Russell (1937) charted some distant migrations by cod beyond their normal ranges, for example, the recovery of a fish tagged at Iceland from the Faroe Islands, etc. The general picture of migration that remains from this period is that fish are at the mercy of the ocean currents and that the exchange between spawning groups might be considerable particularly as cod from one end of the ocean are indistinguishable to a taxonomist from those at the other. Another facet of the same general picture was that fish larvae were supposed to be carried by the currents beyond the nursery grounds and were lost and that this phenomenon accounted for observed differences in the recruitment of year classes.

In the last decade or so this general picture of fish being at the mercy of a variable current system has changed somewhat. Harden Jones (1968) pointed out that fish in midwater rarely had any external referent and that they may make some important migrations merely by drifting in a current or counter-current. He pointed out that Meek's words 'contranatant' and 'denatant' never had any connotation of orientation but merely described the movement that had to be made, if the fish were to return to their spawning grounds or were to drift away in the same currents that carried the larvae. Further, in recent years, it has become clear that in temperate waters at least, the peak date of spawning is fixed (Cushing 1971c) and that the same spawning ground is used year after year. Hence the migration circuit has qualities of regularity that were not suspected by fisheries biologists some decades ago but must have been known to fishermen. The idea that fish might return to their native grounds to spawn in the sea is no longer treated with suspicion, particularly since the extensive work on the Pacific salmon has revealed that such mechanisms must exist. Another point is that if spawning grounds and nursery grounds are at fixed stations along a regular current, then the decrease in stock level by variation in current strength might be very much less than originally thought and therefore the source of recruitment variability must be sought elsewhere. Indeed, the trend today is to search for the behavioural or physiological mechanisms by which fish might be thought of as joining or leaving a particular current at points convenient to maintain the regularity of their migration circuit.

If the circuit is as regular as suggested, then each stock might be isolated from its neighbours, because the apparent mechanisms of exchange as shown by the rare losses of tagged fish are illusory. As shown in Chapter 13, the chances of mixture between the stocks of North Atlantic cod are very low indeed; this conclusion is in very sharp contrast to the concept implicit in the work of earlier fisheries biologists that some exchange between stocks is almost inevitable. Indeed, that was the conclusion drawn from Russell's famous diagram of the migration of tagged cod in ones and twos from their expected grounds. Thus with this increase in knowledge our concepts of fish migration have changed rather radically from the time when a stock was considered to be a unit that had achieved a considerable degree of reproductive isolation. Although such mechanisms may not exist on such a broad scale amongst other groups of animals in the sea, the bland uniformity of the environment and its heavy turbulence should not deceive us into believing that they do not exist at all.

FISH POPULATION DYNAMICS

There have been three stages in the development of the study of fish-population dynamics. The first was the statement by the old masters, Garstang (1900), Petersen (1893) and Heincke (1913), that there were two courses of action necessary. The first was to prevent an excess of fishing that might damage the incoming-year classes. This should be limited by law. The second course was that open to the fishermen themselves; to let the little fish grow rather than be caught at too early an age. Tagging experiments were used to estimate death rates and the spread by migration, and from them Heincke (1913) was able to present solutions to the 'small plaice' problem: in summer in the first decade of the century trawlermen working close to the coasts of Holland and Germany discarded six times more than they caught, because the fish were too small.

The next stage in the understanding of population dynamics was the application of the logistic curve by Graham (1935), in which the net rate of increase was put as proportional to the rate of fishing; consequently, the build-up of stock compensates for increased fishing, provided, of course, that fishing is not allowed to get too intense. This convenient axiom allowed Graham to shelve the problem of recruitment and to turn to the problem of growth overfishing, the problem of how to allow the smaller fish to put on all the weight they could during their adult lives before being caught (some fish can grow by more than an order of magnitude after maturation). In practical terms this meant enlarging the meshes in the cod ends of the trawl in order to let the little fish escape and put on weight before they were caught again.

Beverton and Holt (1957) solved the problem of growth overfishing in model terms, expressing catches as yield per recruit as a function of fishing mortality. Although it evaded the possibility of recruitment failure the system of models developed from this idea allowed management to make decisions very quickly, whilst Graham's method (or its successor due to Schaefer) needed perhaps a decade of observations to provide the basis for the same sort of decision. However, the use of growth and death rates by age independently of stock weight or numbers represented a considerable scientific advance; indeed, the method resembles the Leslie matrix (1948), which has been used in other population studies in which the animals can be aged properly.

In a quite different direction, Ricker (1954, 1958) defined the dependence of recruitment on parent stock in terms of density dependent and density independent mortality. The variability of recruitment about any curve calculated on this basis is very high and the method suffers from the same defect as the Graham–Schaefer model: that many observations are needed before any management decision can be made. What is needed now is a step similar to that made by Beverton and Holt to construct a model from independently determined parameters. Inevitably, this is rather difficult because the nature of both forms of mortality in Ricker's equation remains somewhat obscure. The dynamics of fish populations are as well developed as those of other animals, perhaps because so many fish are counted and weighed upon the quays, because many can be readily aged and because they live for as long as two decades. The models are now well developed and the changes in a number of cod stocks in the North Atlantic can be simulated well enough for prediction (Clayden 1972).

Conclusion

From its roots in natural history, on the Victorian sea-shore or from the collections made on the Victorian expeditions, the ecology of the sea has developed into the study of populations and ecosystems. Although an ecosystem can be considered as a productive engine, it is perhaps no more than an assembly of populations. What makes populations in the sea interlock is the enormous loss of individuals during the life history, which may stabilize the separate populations and contribute to the appearance of stability in the ecosystem. It is in the use of models in one form or another that marine ecology progresses at the present time and indeed in fisheries science the same sort of models are used for practical and predictive purposes. In the short history of the development of the concepts given above, within limited fields each section traced the unfolding of successive models; even in the study of migration, a profound change can be seen from the early model of indistinct movements to the later one of somewhat ordered migration within the vagaries of the ocean currents.

A distinction has been drawn between the study of populations and that of ecosystems, though perhaps unjustifiably because the point was made that an ecosystem cannot be more than the sum of its component populations. Perhaps the real distinction lies in the detail used. Nobody who ventures to study an ecosystem can afford to examine any population very closely, because there are so many. Conversely, anyone who has looked at a single population for a long period cannot take his mind away from the interesting detail to look at irrelevancies outside his chosen and restricted field. Perhaps both outlooks are a little blinkered, because the ecologist who studies the single population must consider the nature of the death that it inflicts and the form of the death that it suffers, which means that three trophic levels are being examined at one and the same time. In a similar way the ecologist who studies the ecosystem and its stability might find it rewarding to look at the stabilization mechanisms that occur in the single populations. That which stabilizes one population may make a competitor temporarily unstable and as noted above, competition is the field of study common to both outlooks.

Physical science developed into a compact and interlocking body of knowledge, when the early models acquired predictive power. Ecology emerges from the precopernican mists and starts to simulate small pieces of housekeeping in the sea. The step from Copernican speculation to Newtonian certainty may take a long time because biology is a science of processes that are in themselves very complicated. The sea is a good place in which to take such steps because of the uniform environment, the simple sampling systems and the scale of distance on which one is forced to work with ships.

PART I. THE SEA AND THE ORGANISMS THAT LIVE IN IT

Any ecological study describes how plants and animals adapt to and insulate themselves from their environment. The sea is a very special environment compared to the land; it is relatively uniform and perhaps even benign. However, sea water is a fluid and has characteristics foreign to that terrestrial environment in which we live. A step in imagination is demanded of us before we can conceive of the sea in which present-day plants and animals live and where most primitive organisms probably evolved. One disadvantage is that we do not always know where to look for the significant boundaries in the sea in the way in which such edges, or ecotones, are part of our living history on the land. However, a great advantage is that our relative detachment from this simple and well-defined environment encourages objective and progressive analysis.

Chapter 2 on physical oceanography describes the properties of sea water necessary for the physical description of ocean currents and the survival of living organisms; but more importantly, it also covers the motions of the sea and its turbulence. This is a mathematical description but we expect biologists who study the sea to follow the arguments. Indeed, it would be difficult to understand some of the biological discussion in later chapters properly without an intuitive grasp of the movement of ocean currents and the diffusive properties of the turbulent waters.

A full account of all the plants and animals in the sea cannot be presented in the short space of a single chapter. There are very few phyla not represented in the oceans and comprehensive lists and inventories would convey little of the taxonomic diversity of life in the sea. Ecologically, such diversity is represented in sizes, in shapes, in life histories and in survival strategies, however, and it is these characteristics that are represented in Chapter 3. Not only has no systematic structure been described, but few pictures are presented of fishes, blue whales, crab zoeae, or naked flagellates. Any ecologist is quite precise in defining the organisms he examines in a taxonomic sense and the necessary textbooks are part of his workaday training.

Chapter 2. Waters of the Sea: The Ocean's Characteristics and Circulation

R. L. Smith

Introduction

The qualities of sea water
Unique properties of sea water
Salinity
Temperature
Density
Sound in the sea
Light in the sea

The dynamics governing motions in the oceans
The equations of motion
Turbulence
The equation of continuity
Some specific examples of dynamic balances

The dynamics governing motions in the oceans *continued*
Geostrophy
The effect of the wind
Inertial oscillations
Vorticity

The circulation and distribution of properties
The wind-driven circulation system
The distribution of properties
The observed distribution of properties in the deep ocean
Vertical distribution of properties
The abyssal circulation.

Epilogue

> He saw that the water continually flowed and flowed
> and yet it was always there; it was always the same
> and yet every moment it was new.
>
> (*Siddharta*, by H. Hesse)

> The atmosphere and the oceans move ceaselessly
> in complex patterns and on awesome scales.
>
> (J. Pedlosky 1971)

Introduction

To the physical scientist the waters of the sea are a dilute electrolytic solution, with slight but significant variations in density and properties, moving within a thin shell on the rotating, gravitating, spherical earth. The physical oceanographer studies the physical state and properties of the oceans, their currents and motions, and the transfer and mixing processes—all of which are basic to an understanding of the ecology of the sea. The marine ecologist has to contend with the complication (unlike the terrestrial ecologist) that the environment is moving, and transporting various properties and biota. These naturally occurring ('geophysical') motions cover an enormous range of scales: from the capillary waves in the cat's paws on wind-rippled water to the major currents like the Gulf Stream.

The contemporary physical oceanographer is awed and intrigued by the complexity of oceanic phenomena and the richness of the spectrum of scales: the complexity of the sea's structure is finer grained—and the variability and intermittency of the processes more dominating than was believed a decade or so ago. We recognize that there are major, large-scale occurrences (e.g. El Niño) that are fundamental to the global oceanic environment, and that significant processes occur on a scale as small as can now be resolved (e.g. temperature microstructure). The controlling mechanisms and the interactions between the various scales are extremely complex and too poorly understood for satisfactory prediction on the local scale—which is what the ecologist, as most of us, would like to have.

The above may serve as an *apologia* for the absence of a totally satisfying picture or simple explanation of the physical ocean. In the relatively few pages that follow in this chapter an attempt will be made to describe some of the properties and processes that would seem important to the ecologist. Some topics will be virtually ignored because the aspects important for ecologists are very well known and the information readily available (e.g. tides in coastal waters), or because the subject is extremely complex and inconclusively known (e.g. the dynamics of the upper couple of metres of the ocean). The emphasis will be on those quantities, processes, and concepts that form the basis of much of our present understanding of the physical environment of the ocean.

The qualities of sea water

Sea water is a solution of virtually all the naturally occurring elements in pure water. It has many of the anomalous properties exhibited by fresh water, and additional other properties because of the materials dissolved and suspended within it. The mass of *dissolved* solids per unit mass of sea water is called the *salinity*.

Salinity, temperature and pressure define the density, and the thermodynamic phase, or state, of a sample of sea water. For most purposes the pressure can be considered to be determined by the weight of water above the sample, and pressure increases by 1 decibar (10^4N m^{-2} or 10 kPa) per meter, or by about 1 atmosphere for every 10 metres. The equation of state is the functional relation between density and salinity, temperature, and pressure. There is no simple analytical form and empirical formulas and tables are used (see Fofonoff 1962).

Chemistry occupies a central niche in oceanography: both the physical oceanographer, concerned with distributions of water masses and inferences about their origin, and the marine biologist, concerned with the fertility of the sea, rely on chemical studies. The chemistry of the sea will receive only implicit discussion in this text and the reader is referred to Horne (1969), Riley and Chester (1971), and Dyrssen and Jagner (1972) for more details and discussion.

Unique properties of water

Water has unique characteristics, anomalous physical–chemical properties compared to other substances, because of its molecular structure and the hydrogen bonding that holds water molecules to one another with tenacity. The anomalously high heat capacity, latent heat of fusion, thermal conductivity and dissolving power are important in biological and physical processes. One of the unique characteristics of pure water is the occurrence of a density maximum at a temperature (4°C) which is greater than the freezing point (0°C). Thus a freshwater lake in winter will cool down to 4°C from top to bottom—but with more cooling the surface layer becomes lighter, does not sink, and eventually will freeze. The water at the bottom of the lake is not the coldest—and the layer of ice on the surface tends to insulate the water beneath. For further discussion of the anomalous and unique properties of water see Sverdrup *et al.* (1942) and Horne (1969).

Salinity

Salinity (S) is usually given in parts per thousand (‰), or grams per kilogram of sea water, and in the open ocean it will usually be found within the range of 32 to 38 ‰. Most properties of sea water are a function of salinity (Sverdrup *et al.* 1942, Neumann & Pierson 1966). For example, the freezing point and the temperature of maximum density both decrease with increasing salinity and at $S = 24{\cdot}7$ ‰ they are equal at $-1{\cdot}33$°C. Thus at normal oceanic salinities vertical convection continues under cooling until the entire water body has been cooled to the freezing point. It is for this reason that the ocean does not freeze as readily or rapidly as freshwater lakes.

The intuitive definition of salinity as the fractional mass of dissolved solids in sea water did not suffice for precise quantitative studies. Early in the twentieth century an international commission established the formal definition of salinity as: the total amount of solid material, in grams, contained in one kilogram of sea water when all the carbonate has been converted to oxide, the bromide and iodine replaced by chlorine, and all organic matter completely oxidized. The history of the salinity concept and definition, and the related scientific and philosophical problems are discussed by Cox (1963) and Johnston (1969).

Although the salinity (i.e. total amount of dissolved salts) is variable, the relative proportions of the major constituents, which comprise more than 99·99 per cent of the dissolved substances in the ocean, are nearly constant. This was recognized by nineteenth-century marine chemists and firmly established by Dittmar's (1884) definitive work using samples from the *Challenger* expedition. The major constituents are given in Table 2.1. Because of the constancy of the major constituents relative to each other, a good

Table 2.1. The major constituents of sea water

Ion	g/kg of water of salinity 35‰*
Chloride	19·353
Sodium	10·762
Sulphate	2·709
Magnesium	1·293
Calcium	0·411
Potassium	0·399
Bicarbonate	0·142
Bromide	0·0673
Strontium	0·0079
Boron	0·00445
Fluoride	0·00128

* Values from Riley & Chester (1971)

measure of any one constituent would be a good index of salinity. Until the 1960s the salinity of a sea-water sample was usually determined by measuring its chlorinity. At present, salinity is determined almost exclusively from electrical conductivity measurements using tables (Cox *et al.* 1967) which preserve a correspondence to determinations based on chlorinity. Electrical conductivity is a function of temperature and pressure as well as salinity, and modern electronic instrumentation capable of determining salinity with a resolution of 0·001 ‰ *in situ* must measure all three parameters (see Perkin & Walker 1972 for computational formulas).

In addition to the major constituents, there is in the sea virtually every natural element, but most are in very small concentrations. The elements and compounds associated with living organisms such as carbon, oxygen, nitrate and phosphate are found in small but highly variable amounts. The trace elements important for biology, e.g. copper, cobalt, etc., have even smaller concentrations. As an example of the relative concentrations of major constituents, nutrient elements, and trace elements, the typical concentration ratios of Na:N(in NO_2, NO_3, NH_4):Co are roughly $10^7:10^3:1$. The surface waters, where photosynthesis occurs, tend to be depleted in the dissolved nutrient elements, and the deeper waters enriched through settling of detrital matter. If the nutrient elements were not returned to the surface layer, the concentration would soon become too low for primary productivity: at the rates of productivity present in the ocean, the depletion of nutrients from the surface layer would lead to a virtually lifeless sea in less than a year (Turekian 1968). Therein lies some of the importance of circulation and mixing to the study of marine ecology.

Temperature

The temperature measured in the world oceans ranges from almost −2°C at the bottom near the Antarctic continent and in the Norwegian Sea to almost 30°C in the equatorial regions. In low and mid latitudes the temperature profile with depth typically consists of

1 an almost uniform (*mixed*) layer, which may be a few to 200 m thick, with a temperature similar to the surface temperature;
2 a permanent *thermocline* to about 500–1000 m depth in which the temperature decreases from that in the mixed layer to
3 the *deep* layer of cold water below 1000 m which has its origin in the high latitudes. In middle latitudes, with the lighter winds and surface heating during the summer, a shallow (20 to 50 m) *seasonal thermocline* may be established.

The temperature is measured routinely to 0·01°C with thermistors or resistance thermometers—or with 'reversing' mercury thermometers attached to the bottles that 'catch' sea-water samples. The bottle is activated to close, and to turn the thermometer upside down (reversing the thermometer and trapping the appropriate amount of mercury) at depth by dropping a weight ('messenger') down the wire. Advances in instrumentation during the past decade allow the oceanographer to obtain essentially continuous profiles of temperature (to a resolution of 10^{-3} to 10^{-4}°C) and salinity instead of several discrete samples obtained with mercury thermometers and water-sampling bottles. This has resulted in the discovery of the existence in the ocean of fine temperature structure (microstructure) in the form of layers or sheets of uniform temperature, centimetres to metres thick, separated by very sharp gradient layers of cm or less thickness. The rate of diffusion of temperature and salinity across the microstructure gradients sets a limit to the smallest significant scales of fluctuation in the ocean. Microstructure may play an important role in the diffusion of substances—and the dissipation of energy in the ocean (Gregg 1973).

Sea water is slightly compressible and this leads to an increase in temperature of water when it is compressed adiabatically (i.e., without exchange of heat). Thus at great depths, the measured temperature may be found to increase with depth. The adiabatic effect of increasing pressure leads to temperature increase of the order of 0·1°C per 1000 m (see Neumann & Pierson 1966, pp. 45–7). *Potential temperature* is defined as the temperature a water sample would have if it were brought adiabatically to the sea surface. The potential temperature profile at great depths is usually slightly decreasing downward.

Density

The relation of density to salinity, temperature and pressure has been the objective of elaborate and highly precise laboratory measurements and accurate tables have been prepared. The measurements and tables are discussed in more detail in Fofonoff (1962), Neumann and Pierson (1966) and Riley and Chester (1971).

The density of sea water (ρ) is always a little greater than unity in the cgs system. The physical oceanographer routinely uses a precision of $1:10^5$, and to avoid having to write long decimals uses σ, based on the specific gravity which is numerically equal to the density (ρ) in CGS units:

$$\sigma = (\rho - 1) \times 10^3. \tag{2.1}$$

The quantity σ_t is defined as the σ of sea water at atmospheric pressure and at the temperature (and salinity) at which it was collected. Sea water is compressible, and at a pressure corresponding to that at the average bottom depth of the oceans, 4000 m depth or 400 atmospheres pressure, sea water is compressed about 2 per cent. Tables of compressibility can be found in Neumann and Pierson (1966). It is useful to define a *potential density* analogously to potential temperature: the density a sample of water would have if brought adiabatically to the surface (atmospheric pressure). Thus both sigma − t (σ_t) and the analogous quantity for potential density (σ_θ) are computed at atmospheric pressure; they differ only in that for σ_t no account is taken of the adiabatic temperature change in ‘bringing’ the sample to atmospheric pressure.

The T–S diagram (Fig. 2.1) is a useful nomogram which allows one to obtain σ_t from salinity and temperature; the family of smooth curves on the T–S plane are loci of constant σ_t—but the usefulness of the T–S diagram is greater than that. Temperature and salinity are physically independent variables but they are not randomly distributed in the world ocean: it is a fact of observation that there are water masses that extend horizontally and vertically with significant total volume and have specific characteristics with respect to temperature and salinity. If you plot the temperature and salinity values from different depths for any region of the ocean, the values of the

T–S diagram generally lie on well-defined curves or narrow bands. The T–S diagram has proved to be a convenient means of examining the bivariant distribution of temperature and salinity, classifying and tracing water masses, and obtaining a qualitative picture of the amount of mixing between water masses of different geographic locations. Characteristics for some water types (water mass of narrow range of temperature and salinity) and masses are given in Table 2.2.

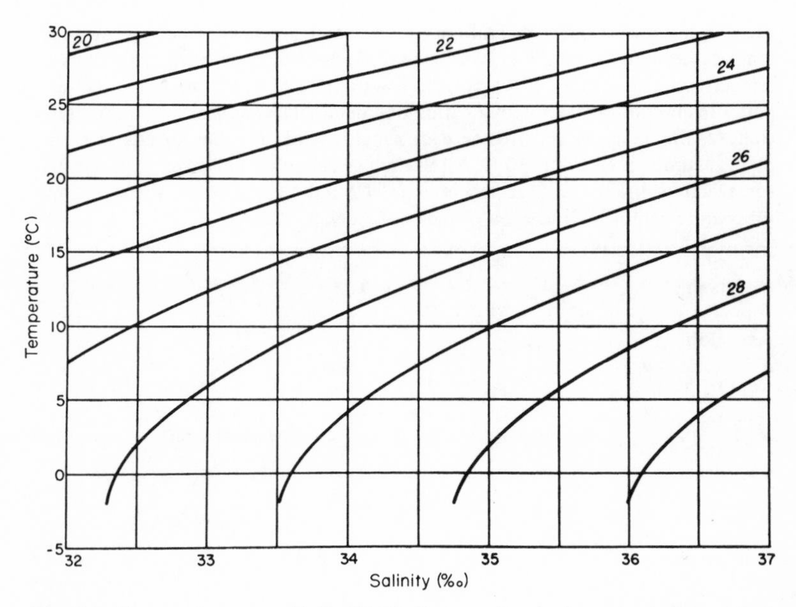

Fig. 2.1. Temperature vs. salinity (T–S) diagram. Loci of constant sigma-t (σ_t) shown.

Stommel (1960) has succinctly stated some considerations that must be kept in mind in using T–S diagrams, and we paraphrase and expand them in the following:

1 Potential temperature and salinity are conservative properties in the ocean (i.e. there are no significant processes by which either property is changed except by mixing), except in the relatively shallow mixed surface layer where evaporation, solar radiation, exchange of heat with the atmosphere, etc., occur. We have noted that sea water is compressible and therefore true temperature is not strictly conservative—but potential temperature is conservative. The adiabatic effect of about 0·1°C per 1000 m is often small compared to the true temperature gradient; thus potential temperature and true temperature are often interchangeable on the T–S diagram. Potential

density (σ_θ) can be found from the potential temperature (θ) and salinity using the T–S diagram: if θ is plotted instead of T, the σ_t loci become σ_θ.
2 The large-scale mixing processes in the interior of the ocean mix both temperature and salinity in the same way. Thus the mixture of two water masses, represented by two points on the T–S diagram (see Figure 2.1), lies along a straight line. Since mixture on the T–S diagram occurs along straight lines, whereas the lines of equal σ_t (or σ_θ) are curved convexly upward, the

Table 2.2. Some water types and masses. A water type has a narrow range of temperature and salinity, and can be characterized by a representative value of temperature and salinity. A water mass is a mixture of water types, and the temperature and salinity vary with location and depth. The temperature-salinity relations for the water masses given here are roughly linear and they are characterized by the salinity at 15°C and 10°C. All the values below are only representative and somewhat arbitrary. See Sverdrup *et al.* (1942) and Neumann and Pierson (1966) for a more complete discussion of the ocean's water masses.

Name	T(°C)	S(‰)
Antarctic Bottom	0·0	34·7
North Atlantic Deep	3·0	34·9
Antarctic Circumpolar	1·5	34·7
Antarctic Intermediate	3·0	34·0
North Pacific Intermediate	4·0	33·5
Mediterranean	12·0	36·5
Red Sea	13·0	37·0
	S‰ at 10°C	S‰ at 15°C
E.N. Pacific Central	34·0	34·5
W.N. Pacific Central	34·25	34·6
E.S. Pacific Central	34·5	35·0
Pacific Equatorial	34·7	35·1
W.S. Pacific Central	34·8	35·5
Indian Central	34·8	35·5
S. Atlantic Central	34·8	35·5
N. Atlantic Central	35·35	36·0

mixing of two water masses of the same σ_t would result in the σ_t of the mixture being greater. This effect is called cabbelling—but, although interesting, is of unknown importance to major oceanic processes.
3 In all the oceans ρ generally increases with depth (i.e., the oceans are stratified) irrespective of the effect of compressibility. A water mass parcel (imagined as a 'solid' body embedded in surrounding water) will remain 'floating' at rest if the density of the embedded mass is equal to the density of

the surrounding water (Archimedes principle). If the fluid parcel is displaced vertically, in a fluid with density increasing with depth, there will be a restoring force tending to return the parcel to its rest position. The 'stability' of a water body can be defined as the relative acceleration that a water parcel experiences when displaced vertically a unit distance from its rest position. The natural frequency of oscillation (in radians/unit time) of the displaced parcel about its equilibrium position (given a small vertical displacement) is the Brunt–Väisälä frequency, N, where:

$$N^2 = \frac{-g}{\rho} \frac{\partial \rho_\theta}{\partial z} \text{ (where: } \rho_\theta \text{ is potential density and } z \text{ positive upwards)} \quad (2.2)$$

The fluid is statistically stable when N is real ($\partial \rho_\theta / \partial z < 0$, i.e. ρ_θ increasing downward). N is one of the most important dynamical characteristics of the ocean. The period ($2\pi/N$) varies from a few minutes in the thermocline to hours in the deep ocean where the water is nearly neutrally stable. In the upper ocean generally the actual temperature gradient is much greater than the adiabatic temperature gradient and hence $N^2 \simeq -g \frac{\partial \sigma}{\partial z} t \; 10^{-3}$. N^2 or N^2/g is frequently used as a measure of static stability. Because of the damping of vertical turbulence by the stability, there is a tendency for most major flows to occur along surfaces of equal potential density. In considering deep flows it is important to use potential density instead of density, which is not conservative under changes of pressure (i.e. depth), to indicate the directions of preferred flow and mixing. The compressibility of sea water is approximately accounted for by use of σ_t although the rigorously correct σ_θ is preferable.

Sound in the sea

Sound is mechanical radiation, a propagating pressure disturbance, and the ocean is an excellent medium for propagating acoustic waves. The speed (c) at which sound waves propagate is given by $c^2 = (k\rho)^{-1}$ where k is the adiabatic compressibility of the water, which like ρ is a function of salinity, temperature and pressure. For $S = 34{\cdot}5$ ‰, $T = 13$°C, and at atmospheric pressure, the speed of sound in sea water is 1500 m sec^{-1} (compared to 340 m sec^{-1} in air). The speed increases with increasing salinity, temperature or pressure (depth): $\partial c/\partial s = 1{\cdot}4$ m sec^{-1} per ‰ S; $\partial c/\partial T = 4{\cdot}6$ m sec^{-1} per C°; $\partial c/\partial z = 1{\cdot}6$ m sec^{-1} per 100 m. Since the speed of sound varies significantly in the ocean, sound undergoes refraction. The concept of an acoustic ray, analogous to a light ray, is used and Snell's law, which relates the angle of incidence and the angle of refraction to the ratio of wave propagation speeds, can be applied. In the thermocline the effect of temperature dominates and the speed of sound decreases with depth, and sound rays tend to 'bend' (refract) downward. In the relatively uniform water below 2000 m, the pressure effect dominates and sound speed increases with depth and the rays tend to 'bend' back upward. Thus the sound energy tends to be trapped in a 'sound channel' and is propagated extremely well over great distances. This effect is used for SOFAR (sound fixing and ranging).

Acoustic radiation, like electromagnetic radiation (e.g. light) is attenuated by absorption and scattering. In absorption, the radiation interacts with the

Table 2.3. Attenuation lengths. For sound (Frosch 1964) and electromagnetic radiation (Liebermann 1962), including light (Duntley 1963), in sea water.

Frequency (Wavelength for light)	Attenuation length (metres)
sound	
100 Hz	$\simeq 1300 \times 10^3$
1 kHz	100×10^3
5 kHz	40×10^3
10 kHz	1×10^3
50 kHz	400
electromagnetic radiation	
1000 kHz	0·25
100 Hz	25
light*	
400 nm	13
440 nm	22
440	12 (Pacific Countercurrent)
440	8 (Caribbean)
440	4 (Galapagos Islands)
465 (blue)	20 (Atlantic near Madeira)
480	28
520 (yellow)	25
560	19
600 (red)	5
700	2

* Values given are for pure water, which is equivalent to clearest ocean water, unless otherwise indicated.

water molecules and the dissolved ions in sea water and is irreversibly altered and the energy transformed into heat, or chemical potential energy, etc. Scattering is any random process by which the direction of radiation is changed without any other alteration. The scattering by organisms accounts for the deep-scattering layer (Hersey & Backus 1962). Absorption of acoustic energy by sea water is a function of the frequency f and absorption is increased as f^2 above 1 kHz. Absorption and scattering decrease the intensity of sound exponentially with distance. The attenuation length represents the distance at which the intensity is decreased to e^{-1}. Table 2.3 gives the attenuation length for sound and for electromagnetic radiation (including light) in the sea.

Because of the long attenuation length for sound, it is useful for detecting both the limits of the sea (the bottom) and objects within it (e.g. fish or submarines). Frequency and wavelength, λ, are inversely related through the relation $c = f\lambda$. Radiation is largely unaffected by objects smaller than the wavelength of the radiation, and hence there is a trade off between the resolving power, or ability to detect an object, and the range of propagation. The reader interested in underwater acoustics is referred to Albers (1967), Frosch (1964), Tucker and Gazey (1966), and Urick (1967).

Light in the sea

Although very small amounts of water appear to be transparent, the ocean absorbs most of the solar radiation incident at the surface within several tens of metres from the surface. It is this absorption of solar energy that supplies most of the energy input to the ocean and that limits most of the primary production to the upper layers of the sea. The average energy input to the sea by solar radiation is about 10^4 times that input by the wind or tides.

The spectrum (distribution of energy as a function of wavelength or frequency) of solar radiation that reaches the sea surface covers the spectral range from 290 to 3000 nm (1 nm = 10^{-9} m) in wavelength, of which the range 350 to 750 nm is visible. The energy has a peak near 480 nm (blue–green) but nearly half the energy is in the infra-red. Some of the light is reflected at the sea surface (only a few per cent for angles of incidence less than 45° from normal) and the rest enters the sea to be refracted, and attenuated by scattering and absorption; the latter is the ultimate fate of most of the light. The reader is referred to Duntley (1963) and Jerlov (1968) for more complete treatments on optical oceanography.

On a clear day in mid-latitudes, at noon, the order of 5×10^4 μw (microwatts) cm^{-2}, or $1{\cdot}2 \times 10^{-2}$g-cal sec^{-1} cm^{-2}, of solar radiation may enter the sea. The light is scattered and absorbed, both by pure water and by the suspended and dissolved matter, in processes that are strongly wavelength dependent. In the sea, light, like sound, is attenuated exponentially with path length. Some attenuation lengths for light are given in Table 2.3. Practically all the infra-red radiation and much of the visible red light ($\lambda > 700$ nm) is absorbed within 1 m, but clear sea water is relatively transparent for violet to blue–green light. The band of radiation from 380 to 720 nm is used for photosynthesis (Strickland, 1958).

A distinction is made between the beam attenuation coefficient (α) and the diffuse attenuation coefficient (k). The beam attenuation coefficient is an 'inherent' property of the water, depending only upon the optical characteristics and is the sum of the absorption coefficient and the *total scattering* coefficient; the diffuse attenuation coefficient would measure the extinction of daylight with depth in the sea, and hence depends on sun altitude, cloud cover, path length, etc., as well as on the inherent optical properties such as α. The amount of light that reaches a depth z in the sea is given by $I(z) = I_0 e^{-kz}$, where I_0 is the intensity at the surface. The important parameter for ecology

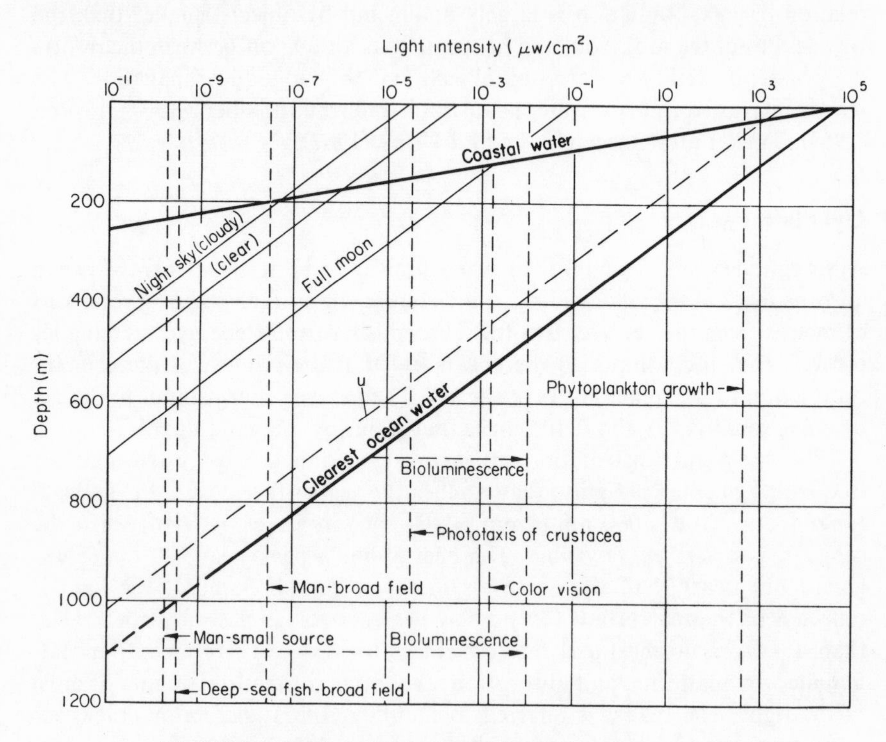

Fig. 2.2. Penetration of light into the sea (from Clarke 1971).

is the amount of solar radiation (the photon flux) in the photosynthetic waveband. For a discussion of measuring procedures, etc., the reader is referred to Tyler (1973). Figure 2.2 gives an indication of light intensity in the sea versus depth. Near the surface, the direction of maximum light intensity is from the direction of the refracted beam of solar radiation, but moves toward the vertical with increasing depth (Duntley, 1963). The amount of upward scattered light is indicated by u in Figure 2.2. The dashed vertical lines indicate various ranges and limits of light intensity levels.

The transparency of the ocean may be defined in terms of k, or, crudely as the Secchi disc depth. The venerable Secchi disc (a flat white plate) of 25 to 50 cm (nominal standard is 30 cm) diameter is frequently used as a simple device to estimate k and depth of the euphotic zone. The depth ($z \cdot_{SD}$) at which a Secchi disc, lowered vertically from the surface, just disappears corresponds roughly to the depth at which the surface light is reduced to about 10 per cent (see Tyler, 1968 for a more complete discussion). Since the light attenuates as e^{-kz}, this implies $k \simeq 2{\cdot}3\ (z \cdot_{SD})^{-1}$, or the Secchi depth is very roughly 2·3 attenuation lengths (attenuation length $= k^{-1}$). A more common, and perhaps more accurate, estimation formula is $k = 1{\cdot}7\ (z \cdot_{SD})^{-1}$ (Sverdrup *et al.* 1942, Strickland 1958).

The dynamics governing motions in the oceans

The motion of a fluid is governed by the conservation laws of momentum (Newton's Second Law) and mass (equation of continuity), the equation of state, and the laws of thermodynamics. These will provide a sufficient number of independent equations to match the unknowns (the velocity and the state variables) in problems of geophysical fluid dynamics but no guarantee of tractability. Our objective here is heuristic and we shall only attempt to point out some of the important dynamical concepts and balances in the momentum equations.

The equations of motion

The momentum equations are simply an expression of Newton's Second Law ($F = ma$) which states that the acceleration of a mass equals the applied force divided by the mass. The equations are given below in a form usual for oceanographic applications (Sverdrup 1957). Ignoring the curvature of the earth, which however becomes important for scales of motion extending over more than about 1000 km, the equations are written in a locally Cartesian coordinate system, at rest relative to the rotating earth. The (x,y,z) axes and (u,v,w) components of velocity are positive eastward, northward, and upward from the mean free sea surface:

$$\frac{Du}{Dt} = \frac{\partial u}{\partial t} + u\frac{\partial u}{\partial x} + v\frac{\partial u}{\partial y} + w\frac{\partial u}{\partial z} = -\frac{1}{\rho}\frac{\partial p}{\partial x} + fv + F_x \qquad (2.3)$$

$$\frac{Dv}{Dt} = \frac{\partial v}{\partial t} + u\frac{\partial v}{\partial x} + v\frac{\partial v}{\partial y} + w\frac{\partial v}{\partial z} = -\frac{1}{\rho}\frac{\partial p}{\partial y} - fu + F_y \qquad (2.4)$$

$$\frac{Dw}{Dt} = \frac{\partial w}{\partial t} + u\frac{\partial w}{\partial x} + v\frac{\partial w}{\partial y} + w\frac{\partial w}{\partial z} = -\frac{1}{\rho}\frac{\partial p}{\partial z} - g + F_z \qquad (2.5)$$

The left-hand side is the fluid acceleration, i.e., the time derivative of the velocity of a fluid element as it moves in space. In a fluid one must distinguish between 'local' change and the changes experienced by an individual fluid element as it moves. Local changes are measured by fixed sensors and are represented by partial derivatives with respect to time (this is called an Eulerian description). The change experienced by the individual fluid element is measured by a sensor moving with the element (Lagrangian description). The total time derivative D/Dt, i.e. the derivative 'following the motion', can be expressed in Eulerian terms by expansion $D/Dt = \partial/\partial t + u\partial/\partial x + v\partial/\partial y + w\partial/\partial z$. The total time derivative of the velocity is the acceleration of the fluid element as it moves, and hence is the acceleration term in $F = ma$.

In the equation of motion, the first terms on the right represent the forces (per unit mass) arising from pressure gradients (which may result from density differences, slope to the sea surface, etc.). The simple form of Newton's Second Law, force = mass multiplied by acceleration, holds only in a 'fixed' or non-accelerating coordinate system. When the law is written in an accelerating system additional terms appear. The second terms are the Coriolis terms, an apparent acceleration arising because the coordinate system is rotating with the earth: objects moving in a straight line relative to a 'fixed' celestial coordinate system appear to be accelerating when viewed from a coordinate system fixed to rotating earth. The Coriolis acceleration (force per unit mass) is equal to twice the vector product of the earth's rotation vector and the fluid velocity vector (see Neumann & Pierson 1966, pp. 125–6). The horizontal components are fv and $-fu$ where $f = 2\Omega \sin \theta$, where $\Omega = 0{\cdot}76 \times 10^{-4}$ radians. $\sec^{-1}$, the earth's rotation rate; θ is the latitude; and f is called the Coriolis parameter. The vertical component of the Coriolis force is negligible in comparison with the gravity term and is thus neglected in the vertical component equation.

The centrifugal effects of the earth's rotation are incorporated in 'gravity', since the free surface of the sea at rest will be perpendicular to the vector sum of the gravitational and centrifugal forces; this defines the local vertical. For most oceanographic cases (an exception being the relatively high accelerations in surface wind waves with periods of seconds) there is near balance between gravity and the vertical pressure gradient, i.e., there is no net force in the vertical:

$$\frac{Dw}{Dt} = 0 = -\frac{1}{\rho}\frac{\partial p}{\partial z} - g, \text{ or } \frac{1}{\rho}\frac{\partial p}{\partial z} = -g, \tag{2.6}$$

This is referred to as the hydrostatic equation and the pressure at any depth can be found by integration: $p(z) = \int_0^{=z} \rho g dz + p(z = 0)$, where $p(z = 0)$ is the atmospheric pressure. Since ρ varies by only a few per cent in the ocean, the pressure increases almost linearly with depth, as was stated earlier. The tide producing forces are small, the order of 10^{-4} dyne gr^{-1} (cm sec^{-2}), compared with hydrostatic pressure gradient term, which is the order of 10^3 dynes gr^{-1} (cm sec^{-2}). The tide producing forces act principally through the horizontal equations. Since tides will not be discussed in this chapter, the tidal terms are not included (see Neumann & Pierson 1966, Ch. 11).

The third term on the right represents the friction forces. For example, in moving water where the velocity varies spatially, frictional stresses are present as a result of momentum transfer between layers of different velocity. In laminar flow, where individual layers of fluid slide over each other, the exchange of momentum is the result of random molecular motion. If there is a velocity shear, say, $\frac{\partial u}{\partial z}$, then a frictional stress $\tau = \mu \frac{\partial u}{\partial z}$ (dynes cm^{-2}) will occur between the adjacent layers. The coefficient μ is the molecular dynamic viscosity and is of the order of 0·01 gr cm^{-1} sec^{-1} for water. The resultant force on a unit volume of fluid is $\mu \frac{\partial^2 u}{\partial z^2}$. In the equations of motion, the effects of molecular viscosity can be somewhat generally written (assuming incompressibility of the fluid) as F_x, etc. $= \frac{\mu}{\rho}\left[\frac{\partial^2 u}{\partial x^2}+\frac{\partial^2 u}{\partial y^2}+\frac{\partial^2 u}{\partial z^2}\right]$, etc. The coefficient μ/ρ has units cm^2 sec^{-1} and is sometimes called the kinematic viscosity.

These equations of motion are known as the Navier–Stokes equations, and were developed by hydrodynamicists in the first half of the nineteenth century. Using them, we should be able, in principle, to describe the motions in the oceans—although the complexity and non-linearity of the system of equations, as well as nature, thwart us.

Turbulence

In nature fluid motions are rarely laminar with sheets of water moving with steady speed. Almost always there are rapid and complex flows or eddies, that can best be described as random, which are superimposed on, and interact with, the simpler mean flow patterns. This irregular condition of flow is called turbulence. A definition of turbulence is not easily given. Some main characteristics of turbulent flow are: (1) the disorder of the flow, (2) the irreproducibility in detail of the flow pattern no matter how carefully conditions are duplicated, (3) the efficient mixing done by turbulence and its ability to transport and exchange momentum and dissolved substances. Turbulence cannot be discussed without statistical methods and the statistical properties of the flow, e.g. correlations, variance, means, are well defined and reproducible. Turbulence is the result of an instability in the flow, i.e. a small random perturbation in the flow continues to grow, involving more of the fluid, rather than diminishing and disappearing. Turbulence in nature is usually the result of either convective instability, in which the kinetic energy of the turbulent motion is derived from the potential energy of the fluid, or shear flow instability, where the turbulence is fed from the kinetic energy of the mean motion. The description and treatment, theoretically and experimentally, of turbulence and the interaction between the various scales of motion are immensely difficult. The reader is referred to Bowden (1964), Phillips (1966), Monin and Yaglom (1971) and Tennekes and Lumley (1972) for a more complete and rigorous discussion than can be given here.

We can obtain some insight into the problem of turbulence by attempting to separate the 'mean' and turbulent fluctuating quantities. Let us write the variables in the Navier–Stokes equations of motion, u, v, etc., in the form $u = \langle u\rangle + u'$, $v = \langle v\rangle + v'$, etc. The brackets $\langle\ \rangle$ represent a time average sufficient to 'average' out the random turbulent fluctuations, and hence the quantities $\langle u\rangle$, $\langle v\rangle$, etc., represent the 'mean' flow components, which may be a function of time (over time scales greater than the averaging time) and position. The u', v', etc., represent the random fluctuating turbulent quantities with zero mean; $\langle u'\rangle$, $\langle v'\rangle$, etc., are zero by definition. Thus the instantaneous flow components, u, etc., equal $\langle u\rangle + u'$, etc.

A complete rigorous treatment of turbulence following this approach requires the use of tensors, which is beyond the scope of this chapter. But let us proceed heuristically and express the acceleration terms $u\dfrac{\partial u}{\partial x}$, etc., in the Navier–Stokes equations in terms of $u = \langle u\rangle + u'$, etc., e.g.:

$$u\frac{\partial u}{\partial x} = (\langle u\rangle + u')\frac{\partial}{\partial x}(\langle u\rangle + u') = \langle u\rangle\frac{\partial\langle u\rangle}{\partial x} + \langle u\rangle\frac{\partial u'}{\partial x} + u'\frac{\partial\langle u\rangle}{\partial x} + u'\frac{\partial u'}{\partial x}. \tag{2.7}$$

If now we take the time average $\langle\ \rangle$ of this equation we obtain:

$$\langle u\frac{\partial u}{\partial x}\rangle = \langle u\rangle\frac{\partial\langle u\rangle}{\partial x} + 0 + 0 + \langle u'\frac{\partial u'}{\partial x}\rangle \tag{2.8}$$

since $\langle u\rangle$ is already averaged and $\langle u'\rangle \equiv 0$, $\langle u'\dfrac{\partial\langle u\rangle}{\partial x}\rangle = \langle u'\rangle\dfrac{\partial\langle u\rangle}{\partial x} = 0$ and $\langle\langle u\rangle\dfrac{\partial u'}{\partial x}\rangle = \langle u\rangle\dfrac{\partial\langle u'\rangle}{\partial x} = 0$. On the other hand, the averages of terms containing products of the fluctuating quantities aren't necessarily zero if the fluctuating quantities are correlated, e.g., in the simple case of $\langle u'\dfrac{\partial u'}{\partial x}\rangle = \dfrac{1}{2}\dfrac{\partial}{\partial x}\langle u'u'\rangle = \dfrac{1}{2}\dfrac{\partial\langle(u')^2\rangle}{\partial x} \not\equiv 0$, since $(u')^2 \geq 0$. If we average the Navier–Stokes equations to obtain the 'mean' equations of motion we obtain for the acceleration, omitting terms that are zero:

$$\left\langle\frac{Du}{Dt}\right\rangle = \frac{\partial\langle u\rangle}{\partial t} + \langle u\rangle\frac{\partial\langle u\rangle}{\partial x} + \langle v\rangle\frac{\partial\langle u\rangle}{\partial y} + \langle w\rangle\frac{\partial\langle u\rangle}{\partial z} + \\ + \left\langle u'\frac{\partial u'}{\partial x}\right\rangle + \left\langle v'\frac{\partial u'}{\partial y}\right\rangle + \left\langle w'\frac{\partial u'}{\partial z}\right\rangle \tag{2.9}$$

Rewriting to express the acceleration of the 'mean' flow explicitly, we have:

$$\frac{\partial\langle u\rangle}{\partial t}+\langle u\rangle\frac{\partial\langle u\rangle}{\partial x}+\langle v\rangle\frac{\partial\langle u\rangle}{\partial y}+\langle w\rangle\frac{\partial\langle u\rangle}{\partial z}=$$

$$-\left\langle u'\frac{\partial u'}{\partial x}\right\rangle-\left\langle v'\frac{\partial u'}{\partial y}\right\rangle-\left\langle w'\frac{\partial u'}{\partial z}\right\rangle \quad (2.10)$$

Thus the terms $\left\langle u'\frac{\partial u'}{\partial x}\right\rangle$, etc., act as a virtual stress generated by the turbulent fluctuations and act on the mean flow. These terms are called the Reynolds stress. The Reynolds stress can be directly measured, but with difficulty, by suitably small sensors with fast response times. Since our measurements and theoretical applications of the equations of motion generally deal with motions which are in effect averages over some scale determined by the phenomenon studied, and since we *cannot* ignore turbulence and the interactions of other scales, it is desirable to attempt to express the Reynolds stresses (or turbulent eddy transports) parametrically as functions of the averaged variables which are measured. This is done using empirically estimated coefficients of eddy transport:

$$-\left\langle u'\frac{\partial u'}{\partial x}\right\rangle, -\left\langle v'\frac{\partial u'}{\partial y}\right\rangle, -w'\frac{\partial u'}{\partial z}\right\rangle, \text{ etc.}$$

become

$$\frac{\partial}{\partial x}\left(A_x\frac{\partial u}{\partial x}\right), \frac{\partial}{\partial y}\left(A_y\frac{\partial u}{\partial y}\right), \frac{\partial}{\partial z}\left(A_z\frac{\partial u}{\partial z}\right), \text{ etc.}$$

where u, etc., are now used to represent the 'mean' or averaged velocity components. We generally do not explicitly indicate averages, e.g. $\langle u\rangle$, but we should implicitly recognize that the variables we measure are 'averaged'.

In the equations of motion these terms supplant, for turbulent flow, the molecular viscosity terms. The turbulent viscosity coefficients, A_x, A_y, A_z, have the same dimensions as kinematic molecular viscosity, $\frac{\mu}{\rho}$, but their value is much larger. Because of the restraining effect of stratification, generally $A_z \ll A_x$, A_y. The typical order of magnitudes of $\frac{\mu}{\rho}$: A_z:A_x or A_y are about 10^{-2}:10:10^6 cm^2 sec^{-1}. Turbulent viscosity is not an intrinsic property of the fluid but varies with location, stability, averaging period, etc. Pedlosky (1971) has described the situation: 'For large-scale motions, the coefficient of viscosity is often assumed to also represent or parameterize the momentum transport of small-scale random eddies. This parameterization [*sic*] of the interaction of the small- and large-scale motions is accomplished by increasing the size of μ from its molecular value, supposedly because of the greater efficiency of smoothing of momentum on a large scale by the transport of great blocks of fluid across the mean momentum gradient. . . . In some cases it is clearly wrong to imagine that small-scale turbulence acts on the large-scale flow as a beefed-up group of molecules. In those cases when it is qualitatively and empirically correct it is still difficult to assign with confidence numerical values to the turbulent viscosity. However, it is also absurd to rely totally on molecular viscosity as the agent for the dissipation of momentum. More sensible results are obtained from the use of the larger turbulent viscosities, but any such results, while viewed with sympathy, must also be viewed with suspicion. Any dynamical theory which depends critically on a specific value of the turbulent viscosity must be considered as having a shaky foundation indeed.'

We shall later see that the same general considerations and problems are encountered in the distribution of substances in the ocean—for turbulence transports not only momentum but also substances—and the same parameterizations are used. One of the important but unsolved problems in geophysical fluid dynamics is that of finding more adequate representation of turbulence.

An example of one important oceanographic phenomenon in which turbulence plays the central role is the growth of the mixed surface layer. A turbulent state leads rapidly to a uniform distribution of properties and the fluid becomes well mixed. The mixed surface layer grows in this manner as the turbulent momentum from wind and waves is transferred downward. In stably stratified water ($N^2 > 0$) a vertically displaced parcel of water has to work against buoyancy forces, i.e. the lighter parcel tends to 'float' above the denser water and hence resists being 'mixed' downward. If the kinetic energy available in the turbulence is sufficient to overcome the stratification, the flow becomes dynamically unstable. The rate at which kinetic energy is transferred from the mean motion scales to turbulent motion is proportional to the product of the Reynolds stress and the vertical gradient of the horizontal velocity field. As discussed above, N^2 (N = Väisälä–Brunt frequency) is a measure of the 'stability' of a fluid due to stratification. A measure of the dynamic stability of a stratified shearing fluid is the Richardson number, $R_i = N^2 / \left(\frac{\partial u}{\partial z}\right)^2 + \left(\frac{\partial v}{\partial z}\right)^2$. A sufficient condition for stability in a shearing flow is $R_i > \frac{1}{4}$ everywhere (Phillips 1966).

The equation of continuity

The conservation of mass states that the density of a fixed volume of a continuous fluid must change if there is a net flux of mass into or out of the volume. The equation of continuity expresses this:

$$\frac{\partial \rho}{\partial t}+\frac{\partial(\rho u)}{\partial x}+\frac{\partial(\rho v)}{\partial y}+\frac{\partial(\rho w)}{\partial t}=0 \tag{2.11}$$

This can be expanded and rewritten as:

$$\frac{1}{\rho}\frac{D\rho}{Dt}+\left(\frac{\partial u}{\partial x}+\frac{\partial v}{\partial y}+\frac{\partial w}{\partial z}\right)=0. \tag{2.12}$$

If we assume, as is valid for most considerations of ocean flow, that the fluid is approximately incompressible, i.e. the change in the density of a fluid element is small compared with its density, then the equation of continuity becomes simply:

$$\frac{\partial u}{\partial x}+\frac{\partial v}{\partial y}+\frac{\partial w}{\partial z}=0 \tag{2.13}$$

Some specific examples of dynamic balances

The present state of oceanography is such that we do not and cannot study the global ocean in all its scales simultaneously—but rather, study specific problems and look for the dynamical mechanisms, processes, and balances important in them. Sophisticated and reasonably successful numerical models, which integrate the relatively complete equations of motions by numerical methods on large computers, do exist and are employed in problems ranging from the general circulation (Bryan & Cox 1967, 1968, 1972) to response of the ocean to a hurricane (O'Brien & Reid 1967). The fact that numerical models are mentioned only in passing in this chapter does not imply their unimportance or any benign neglect by the author. The complication and necessary attention to computational procedure does, however, sometimes obscure the basic physics and dynamical processes to the uninitiated. We shall primarily concern ourselves, in this introductory chapter, with some simple approximations to the equations of motion that reveal some features of the dynamics of the oceans.

The various terms in the equations of motion are generally not of the same order; which dominate and which can be neglected depends on the scales of the motion. Scale analysis is a useful procedure for indicating the important terms (see Kraus 1972, Section 1.4; Pedlosky 1971). For example, an indication of the importance of the Coriolis term is given by the Rossby number: $R_0 = VL^{-1}\,\Omega^{-1}$ where V is the characteristic horizontal velocity scale, L the characteristic horizontal length scale, Ω the earth's rotation rate. If $R_0 \ll 1$ then the Coriolis force is important.

GEOSTROPHY

An important feature of the large-scale motions (say $L \gg V\Omega^{-1} \sim 10$ km) in the ocean is the near balance between the horizontal components of the pressure gradient and the Coriolis force. The acceleration and the other terms are generally small in comparison, and the geostrophic (from the Greek for 'earth-turned') approximation reduces the equations to:

$$fv = \frac{1}{\rho}\frac{\partial p}{\partial x} \tag{2.14}$$

$$-fu = \frac{1}{\rho}\frac{\partial p}{\partial y} \tag{2.15}$$

$$-g = \frac{1}{\rho}\frac{\partial p}{\partial z} \tag{2.16}$$

As a comparison of magnitude we note that in mid-latitudes $f \simeq 10^{-4}$ sec^{-1}, and u, $v \simeq 10$ to 100 cm sec^{-1} (and greater in the Gulf Stream); thus the magnitude of the horizontal terms are 10^{-3} to 10^{-2} cm sec^{-2}. If a flow of 50 cm/sec (1 knot) is reduced to 0 in a week, the acceleration $\left(\frac{Du}{Dt}, \frac{Dv}{Dt}\right)$ is of magnitude 2×10^{-4} cm sec^{-2}. The tidal forces, as we have already mentioned, are also of order 10^{-4} cm sec^{-2}.

Flows which satisfy the above equations are called geostrophic. The equations show that as a consequence of the Coriolis force the horizontal geostrophic velocities are perpendicular to the horizontal pressure gradient; the horizontal flow is thus along lines of constant pressure. This relationship is used in weather maps—and is of wide practical use in estimating the horizontal velocities in the ocean. The details of the computational process are given in Sverdrup *et al.* (1942) or Neumann & Pierson (1966). The procedure follows from noting that the set of above equations can be manipulated to:

$$\frac{\partial(\rho f v)}{\partial z} = -g\frac{\partial \rho}{\partial x} \tag{2.17}$$

$$\frac{\partial(\rho f u)}{\partial z} = +g\frac{\partial \rho}{\partial y} \tag{2.18}$$

These equations are the so-called thermal wind equations of meteorology: a horizontal density gradient implies a velocity gradient in the vertical. It is apparent that if the density of sea water as a function of depth is known along a section, the vertical gradient in the velocity perpendicular to the section could be obtained. If one knows, or can assume, a velocity at some depth (usually one assumes a 'level of no motion' sufficiently deep that low velocities are a reasonable assumption), then u and v can be found by numerical integration of the thermal wind equations. Thus, it is not necessary to measure the horizontal pressure gradient directly, which cannot be done with sufficient precision, but only the density field, which can be measured quite accurately, in order to calculate geostrophic velocities. This (and ships' drift) has been the basis of much of our knowledge of the large-scale ocean circulation—and the reason for much of the hydrographic sampling to obtain the density structure of the oceans.

The total pressure field in the density stratified ocean can be considered to be composed of the relative field of pressure due to differences in the density field, and the 'slope' field of pressure due to the slope of the sea surface. (If there is a level of no motion in geostrophic flow it implies the density field has compensated for the sea surface slope. One can estimate the topography of the sea surface this way, and it is referred to as the dynamic topography.) The terms barotropic and baroclinic are sometimes used to refer to velocity fields that are independent of depth, and a function of depth, respectively. Strictly (and etymologically correctly) barotropic implies that the isobaric (equal pressure) and isosteric (equal density) surfaces coincide, and baroclinic implies the surfaces intersect.

THE EFFECT OF THE WIND

When the wind blows over the ocean it exerts a force on the sea surface in the direction of the wind, i.e. momentum of the wind is transferred to the ocean, tending to drag the surface water with the wind. The exact process, or mechanism, by which this is done is complex and incompletely understood. The wind stress (force per unit area) could be obtained by measuring the Reynolds stress (turbulent transfer of momentum) at the air–sea boundary. Measuring the Reynolds stress is difficult and it is not done routinely. Empirically, it has become apparent that a good estimate of the wind stress can be made using a square law and a drag coefficient determined from measurements of the Reynolds stress. By this law, the wind stress vector has the direction of the wind and is proportional to the product of the air density, ρ_a, a dimensionless drag coefficient, $C_D \simeq 1{\cdot}5 \times 10^{-3}$, and the square of the wind speed:

$$\tau = \rho_a C_D \| V_{\text{wind}} \| V_{\text{wind}}$$

Hellerman (1967) gives seasonal average wind stress for the world ocean.

The work of Ekman (1905) provided the basis for our understanding of the effect of wind stress on ocean circulation. Nansen, during the drift of the *Fram* across the polar sea in 1893–6, observed that the ice drift was at 20° to 40° to the right of the wind. He attributed this to the effect of the Coriolis force, and inferred that the water below, dragged along by the ice or water above, must be deflected even further to the right. Ekman, motivated by these observations, produced the theoretical explanation in his classic paper of 1905. He assumed a balance existed between the Coriolis force and the frictional forces resulting from the vertical turbulent transport of momentum from the wind and down into the sea. Using a constant eddy viscosity, the equations of motion are simply:

$$0 = fv + \frac{A}{\rho}\frac{\partial^2 u}{\partial z^2} \tag{2.19}$$

$$0 = -fu + \frac{A}{\rho}\frac{\partial^2 v}{\partial z^2} \tag{2.20}$$

A simple elegant solution is obtained (Ekman 1905, see, e.g., Kraus 1972) by requiring that the wind stress, τ (force per area), matches the shear stress in the fluid at the surface. The surface boundary condition becomes:

$$\tau_x = A\left.\frac{\partial u}{\partial z}\right|_{z=0}; \ \tau_y = A\left.\frac{\partial v}{\partial z}\right|_{z=0} \tag{2.21, 22}$$

The velocities and velocity gradients are assumed to vanish at great depth.

The resulting surface velocity is 45° to the right (left) of the wind direction in the northern hemisphere (southern hemisphere) and decreases exponentially with depth with the direction rotating to the right (left) with increasing depth. This logarithmic spiral is called the Ekman spiral. The surface speed is $\tau(fA\rho^2)^{-\frac{1}{2}}$ and has decreased by $e^{-\pi}$ at $z = -\pi(2A/f)^{\frac{1}{2}}$. That depth is called the Ekman depth or depth of frictional influence, and the layer from the boundary to that depth is called the Ekman layer. An Ekman layer will occur along any boundary which exerts a frictional drag on the fluid.

A very important result can be obtained by integrating the above equations of motion with depth to obtain the net mass transport due to the wind (the Ekman transport);

$$M_x = \int_{z=-D}^{0} \rho u dz = \frac{\tau_y}{f} \tag{2.23}$$

$$M_y = \int_{z=-D}^{0} \rho v dz = -\frac{\tau_x}{f} \tag{2.24}$$

The lower limits of integration are formally indicated as $-D$, where D is some depth sufficiently great that the stresses (velocity gradients) vanish. D can be taken as the Ekman depth, since the velocities decay exponentially. The Ekman transport is directed 90° to the right of the wind in the northern hemisphere (and to the left in the southern hemisphere). The total flow in the directly driven Ekman layer is relatively small compared to the flow in the major ocean currents. Nevertheless, the wind is a major driving force for ocean circulation; the wind redistributes the mass in the ocean to set up the density differences and pressure gradients. Thus, the wind indirectly drives major ocean circulation. Wherever and whenever the wind stress varies spatially, convergence, or divergence, of the mass transport in the surface layer will occur, causing a 'pushing' downward, or an 'upwelling' of the subsurface water.

The divergence of the mass transport along coastal boundaries results in the upwelling of cold (and nutrient rich) water off the coasts of Peru, south-west and north-west Africa, and California and Oregon. In these regions the prevailing winds, for months or longer blow nearly parallel to the coast in an equatorward direction. The Ekman transport moves the surface water away from the coast and the water must be replaced from depth. This water is cooler and richer in nutrients; biologically, the coastal upwelling regions are among the most important regions in the oceans. As Ryther (1969) has discussed, the high primary productivity resulting from plentiful nutrients and the shortness of the food chain in the upwelling regions may enable half of the world's fish supply to be produced there. Bakun (1973) has computed an 'upwelling index' for fisheries applications that is based on the Ekman transport. The reader is referred to Smith (1968) for a more extensive review of the physical oceanography of upwelling. More recently Dietrich (1972) has edited a collection of papers on upwelling and its consequences, and O'Brien (1975) has reviewed the status of models of coastal upwelling. Considerable work is being done to develop numerical models of coastal upwelling circulation (O'Brien & Hurlbert 1972, Thompson & O'Brien 1973, McNider & O'Brien 1973).

INERTIAL OSCILLATIONS

Consider setting a parcel of water in motion: since whenever a fluid has motion the Coriolis force acts (and if no other force continues to act after a fluid has been set in motion) the momentum equations are simply:

$$\frac{Du}{Dt} \sim \frac{\partial u}{\partial t} = fv \tag{2.25}$$

$$\frac{Dv}{Dt} \sim \frac{\partial v}{\partial t} = -fu \tag{2.26}$$

which can be combined to give a set of equations:

$$\frac{\partial^2 u}{\partial t^2} + fu = 0; \frac{\partial^2 v}{\partial t^2} + fv = 0 \text{ with solutions } u,v \sim e^{ift} \qquad (2.27, 28)$$

No work can be done by the Coriolis force since it acts at right angles to the motion—but the parcel continues to accelerate in a direction perpendicular to parcel's velocity. Hence, in the northern (southern) hemisphere the parcel moves clockwise (counter-clockwise) in a horizontal circle of radius $(u^2 + v^2)^{\frac{1}{2}} f^{-1}$ with a period $2\pi/f$ (where f is in radians per unit time), called the inertial period. Thus, there is a natural frequency of oscillation in the horizontal flow due to the Coriolis force. In general there are other forces acting which nearly balance the Coriolis force so *pure* inertial oscillations are rarely observed alone. The inertial oscillation is, however, the oceans' 'characteristic' response to a transient horizontal impulse. Current meter records in the ocean often show rather striking evidence of oscillations at frequencies close to f, the local value of the Coriolis parameter.

Vorticity

Vorticity is a concept of central importance in large-scale ocean motions. Fluid vorticity is equivalent to angular velocity: if a small fluid sphere with vorticity ω were instantaneously solidified, its angular velocity (radians per unit time) would be $\omega/2$. The strong stratification in the ocean tends to suppress vorticity about a horizontal axis but the component of vorticity about the vertical axis tends to be conserved. A parcel of fluid on the earth has a *planetary* vorticity equal to the Coriolis parameter, f, arising from its rotation with the earth—and it may also have a relative vorticity, ω, given by

$$\omega = \frac{\partial v}{\partial x} - \frac{\partial u}{\partial y} \tag{2.29}$$

where u and v are measured in the reference frame relative to the earth. A paddle wheel in the fluid with its axis vertical will rotate with angular velocity $\omega/2$ if the fluid has a relative vorticity (ω) at that point. The vorticity is positive if the wheel rotates counter-clockwise (cyclonically). In the absence of friction and any change in vertical extent of the fluid parcel, $\omega + f =$ constant. For example, if fluid in the northern hemisphere moves south toward the equator, f decreases, and so the relative vorticity, ω, increases—or, as descriptively put by Longuet-Higgins (1965): 'Roughly we may say that the fluid finds the Earth "spinning more slowly" beneath it and so appears to spin faster (i.e. in an anticlockwise sense) relative to the earth.'

It is often useful to consider the dynamics of flow in terms of vorticity. The vorticity equation can readily be obtained from the momentum equations and a useful approximate form is:

$$\frac{D(\omega+f)}{Dt} = f\frac{\partial w}{\partial z} + \left(\frac{\partial F_y}{\partial x} - \frac{\partial F_x}{\partial y}\right) \tag{2.30}$$

Since f is only a function of latitude, the term $\frac{Df}{Dt}$ is frequently written βv, where $\beta = \partial f/\partial y$ and the approximation made from f varies linearly with latitude.

$$\frac{D\omega}{Dt} + \beta v = \frac{f\partial w}{\partial z} + \left(\frac{\partial F_y}{\partial x} - \frac{\partial F_x}{\partial y}\right) \tag{2.31}$$

The term βv expresses the change in the planetary vorticity a parcel of fluid would experience if it moves north or south. The $f\frac{\partial w}{\partial z}$ term expresses the change in vorticity due to a change in the vertical extent of the parcel of fluid: The equation of continuity requires a horizontal convergence if $\frac{\partial w}{\partial z} \neq 0$, and hence the 'moment of inertia' of the fluid changes—and the vorticity changes to conserve angular momentum. The $\frac{\partial F_y}{\partial x} - \frac{\partial F_x}{\partial y}$ represent the effect on vorticity by frictional effects, i.e., the wind stress on the surface of the ocean. Smith (1968) discusses the use of vorticity in upwelling dynamics and in the next section we shall discuss its central role in wind-driven circulation theory.

The circulation and distribution of properties

The sun is ultimately responsible for the oceanic circulation—it acts indirectly through the differential heating of the atmosphere which causes the winds to blow—and acts directly by heating the ocean with solar radiation. The mechanisms and the resulting motions are coupled and interactive but it has been conceptually useful to separate the circulation of the ocean into the *wind-driven* and the *thermohaline*. The wind is the predominant force driving the circulation in the upper kilometer of the ocean. The thermohaline circulation is driven by density differences resulting from the heat fluxes and salt fluxes, due to evaporation and precipitation, at the surface. The deep circulation is the result of global thermohaline processes: cooling and freezing occurring at high latitudes, and warming and evaporation at lower latitudes.

Our general knowledge of the surface circulation is derived largely from ships' drift observations (the difference between the course and speed set, and that made good) and from geostrophic calculations based on the observed density distribution. Below a kilometre or so the density gradients become too small to give meaningful geostrophic estimates and the abyssal circulation must largely be inferred from the distribution of the natural tracers provided in the ocean: temperature, salinity, dissolved oxygen, etc. It is convenient for theoretical purposes and from the point of the observational methodology to discuss separately the near surface wind-driven circulation, and the abyssal thermohaline circulation.

The wind-driven circulation system

On either side of the equator the winds have a tendency to blow from east to west—these are the trade winds. Further north and south (about 40° latitude) the winds blow from west to east—these are the westerlies. (In meteorology it is customary to refer to the direction *from* which winds blow; in oceanography, one gives the direction *toward* which currents flow.) The major ocean current pattern is a large circulation gyre—clockwise in the northern hemisphere, counterclockwise in the southern hemisphere.The major ocean current pattern follows generally the pattern of the winds: westward near the equator, eastward near 40°.

There are, however, striking discrepancies between the winds and the currents: just to the north of the equator, in the region of the weak westward doldrums, there is an equatorial counter current (eastward) between the north and the south westward equatorial currents; even more striking is the fact that the strongest major currents all occur along the western boundary of the oceans. These western boundary currents, e.g. the famous Gulf Stream in the north Atlantic and Kuroshio in the north Pacific, are narrow, strong (4 knots, or 200 cm/sec) currents flowing generally poleward. They differ markedly from the broad, diffuse, and slow (20 cm/sec) equatorward flow along the eastern boundaries, e.g. the California Current. There is no such asymmetry in the general wind pattern, and there was no convincing explanation for these features of the oceanic circulation until the late 1940s (see Stommel 1960, for a complete and excellent discussion).

The explanation is easiest in terms of vorticity. It is a basic fact of observation that over the greater part of the ocean the mean currents are slow and, therefore, for the large-scale circulation the relative vorticity ω is negligible compared with f. If the vorticity equation is integrated from the surface to the bottom, which is assumed to be at the same depth everywhere, the integral of $\frac{\partial w}{\partial z}$ is zero, since the mean vertical velocity at the sea surface and bottom must be zero, and the following integral form of the vorticity equation is obtained:

$$\beta M_y = \frac{\partial \tau_y}{\partial x} - \frac{\partial \tau_x}{\partial y} + \text{frictional effects} \tag{2.32}$$

$M_y = \int_{\text{bottom}}^{\text{surface}} \rho v dz$ and τ_x, τ_y are the wind stress components acting on the surface. Only flow in the meridional direction experiences a change in planetary vorticity and thus only M_y (and not M_x) appears explicitly in the vorticity equation.

It is generally assumed the flow at depth is small and the stress along the bottom negligible. In the absence of frictional effects, which may be expected to be primarily important along boundaries, a balance exists between the rate of change of the planetary vorticity and the input of vorticity by the wind:

$$\beta M_y = \frac{\partial \tau_y}{\partial x} - \frac{\partial \tau_x}{\partial y} \tag{2.33}$$

The right-hand side of the equation is the curl (in the terms of vector analysis) of the wind stress. This relation was first noted by Sverdrup (1947). Applying it and the equation of continuity in the equatorial regions, where the wind stress varies with latitude but is mainly westward, he found the mass transport in the east–west direction M_x is proportional to $\frac{\partial}{\partial y}\left(\frac{\partial \tau_x}{\partial y}\right)$. Hence, the eastward equatorial countercurrent developed in the region of minimum wind stress (the doldrums).

Stommel's (1948) explanation of the existence of intense western boundary currents and Munk's (1950) more general theory of the wind-driven ocean

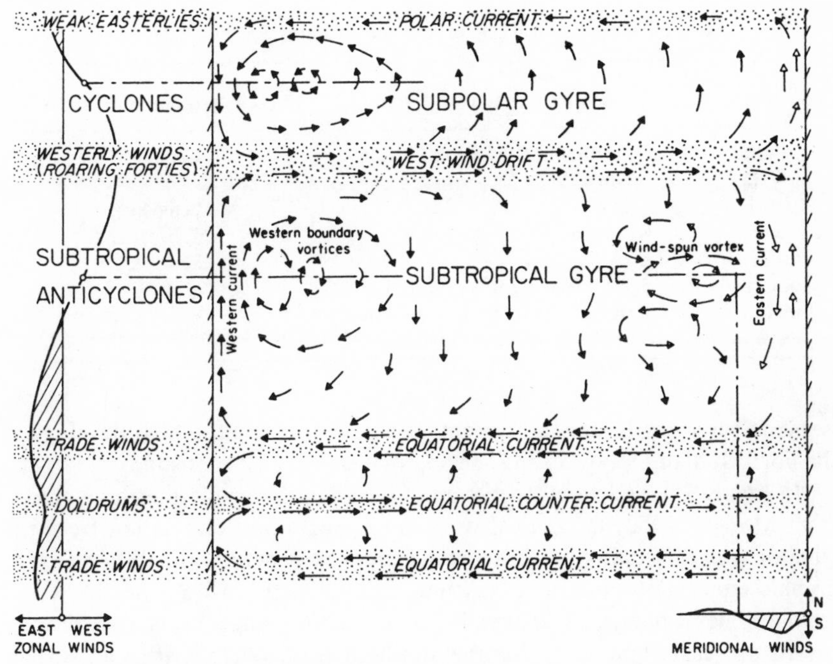

Fig. 2.3. Generalized schematic of the wind driven circulation (from Munk 1950) in ocean extending from equator to polar latitudes.

circulation also follow from the vorticity relation. Consider for example the north Atlantic, where the wind stress curl is negative (clockwise) between 20°N and 50°N, and $\beta = \partial f/\partial y$ is positive. Hence, from the above equation, the mass transport, M_y (which represents the net northward flow in a column of water) must be negative (southward) over nearly all the north Atlantic. It is obvious that somewhere this must break down: there must be a return northward flow somewhere so the total mass is conserved, and, furthermore, the total vorticity must be conserved since the wind doesn't spin up the oceans clockwise ever faster and faster. If an intense northward flow exists along the western boundary, both the necessary mass balance can be satisfied and counterclockwise (positive) vorticity can be supplied by frictional forces due to the 'drag' along the western boundary. Since the frictional forces along a boundary oppose the current, an intense northward current along the eastern boundary, instead of along the western boundary, would generate clockwise (negative) vorticity.

Similar reasoning applies in the southern hemisphere and leads again to equatorward flow generally, with narrow western boundary current flowing poleward (south). The monsoon regime in the Indian Ocean results in a

Table 2.4 Geographical names of some currents (See **Fig. 2.3**)

Location	Western boundary current	Eastern boundary current
N. Atlantic	Gulf Stream (including Florida Current)	Canary
S. Atlantic	Brazil	Benguela
N. Pacific	Kuroshio	California
S. Pacific	E. Australian	Peru
N. Indian	Somali (during S.W. monsoon)	—
S. Indian	Agulhas (including Mozambique)	W. Australian

'seasonal' western boundary current, the Somali current, which develops in the north Indian Ocean during the summer south-west monsoon.

In Figure 2.3 a schematic diagram of the wind-driven circulation is shown from Munk's (1950) theoretical work. The generalized circulation from the equator (the thermal equator and doldrums are generally north of the geographic equator) to the polar latitudes is applicable to both northern and southern hemispheres. The upside-down mirror image gives the 'proper' (north up, east right) map for the southern hemisphere, with the western boundary currents flowing poleward and the eastern boundary currents equatorward. In the southern hemisphere the west wind drift expands to replace the subpolar gyre with the eastward Antarctic Circumpolar Current. The width of the arrows is an indication of the strength of the currents, which result from zonal winds (filled arrowheads), meridional winds (open arrowheads), or both (half-filled arrowheads). The geographical names corresponding to the western and eastern boundary currents are given in Table 2.4. The strongest currents are the Gulf Stream, Kuroshio, and the Somali current systems with transports of the order of 60×10^6 m^3 sec^{-1}. All the eastern boundary currents are associated with major coastal upwelling which occurs under the prevailing equatorward winds along the eastern boundary. The eastern boundary currents are broad (1000 km width) compared with western boundary currents (100 km width), and have appreciably smaller transports (15×10^6 m^3 sec^{-1}); Wooster and Reid (1962) discuss eastern boundary currents in some detail. The general features of the large-scale circulation above 1000 m are thus understood in terms of the wind stress distribution.

The distribution of properties

Knowledge of the distribution of oceanographic parameters, such as temperature and salinity, and dissolved constituents can be used to infer the flow patterns and mixing processes occurring in the ocean. The equation governing the concentration (s: quantity per unit volume) of some quantity at a point is:

$$\frac{\partial s}{\partial t} = -\frac{\partial}{\partial x}(su)-\frac{\partial}{\partial y}(sv)-\frac{\partial}{\partial z}(sw)+\mu\left[\frac{\partial^2 s}{\partial x^2}+\frac{\partial^2 s}{\partial y^2}+\frac{\partial^2 s}{\partial z^2}\right]+R \qquad (2.34)$$

where μ is the appropriate molecular diffusion coefficient and R the source or sink for non-conservative properties, e.g. radioactive decay of radioactive constituents or consumption of dissolved oxygen. The terms $\frac{\partial}{\partial x}(su)$, etc., represent the advective flux—the flux of the quantity due to all the scales of the fluid flow—and hence the same considerations that led to eddy or turbulent viscosity apply. We measure some mean or averaged $\langle s \rangle$ and velocity components $\langle u \rangle, \langle v \rangle, \langle w \rangle$, but the net flux is $\langle su \rangle = \langle s \rangle\langle u \rangle + \langle s'u' \rangle$ where s', u', etc., are the turbulent or random fluctuations. The turbulent fluctuations have individual zero mean, but non-zero mean products $\langle s'u' \rangle$, if the fluctuations are correlated, and hence result in a net transport or flux. These eddy or turbulent transports are expressed parametrically in terms of measured average variables, as in the equations of motion and with the same reservations, by the introduction of eddy or turbulent diffusion coefficients (where s now is the 'mean' or averaged value):

$$\langle s'u' \rangle = -K_x \frac{\partial s}{\partial x}, \; \langle s'v' \rangle = -K_y \frac{\partial s}{\partial z}, \; \langle s'w' \rangle = -K_z \frac{\partial s}{\partial z} \qquad (2.35)$$

The turbulent diffusion coefficients have the same dimensions as the molecular coefficients but are much larger. In general $K_z \ll K_y$ or K_x since the vertical density stratification inhibits vertical motion. In terms of measured mean concentrations, the molecular diffusion terms are negligible compared with turbulent diffusion terms. The concentration equation, i.e. the advection–diffusion equation for the distribution of properties, can be written for an incompressible fluid in terms of 'averaged' s, u, v, and w:

$$\frac{\partial s}{\partial t} = -u\frac{\partial s}{\partial x}-v\frac{\partial s}{\partial y}-w\frac{\partial s}{\partial z}+\frac{\partial}{\partial x}\left(K_x\frac{\partial s}{\partial x}\right)+\frac{\partial}{\partial y}\left(K_y\frac{\partial s}{\partial y}\right)+\frac{\partial}{\partial z}\left(K_z\frac{\partial s}{\partial z}\right)+R \qquad (2.36)$$

Thus the concentration at a point may vary because of flux of the quantity with the mean flow if a gradient of concentration exists, because of turbulent diffusion, or as a result of sources or sinks of the quantity.

If the first three terms on the right-hand side (the advective fluxes) are written on the left-hand side, the left-hand side becomes equal to $\frac{Ds}{Dt}$, the concentration following the motion. In other words, the equation then expresses the change in concentration experienced by a sensor advected with the flow. If the quantity is conservative, then $R \equiv 0$, and the concentration following the motion will be affected only by turbulent diffusion. Away from the sea surface, salinity and potential temperature are conservative properties (true temperature is not because of the adiabatic effects). The concentration of radioactive quantities decrease at a rate proportional to the concentration ($\frac{\partial s}{\partial t} = -\lambda s$ if s is distributed uniformly, i.e. $s\,(x,y,z) =$ constant). Estimating the form or rate, for the source–sink term for stable (non-radioactive) non-conservative constituents may be difficult.

THE OBSERVED DISTRIBUTION OF PROPERTIES IN THE DEEP OCEAN

The deep circulation is usually attributed to thermohaline processes: the differential heating between the equator and the polar regions, evaporation, the freezing of ice, etc. Once a water type is formed at the surface it seeks a level in the ocean based on its density or, more accurately, its potential density. Because both potential temperature and salinity are conservative properties one can trace back from the distribution to the origin of the water mass using the advection–diffusion equation and T–S curves as a guide. Both the intermediate depth (1 km) waters and the deep abyssal waters are formed this way. Antarctic Intermediate water and Mediterranean water (which sinks after exiting the Mediterranean at the Straits of Gibraltar) are examples of intermediate depth water which can be traced over most of the southern hemisphere of the world ocean and over the Atlantic, respectively. The North Pacific Intermediate water may not be formed at the surface, but may acquire its properties by mixing at depth (Reid 1965). The water of greatest density in the ocean is formed at high latitudes, and sinks and fills the deepest parts of the ocean basins. The two main water masses constituting the abyssal circulation of all oceans are both formed in the Atlantic: In the Weddell Sea near the Antarctic continent and in the Norwegian–Greenland sea region of the North Atlantic.

Water in the Weddell Sea is cooled in the Antarctic winter and sea-ice is formed, which is relatively fresher, leaving a cold, saltier water which mixes with circumpolar water and sinks, spreading northward, and also eastward around the Antarctic continent into all oceans. This Antarctic Bottom Water (AABW) is the coldest water in the deep ocean and the spreading can be seen by the distribution of potential temperature in all ocean basins deeper than 4 km. Some AABW is probably also formed in the Ross Sea and near the Adelie coast, but the major source is the Weddell Sea (Committee on Polar Research 1974). The North Atlantic Deep Water (NAD) is formed principally in the Norwegian Sea, when the warm, saline water from the western North Atlantic is brought north into the Norwegian Sea and is cooled at the surface in winter. The cold NAD enters the North Atlantic by overflowing at depth through the Denmark Strait and Iceland–Scotland Ridge (Worthington 1970). NAD has a higher temperature and salinity than AABW and is slightly less dense and tends to spread above the AABW, filling most of the Atlantic between 1500 and 3000 m. It has an oxygen maximum which can be traced to nearly 50°S. How the deep water is distributed to the rest of the world ocean is the topic of the section on the abyssal circulation.

VERTICAL DISTRIBUTION OF PROPERTIES

The vertical distribution of properties in the ocean sandwiched between the permanent thermocline and the abyssal bottom kilometer has received considerable attention (Wyrtki 1962, Munk 1966, Craig 1969). The return of the abyssal water to the surface layer must occur, completing the circulation back to its source. Age dating by ^{14}C puts the age of the abyssal waters generally the order of hundreds of years (Bien *et al.* 1965). A related question is how at the permanent thermocline in the low and middle latitudes is maintained in the presence of the net heat flux down into the surface layers from the atmosphere. In the interior region below the thermocline, the horizontal gradients are generally small, and, to a first approximation, horizontal diffusion and advection may be ignored locally in comparison with those in the vertical direction. Assuming steady state exists:

$$\frac{\partial s}{\partial t} = 0 = -w\frac{\partial s}{\partial z} + \frac{\partial}{\partial z} K_z \frac{\partial s}{\partial z} + R \tag{2.37}$$

If we consider the distribution of potential temperature or salinity, and assume K_z is constant, we obtain

$$\frac{d^2 s}{dz^2} = \frac{w}{K}\frac{\partial s}{\partial z} \tag{2.38}$$

and can fit the exponential solution to observed distributions and obtain an estimate of $\frac{w}{K} \simeq 10^{-5}$ cm^{-1}, from either salinity or potential temperature profiles (Munk 1966). If molecular processes dominated, i.e. if the vertical diffusion was by molecular diffusion rather than turbulence, the estimate of w/K (or rather in that case, w/μ) would be different for salinity and temperature since the coefficient for molecular diffusion of salt is 10^{-2} that of heat (thermal conductivity). If the transport is by turbulent fluxes, the 'eddy' coefficients should be the same—which is what is found.

If we have the vertical distribution of a radioactive constituent, then it should obey the equation:

$$\frac{\partial^2 s}{\partial z^2} - \frac{w}{K}\frac{\partial s}{\partial z} = \frac{\lambda}{K} s \tag{2.39}$$

This equation has a solution in terms of the parameter w^2/K. The fit to the observed distributions gives a characteristic time K/w^2 of the order of 10^{10} sec (300 yrs). From estimates of w/K and w^2/K we conclude w is of the order of 1×10^{-5} cm sec^{-1} and K the order of 1 cm^2 sec^{-1} (Munk 1966). (Molecular diffusion coefficients are the order of 10^{-5} cm^2 sec^{-1} for salt and 10^{-3} for heat). The slow upwelling of the cold abyssal water through the thermocline returns the abyssal water to the surface layer and also balances the downward diffusion of heat, thus maintaining the thermocline.

The same vertical diffusion—vertical advection model has been used, with an oxygen consumption function R to explain oxygen minimum layer (Wyrtki 1962), and by Craig (1969) in discussing the ^{14}C distribution in the deep ocean.

The abyssal circulation

The present theory and understanding of the abyssal circulation makes use of the observed distributions of properties discussed above. There is no thermohaline theory yet that explicitly uses the details of the presumed driving mechanism of differential solar insolation between the equator and the pole. The most complete and generally accepted theory of the abyssal circulation (Stommel & Arons 1960) makes use of indirect arguments to obtain the abyssal circulation pattern.

It starts with the recognition that there exist the two localized sources of deep water for the world ocean and that there exists in lower and mid-latitudes a relatively sharp temperature gradient (the permanent thermocline) between the warm surface waters and the cold abyssal waters. To maintain a steady state in the presence of the downward turbulent heat flux, it is assumed there is a relatively steady and horizontally uniform upward flux of cold water. In the previous section we saw this required an upward velocity of 10^{-5} cm sec^{-1} (3 or 4 m yr^{-1}). The deep circulation is, in effect, driven by the horizontally uniform upward flux of water at the base of the permanent thermocline, and this flux serves as the sink of the abyssal water which has sources in the high latitudes of the North and South Atlantic.

Let us consider the vorticity equation again, and note the analogous role of the $f\,\partial w/\partial z$ term to that of the wind stress, through the term $\left(\frac{\partial F_y}{\partial x}-\frac{\partial F_x}{\partial y}\right)$. When the depth of a column of fluid is increased, there is a horizontal convergence due to mass continuity. Conservation of angular momentum requires the angular velocity of the column of fluid to increase: the relative vorticity (and total vorticity) is affected by the vertical gradient of vertical velocity, i.e. the stretching or shrinking of the column of fluid. If the total vorticity is to be conserved, a change in relative vorticity must be accompanied by meridional flow, which allows the planetary vorticity to change oppositely. Thus, a stretching or shrinking of a water column is accompanied by meridional flow.

It is assumed that over the large part of the deep ocean the currents are slow and mainly geostrophic, and thus the relative vorticity is negligible compared to f and the stresses at depth are negligible. Below the depth of the wind's frictional influence the vorticity equation becomes:

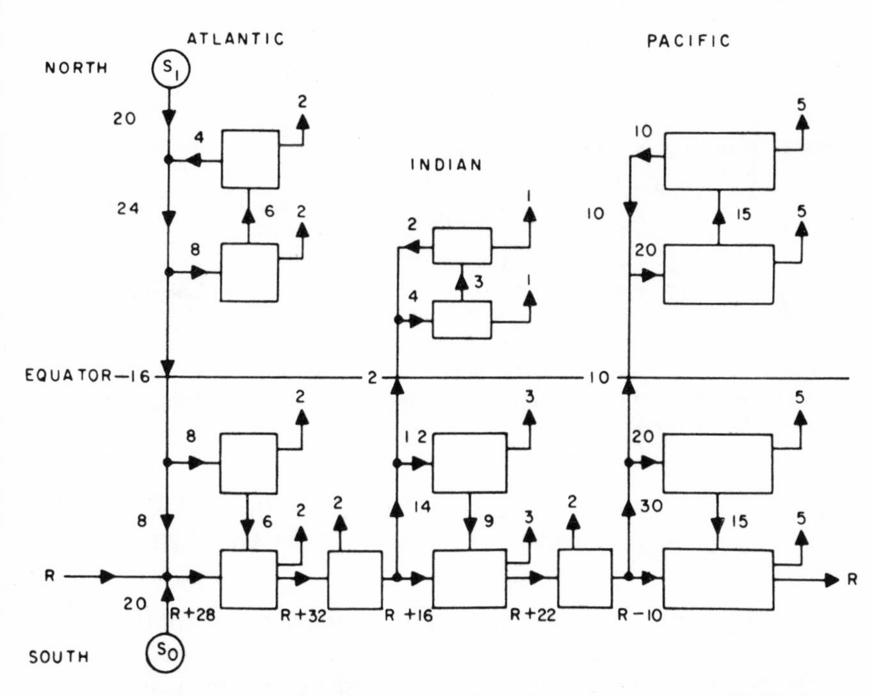

Fig. 2.4. Schematic of abyssal circulation from Stommel and Arons (1960). S_I and S_o represent sources of North Atlantic Deep and Antarctic Bottom Water, respectively, and R the circulation in the Antarctic Circumpolar Current. Transport into/out of various 'basins' and of upwelling (upward arrows) is in Sv (Sverdrup) units (10^6 m^3 sec^{-1}).

$$\beta v = f \frac{\partial w}{\partial z} \tag{2.40}$$

Integrating from the bottom to the base of the permanent thermocline:

$$\beta M_y = f w_t \tag{2.41}$$

where w_t is the vertical velocity at the base of the thermocline, which we concluded earlier must be positive. Therefore, the deep flow (below 1000 m or so) over most of the ocean must be poleward. As in the wind driven circulation, it is necessary to include a return flow, and again the current is on the

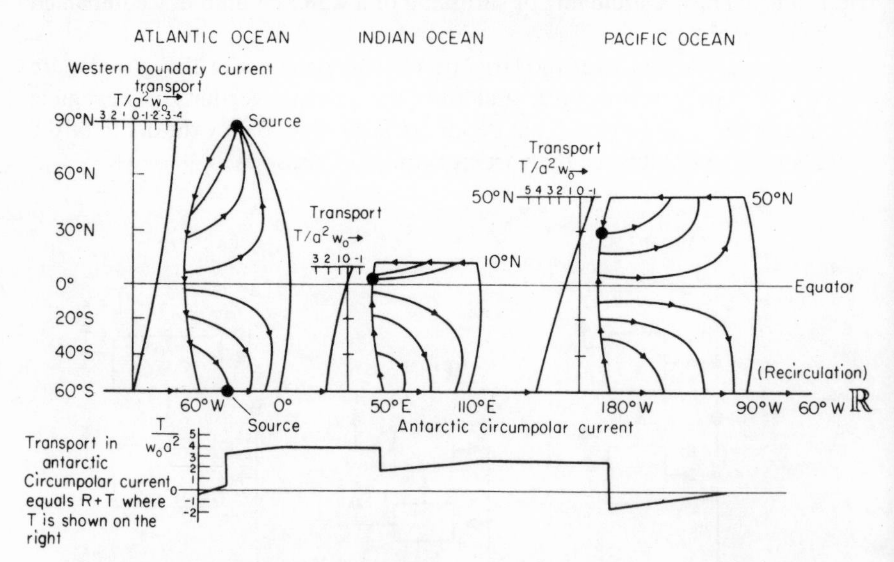

Fig. 2.5. Conceptual model of the abyssal circulation of the world ocean from Kuo and Veronis (1970). Transport is in units of $w_o a^2$; w_o = upwelling velocity, a = radius of earth.

western side of the ocean. In Figures 2.4 and 2.5 are shown conceptual flow patterns for the abyssal circulation from Stommel and Arons (1960) and Kuo and Veronis (1970). In words:

1 The deep water, which originates in the poleward regions of the Atlantic is carried in intense deep western boundary currents. The abyssal waters in the interior ocean basins are transferred from these western boundary currents. There exists observational evidence of the existence of abyssal western boundary currents (e.g. Warren & Voorhis 1970; Warren 1971).

2 In the interior, the abyssal flow is eastward and poleward everywhere and is nearly geostrophic.

3 The abyssal water is returned to the upper layer by a very low vertical velocity throughout the interior ocean basin. In these pictures it is the *net* transport of the ocean below the permanent thermocline that is shown—and intermediate structure (or counter flows) are 'averaged' out.

Kuo and Veronis (1970, 1973) have used the horizontal (two dimensional) form of the advection–diffusion equation governing the distribution of properties to interpret the distributions of tracers in terms of the circulation. They assume a local balance between vertical convection and diffusion. The optimum pattern obtained from the numerical model compared with the deep observed oxygen distribution for the world ocean gives K_x, $K_y \simeq 6 \times 10^6$ cm^2 sec^{-1}, $w_t \simeq 1{\cdot}5 \times 10^{-5}$ cm sec^{-1} and oxygen depletion of 2×10^{-3} ml l^{-1} yr^{-1}, and the recirculation around the Antarctic to be 35×10^6 m^3 sec^{-1}. The concentration of O_2 and, presumably, all other properties, is the result of *both* advective and turbulent diffusive processes.

More complex models which attempt to interpret the distribution of properties or tracers in terms of the circulation have been done: e.g. the three-dimensional system is numerically integrated by Holland (1971). At the other extreme many simple box models have been used to obtain a 'residence' time for the abyssal oceans (Bolin & Stommel 1961, Keeling & Bolin 1967, Wright 1969).

Epilogue

This chapter started with an *apologia*: our understanding of the ocean, in its richness and complexity of phenomena, is incomplete. Some aspects that seemed basic and comfortably well established were then pursued in the following pages. It is appropriate to end with a *mea culpa*: important topics have been neglected—and to amend this with references to those topics.

Exchange of momentum and mass occurs across the boundaries of the ocean; some topics are: the heat budget of the ocean (Neumann & Pierson 1966, Chapter 9), the atmosphere–ocean interaction (Kraus 1972), and the flux of materials from the continents (Hood 1971). The dynamics of the upper tens of metres, where most of the primary production occurs, is clearly of considerable importance but terribly complex and inadequately understood. The present knowledge is well summarized in Phillips (1966) and Kraus (1972). Estuaries, too, are of great ecological importance and deserve a chapter by themselves. Fortunately, the reader can find a virtual compendium on estuaries in the volume edited by Lauff (1967). Diffusion and mixing processes in the ocean warrant more discussions and the reader is referred to the excellent and recent discussion by Okubo (1971).

Although it was stated that oceanic phenomena are variable and intermittent, the discussions shied from confronting that fact and concentrated on a smooth mean picture. The ocean is in fact characterized by great variability on time scales from hours to millenia (Namias 1972). Fluctuations occur with both rather constant periods, e.g. tides (Neumann & Pierson 1966, Chapter 11), and with variable periods, e.g. internal waves whose frequencies may vary between the local Väisälä–Brunt and the inertial frequencies (Phillips 1966) and Rossby or planetary waves with periods of days and longer (Longuet-Higgins 1965). There are non-linear interactions between processes at different frequencies. The problems of handling the non-linear interactions and time-varying forcing are not trivial. The fluctuations of the currents themselves influence, and may even drive, the mean motion through rectified transports of momentum and vorticity. This idea is at the centre of the planning of recent and future major experiments (MODE-I Scientific Council 1973). In many parts of the ocean the time varying part of the current is one or two orders of magnitude greater than the mean.

The variability of physical processes in the ocean may greatly affect the ecology. Thus, for example, in coastal upwelling regions an understanding of the ecology may require knowledge of the response of the circulation over the continental shelf to the diurnal fluctuations in the sea-breeze (Halpern 1974), to the several day-wind variations occurring on a synoptic scale (Huyer *et al.* 1974) and to the irregular decadal catastrophic fluctuations of 'climate' that produce an 'el Niño' (Idyll 1973).

The present state of physical oceanographic knowledge was succinctly summarized by Stewart (in Dyrssen & Jagner 1972): 'Physical oceanographers (and meteorologists) have now a fairly satisfactory grasp of the mechanisms and nature of the average circulation of the ocean—both as a wind-driven system and as a thermo-haline system. Many details have yet to be incorporated into these concepts, however, and many important processes are still very crudely parameterized. Some ideas are also available about the time-dependent motions—but there we are on far less-secure ground and the relation between observations and theory is tenuous indeed. About the very important and very complex upper couple of meters of water we know far less than we need to know.'

Chapter 3. Plants and Animals of the Sea

T. Wyatt

Introduction

The physiognomy of survival
Body size
    Body size and mortality in planktonic organisms
    Body size in benthic forms
Colour
    Plants
    Animals

The physiognomy of feeding
Filter feeding

Life cycles
Egg and larval size and reproductive strategies
Larval life

Life in the photic zone
Littoral and sublittoral environments
Pelagic environments

Plants and animals of the sea

Introduction

Ecology is largely a study of population numbers, and of the processes which bring about fluctuations in these numbers. Each individual in a population interacts with other members of that population, with other species, and with the environment. These interactions taken together can be seen as the ecological roles of the individuals. In this chapter, an attempt is made to provide a general view of the kinds of organisms found in the sea, and the ways in which they play their ecological roles. We are concerned here with two main themes; firstly, the physiognomic characteristics of groups of organisms found in different marine habitats, which cut across the boundaries of systematic biology, and often tell us a great deal about the kinds of lives their possessors lead, and secondly with life histories. These two themes are closely intertwined, and are receiving increasing attention as ecologists sense more clearly the need for an understanding of community structure additional to that based on production and ordination studies.

There are perhaps 200,000 different species of plants and animals which live in the sea, and these can be classified along a number of axes which stress their static and dynamic relations within the marine province as a whole. The more descriptive divisions of the inhabitants of the marine environment naturally preceded historically those based on awareness of the dynamic processes which result in community integration. One of the oldest distinctions is that made between pelagic life which lives freely in the water, and benthic life which lives on or in the sea bed. Only about 2 per cent of marine organisms are pelagic, and the remaining 98 per cent benthic. The vast majority of the latter are found in shallow or intertidal depths in tropical seas, and less than 0·5 per cent in the hadal zone (deeper than 6000 m) of the deep ocean. This kind of statistic serves not only to reduce the difficulty of comprehending as a whole the variety of marine life, but also provides insights into the ecological differences between the subdivisions of the marine environment. When taken together with taxonomic information, a good deal can be surmised about the nature of different marine habitats. This approach to descriptive ecology is considered in more detail later in this chapter.

The different environments of the ocean inhabited by pelagic and benthic organisms have been named on the basis of depth and distance from shore. Thus we have an epipelagic zone, more or less equivalent to the photic zone, reaching from the surface down to about 100–200 m. Below this are mesopelagic (200–1000 m) and bathypelagic (1000 m to bottom) zones. The latter is divided into bathyal, abyssal and hadal zones. The 4°C isotherm is an important faunistic boundary, and may be taken to divide the bathypelagic and abyssopelagic zones. This isotherm lies much deeper in the Atlantic Ocean (about 2000 m) than in the Indian and Pacific (1000–1500 m). The upper limit of the hadopelagic fauna is at about 6000 m. Pelagic life is also said to be neritic, i.e. found in the water over the continental shelf, or oceanic, beyond it. Benthic environments are supra-littoral, above mean high water, littoral, between high water and low water, sublittoral, to the edge of the continental shelf, and beyond here bathyal, abyssal, and hadal, corresponding in depth to the pelagic zones and topographically to the continental slope, abyssal plain, and trenches of the deep sea. As in the pelagic region, the 4°C isotherm separates bathyal and abyssal animals. A large number of other terms, some synonymous with these, others slightly different, and yet others which make further subdivisions, will be found of use where more detailed description is required. Many of these have been reviewed by Hedgpeth (1957). Each of these zones has a characteristic fauna, and in the photic zone, a corresponding flora. Some brief notes on these are given below.

No classification of the littoral environment so far devised does justice to the amazing variety of habitats found on rocky shores. The effects of heavy surf or of shelter, of grooves and gulleys and changes of slope, of erosion and seepage and sedimentation, of exposure to sun and wind and freshwater influences, create an enormous number of habitats, each and every one of which seems to harbour a specialized fauna.

More recently coined divisions of marine organisms reflect an awareness of community dynamics. Terms which are widely used indicate the trophic levels occupied by organisms—producers, herbivores, carnivores, decomposers—or, more briefly, autotrophs and heterotrophs. These form the subject matter of Chapter 14. A recent approach to ecological classification sees organisms in terms of strategic axes (MacArthur 1962). An organism makes a 'choice' in an evolutionary sense, for example, between being large or being small, or between being more or less fecund. These strategic axes are interrelated, and the position of an organism with respect to any one of them puts certain constraints on its freedom of choice with respect to other axes. A description of an organism on the basis of such axes will provide an effective definition of its ecological role.

Despite all efforts to survive, the natural destiny of almost all animals and plants in the sea is to be eaten before maturity is attained. It is only the few which live long enough to perpetuate their species. The best that most species can do is remain one step ahead of those which prey on them, and so avoid extinction. So eating and being eaten are central to the whole science of ecology, and it is these two topics which are emphasized here.

The physiognomy of survival

Body size

The importance of body size is implicit in several terms used broadly to classify marine organisms. The terms plankton and nekton are used to distinguish between those pelagic organisms which are more or less at the mercy of water currents and turbulent mixing processes, and those which can swim against them. The distinction is largely one of size. The plankton in turn is classified into a number of classes based again on size, and these are listed in

Table 3.1. Classification of plankton by sizes. Pérès and Devèze use the term megaplankton for very large plankton animals such as scyphozoans, siphonophores. Dussart uses the term 'ultramicroplankton'. Dussart regards meso- and macro-plankton as synonymous.

Size class	Margalef 1955	Peres & Deveze 1963	Dussart 1965
Ultraplankton	$<5\ \mu$	$<5\ \mu$	$<2\ \mu$
Nanoplankton	5–50 μ	5–60 μ	2–20 μ (note 2)
Microplankton	50–500 μ	60–100 μ	20–200 μ
Mesoplankton	0·5–1·0 mm	1·0–5·0 mm	0·2–2·0 mm (note 3)
Macroplankton	>1 mm	>5·0 mm	—
Megaplankton	—	(note 1)	>2·0 mm

Table 3.1, together with the size ranges assigned to them by various authors. These categories have important practical implications, in that each requires different sampling techniques, and they also offer a crude guide to the kinds of organisms likely to be encountered. Thus the ultra- and nanoplankton consist very largely of unicellular algae, together with bacteria, fungi, and some invertebrate eggs. The micro- and mesoplankton contain many larval and juvenile forms, especially of crustaceans, fish eggs, protozoans, and appendicularians. The macroplankton is composed of crustaceans, fish larvae, chaetognaths, ctenophores, thaliaceans, etc. Very roughly, these three groupings correspond to producers (and decomposers), herbivores, and carnivores, and hence possess some ecological validity in addition to their practical use.

The nekton consists of large crustaceans, squids, fish, sharks, and mammals, and very specialized members of some other groups more characteristically benthic. Marine birds are perhaps a rather special category of nekton too. A distinction is sometimes made between micro- and macronekton. Similar terms are used with reference to the in-faunal benthos, and the terms microbenthos, meiobenthos, and macrobenthos are in common usage. Unlike the plankton however, these categories have no obvious ecological meanings, in terms of trophic relations. Macrobenthos refers to animals which move by swallowing the substrate, or by displacing it. The meiobenthos is synonymous with the interstitial fauna, and is composed of animals from 50 to 500 μ long, which live in the spaces between sand or mud particles. The microbenthos consists of animals smaller than 50 μ, mainly protozoans.

BODY SIZE AND MORTALITY IN PLANKTON ORGANISMS

Since mortality is in general density dependent (see Chapter 13), and since small organisms are more abundant than large ones (Elton 1927), mortality is necessarily higher in small organisms than in large ones. Smaller organisms should therefore reproduce more rapidly than larger ones, unless particular ways have been evolved to offset this apparent disadvantage. Within this framework it is possible to visualize a number of different strategies available to populations in their evolutionary search for survival.

The spines of diatoms, radiolarians, decapod larvae, and many other planktonic groups, are generally thought of as devices to assist flotation. But in a turbulent environment, these processes are as likely to carry their bearers downwards as upwards. Besides, very small diatoms often have spines while their sinking rates would be negligible without them, and the largest species generally lack them. The possession of spines however increases the effective dimensions of the body *vis-à-vis* predators of the raptorial kind, and hence lowers the mortality which might otherwise be anticipated from this source on

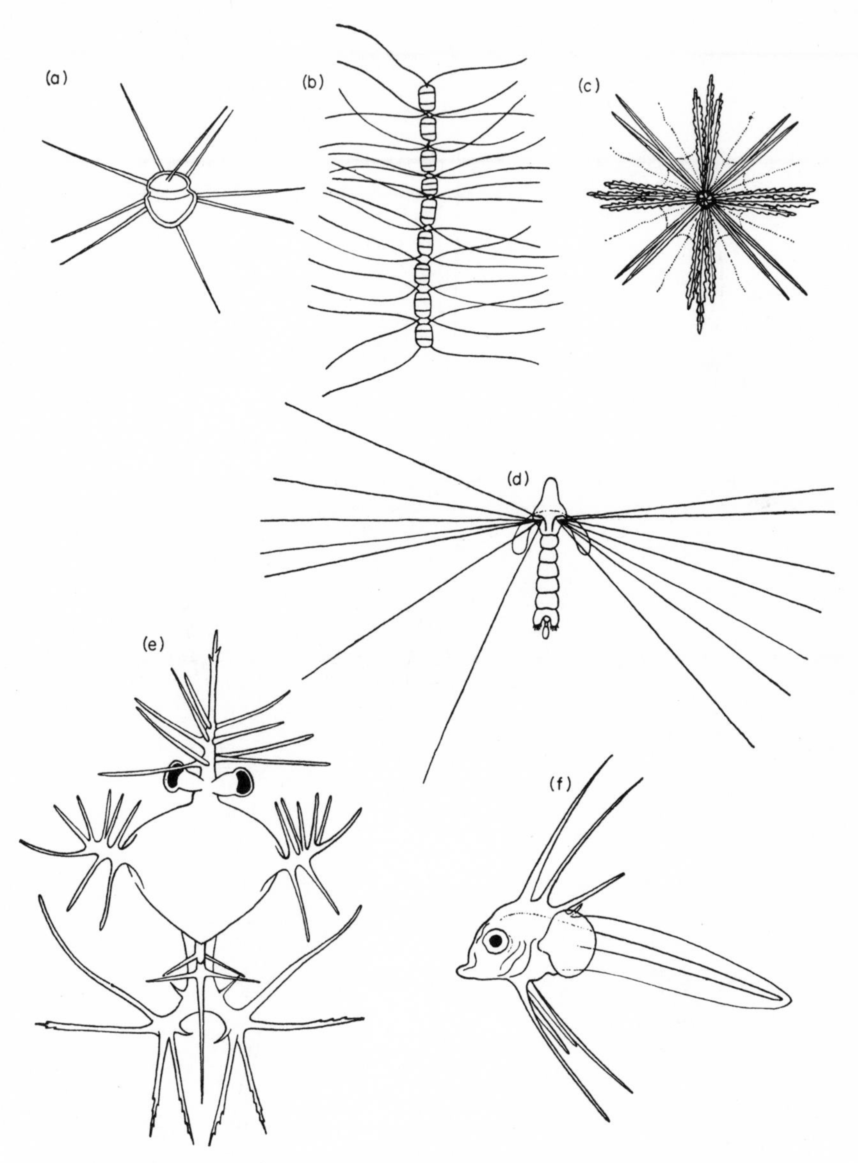

Fig. 3.1 Plankton organisms which have an increased effective size *vis à vis* predators. (a) *Microcanthodium setiferum* (dinoflagellate); (b) *Chaetoceros contortum* (Schütt) (diatom); (c) *Conacon foliaceus* (Haeckel) (radiolarian); (d) larva of *Rostraria* (polychaete); (e) zoea stage of *Sergestes* (penaeid prawn); (f) larva of *Lophius piscatorius* (Linnaeus) (teleost fish).

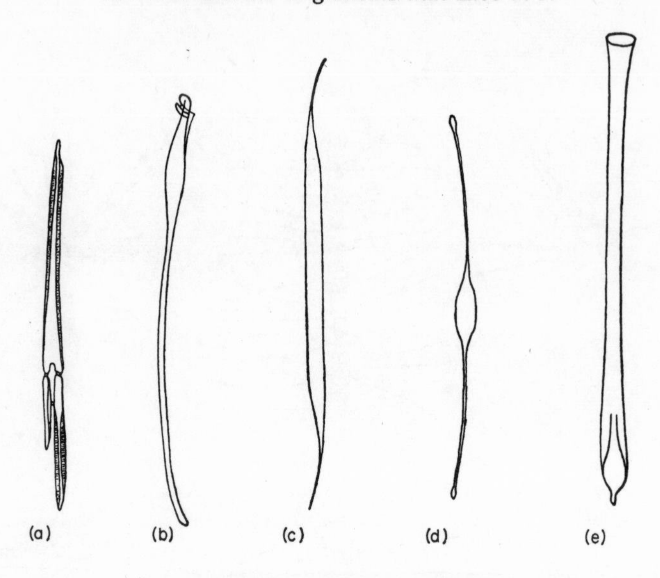
(a)
(b)
(c)
(d)
(e)

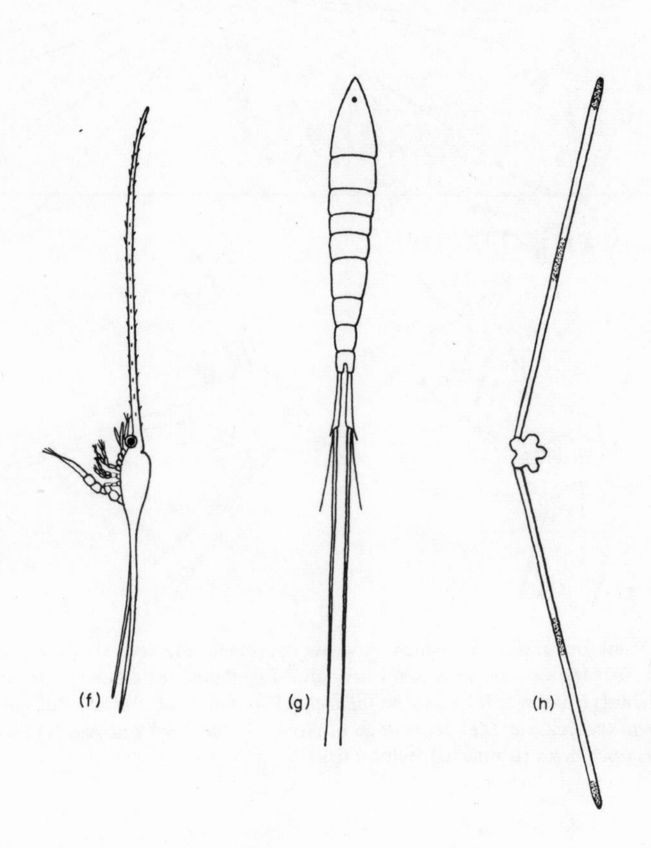
(f)
(g)
(h)

the basis of absolute size alone. Some examples of spiny plankton organisms are shown in Figure 3.1. *Chaetoceros contortum* (Schütt) is a widely occurring diatom in which the spines increase the cell diameter with very little increase in biomass. It has recently been shown experimentally that freshwater rotifers can respond to the presence of a predator by an increase in spininess (Pourriot 1974). If such devices do indeed reduce size-dependent mortality, then reproductive effort can also be lower, since in the long term mortality and fecundity must be equal if population numbers are to remain unchanged. In turn, this may allow the energy demands of other activities as a proportion of the total energy budget to rise. Other organisms in Figure 3.1 are larval or juvenile forms, and here it is the reproductive effort required of the parents which may be lessened.

Several other features of planktonic organisms may be seen in the same light. The greatly elongated cells of the diatom *Rhizosolenia* and some species of *Nitzschia*, shown in Figure 3.2 spring to mind. Also seen in this figure are organisms of similar habit from several other planktonic groups. The saucer-like extension of *Planktoniella sol* (Wallich), the formation of chains, as seen for example in *Thalassiosira gravida* (Cleve) and *Cerataulina*, and aggregation in bundles (e.g. *Trichodesmium*) or in jelly-like masses (e.g. *Phaeocystis*) may all be ways of securing a larger size relatively economically. Examples are shown in Figure 3.3.

Real rather than simulated size is an obvious way to escape predation. Diatoms again provide examples. *Coscinodiscus* is a familiar genus containing species up to 400 μ diameter, and the very large *Ethmodiscus rex* (Wall) of the Sargasso Sea may reach over 1 mm in diameter, which makes it larger than many planktonic herbivores. A number of planktonic groups seem to have specialized in large size without a concomitant increase in living material by the production of 'jelly' which consists mostly of water. Medusae, siphonophores, ctenophores and tunicates are examples. In these groups the dry weight is generally less than 5 per cent of the wet weight, and frequently less than 1 per cent. In most groups of planktonic animals, the dry weight is usually above 10 per cent of the wet weight.

BODY SIZE IN BENTHIC FORMS

Some features of benthic animals are clearly related to protection from

---

Fig. 3.2 Examples of elongated plankton organisms. (a) *Ceratium belone* (Cleve) (dinoflagellate); (b) *Amphisolenia spinulosa* Kofoid (dinoflagellate); (c) *Rhizosolenia hebetata* (Bail) (diatom); (d) *Nitzschia closterium* (Hasle) (diatom); (e) *Salpingella glockentogeri* Brandt (tintinnid); (f) metazoea stage of *Porcellana platycheles* (Pennant) (anomuran crab); (g) *Macrostella gracilis* (Dana) (copepod); (h) ophiopluteus of *Ophiothrix fragilis* (Abildgaard) (ophiuroid).

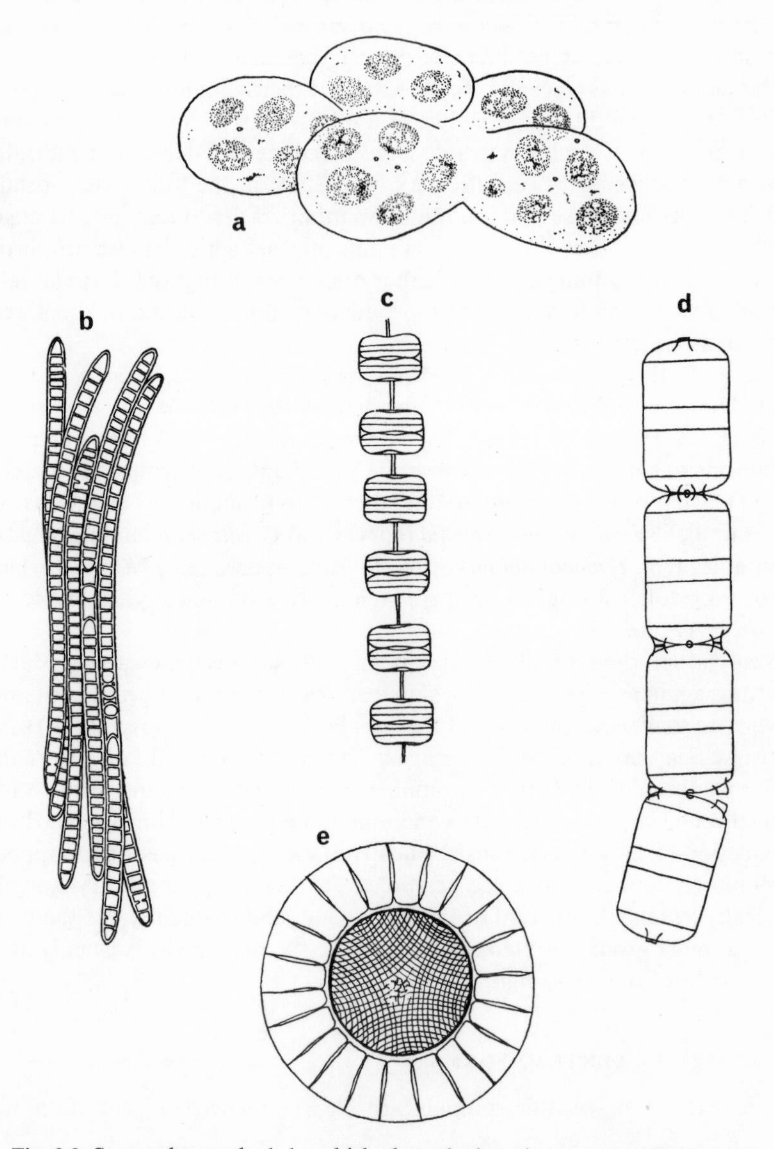

Fig. 3.3. Some other methods by which phytoplankton increases its effective size. Compare with Figure 3.1. (a) *Phaeocystis* aggregated in jelly-like masses (each individual dot represents a single cell); (b) *Trichodesmium erythraeum* (Ehrenbaum) filaments joined together to form a bundle; (c) *Thalassiosira gravida* (Cleve), and (d) *Cerataulina bergonii* (Pérag) forming chains; (e) *Planktoniella sol* (Wallich) with an equatorial plate surrounding the cell.

predators, and often seem less subtly employed than has been suggested for planktonic organisms. The sharp and rigid, and sometimes poisonous spines of many echinoderms and fishes are frankly dangerous, and correspond more closely to our own ideas of protective armour. But on the whole, the feeding mechanisms of benthic animals are more elaborate than those found amongst pelagic animals, and one gains the impression that the rather simple strategies evolved to reduce predation in the pelagic environment may be less adequate on the bottom. Alternatively, the constraints on benthic life may be less severe—for example, the need for buoyancy is absent—and hence a greater variety of strategies is possible. This may account for the enormous preponderance of benthic species. Nevertheless, a refuge is gained in size; Ansell (1960) for example has shown that the venus clam (*Venus striatula* (Da Costa)) becomes immune to attack by the boring snail (*Natica alderi* (E. Forbes)) above a certain size. At the same time, the smallest clams cease to be eaten by the larger snails. Similarly, the predatory snail (*Navanax inermis* (Cooper)) is restricted to a certain size range of prey with upper and lower limits, when itself of a certain size (Paine 1965). Further examples of this type of relationship were the subject of field experiments conducted by Dayton (1971). Gigantism is found among several groups of deep-sea animals. *Colossendeis* (a pycnogonid), *Geryon* (crab), *Storthyngura* and *Bathynomus* (isopods), and several echinoids and pennatularians provide examples. This feature has been explained on the basis of low temperature retarding the onset of sexual maturity and prolonging the period of development, and as a response to the effect of hydrostatic pressure on metabolism (Bierstein 1957). But it is not a feature of the entire deep-sea fauna, so neither of these hypotheses is of general value.

Colour

PLANTS

The colours of marine algae depend to a large extent on their complement of pigments, and have given rise to the common names of some major classes of algae (green, yellow-green, brown, red, blue-green). The selective absorption of different wavelengths of light seems to be one factor dictating the ecological zonation of algae, which is expressed in the temperate littoral and sublittoral zones. Thus red algae are more abundant in the lower euphotic zone where only the shorter wavelengths penetrate. This light would be largely reflected were their chlorophyll unmasked by the red pigment phycoerythrin, and hence lost. But red algae are also found high in the euphotic zone as on the algal ridge of Indo-Pacific coral reefs, so that phycoerythrin may confer other advantages unrelated to light quality. Also, some green algae are found in relatively deep water (Feldmann 1937), and the study of complementary chromatic adaptation is no simple matter (Rabinowitch 1945).

ANIMALS

Red animals of course appear black in the absence of red light, and the red pigments so characteristic of the mesopelagic fauna must make them more or less invisible in their natural habitat, and so conceal them. Even in shallow water, red pigment may serve only to emphasize some other feature, which in full daylight may be quite inconspicuous. Two photographs by Eibl-Eibesfeldt (1965) of the bright-red dragonfish (*Pterois volitans* (Linnaeus)) demonstrate that where it lives, the most prominent feature is a group of small white spots on and near the pectoral fins. Similarly, it is only the light organs of bathypelagic fishes which can be seen against their red, brown or black skins. The specific colours of animals from the poorly lit or unlit parts of the sea then may be of no special significance beyond concealment. In shallow water though, while red is not well seen, yellow, blue, and white show up well, and alone or combined in patterns, are used in territorial, sexual and perhaps didactic display in fishes and in other forms of intra- and inter-specific behaviour.

These brightly coloured fish must have other means of escaping from their enemies. Many animals, like nudibranchs, are brightly coloured as a warning of their distasteful, or even poisonous nature. Yet the often brilliant colours of many quite innocuous and apparently helpless intertidal animals remain a puzzle unless mimicry is involved. Colour is also used as a cryptic device in shallow water. Animals which live on sandy or muddy bottoms can usually adjust their colour to that of the substrate, and can also obscure their outline as well by selective control of their chromatophores. Similar techniques are used by animals which live amongst rocks or corals, or amongst algae. Some cryptically coloured fishes conceal brilliant pigment patterns needed for sexual display inside their mouths, or on their folded dorsal fins. Many examples are found among the blennies and gobies. One of the most striking examples of cryptic coloration is found in the fauna associated with floating *Sargassum*, much of which has the same bright yellow-brown of the seaweed itself. Cryptic patterns in fishes are sometimes effective only when shoaling, when they serve to break up the outlines of individual fish, and hence presumably make attacks by predators less accurate. The resident planktonic fauna of the euphotic zone is characteristically transparent.

At the immediate surface, in the top few centimetres of the euphotic zones, where the light intensity is highest, many animals contain blue pigment. The siphonophores *Physalia*, *Velella*, and *Porpita* and the gastropods *Glaucus* and *Ianthina* are well-known members of this fauna, and have air-filled floats. They are collectively called pneuston. A second group, the hypopneuston, includes pontellid copepods, decapod crustaceans like *Parapeneus longipes* (Alcock) and larval squids and fish. Some species of the copepod *Sapphirina* contain no blue pigment, but iridesce in the blue part of the spectrum. The function of this blue colour is not known, but it may again be cryptic. Zaitsev (1961) has noted that the fish *Mugil auratus* (Risso) is silvery-white when the sea surface is foam-flecked, and blue when it is calm.

Between the surface and about 500 m, the fish fauna is mainly silver or grey in colour and darker above than below. As in other layers, it appears that the characteristic colour of this zone is cryptic. It is in this zone too that bioluminescence is such a marked feature of the fauna, and it has been demonstrated that the lantern fish *Tarletonbeania crenularis* (Jordan and Gilbert) can match the downwelling light through control of its ventral photophores, and hence be difficult to see from below (Lawry 1974). The anatomical structures on which this sytem relies are widespread in bathypelagic fishes, and the light from photophores has the same wavelength as the ambient light (Nicol 1960), so that the system may be a very effective form of camouflage.

The physiognomy of feeding

Filter feeding

Many methods of capturing food are found in marine animals, but by far the commonest is filter feeding. This takes many forms, but one, ciliary feeding, is found in almost every group of invertebrates, and in all members of some. Sponges, polyzoans, bivalve molluscs, brachiopods, and ascidians are all exclusively ciliary feeders, and while the feeding apparatus may vary widely in structural details between groups, the principle of operation is in all cases similar. Many sedentary polychaetes (*Serpulidae* and *Sabellidae*—fan worms) and some gastropods (e.g. Calyptraeidae (slipper limpets) and Turritellidae) provide other examples. The filter itself separates inhalant and exhalant currents of water, and removes the suspended material in the process. Three separate sets of cilia are generally required, to create the current, to collect the food, and to convey it to the mouth. Mucus secretion usually plays a prominent role in the last two processes. A second method of filter feeding makes use of specialized limbs fringed with setae, and is the most widespread feeding technique of barnacles and planktonic crustaceans. In these the filtering apparatus is extended slowly and flexed rapidly, and so creates a current of water past the mouth. Mucus again often plays an important role. The effectiveness of filter feeding may be magnified in sedentary colonial animals by their habit of forming branching tree-like forms which vastly expand the volume of water they can control. Branching colonies of polyzoans for example are especially common in deep water where currents are weak, while encrusting forms are the rule where tides and wave action are strong. Some encrusting polyzoans and ascidians grow on the fronds of algae, and hence gain the advantages of being swept back and forth through the water. The 'displaced benthos' of floating *Sargassum* weed is also in this category, as are ellobiopsids, branching protozoan colonies which grow on pelagic crustaceans.

Coelenterates, and ctenophores, which are all carnivorous, obtain their food, either zooplankton or small nekton, with tentacles. These extend passively in the water, and only contract towards the mouth when food organisms have been encountered and subdued by the stinging cells. As in polyzoans, many colonial coelenterates form tree-like colonies which increase the volume of water available to them. Many hydroids, scleractinians, and gorgonians provide beautiful examples. The pelagic *Physalia* can extend its tentacles up to 30 metres to achieve the same end. Sea-lilies and feather-stars (stalked and sessile crinoids) employ a similar method of feeding, though the food is conveyed to the mouth by tube-feet in these groups. They have poison glands on the pinnules which paralyze their prey. A few brittle-stars (ophiuroids), also feed in this way, though without the benefit of narcotizing agents. The temporary rhizopods of radiolarians and foraminiferans may also perhaps be compared with the tentacles of coelenterates.

Other predatory techniques, like ciliary feeding, are very widespread. Suctorial feeding, preceded by drilling or piercing a hole in the skeleton or integument of the prey, is widespread in sea-slugs (nudibranchs) and other gastropods which feed on sponges, polychaetes, barnacles, other molluscs, etc. This method is common among arthropods, notably in the cyclopoid copepods known as siphonostomes, in a few ostracods, and in some anemones, holothurians, and tunicates. Members of the interstitial fauna which live in the spaces between sand grains are also suctorial, in this case on the diatoms and dinoflagellates which form a scum on the surface. These include nematodes, gastrotrichs and tardigrades. Many animals simply gulp their food whole which clearly requires little in the way of specialized apparatus beyond efficient jaw mechanisms and strong muscles. A great variety of fish belong here. Others catch their prey with limbs specialized for the purpose. Large maxillae are the keynote of predatory copepods like *Candacia* which can feed on fish larvae and chaetognaths. Yet others push food into their mouths with appendages of various kinds, like many carrion- and detritus-eating crabs, and ophiuroids with their five triangular jaws. Mud-eating, as in the earthworm, is found in many polychaetes and molluscs, in holothurians, and in some ophiuroids. The general degree of selectivity practised by mud-eaters can often be inferred from the appearance of the digestive tract. In Capitellid and Maldanid polychaetes, spatangids (holothurians), many asteroids, and sipunculids like *Aspidosiphon*, large volumes of mud of poor food value are consumed indiscriminately, and the guts are consequently very large relative to the size of the animals. Other polychaetes, the isopod *Eurycope*, and nuculid and tellinid bivalves eat only the organically rich upper layers of mud, and their guts are less striking. Scraping and browsing techniques on algal films or macroscopic algae are widespread in limpets, chitons, sea-hares and other molluscs, in sea-urchins (echinoids), in amphipods, and in the copepod *Balaenophilus*, which lives on whales and browses on the algae which grow on their skins. Finally, though by no means exhaustively, we may recall the feeding technique of asteroids in which the stomach is everted, and the prey digested outside the body. Polyclads can evert the pharynx to the same end. It is a short step from some of these feeding methods, especially in the case of small predators feeding on large prey, to a state of parasitism.

It has not been possible in this section on adaptations to avoiding capture, and to obtaining food, to describe, or even mention the many different strategies which organisms have evolved in response to these two major problems of life. But when a broad view is taken, the number of solutions is relatively small compared with the number of species, and lends some comfort to the hope expressed in the introduction that the task of comprehending marine life as a whole is manageable. A detailed examination of the feeding and digestive organs can with experience provide valuable information about the food habits of animals, leading to a fuller understanding of their ecological roles.

Life cycles

Egg and larval size and reproductive strategies

Reproduction counters mortality. The two processes must evolve hand in hand if population stability is to be maintained. Any strategy which reduces mortality at any phase of the life cycle allows a reduction in reproductive effort. Conversely, reliance on small unprotected eggs, so common in teleost fishes will generally require a massive expenditure of energy (just as in creatures of small adult size), since fecundity will then be at a premium. In general, the former strategy is favoured in stable and hence predictable environments where mortality is more strictly density dependent, and the latter in unstable or turbulent environments where stochastic processes play a more important part (MacArthur 1962). High fecundity is often a notable feature of parasites for much the same reasons. Within this broad division, further choices are possible, and we can recognize a spectrum of approaches to the problem of mortality ranging between the two extremes.

The life cycle begins with reproduction. Each of the environments distinguished in the introduction offers certain advantages and disadvantages to eggs and young. Herring (*Clupea harengus* (Linnaeus)) and sand-eel (*Ammodytes marinus* (Raitt)) for example, both pelagic fish, and almost unique at any rate among commercially important species, lay their eggs on the sea-bed in sticky masses. The famous grunion (*Leurestes tenius* (Ayres)) bury their eggs beneath the sand at high-water mark. But these fish are exceptional. The vast majority of both pelagic and benthic animals produce pelagic eggs and larvae. The great majority of bottom-living animals produce free-swimming larval stages which at times form a very important constituent of the pelagic fauna in coastal waters. These eggs are characteristically small and contain relatively little yolk. When they hatch, therefore, the larvae must feed themselves, and their production in middle latitudes usually seems timed to coincide with the season of peak abundance of suitable food. It may also be timed to allow the metamorphosed larvae to settle on the bottom at a time when their potential predators are in non-feeding condition (Thorson 1957). The extremely high mortality suffered by small pelagic eggs and larvae is indicated by the fecundity of their parents which frequently runs into the tens of millions, or even over 100 million as in the oyster. The causes of this mortality are very difficult to estimate quantitatively, but several different physical and ecological processes are known to be important (see below).

Animals which produce eggs with a lot of yolk are much less fecund. This strategy reaches its climax in the sharks, some of which produce as few as a dozen eggs each year, and in some instances breed only in alternate years. Some also are viviparous. In Arctic waters, almost all invertebrate eggs are large and yolky, even though their close relatives in temperate regions produce small eggs. Species of *Cardium* and *Modiolaria* illustrate this difference. The Arctic representatives of these genera are also protandrous hermaphrodites, while the temperate species are dioecious. The significance of this observation is obscure, but may be related to the nature of the production cycle in high latitudes. In invertebrates with non-pelagic eggs, there is frequently some form of brood protection. Retention of the eggs in strings or capsules or special cavities is common. Many crustaceans have special brood pouches between their abdominal limbs, where the larvae find protection for some time after hatching. Many larval asteroids find security by living on the dorsum or in the ambulacral grooves of their parents. Some polychaetes and anemones show the same adaptations. These features are again characteristic of animals of high latitudes, suggesting that the problem of timing the spawning of pelagic eggs in relation to the brief production cycle has not been solved.

Larval life

Sedentary animals stand in greater need of a free-swimming phase in their life history than motile kinds if they are to colonize new environments. This is true of all benthic plants as well. Eggs, larvae and spores constitute the dispersal phase in most cases, and water movement the vehicle. Alternation of fixed and motile generations, as in many coelenterates, achieves dispersal too.

The larval phase is generally ended by a profound morphological reorganization known as metamorphosis. So different are larval and post-larval forms, those of the same species have in the past frequently been assigned to different genera. It is from this confusion that we have received the common names of different kinds of developmental stages like nauplius, zoea, and megalopa among crustaceans, auricularia and pluteus among echinoderms, and so on. During metamorphosis, all the adult organ systems are acquired in definitive form, though in many parasites, there may be further dramatic morphological changes later in the life cycle.

Following larval life, there is a period of growth before maturity is attained, the duration of which will depend to some extent on the adult size, but also on such factors as light, temperature and food supply. Chapter 11 is devoted to the subject of growth of fishes. It is apparent in many fish and crustaceans that the timing and location of spawning are such that the young, following their larval life spent drifting in the plankton, will reach regions where conditions are suitable for survival and growth. This topic is discussed in Chapter 13. Equivalent processes must occur in many deep-water populations whose larvae and juveniles live at lesser depths. Thorson (1950) reviewed the evidence that the larval bottom-dwelling invertebrates can delay their metamorphosis within limits until such conditions are found. They can therefore adapt to variations in hydrographic conditions, and select the kind of substrate they require to settle on for their further development.

Life in the photic zone

Littoral and sublittoral environments

The inshore environments, and especially the intertidal region, are, as already indicated, extremely complex, and are the habitat of the vast majority of marine plant and animal species, and of the greatest concentrations of individuals. Though there is of course much variation from one region to another, a total biomass in excess of 1000 g/m^2 is typical of the fauna at low water. At a depth of 200 m, a figure of 50 g/m^2 can be expected. In the deep sea, very low standing stocks are found, generally less than 1 g/m^2, and less than 0·1 g/m^2 in abyssal depths. These figures indicate the importance of the inshore region despite its small extent compared to the total area of the sea. Its accessibility and its richness and diversity have ensured for it the attention of many naturalists, and several excellent books are available which describe life just beyond the shore. Those by Morton and Miller (1968) and Lewis (1964) are especially recommended. Though both are concerned with particular regions (New Zealand and Britain respectively), the principles of the zonation of plants and animals in response to light, immersion and exposure, substrate type, and so on are of world-wide significance (Stephenson & Stephenson 1949).

The benthic flora may be divided into three main groups, the macroscopic and microscopic algae, and the marine phanerogams. The large algae may be divided into perennial species in which either the whole plant persists for several years, or in which only the basal part persists to produce new fronds each year, and the annuals which either pass through several generations in a year and produce eggs or spores which germinate immediately, or which spend part of the year as resting spores or as inconspicuous protonemata. The plants are usually either erect or encrusting, though sac-like and other forms occur, and their development is very labile in response to the particular characteristics of the site. Plant form is also greatly modified by browsing herbivores. The microscopic algae of the littoral are mainly diatoms, blue-greens, and a variety of flagellates. These form films on the substrate and on the surfaces of other organisms. A composite film of blue-green algae, diatoms and fungi, all embedded together in a gelatinous matrix, is a conspicuous feature of high levels on some shores. Other species form miniature branching colonies. On soft shores of mud or sand, this algal film is also present. On mud it increases the resistance of the surface to the scouring action of currents. Macroscopic algae are relatively unimportant on soft shores, and it is in such situations in temperate regions that marine grasses are richly developed. In the tropics, these grasses are replaced by mangrove vegetation.

Very briefly, three main zones are recognized on consolidated shores in the temperate regions of both hemispheres. These are an upper zone where a few green algae (especially *Enteromorpha*) and lichens occur, and an associated fauna of littorinids, a middle zone of Fucaceae with barnacles and limpets, and a lower zone of Laminariales, with many smaller algae, where sea-urchins, mussels and anemones are prominent. The control of these zones seems to lie as much in the hands of predatory animals as on tidal influences. Paine (1966, 1971) has shown that the major patterns of zonation, as well as the spatial heterogeneity and species richness, on the coast of Washington, are controlled by *Pisaster ochraceus* (Brandt), and that species of *Acmaea* influence the distribution and abundance of both the algae and barnacles of the middle zone. Similar principles are emerging from studies in other areas, though the actors are often quite different (e.g. Mann & Breen 1972, Kitching & Ebling 1961). The widely publicized depredations of *Acanthaster planci* Linnaeus on some coral reefs, and the replacement of the hard corals by soft corals and algae, are another example of the importance which ecological factors have. On the other hand, random physical events such as the occurrence of a hard frost at low tide immediately following larval settlement, or the destruction of parts of mussel beds by driftwood (Paine 1969), can lead to the establishment of communities quite different from those which might be anticipated on non-historical premises. So any attempt to account for the patterns of plants and animals in the littoral region must take account of both predictable and unpredictable physical events, the trophic relations of the inhabitants, and perhaps more complex interspecific relationships also (Dayton 1973). The same injunction applies to all communities, though the relative importance of the different determining factors may alter.

Pelagic environments

The photic zone beyond the coastal fringes is occupied by a fauna, and fauna relatively impoverished by comparison with the intertidal zone, especially if we discount the multitudes of larval stages of the benthos. About ten groups of unicellular algae contribute to the phytoplankton, but of these only three, diatoms (Bacillariophyceae), dinoflagellates (Dinophyceae), and yellow-browns (Chrysophyceae) are found regularly and abundantly. A single genus of blue-green algae, *Trichodesmium*, is very important in the tropics. Species from other groups may give rise to extensive blooms from time to time in some areas, but cannot be reckoned as major contributors to primary production on a global scale. They may nevertheless form the basis of important food chains locally. From a practical point of view, these rarer forms often provide useful indicator species, and more remotely, may help us to interpret the nature of extinct environments.

Each important algal group tends to be predominant at characteristic times of the year. Very broadly, diatoms have seasonal standing-crop maxima in spring and autumn in middle and high latitudes, while lower numbers occur during the intervening months. Within these maxima, a succession of different species can generally be recognized, each of which rises and falls in numerical abundance within a short period of time. It is the integrated result of this succession which is recognized as a bloom. In lower latitudes, diatoms are usually most abundant in 'winter'. Blooms of dinoflagellates are more characteristic of summer and autumn, especially in warm temperate and subtropical seas. From time to time in these regions, they may give rise to massive and occasionally destructive outbreaks of red tide. In tropical and subtropical waters dinoflagellates are usually the most abundant phytoplankton group numerically. Most are autotrophic, but other modes of nutrition are of wide occurrence in this group, though detailed information is lacking. Dinoflagellates in the terminal stages of succession are typically of very unusual form, and have reduced chromatophores. In some, photosynthetic pigments are absent. Much of the literature on marine successions has been summarized by Margalef (1963).

The meaning of algal succession in the pelagic environment is as yet not very clear. Some of the processes which control succession stem from the dynamics of food chains and are mentioned in Chapter 14. But successions in mixed algal populations occur in the absence of higher trophic levels, and must reflect the abilities of different species to respond to seasonal changes in the availability of light and nutrients, and to changing patterns of turbulence. It seems likely for example that some algae are much better than others at storing phosphate, and that as a consequence their growth rate is not inhibited when the phosphate concentration in the water falls. Each species may be adapted in a like manner along a variety of different axes related both to the environment itself, and to trophic relationships. Margalef (1967) has reviewed some of these problems. Since succession forms a link between the population dynamics of single species and community dynamics as a whole, it should prove a rewarding area for future study. The respective roles played by, say, turbulence and grazing in determining the species composition of a seawater sample are analogous to the processes discussed earlier in relation to rocky shore ecology, but the time scales on which these forces operate are quite different, and are measured in days and weeks rather than months or even years. In the long run, it should be possible also to relate the physiognomic characteristics of algal species, their shapes and sizes, the form of their skeletons, their particular complements of pigments, their ability to join into chains or other associations, their nutritional characteristics, and so on, to the environment in which they occur, and to the part they play in succession.

The herbivorous zooplankton consists largely of protozoans, the larval and juvenile stages of pelagic crustaceans, especially copepods, and appendicularians. The protozoans (tintinnids and radiolarians, and a few species of foraminiferans are the most abundant kinds) are mainly floating forms, but epiphytic and epizoic species (mainly peritrichous and chonotrichous ciliates) also occur. Many of the acantharian radiolarians of tropical waters contain symbiotic dinoflagellates (zooxanthellae) like those of hermatypic corals and other reef animals. Some radiolarians grow to a very large size, up to several millimeters in diameter, and can be even more impressive when they adopt the colonial habit. *Collozoum* for example can reach a colony diameter of 20 cm. Copepods, as a general rule, predominate in all zooplankton samples, and usually comprise 70–90 per cent of the fauna. Two distinct types of life history can be recognized in pelagic herbivorous copepods. There are those in which breeding is confined to periods of phytoplankton abundance like the well-known *Calanus finmarchicus* (Gunnerus), and those which reproduce throughout the year but whose egg production is dependent on the availability of food. *Oithona similis* (Claus) and some *Acartia* species belong to this second group. Heinrich (1962) recognizes a third group in which egg production occurs both in and out of periods of phytoplankton growth, but nevertheless during definite seasons. He cites a number of offshore and oceanic species as examples. These differing strategies may be related to the production cycles in their regions of occurrence, but the relationship, if any, is not a simple one, since species occurring in the same region (and often closely related) fall into different categories. There is perhaps a greater preponderance of continuously reproducing copepods in seas where shallow depths allow a longer period of primary production, but a full understanding of these observations, in other groups too, may ultimately depend on a detailed knowledge of food preferences in relation to phytoplankton species composition.

Succession is not an obvious feature of zooplankton populations, though it is generally true in fluctuating environments that predatory forms succeed herbivores. As already pointed out, many species have fixed and different spawning seasons, the sequence of which effectively simulates succession. It is not even clear at present whether the latter kind of succession, genetically determined, is usefully distinguished from the more opportunistic kind of succession observed in the phytoplankton. If such a distinction were to be made, it would presumably serve to define a further axis along which the ecological strategies of different organisms could be described.

This chapter has sought loosely to find ways in which the confusion and abundance of marine life can be viewed as a whole, and to explore some avenues along which we might hope to reach a fuller understanding of marine ecosystems. Subsequent chapters of this book describe different aspects of such systems in more detail, and it is hoped that this general account will provide an introduction to them. The main theme throughout is the regulation of numbers. It is this regulation at the population level which ultimately lies behind the stability of ecosystems, and which requires to be analyzed in full as a prelude to a predictive and general science of ecology.

PART II. THE STRUCTURE OF LIFE IN THE SEA

Life in the sea is dominated by the penetration of light into the water. Photosynthetic activity is limited to the depth of the euphotic layer and all the plant food is produced there, where indeed much of it is consumed. More important, below the euphotic layer the numbers of animals decline exponentially with depth, so in very deep water animals are very sparse, although relatively large. The upward motion of sea water is 3–5 orders of magnitude less than the horizontal flow, and the resupply of nutrients to the euphotic zone is another constraint on the maintenance of life in the sea.

In the horizontal plane, the surface waters flow continuously in the great structures of ocean currents as described in Chapter 2. There tend to be recognizable oceanic structures such as upwelling areas or the North Sea swirl, which are also biological units, based on the current systems, on diffusive processes and upon the vertical structure of life in the sea. Fish populations tend to express some biological rules in somewhat summary form; for example, populations that are genetically distinct tend to be restricted to defined oceanic structures. It is likely that the animals make use of the mix of structure to survive and to obtain reproductive isolation, for example, *Euphausia superba* (Dana), which migrates seasonally north and south at different depths within the Southern Ocean.

It is very hard to describe the structure of life in the sea in any comprehensive way because many details cannot be grasped on the scale which work from ships allows. There is a sense in which the three chapters in this section indicate a text that might be written in the future when the detailed structure has been described more comprehensively. Chapter 4 contrasts marine and terrestrial ecosystems and looks ahead to models based merely on the sizes of animals. In a later section, we shall examine how fishes exploit successive size distributions as they grow. Because they also grow considerably during their lives, the point is made obvious that the same principles may apply to invertebrate populations as well. This common characteristic of increasing size may give rise to a generality about size which some current models do not allow.

Chapter 5, on patchiness, examines a different sort of structure. Because the sea is a diffusive fluid some of the potential variability in the system must become damped. The important point here is that animals (or plants) of different sizes are affected by different scales of the diffusive process. Hence, variability in the vital population parameters of animals of different sizes (which live at different distances apart) may well be damped in different ways. Any damping of variability might also assist the stabilization of numbers and the question arises to what degree the animals exploit the diffusive properties of the medium to obtain numerical control in the face of a variable environment.

Chapter 6, on vertical migration, describes a function, but also illuminates the dependence of animals of different sizes upon the algae of the euphotic zone, not only diurnally but also seasonally. To the limits of light's penetration, animals' size tends to increase with day depth, yet all the array of sizes reach the surface layers at night. As they migrate, the animals are subjected to diffusive processes of different orders at different depths, yet the same processes are significant to both prey and predator on different scales, from aquatic microstructure to permanent thermocline. It is this problem of scale resolution and spatial integration by larger organisms of the temporal variability of smaller organisms that limits our present knowledge of food-chain dynamics in the sea.

This brief study of the structure of life in the sea thus really concerns the sizes of organisms and how their distributions depend upon the penetration of light and on the diffusive properties of the water. In the subsequent section on functions within the marine ecosystem, many of the interrelationships between sizes and groups of plants and animals will become illuminated. In essence, we shall return to Ivlev's principle that for each predator there is an optimum size of prey. This is the first and most important stage in understanding the structure of life in the sea.

Chapter 4. The Structure of Life in the Sea

T. R. Parsons

A comparison of terrestrial and pelagic ecosystems

The marine pelagic ecosystem

The Peruvian upwelling ecosystem—a case study

Environmental and size constraints on phytoplankton

Environmental and size constraints on zooplankton

Some additional factors affecting the structure of pelagic ecosystems

Summary

A comparison of terrestrial and pelagic ecosystems

Existence in a terrestrial community is first and foremost an ability to survive the strain of gravity; in contrast pelagic organisms grow up in the amniotic cradle of the sea. Thus, competition between plants and animals in a terrestrial ecosystem requires heavy energy investment in structural material such as wood and bone. Among pelagic organisms the need for large bones is greatly decreased and cellulose is a very minor constituent of the marine biota; pelagic organisms are suspended in sea water and are close to being neutrally buoyant in their life medium.

When movement is involved, a further exaggerated energy requirement is needed by terrestrial animals as each step requires their mass to be lifted against the force of gravity; again in contrast the business of swimming requires approximately an order of magnitude less energy (Gold 1973). For example, the energetics of exercise is a very appreciable debt in terrestrial ecology while the swimming of copepods during vertical migration has been determined to require less than 0·3 per cent of their basal metabolic rate (Vlymen 1970).

From these observations it follows that the major difference in the structure of terrestrial and pelagic ecosystems lies in the store house of refractory structural materials in the former and its virtual absence in the latter. In the pelagic environment of the sea the maximum pressure for survival is exerted between the organisms themselves. It is further apparent that the medium in which they live allows for continual contact between predator and prey with little refuge.

The predominant organic material of the sea from phytoplankton to whales is protein. In terrestrial environments the predominant organic material is carbohydrate. This basic difference in composition also reflects the very different functions operating in the two communities. A carbohydrate-predominated ecology is rich in stored energy but slow in growing; a protein-predominated ecology uses its energy to grow and there is no equivalent energy storage depot. The business of rapid growth in terms of generations, or growth up to size large enough to avoid most predators, is the primary concern of pelagic organisms. These two basic strategies of survival in the pelagic environment appear to operate at different ends of the food chain. Thus, for primary and secondary producers the greatest strength for their survival is to be small but numerous so that in the absence of cover in the sea their presence is difficult to detect. In addition, their numbers are maintained by growth rates so rapid as to be measured in hours or days. At the other end of the food chain security from predation is achieved through growing large. Thus an animal (e.g. a cod-fish) which spans both the small organism plankton community and the large organism, nekton community, requires a reproductive strategy of millions of small eggs from which only two adults survive to maintain the status quo. From these two strategies it can be concluded that the two most important parameters governing life in the pelagic environment of the sea are the growth rate and size of organism within the ecosystem.

The marine pelagic ecosystem

In the simplest terms, one may consider a marine pelagic ecosystem as a food chain in which the phytoplankton are eaten by zooplankton, which are in turn eaten by fish. This 'food chain' description appears satisfactory if one is considering not more than one trophic level at a time, and then over relatively long time-intervals and large areas. Models based on the food-chain concept involving single trophic levels have been reasonably successful in explaining the gross features of production (e.g. Riley 1946, Steele 1962). Yet in spite of this success it has been apparent for some time that these explanations do not in themselves contain sufficient information for use in such practical problems as fisheries management. A particular example of this was the disappearance of the Plymouth herring fishery (Russell 1939). The lack of herring was associated with other biological events such as a decrease in inorganic phosphorus, the replacement of one species of chaetognath (*Sagitta elegans* (Verrill)) by another (*S. setosa* (Müller)), a reduction in the amount of macroplankton and an increase in the number of pilchard eggs (Cushing

1961). While scientists have speculated on the cause of these events, a clear relationship is lacking. It appears, however, that with the reversal of the whole trend over a period of 40 years (Russell *et al.* 1971) the observed changes were more directly associated with the structure of the ecosystem than with its total productivity.

The inability to solve problems with food-chain production models lies not with the explanations themselves but is probably more closely associated with two additional factors, both involving the structure of the ecosystem. The first of these is the time/space scales on which the descriptions are based and the second is the oversimplification of the food-chain concept which is used in deriving the model.

Time/space scales currently employed in descriptions of marine ecosystems are usually too large for the solution of many practical problems. Questions related to fish migrations, schooling and larval survival have time and space scales which are very often much smaller than those employed in production models. For example, schooling of fish may be a phenomenon lasting from hours to days; larval fish survival is critical over a period of a few days and distances of several metres. When these time/space scales are considered it is apparent that the uniformity of physical/chemical effects in time and space and absence of species differences in marine populations which have been assumed for the production/food-chain models are no longer applicable. It has been recognized for some time in fact, that biological data have a high degree of variability which is often neglected in order to formulate a general relationship. For example, the distribution of phytoplankton in a mixed water column has been assumed to be uniform (e.g. see Sverdrup's 1953, critical depth model). In fact it has been shown (e.g. Ryther & Hulbert 1960) that in a homogeneous water column to 150 m, in which temperature and salinity differed by less than 0·05°C and 0·05 per cent, respectively, differences of several orders of magnitude occurred in the concentration of phytoplankton species at different depths. Similar studies on the vertical distribution of plants and animals in association with known discontinuity layers show even greater differences in the vertical distribution of the biota. As examples, abundant zooplankton concentrations have been found in association with the oxygen minimum in the eastern Pacific (Longhurst 1967); a deep chlorophyll *a* maximum has been shown to occur in association with the nitrite maximum in the Pacific (Anderson 1969, Venrick *et al.* 1973); time variations in nutrients at a fixed point have been shown to be correlated with the height of internal waves (Armstrong & LaFond 1966) and similar non-random distributions in plankton concentrations have been shown to occur in areal distributions (e.g. Bainbridge 1957, Cassie 1960, Wiebe 1970). The exact cause and extent of these non-random distributions of plankton in the sea are discussed in the next chapter (see Chapter 5). For the purpose of discussing the structure of life in the sea it is only necessary at this point to recognize that the plant and animal plankton communities are built on a microscale of associations which are quite different from the macroscale considered in food-chain/production models. Thus much of the data which have been collected on the plankton community are only useful in showing the gross features of production; the same data can seldom be used for detailed trophic models because the degree of contagion or patchiness of the plankton distributions have not been measured.

Ivlev (1961) was the first to emphasize the importance of food aggregation in fish feeding. While his data applied only to benthic feeders, it is quite apparent that the same situation applies to plankton feeders. As an example of this Kumlov (1961) showed that the average concentration of plankton reported by scientists using plankton nets was not sufficient to support the metabolic and growth needs of baleen whales in the North Pacific. The explanation of this inconsistency was given by Barraclough *et al.* (1969) who used a high frequency echo sounder to show that the zooplankton occurred in very dense layers in which the concentration of animals was many times the reported average concentration. The latter had been obtained traditionally by towing nets over some distance; this effectively integrates patches of plankton into an average concentration which is unrepresentative of the true plankton concentration at any point. The association of herbivorous micro-zooplankton with the 'pigment layer' in the sea (e.g. Beers & Stewart 1970) is another example of where patchiness plays an important role in trophodynamic relationships of plankton communities.

The second important consideration in obtaining a better description of the structure of pelagic communities is to replace the concept of a 'food chain' with that of a 'food web' (Steele 1965). The latter analogy suggests that there are alternative pathways for the flow of energy in the marine environment. The presence of these alternative pathways is generally held to give greater stability to any biological system. This concept was strongly held by some terrestrial biologists who considered that stability could be maintained through the diversity of organisms available in an ecosystem (e.g. MacArthur 1955). From this it followed that less diverse communities were considered less stable. However, if we regard stability as the ability of an ecosystem to maintain itself in spite of small perturbations, then the means of maintaining itself may not always be through the diversity of organisms present. For example, a predator which is about to exhaust its food supply of a particular prey may turn to an alternative prey if the ecosystem has a sufficient diversity of prey organisms. However, another pathway to stability in this predator/prey situation might be for the predator to migrate vertically into colder water where it could hibernate (or aestivate) until its particular prey had again increased in quantities sufficient for consumption. Thus, in this case temperature and migration would be the chief components in stabilizing the community. Within the food web, therefore, one has to consider not only the variety of different organisms present but also the differences in their physiological, behavioural and genetic properties, depending on the time scale of events under study. It is in fact these alternatives in both intra- and inter-organism response which offer the most probable general explanation for the structure and stability of the marine pelagic ecosystem.

The Peruvian upwelling ecosystem—a case study

Off the coast of Peru, Beers *et al.* (1971) and Ryther *et al.* (1971) have reported on two quite separate and independent studies of plankton production in the Peruvian upwelling. A summary of their findings is given in Table 4.1 and the

Table 4.1. Plankton populations off the coast of Peru. Data from Ryther *et al.* (1971), Beers *et al.* (1971).

Study date:	March–April 1966	June 1969
Approximate location	15° S, 75°40′ W	11° S, 78°20′ W
Euphotic zone (m)	11 to 28	18 to 42
Mixed layer depth (m)	*c.* 20	*c.* 25 to 50
Chlorophyll (mg/m^2)	50 to 600	39 to 60
Surface nitrate (μgat/l)	5 to 15	10 to 20
Photosynthetic rate ($gC/m^2/day$)	1·6 to 11·2	0·83 to 1·23
Major phytoplankton species (approximate size)	Diatoms: *Chaetoceros debilis* and *C. lorenzianus* (8–36μ and 7–48μ)	Flagellates: Cryptomonads and chrysophytes (2–4μ and 5–7μ)
Major herbivores	Anchovy	Ciliates (Oligotrichs and tintinnids)

results of these two investigations appear to give very different descriptions of the same water mass. However, there are good ecological reasons for this difference in the data contained in these two papers and it is these differences which have very wide ecological implications in an examination of the structure of pelagic ecosystems.

While the two locations are not precisely the same, both occur in the general region known as the Peruvian Upwelling under oceanographic conditions which were remarkably similar in the two years (Walsh, 1975). The principal effect of the two-month difference in season between the studies was that the winter water (June) was more strongly upwelled (2×10^{-2} cm/sec) and less stable than the water observed in the period March/April. However, these relatively small differences in season and area were sufficient to cause very great differences in the structure of the food web in the two areas. In the study performed by Beers *et al.* (1971) the authors were impressed by the large number of ciliates which were the principal herbivores. In contrast, Ryther *et al.* (1971) enter into a discussion at the end of their presentation on the role of anchovy as the principal herbivore in their study area. Obviously, there is a very great difference in the size of these two herbivores and in the structure of the two food webs. It is these differences on a local scale which serve to illustrate alternative ecological responses of the food web to differences in the physical/chemical environment.

Differences in total primary productivity in the two studies can be accounted for in terms of known relationships between the amount of chlorophyll *a* in the water column and the depth of mixing. In the latter case, when the depth of mixing extends below the euphotic depth, the productivity is decreased by the average amount of light available for primary production (e.g. Cushing's D_c/D_m correction, 1959). Production would appear to be decreased by this process in the study conducted by Beers *et al.* (1972) but not in the study conducted by Ryther *et al.* (1972). Further, the quantity of chlorophyll in the water column determines the net photosynthesis per m^2 assuming conditions of constant daily radiation and in the absence of appreciable quantities of other light absorbing substances (Takahashi & Parsons 1972). The net results of these effects would account for the approximate fourfold difference in primary productivity observed in the two environments.

The second difference in these two environments is the species and size range of primary producers present in the two water columns. The water column with the deeper mixed layer (which the authors also describe as lacking in stability) contained small flagellates (2–4 μ and 5–7 μ) while the water column having a shallow mixed layer was predominated by relatively large, chain-forming diatoms. The difference in the concentration of nutrients (as reflected in the nitrate values—Table 4.1) was not sufficiently great to affect the size of organism. On the other hand, the amount of light available to the primary producers would have been very much lower in the water column studied by Beers *et al.* (1971) compared with that studied by Ryther *et al.* (1971). The effect of light and nutrients in determining the size of phytoplankton cells is discussed in the next section. The important point at present is that the combined effect of low primary production, low phytoplankton standing stock and small phytoplankton cells reported by Beers *et al.* (1971) was in sharp contrast to the high primary productivity, high phytoplankton standing stock and large phytoplankton cell size reported by Ryther *et al.* (1971). These differences lead to the third contrast which is in the size of the major herbivore species in each area (Table 4.1). The success of anchovy feeding in Ryther's study compared with the success of ciliate feeding in Beer's study must be further linked to the food web through a consideration of the food available and the energetics of these two herbivores.

The overall effect on fish production in these two areas is to reduce the production of a terminal resource (anchovy) by *c.* 96 per cent in the water studied by Beers *et al.* (1971). This follows from the assumption that the ciliate population could be fed off directly by anchovy while the flagellates were too small to be retained by the gill rakers of the anchovy. If an ecological transfer efficiency of 20 per cent is assumed for each trophic level, then the transfer of food to a higher trophic level is given as

$$P = BE^n \tag{4.1}$$

where B is the primary production, E is the ecological efficiency, and n is the number of trophic levels. Using the above equation the terminal production (P) of anchovy in the case of Ryther's study would be 1 $gC/m^2/day$ ($n = 1$) compared with 0·04 $gC/m^2/day$ ($n = 2$) in Beer's study (assuming $B = 5$ $gC/m^2/day$ and 1 $gC/m^2/day$, respectively).

The question in this case study of two alternate pathways in the food web of the sea is to examine what kinds of biological mechanisms are important in determining the reaction of organisms to differences in the physical/chemical environment. A possible general answer to this question may be found in considering marine pelagic organisms on a basis of size selectivity at both the primary and secondary trophic level.

Environmental and size constraints on phytoplankton

In studies on the proportion of large (net) phytoplankton to small (nano-) phytoplankton in the California current system, Malone (1971) was able to show that the ratio net/nano plankton increased under grazing pressure from herbivorous zooplankton. Using the concentration of nitrate as an index of upwelling and the ratio phaeopigments/chlorophyll α as an index of zooplankton grazing, the author obtained a statistically significant relationship at the 0·05 level such that

$$\text{net/nano} = 1{\cdot}76 + 0{\cdot}003\,(NO_3 - N) - 2{\cdot}53\,(\text{Phaeo/Chl}\ \alpha) \tag{4.2}$$

where nitrate concentration is expressed in μM and net/nano is the ratio of chlorophyll α per m^2 of net plankton to nanoplankton.

Semina (1972) recognized that definite distributional patterns existed in the size of phytoplankton species in the Pacific Ocean. The author described seven areas of different phytoplankton cell size and these approximated known hydrographic differences in the water masses. From a study of the physical/chemical properties of these areas, Semina (1972) concluded that phytoplankton cell size depended on (1) the direction and velocity of vertical water movement, (2) the density gradient in the main pycnocline, and (3) the phosphate concentration. This appears to be a useful grouping of determinate factors but it does not include light or zooplankton grazing. Also it may be preferable to examine size relationship in terms of nitrate limitation rather than in terms of phosphate limitation. This is because nitrogen appears to be more often the rate-limiting element in marine environments (Ryther & Dunstan 1971) and more data are available on the kinetics of nitrate uptake.

Using a wider grouping of determinate physiological and environmental factors Parsons and Takahashi (1973) examined the growth constants of two different-sized phytoplankton under various environmental conditions. The equation used in these studies was

$$\mu = \mu_{max}\left\{\left[\frac{\langle I\rangle}{K_I+\langle I\rangle}\right]\left[\frac{[\mathrm{N}]}{K_N+[\mathrm{N}]}\right]-\frac{S-U}{D}\right\} \tag{4.3}$$

where μ is the growth constant of the phytoplankton as defined by the equation $n_t = n_o \exp(\mu t)$ for a population of phytoplankton (n_o) and after some time interval (t) when the population increases to n_t. $\langle I\rangle$ is the average light intensity, [N] is the concentration of nitrate, S is the sinking rate, U the rate of upwelling, D is the depth of the mixed layer, and K_I and K_N are physiological constants characteristic of particular phytoplankton species. The term $\langle I\rangle$ is determined from the expression

$$\langle I\rangle = \frac{I_o}{kD}(1-e^{-kD}) \tag{4.4}$$

where I_o is the photosynthetic radiation at the sea surface and k is the extinction coefficient of the water column.

Using equation 4.3 it was possible to show that appreciable growth of large-cell phytoplankton would occur in temperate coastal estuarine environments during the spring, in tropical upwelling areas throughout the year and in Antarctic upwelling areas during the summer. These theoretical results are in general agreement with observations in the literature which show that large diatoms predominate in estuarine environments during the spring (e.g. Parsons *et al.* 1969) and that large phytoplankton also predominate during maximum upwelling in the tropics (e.g. Ryther *et al.* 1971) and in the Antarctic during the summer (e.g. Marr 1962). On the other hand small-cell phytoplankton are known to predominate in the oceanic subarctic Pacific (e.g. McAllister *et al.* 1960, Parsons 1972) and in the Sargasso Sea (e.g. Hulbert *et al.* 1960). The latter cell sizes are also predictable in terms of the nine parameters employed in equation 4.3.

As a generalization of the effect of light and nutrients on the cell size of phytoplankton, Parsons and Takahashi (1973) presented a three-dimensional model (Fig. 4.1) which shows the ratio of the growth rate of a large cell (*Ditylum brightwellii* (West)) to a small cell (*Coccolithus huxleyi* (Lohm)). In this figure it is apparent that growth rates of large-cell organisms only exceed that of small-cell organisms in bright light and nutrient environments. Not

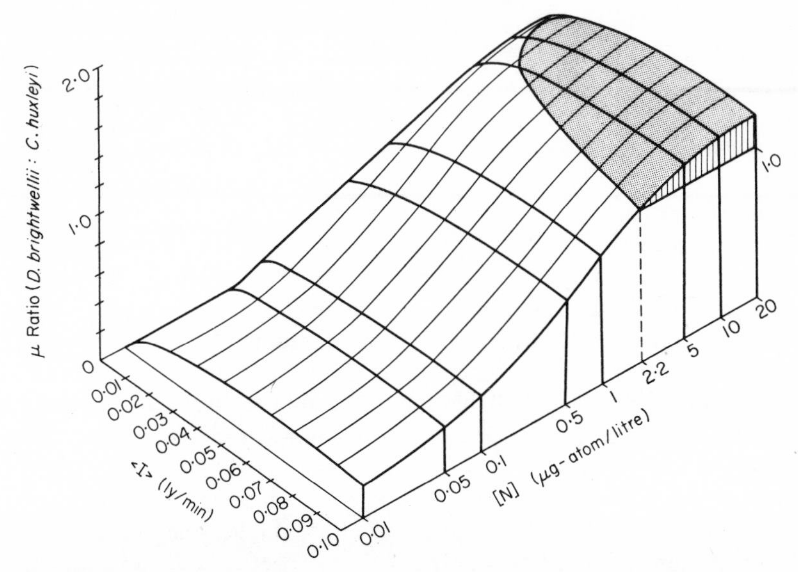

Fig. 4.1. Three-dimensional graph showing the effect of light and nutrient on the ratio of the growth rate of a large phytoplankton cell (*Ditylum brightwelli*) to that of a small phytoplankton cell (*Coccolithus huxleyi*). Values of μ ratio > 1 indicate a predominance of *D. brightwellii* over *C. huxleyi* (From Parsons & Takahashi 1973).

included in this diagram, however, are the sinking and upwelling rate terms relative to the depth of the mixed layer. In addition there are some morphological adaptations among the phytoplankton which could be cited as exceptions to the data in Figure 4.1. For example, the role of motile phytoplankton (e.g. flagellates) may be particularly important in some environments since motility enables a phytoplankton to improve its ecological position. As a flagellate moves through the water it increases its exposure to different water and therefore improves its ability to take up low concentrations of nutrients. In low-light intensities flagellates are also at an advantage since they can move towards the source of light and again make the most of a growth-limiting situation.

Another size restraint on the phytoplankton which does not occur in equation 4.3 is that the standing stock will depend to some extent on the size selectivity of the herbivorous grazers. This effect has been demonstrated both in the laboratory (e.g. Parsons *et al.* 1967, Poulet 1973) and in the environment (Malone 1971). In general it appears from these and earlier experiments (Mullin 1963, Conover 1966) that large phytoplankton cells are grazed more efficiently than small cells. In many cases, however, there may be an upper and lower prey size threshold for a filter feeding animal. Burns (1968) showed that the upper-size threshold of the freshwater zooplankton *Bosmina* and *Daphnia* was related to the body size of the organisms; the lower threshold size in a filter feeding crustacean can be determined from the spacing between the setules of the animal's setae (see review of Jørgensen 1966). More detailed experiments on phytoplankton-size selection by copepods have been carried out by Wilson (1973) who deduced that a herbivorous zooplankton (*Acartia tonsa* (Dana)) might have an ability to 'scan' the size spectrum of particulate material in the sea and locate peaks of abundance; grazing was then reported to be most efficient on the particles in size groups slightly larger than the average size or 'peak' in the spectrum of particle size distribution. Poulet (1973) showed that the marine copepod, *Pseudocalanus minutus* (Kröyer), feeding on natural phytoplankton populations could change its feeding to different parts of the phytoplankton-size spectrum depending on the concentration (in ppm and not number) of particles available. Thus, the animal would tend to graze out any peaks of phytoplankton sizes and yield a more or less even distribution of phytoplankton biomass. In this case grazing selectivity within a limited size range of particles normally consumed by a particular species of zooplankton would appear to be a function of concentration. On the other hand, the size of phytoplankton cells over the whole *in situ* size spectrum of food items appear to control grazing selectivity between species of zooplankton. Thus, in spite of the different growth rates of phytoplankton and temporary differences in the peak abundance of large and small species, the long-term effect of zooplankton grazing should be to give an even size distribution of phytoplankton in terms of the biomass of organisms in any one size category (e.g. as suggested by Sheldon *et al.* 1972).

Environmental and size constraints on zooplankton

Some of the properties referred to above with respect to the advantages and disadvantages of size on the growth of phytoplankton can be repeated with respect to properties governing zooplankton survival. Thus, the tendency for small organisms to sink more slowly than large organisms is also true for zooplankton and various anatomical modifications to retard sinking are apparent in some species. On the other hand, there are specific properties of size which are peculiar in an ecological sense to the zooplankton community.

Probably the most important size property of the zooplankton is the difference in food requirements per unit body weight of small and large animals in different environments. A recent discussion of this subject has been given by Ikeda (1970) and data in Table 4.2 have been largely derived from his findings and a summary of filtering rates given by Jørgensen (1966). In reporting these data it is recognized that scientific opinion may vary with respect to absolute values of some of the derived data. For the purposes of

Table 4.2. Relative food requirements of different sized zooplankton in different environments based on data given by Ikeda (1970).

4.2.1. Food requirements for boreal species of different size assuming growth (7 per cent) day) and assimilation efficiency (80 per cent).

Animal wet wt. (mg)	Metabolism	Growth	Total	Food intake	Growth efficiency (K_1%)
	(Per cent body weight per day)				
0·005	27	7	34	42	17
0·05	13	7	20	25	28
0·5	6·4	7	13·4	16·8	42
5·0	2·8	7	9·8	12·3	57

4.2.2. Metabolic requirements for 5 mg animal in different habitats.

	Boreal (8°C)	Temperate (17°C)	Tropic (29°C)
Percentage body weight required for metabolism/day	2·75	7·6	20

4.2.3. Concentration of particulate organic carbon required to support the growth and metabolic requirements of different sized zooplankton.

Animal wet wt.	Food intake (% b.w./day)	Animal C (μg)	Food intake (μgC/animal/day)	Filtering rate (ml/day)	Food conc. (μg C/l)
0·005	42	0·375	0·158	0·2	790
0·05	25	3·75	0·94	2·0	420
0·50	16·8	37·5	6·3	20·0	315
5·00	12·3	375	42·3	200·0	215

this discussion, however, the importance of Table 4.2 lies in the *relative* values.

The range of animal sizes given in Table 4.2 is approximately that of a small nauplius up to a large adult copepod. By comparison with the size of organisms studied by Mullin and Brooks (1970), a *Calanus helgolandicus* (*pacificus*) (Brodskii) Nauplius V would correspond approximately to the 0·005 mg wet weight animal, and a male Copepodite V of the same species would correspond approximately to the 0·5 mg wet weight animal. Alternatively, the data can be regarded as representing size differences of adult copepods such as between *Pseudocalanus minutus* and *Calanus cristatus* (Kröyer). From data in Table 4.2.1 it is apparent that smallness is a disadvantage due to the greater proportion of food intake which is required for metabolism over growth. This is reflected in the growth efficiency (K_1) which shows a threefold increase from 17 to 57 per cent over the size-range of animals. Growth efficiencies reported by Mullin and Brooks (1970) for the different growth stages of *Rhincalanus* and *Calanus* generally show a similar increase in growth efficiency with size from *c.* 20 to 50 per cent. This relationship is only true, however, so long as the animal is growing exponentially; if growth rate declines with age it is possible to obtain a decrease in K_1 with size as shown by Shushkina *et al.* (1968) using the copepod, *Macrocyclops albidus* Jur. However, Makarova and Zaika (1971) have shown that K_2 (growth efficiency based on assimilated ration) for a single species of copepod (*Acartia clausi* (Giesbrecht), data from Petipa 1966) first increased and then decreased

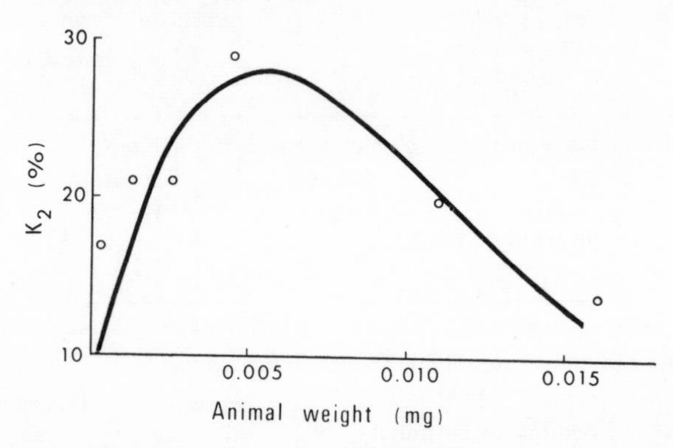

Fig. 4.2. Changes in the growth efficiency (K_2) during the growth (mg dry wt.) of *Acartia* (From Makarova & Zaika 1971).

with age (Fig. 4.2). Thus, the data in Table 4.2 are consistent with the left-hand side of the graph in Figure 4.2 and unless there is a very large increase in the assimilation efficiency, it may be concluded that during the exponential growth phase of zooplankton there is a metabolic advantage to being large.

The disadvantage of being small in terms of growth efficiency may be in part offset by a greater reproductive ability among small animals. Thus, according to Khmeleva (1972) the amount of energy devoted to reproductive tissue ('generative growth') is much greater per unit body weight among small crustaceans compared with large crustaceans. This means that for an equal biomass of small and large adult zooplankters, the amount of tissue involved in egg production will be greater in the case of small animals than in large animals.

In Table 4.2.2 it may be seen that the same-sized animal in different habitats (boreal, temperate and tropic) require much more energy for metabolism in warm (29°C) waters than in cold (8°C) waters. A further difference related to the temperature is that the Q_{10} for large (5 mg) animals has been shown to be *c.* 2 while the Q_{10} for a small animal (0·05 mg) is *c.* 3 (Ikeda 1972). Therefore, the effect of increased temperature is to create a much larger metabolic debt per unit biomass in the smaller of two animals experiencing the same temperature change—this effect being most pronounced in a high temperature environment (tropics) compared with a low temperature environment (boreal).

In Table 4.2.3 the difference in food requirements per unit body weight derived in Table 4.2.1 have been matched with the approximate filtering rate of the different-sized organisms to give the concentration of particulate organic carbon necessary to allow for maintenance and growth (7 per cent per day) of the different-sized animals. The data show that large zooplankton can feed at much lower concentrations of phytoplankton than are available to small zooplankton. It is also apparent that the concentrations (last column of Table 4.2.3) appear to be very high when compared with the levels of particulate organic carbon reported in most oceanic areas. They do not, however, appear as high when one examines the levels of phytoplankton employed in culture experiments in which animals are raised through several generations. For example, the range of cellular organic carbon used in culture experiments by Mullin and Brooks (1970) was from 99 to 614 μgC/l; Paffenhöfer (1970) used different algal foods ranging in concentration from 28 to 800 μgC/l; Corkett (1970) employed concentrations of a flagellate in excess of 500 μgC/l for breeding and rearing calanoid copepods. From this it may be assumed that the *in situ* concentration of chlorophyll α must be very much higher than is generally reported in data collected with standard water bottles. Evidence that high concentrations of chlorophyll occur in micro-layers in the ocean has been given by Strickland (1968) who reported maximum chlorophyll values in a continuous depth profile to be three times the average concentration obtained from bottle casts. With the use of finer instrumentation for measuring local chlorophyll concentrations it is probable that even higher concentrations of chlorophyll may be encountered. At least from a physical point of view it is apparent that very sharp microboundaries exist in the sea but at present there is no biological detector for examining these microgradients (LaFond 1963, Osborn 1974).

In Table 4.2.3, the filtering rate of organisms has been given as being in direct relation to the animal's body size. However, Brooks and Dodson (1965) have suggested that the filtering surface of a filter feeding organism should increase as the square of an animal's linear dimensions. This would indicate that large organisms should be capable of filtering proportionally more water than small organisms. If this is correct, the effect would be to further separate the range of food concentrations required by different-sized animals given in the last column of Table 4.2.3. However, another consideration is the ability of organisms to retain particles. Among copepods this is a function of the spacing of setules on the second maxilla—since these spaces are proportional to the linear dimensions of an organism, it appears that in general small organisms should be capable of filtering smaller particles than large organisms. However, when animals change to raptorial feeding (i.e. seizing their prey), the feeding process becomes more a function of how fast the animal can swim in order to gather food and how well it can detect and capture its prey; these latter properties would generally favour large animal feeding.

In conclusion, it appears that there are a number of physiological/ecological reactions which generally favour the development of large zooplanktonic organisms. However, two mechanisms in favour of small organisms are their high reproductive capacity and any effect of a planktivorous feeder which might preferentially eliminate large zooplankton (e.g. discussion by Brooks & Dodson 1965). In addition, there are further ecological consequences of having populations of small or large zooplankton which can affect the whole metabolic structure of the community. This is apparent, for example, in the effect of the higher metabolic needs of small zooplankton compared with large zooplankton. While this is a disadvantage in terms of growth efficiency of the zooplankton community, the lower efficiency actually results in a greater release of metabolites including the regeneration of nutrients from phytoplankton (Johannes 1964, 1965). Therefore, in an ecological sense, small zooplankton may have a greater stabilizing influence than large zooplankton, on a phytoplankton/zooplankton community.

Taniguchi (1973) has discussed the ecological implications of plankton size and metabolic rate in two widely different habitats, the North Equatorial Current and the Bering Sea. In these two environments the author shows that the body size of herbivorous zooplankton is smaller, and the total biomass of herbivores sustained by the *in situ* primary production is less in the tropics compared with the Bering Sea. From these observations it is apparent that the small-sized tropical herbivores have a greater energy requirement than boreal populations, and that this results both from their small size and the high temperature of their environment. The paradox of this apparent inefficiency of having small zooplankters in oligotrophic tropical waters is offset by the fact that the small zooplankton grow faster and reproduce quicker in the tropics than large species in a boreal habitat. Therefore, over a lifetime the amount of food consumed by small tropical zooplankton is less than the amount of food consumed by large, boreal zooplankton. The latter have to carry their food reserves with them in the form of extra body weight in order to survive long (winter) periods of no food. During the latter periods, two factors favouring survival of large boreal species will be the lower body weight specific metabolism and the animal's lowered metabolism due to colder temperatures (Tables 4.2.1 and 4.2.2).

Some additional factors affecting the structure of pelagic ecosystems

An important observation in the study of pelagic foodwebs is that the ecological transfer coefficient (E in equation 4.1) may decrease both with an increase in the *amount* of production (Cushing 1971) and the trophic *level* of production (Ryther 1969).

Cushing's (1971) data show that transfer coefficients between primary and secondary producers decrease from *c.* 18 per cent to 4 per cent with increased primary production from *c.* 0·1 to 2 g C/m^2/day. Physiological evidence in support of Cushing's data can be found in the work of Paloheimo and Dickie (1966) who showed a decrease in growth efficiency (K_1) with increasing ration among fish. The general relationship was formulated such that

$$\log K_1 = -a - bR \tag{4.5}$$

where a and b are constants depending on the type of animals and their prey, and R is the ration. Metabolically, this relationship implies that as an animal consumes more food (presumably at higher-prey densities) its increase in weight is less per unit food consumed than at low-prey densities. This relationship has been discussed by Sushchenya (1970) with respect to microcrustacean herbivorous feeding. Using Richman's (1958) data for *Daphnia pulex* (Degeer), Sushchenya (1970) showed that the same formulation could be applied to zooplankton and that the change in growth efficiency from a low ration ($K_{1\,max}$) to high rations was from 60 to 6·6 per cent. Thus, although a high ration at high food concentration is advantageous to an organisms' survival as far as its ability to grow is concerned, the increased growth with ration is achieved at greater expense. This greater expense may be due either to greater metabolic cost (as indicated by Paloheimo & Dickie's data) or to lower assimilation efficiencies, as indicated by Shushkina *et al.* (1968). Ecologically, the effect of Cushing's (1971) observation is to decrease the expected increase in yield of tertiary producers in a food web for a given unit increase in primary or secondary producer.

The second factor which may further decrease potential production is a decrease in the ecological transfer coefficient at higher trophic levels. There does not seem to be a great deal of data supporting this contention but it is generally assumed that a fish (such as a tuna) which has to spend a lot of time hunting its prey will use more metabolic energy in feeding per unit biomass than, say, a zooplankton who may only have to float in the water and filter particles in order to get its food. If this is true, then the value of E (equation 4.1) for the primary/secondary trophic level may be at least 20 per cent while for large carnivores it may be less than 10 per cent (Ryther 1969). However, from a number of recent studies it is apparent that the metabolic cost of swimming is quite small (Vlymen 1970, Gold 1973). On the other hand, one may speculate that the metabolic cost of pumping water against a screen in order to filter out small particles may be relatively high. There is, therefore, considerable uncertainty and some contradictory assumptions as to what values to assign to the ecological transfer coefficient at different trophic levels.

Another very important aspect in the structure of pelagic food webs is the ability of an animal to change its position in the trophic hierarchy of the sea. In a simple case it is quite apparent that a salmon which changes from feeding off squid to feeding off lantern fish is shortening its food chain by one step (since the squid also feeds off lantern fish). The difficult point to determine in the 'food web' of the sea is at what point does an animal change its food supply and what is the overall effect on the structure of food web?

Parsons *et al.* (1967) showed that euphausiids in a coastal area were herbivores while Lasker (1966) showed that the same animals are carnivores in oceanic waters. The choice to a euphausiid of when to be a herbivore and when to be a carnivore must be in some way related to the cost in energetics of two types of feeding; one choice being raptorial feeding in which the euphausiid swims through a volume of water in order to capture a relatively large prey (e.g. a copepod); the second choice being for the animal to filter phytoplankton. This problem has recently been discussed by Poulet (1973) in studies on the feeding of *Pseudocalanus minutus* and its choice of large food items (by raptorial feeding) over small food items (by filtering). The author found that when the biomass of particles 1·58 to 22·6 μ diameter was greater than 50 per cent of the total concentration of particles (up to 144 μ diameter) the animal fed off the smaller particles. When the particle distribution from 1·58 to 144 μ diameter was more or less evenly distributed, the animal grazed off large particles; or, if there was a peak in a size category above 22·6 μ, the animal would also select for the large particles. The total particle concentrations used in these experiments was between 0·8 to 4·8 parts/10^6 or from *c.* 40 to 250 μgC/l.

From the above experiments it is apparent that the structure of food webs within which an animal (such as a salmon, euphausiid or microcrustacean) may change its trophic position is probably dependent on the relative biomass of different prey items offered within the size feeding range of the animal. Specialization among animals towards one form of feeding may also be a factor within groups. For example, among euphausiids, deep ocean forms, which must depend on being largely raptorial feeders, have specialized 2nd or 3rd thoracic legs which terminate in grasping claws or small chelae; in shallow-water species the thoracic legs are more uniform and as such more adaptable to making a filtering basket (Nemoto 1967).

Summary

The measurement of plankton communities in terms of the total biomass and productivity of primary and secondary producers has not been found to be sufficiently informative for the solution of some practical problems in fisheries and pollution research. In addition to assessments of total biomass and productivity, greater detail is now required on the complexity of the food web and the interaction of organisms. From this information it may be possible to assess the stability of marine ecosystems and to analyse their sensitivity to natural and unnatural (e.g. pollution) change.

In the previous sections an attempt has been made to discuss the structure of the pelagic food web in terms of the physiological response of planktonic organisms to their environment. While the taxonomic position of each organism and its response to the environment would yield the kind of information required for studies of marine food webs, it becomes impractical even to suggest that individual species reactions should be employed for trophodynamic studies throughout the hydrosphere. As a compromise to this dilemma it has been suggested (e.g. Parsons 1969) that the size spectrum of organisms in the plankton community might serve as a useful analytical basis for obtaining more information on plankton community structure. Differences in the physiological reaction of different-sized organisms have been discussed above in order to illustrate the potential possibilities in this approach. In particular two examples of very different food webs taken from the same water mass off the coast of Peru have been used as an illustration of how the size of organisms produced at the primary trophic level can affect the structure of the food web. Other examples of differences in the structure of food webs can be found in a comparison of the size spectrum of organisms in the Sargasso Sea, north-east subarctic Pacific, estuarine waters, and tropical and subtropical seas.

Additional components required in the analysis of community structure have been referred to in terms of the patchiness of plankton distributions, ecological transfer efficiencies, variations in the trophic position of organisms and in mechanisms which prevent the collapse of ecosystems through alternative strategies. Among the latter are the diversity of prey species, threshold feeding concentrations, vertical migrations and the synchrony of biological events. Feed-back mechanisms involving the regeneration of nutrients and the storage of energy in the particulate and soluble organic detritus of the sea are further factors which help to stabilize the structure of marine ecosystems. The latter subject has been discussed in a number of recent reviews (e.g. Nishizawa 1969, Riley 1970).

Chapter 5. Patchiness

J. H. Steele

Introduction
Plankton patches
Mixing and dispersion
Vertical processes
Special physical conditions
Lateral mixing
Phytoplankton blooms
Zooplankton patchiness
Population stability
Patchiness and fish
Summary

Introduction

The variability of an ecosystem in space and time is usually one of its most important features, influencing both practical problems of sampling and conceptual questions about its structure. Patchiness or spatial heterogeneity can occur on nearly every scale of measurement and must depend on the nature of the response of organisms to their aquatic environment. Sea water is in continual motion and these motions occur at all scales from molecular movement to major ocean gyres. Some movements can appear intense; the breaking waves in a gale, or the Gulf Stream. Others seem relatively calm; the abyssal waters or the Sargasso Sea. But none are completely still. Every organism living in the water is subject to this movement and must either move with it or, in a general sense, navigate through it.

The largest animals, whales, tuna, salmon, can migrate on a scale comparable to that of a whole ocean. Smaller fish such as herring can move in a consistent pattern around parts of an ocean such as the Norwegian Sea or the North Sea (Harden Jones 1968). We expect these organisms to congregate in different areas at different times of year—to be patchy—even though we have no real understanding of how they achieve this. Organisms living on the bottom, the benthos, also display patchiness at different scales (Thorson 1957, Buchanan 1967) but this is generally associated with differences in the character of the bottom on which they live. Plankton, on the other hand, are by definition supposed to be at the mercy of the movements of the sea. Zooplankton such as copepods can migrate vertically but any directed horizontal motion, except at very small scales, would appear to be impossible. Some phytoplankton, the flagellates, also can migrate vertically but many of the major groups, particularly the diatoms, can only sink or rise slowly by changes in buoyancy. Thus, we can expect to find vertical stratification of plankton dependent on the behaviour of the organisms themselves, but horizontal variations should depend on physical factors and so might be expected to be similar in scales and patterns to parameters such as salinity or chemically important aspects like the essential nutrients.

In fact, both the microscopic plants and the small zooplankton seem to display much greater variability than the environment in which they live. The evidence for this at a wide range of scales is still not as definite as one would wish, but will help to define the problem. The corresponding scales of water movement indicate the probable forces which can disperse and mix these populations. To study the relation of the physical and biological processes, it is necessary to try to model their interactions and this will provide some idea of how patches may be formed and of their importance to the stability of these planktonic ecosystems.

Plankton patches

Large-scale plankton distributions have been described elsewhere in this volume. Many of the more common species have geographical ranges extending over thousands of kilometres. At the limits of their range their presence may be due to water movements carrying them from suitable to unsuitable environments so that these extremes do not represent viable populations (e.g. Omori 1970). If one excludes these boundary areas, then the common species are found effectively everywhere within their range. In summer in the northern North Sea, any plankton haul sampling about a hundred cubic metres will contain *Calanus finmarchicus* (Gunner). In this sense, the common species are uniformly distributed within their geographic range. To obtain information on presence or absence one would need to sample volumes of a litre or less. Thus, variability in distribution is normally a matter of quantitative changes and the observed differences are very dependent on the 'scale' of the sampling devices.

For the phytoplankton, samples are usually taken in fractions of a litre since this volume is more than sufficient for counts of the main species. Alternative methods of estimation involve either pigment determinations or estimates of production rate using the ^{14}C method. All these have been used to study small-scale variations. The general conclusion from statistical analyses of these data is that phytoplankton are 'overdispersed', usually defined as positive departure from a Poisson distribution. As Cassie (1963) points out, this statistical terminology relates in some senses to 'aggregation' or 'clumping' but does not necessarily correspond to physically observable aggregations or clumps. Nor does it indicate any active response of the plants to their environment. Bainbridge (1957) described observations of physically distinguishable patches of phytoplankton either observed directly from ships or apparent in the counts of particular phytoplankton species along lines. These patches in the open sea have been reported at a wide range of scales from a few metres wide to areas with diameters of hundreds of kilometres. However, according to Bainbridge, there appear to be two main categories; strips a few metres wide but hundreds of metres in length; and much larger patches, roughly elliptical, with a mean diameter of, very approximately, 50 km. More recent work has used pigment concentration to study spatial heterogeneity. Platt *et al.* (1970) observed variations in a semi-enclosed bay in Nova Scotia with a diameter of about 10 km and concluded that there was a series of discrete size scales of variability indicating patchiness. The main patch diameter was between one and three km.

Continuous measurements from a ship (Lorenzen 1971) showed that in the subtropical Pacific there are large areas of open ocean with uniformly low chlorophyll values showing negligible variability. In these regions any chlorophyll changes which occurred were associated with horizontal tem-

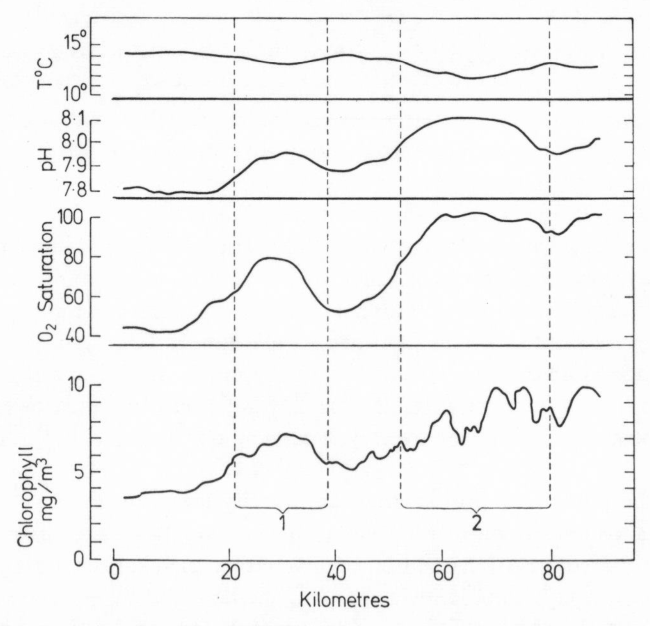

Fig. 5.1. Variations in properties encountered near shore in the Peruvian upwelling area (Lorenzen 1971).

perature discontinuities. In nearshore waters off Peru where the chlorophyll levels were higher, these levels were also much more variable (Fig. 5.1). Oxygen and pH values in the water increase as a result of photosynthetic activity and so these measurements gave some indication of the 'history' of the water. From these data, Lorenzen deduced that the ship's track had cut across two 'parcels' of upwelled water in which phytoplankton blooms had occurred on a scale of, approximately, 20 km.

For zooplankton, it is not normally possible to study microdistribution, since nearly all samples are taken by nets hauled through the water for distances ranging from tens of metres to kilometres. Studies of such replicate tows indicate the same general statistical feature of over-dispersion at this scale (Barnes & Marshall 1951). Equipment has been developed to collect samples at intervals of a few metres (Longhurst *et al.* 1966). The results suggest the superposition of small-scale variability, at scales less than 20 km, on larger patches of kilometre dimensions (Wiebe 1970). However, the small-scale patches may be artifacts of the instrument (Haury 1973). Thus, the main evidence for zooplankton patches is confined to the larger scales.

Cushing (Cushing 1955, Cushing & Tungate 1963) has made detailed studies of phytoplankton and zooplankton patches in the North Sea. The areas surveyed were usually about 100–200 km squares containing 30–40 sampling positions. The main feature of the results, relevant here, is that normally the distributions of both phytoplankton and zooplankton species could be contoured, showing gradients (or 'patches') of the same scale as the area sampled. Figure 5.2 gives two examples from Cushing's work. From mid-March to early June 1954, Cushing and his colleagues followed a patch of *Calanus* which appeared to remain intact throughout this period. These very detailed studies give some idea of the time and space scales of patchiness that can occur with zooplankton.

Using a similar approach in the northern North Sea, I observed the simultaneous distributions of chlorophyll pigment in the upper layers and the zooplankton dry weight under a m^2. Figure 5.3 shows the results from two surveys approximately two days apart. The distributions did not appear to by related to any physical factors such as salinity. Also it can be seen that the concentrations appear to be inversely related. A detailed analysis of these data (Steele 1974a) suggests that this inverse relation between pairs of values for phytoplankton and zooplankton might be expected to be found during a sequence in *time*, as a zooplankton population grows and grazes down the phytoplankton. The problem raised by these data is that they occur as nearly simultaneous distributions in *space*. Further, the variation in zooplankton biomass within this small area is of the same order as that found in surveys with more widely spaced stations covering the whole of the northern North Sea.

These results illustrate the various problems associated with the variability

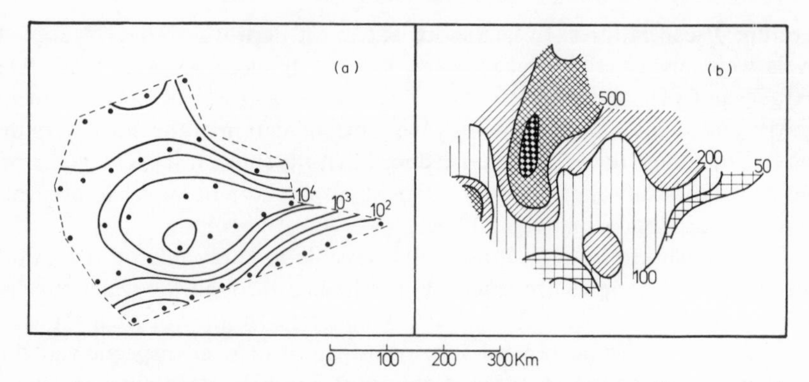

Fig. 5.2. Distribution of (a) a diatom *Thalassiosira gravida* and (b) a copepod *Calanus* Stage V in the southern North Sea (Cushing 1955).

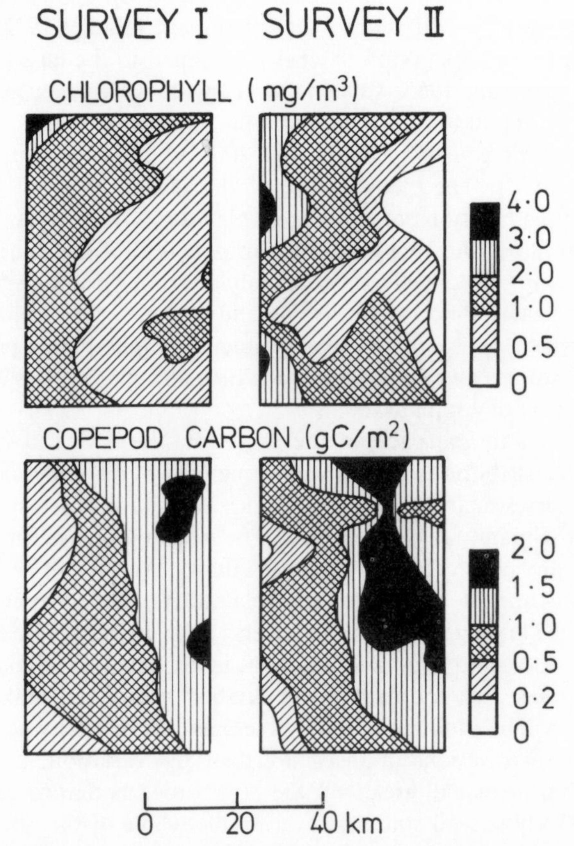

Fig. 5.3. Distribution of chlorophyll α and copepod carbon on two surveys in the northern North Sea, two days apart (Steele 1974a).

of planktonic populations. Underlying the purely biological aspects, such as growth, grazing and reproduction, which can determine part of these variations, there is the question of the effects of water movement in continually redistributing these populations. At the largest scales ocean currents can be important, but at the small scales the turbulent mixing of the water is a dominating factor.

Mixing and dispersion

The smallest scales of motion will involve molecular diffusion but this is negligible compared with turbulent mixing processes that occur horizontally and vertically. These processes can be considered to arise from eddies existing at all scales and these eddies impart fluctuations to the water movement. Usually, these fluctuations appear to be random but in some cases the eddies can have an observable physical reality, such as a large ocean gyre, or the circulation behind a headland as the tide flows past it. In general, however, the 'eddy' concept is used to separate two components in any set of measurements of water flow; the mean flow and the randomly fluctuating elements in these measurements. In any set of measurements, the separation of these components is very dependent on the time and length scales. Mean flow on one scale can appear as eddy motion on a larger scale. Thus, we use the idea of a spectrum of turbulence where the amount of fluctuation is related to the scale on which it is measured. The physical realities underlying this come from the fact that large eddies can derive their energy from external sources such as wind or tide. For example, the very largest eddies, the oceanic gyres, are driven by the general atmospheric circulation (Stommel 1958). Smaller eddies feed on these and in turn break down into still smaller motions until the energy is finally dissipated through viscosity. The cascade of energy down through smaller and smaller scales is the best available picture of how turbulence is generated. This process is known as eddy diffusion because of certain similarities to molecular diffusion. Mathematically, the form of the equations representing these two types of motion are essentially the same and both depend on a diffusion coefficient. In molecular diffusion the coefficient is a constant for any particular system such as heat conduction along a rod but, for the reasons given, in turbulent diffusion the coefficient, k, is a function of the scale, l.

The major differences in k are found when we compare vertical and horizontal mixing. Horizontal mixing appears as much greater than vertical mainly because the horizontal dimensions are so much larger but also because the energy available for the eddies is in the horizontal currents generated by wind and tide. Since our interest is in patchiness, it is the lateral mixing which is of most concern. However, the vertical and horizontal processes are closely linked and both must be considered when we look for explanations of biological patterns.

Vertical processes

Vertical mixing can occur when the upper layers are denser than those below producing convective overturn. This happens in winter as a result of cooling at the sea surface creating a water column of uniform density to a depth well below the lighted zone. On the other hand in spring, summer and autumn, except in shallow inshore areas, it is the existence of a thermocline giving lighter on top of denser water which establishes an upper stable layer and allows plant growth to proceed in this well lit (euphotic) zone. At the same time the utilization of nutrients by plant production and their transfer through the food chain to deeper water will eventually strip this upper layer of nutrients and so inhibit production unless there is some supply from below by vertical mixing. Thus, we are concerned with the balance between the establishment of vertical gradients through heating and the breakdown of such gradients by vertical mixing. It has been shown that this balance depends on the Richardson number, R_i. This number is a function of the changes with depth of density and of horizontal velocity which can be expressed as

$$R_i = \frac{g}{\rho} \frac{\text{vertical density gradient}}{(\text{vertical velocity shear})^2} \tag{5.1}$$

where g is gravity and ρ is density. If R_i is larger than a critical value (approx. 0·2) then mixing does not occur. In many cases where mixing appears to be inhibited, this is due to very large density gradients produced by low salinity in the upper layers rather than from low velocity shears. In general, the strength of the mixing depends on the difference in horizontal water velocities between the upper and lower layers. In shelf areas tidal currents are a major factor in generating such shears. Beyond this, the effects of surface movements produced by strong winds, in breaking down thermoclines, are well known.

Thus, vertical mixing and horizontal movements are closely linked. Further, what we see as apparent horizontal dispersion may also be in part, if not wholly, a combination of these same processes. Bowden (1965) and Kullenberg (1972) have shown that the observed rates of lateral diffusion can often be obtained without invoking a specific horizontal diffusion term. Kullenberg calculated that the lateral spread of dye patches in shelf areas subject to tidal oscillation may be explained by vertical mixing and horizontal velocity shear. For some processes it may not be important whether one uses the simpler idea of lateral diffusion instead of this more complicated explanation. But for plants which exhibit marked vertical variations in concentration and for animals which can migrate vertically, satisfactory explanations of lateral patchiness may be highly dependent on the physical basis invoked for the diffusion processes. At present, however, nearly all the available physical results are given in terms of purely horizontal diffusion coefficients and when these are used in relation to biological processes it is necessary to remember that they are simplifications of very complicated three-dimensional events.

Special physical conditions

One special physical phenomenon is important for small-scale patchiness. Under certain conditions, usually associated with weak to moderate winds, the near surface waters form long parallel vortices with pairs of vortices rotating in opposite directions. The spacing between vortices is usually about 20–100 m. Between vortices there are alternate regions of downwelling and upwelling. The former can accumulate buoyant material and the latter can concentrate organisms sinking or attempting to swim downwards. This 'Langmuir Circulation' (Langmuir 1938) is considered to be responsible for the long narrow patches described by Bainbridge.

A similar type of downwelling circulation may occur at boundaries

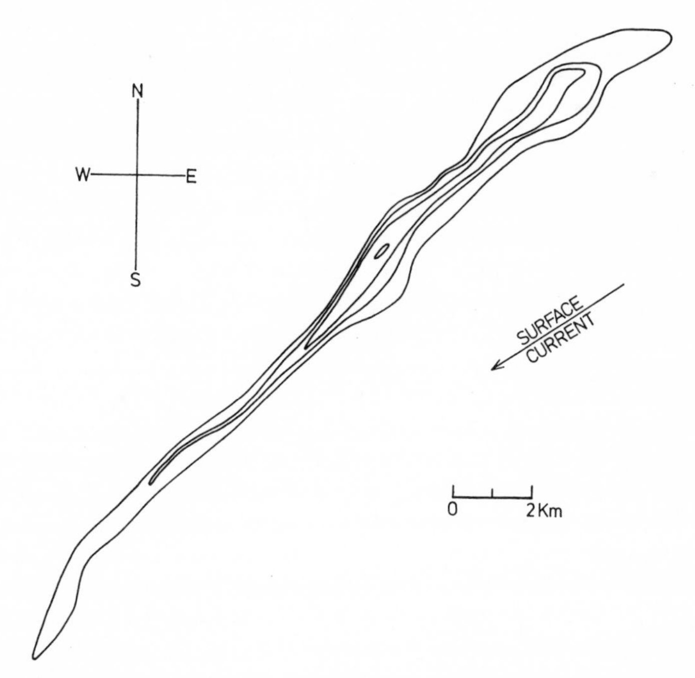

Fig. 5.4. Extreme elongation of a dye patch observed in the Norwegian Deeps two days after release.

between waters of different salinity such as is found when river or estuarine water flows into the sea. This circulation can serve to maintain the sharpness of the boundary and also to concentrate particulate material along the edge. On a still larger scale there are areas of rapid transition of physical and chemical features in the open sea and these special conditions can lead to marked changes in the plankton over distances of a few kilometres. At such boundaries there are usually marked changes in vertical structure and there must be severe inhibition of lateral mixing across the boundary. Off the Norwegian coast, where the Baltic outflow of less saline water is found, there are usually very marked horizontal changes in hydrographic structure and large changes in phytoplankton are often associated with this (Steele 1961). A release of Rhodamine dye in this area, Figure 5.4, shows how the mixing or spreading of the dye appears to occur almost entirely in one direction.

It is apparent that many of the smaller and medium-scale fluctuations in plankton abundance can be related to physical variations but in areas where there are no obvious physical boundaries, patchiness is still observed. What will be the effect of lateral mixing in these less restricted conditions?

Lateral mixing

Much of our information on lateral mixing rates has come from experimental release of Rhodamine dye in surface waters. The dye is injected into the sea as nearly as possible at one point and its spread is followed for days, or sometimes for weeks when a very large quantity of dye has been used (Weidemann 1973). The dye never spreads in regular concentric circles. An example from the Clyde Sea Area, Figure 5.5, gives some idea of the type of distributions which are observed. Usually the general configurations are roughly elliptical but not normally as extreme as that given in Figure 5.4. By measuring the areas within the contours for different concentrations and treating these areas *as if* they were circular, it is possible to calculate apparent lateral diffusion coefficients. By relating the coefficients to the radius of these hypothetical circles, diffusion can be expressed as a function of scale. Natural phenomena, such as the spreading of highly saline water from the Mediterranean into the Atlantic, have been used to calculate coefficients for very large scales (Joseph & Sendner 1958). There is a considerable literature on these dye experiments and on the mathematical formulations which may fit the results. These have been summarized by Okubo (1962, 1971) and the main features presented as a relation of k to scale l, taken to be, approximately, the diameter of a circular patch, Figure 5.6. The scatter of the observations may reflect the fact that dispersion could be dependent on wind strength, sea state or particular hydrographic conditions at the time of each experiment.

This type of experiment seems particularly relevant to the comparable

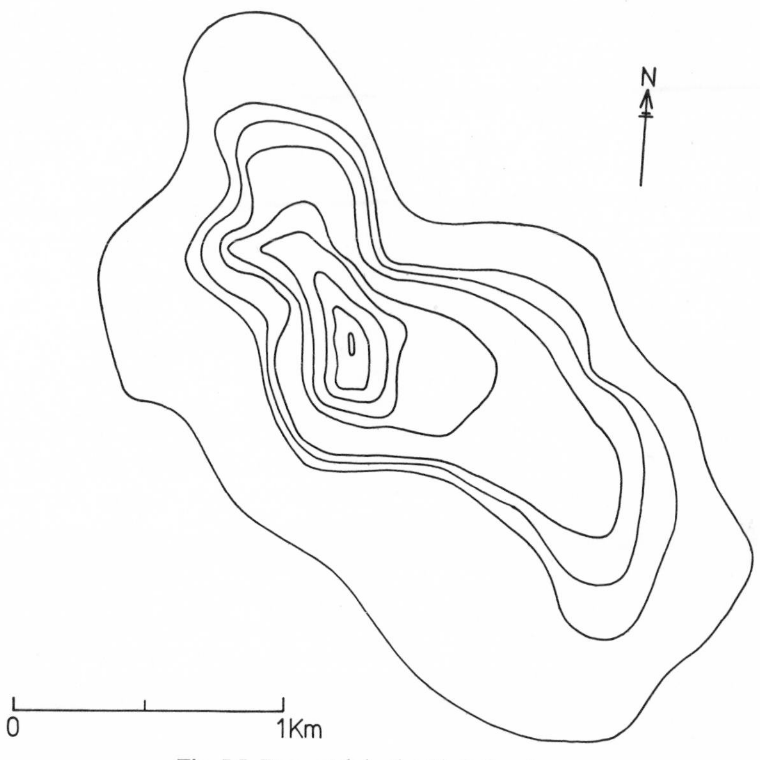

Fig. 5.5. Dye patch in the Clyde Sea Area.

problem of patchiness of plankton. If phytoplankton or zooplankton were considered to be 'inert' and neutrally buoyant, then any high (or low) concentration would be dispersed in this way. However, since they are growing or reproducing, and since animals are eating the plants, these factors will complicate the effects of dispersion. The simplest case to consider first is the growth of phytoplankton on its own. The original treatment of this problem (Kierstead & Slobodkin 1953) dealt with red tides where this is a reasonable, or at least, a possible assumption. Also, at the beginning of a spring outburst, when plant growth begins but herbivore grazing will be at a low level, this can be taken as a first approximation. Kierstead and Slobodkin considered the problem of a circular patch of diameter, l, with a growth rate, a. They showed that if l is small the patch will leak away due to the effects of mixing. However, if l is large enough then the growth within the patch will more than counteract the effects of dispersion, and the concentration of phytoplankton within the patch will increase. They calculated the critical diameter l_c above which this increase would occur as

$$1_c = 4{\cdot}8\sqrt{k/a} \qquad (5.2)$$

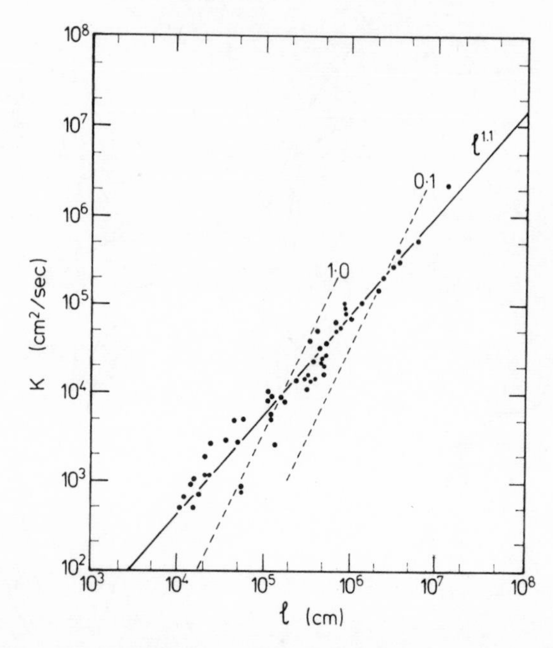

Fig. 5.6. Observations of diffusion coefficient k in relation to scale, l (Okubo 1972). The dashed lines indicate the relation of critical length, l_c to k for two division rates.

For any plant growth rate this defines a relation between l_c and the diffusion coefficient k. For two division rates of 1·0 and 0·1 per day (or doubling time of 1·0 and 10 days) these relations are shown in Figure 5.6. Using the solid line as an average for the observed relation between l and k, the minimum critical size of a patch would go from 2 km to 50 km as the division rate goes from a high to a low value. This is one general conclusion about patch size that can be taken from this relation. The other main conclusion, evident from Figure 5.6, is that, although these lines represent the average conditions, there are likely to be large variations from the average. This is due to the small angle of the intersections of the dashed lines with the solid line and the relatively large scatter of the observations of diffusion coefficients. In particular, if the local conditions in any area restrict lateral mixing, then critical patch size is very much reduced.

Given this limitation, however, certain conclusions about patchiness would still follow. The relation of scale to diffusion coefficient does not imply that 'variability' occurs only above the critical length scale but that below this scale, variability in plant concentration should be a consequence of, and related to, the physical characteristics of the water. Evidence in support of this has been provided by Platt (1972) who measured continuously the fluorescence of chlorophyll pigments at a fixed position in the Gulf of St Lawrence for a period of three days. By relating the time scales to the mean flow of water past this position, Platt showed that the distribution of the variance of the phytoplankton abundance had the same form as that expected from the theory of turbulence over the length scales 10 m to 1 km. As Platt says, this leads to the 'hypothesis that, at least in a two-dimensional world, the local concentration of phytoplankton is largely controlled by turbulence, and not by any dynamic attributes of the organisms'.

These theoretical considerations support the general observations described earlier, that plankton patches in the open sea appear to occur at scales the order of 10 km–100 km. The theory, up to this point, has dealt only with the growth of phytoplankton and not with the interactions of plants and herbivores. This will be considered later but it is useful to look at the possible applications to particular problems.

Phytoplankton blooms

The problem of 'red tides' concerns intense blooms of certain dinoflagellates which, when fed on by shellfish, can render these toxic. There is an extensive literature on the subject (see Adachi 1972 for references; Quayle 1969, Hickel *et al.* 1971) but no clear evidence of the factors which permit these sporadic outbreaks to occur. It appears that they are normally associated with calm weather and often with lowered salinities near the surface. Such factors, on the basis of the physical concepts which have been discussed, could decrease mixing and so increase the likelihood of patches developing. This is the basis for the original ideas of Kierstead and Slobodkin. However, these flagellates, being motile, can concentrate by vertical migration and this could be enhanced by relatively calm conditions. The particular physical environmental conditions may be a necessary condition but are certainly not sufficient to cause a 'red tide' (Prakash & Taylor, 1966). Nor, apparently, is an initial high nutrient concentration essential (Hickel *et al.* 1971). A further factor concerns grazing. If there were no grazing on a spring bloom it would reach the densities often observed in red tides (Cushing 1969, Steele 1974a). Thus, grazing normally prevents the occurrence of such high concentrations and so a causative factor in these flagellate blooms may be the absence from, or avoidance of, these blooms by suitable grazers.

This brief outline is intended to point to the probability that the inhibition of mixing is only one factor, and biological events may be as important, particularly the other main limitation on phytoplankton, grazing.

Upwelling is another cause of intense blooms of phytoplankton for which, again, there is an extensive literature (see Cushing 1969). Distributions off West Africa, Figure 5.7 (Hart & Currie 1960) and off Peru (Walsh *et al.* 1971),

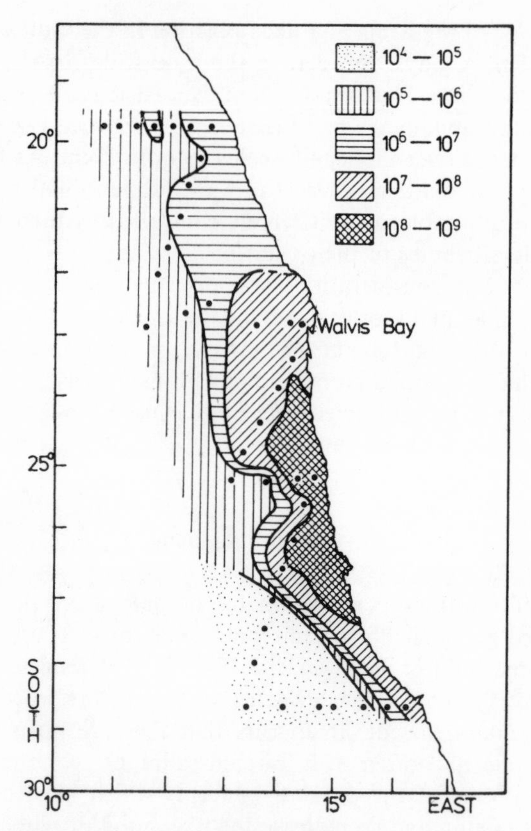

Fig. 5.7. Distribution of the microplankton (numbers per net haul) in an upwelling area off the African coast (Hart & Currie 1960).

Figure 5.1, suggest blooms with lengths of 50–100 km and least diameters considerably less. The development of phytoplankton blooms in particular areas may depend as much on local topography as on the general effect of wind stress producing an offshore water movement near the surface (Yoshida 1967). Features such as headlands may affect the character of mixing in the waters nearby and so be conducive to the development of patches at the right scale for phytoplankton to bloom and for the blooms to persist for days or weeks. However, Walsh also pointed out that, in computer simulations of the sequence of nutrient upwelling—phytoplankton production—and grazing by herbivorous fish or zooplankton, it was essential that there be a lag in the onset of grazing. Otherwise the high concentrations of phytoplankton were not simulated. Thus, again, the physical basis for patchiness may be necessary, but other biological factors are implicated.

A similar local phenomenon was observed in a part of the Clyde Sea Area,

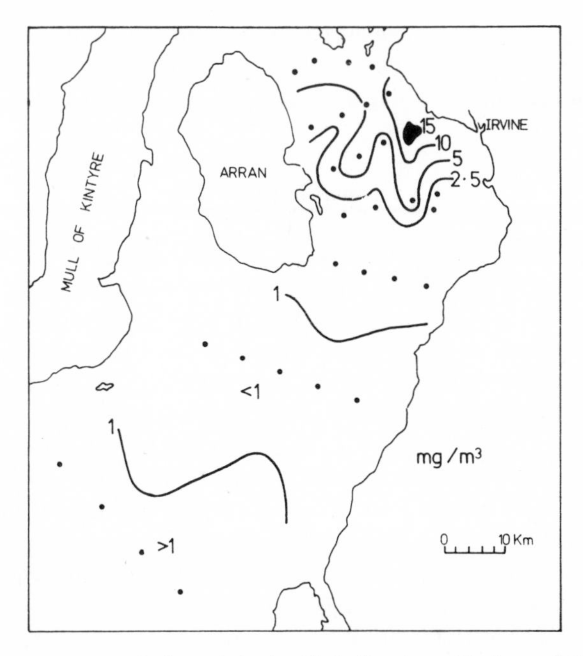

Fig. 5.8. A chlorophyll patch in the Clyde Sea Area (Steele *et al.* 1973).

Figure 5.8. The occurrence of this patch probably resulted from the movement of deeper nutrient-rich water from offshore into the shallower waters of this bay. This may have been supplemented by nutrients from rivers in the area but from observations on other occasions it seems likely that this source alone produced much smaller and very temporary areas of higher chlorophyll values, closer to 'variability' than 'patchiness'. Once more, there is the question of the effects of grazing. Although they were carried out at a subsequent time, feeding experiments with zooplankton indicated that their filtering rates in these inshore waters may be much lower than normal (Steele *et al.* 1973).

These examples of rather special conditions would indicate two common features. Firstly, the scales are, very broadly, of the order of 10–100 km which, again very approximately, is in agreement with the theory outlined earlier. Secondly, other biological conditions are necessary for the potential bloom to actually develop—particularly low or zero grazing. These conclusions may have some relevance to problems of contaminants added to the sea, or to large lakes. The main problems often arise from the enhancement of phytoplankton growth. These illustrations would suggest that for an open coast, effective enhancement of growth would need to occur over length scales of the order of 10 km or greater. Further, even on these scales, actual major increases in phytoplankton should not occur unless there is also a marked decrease in the grazing capability of the zooplankton. Such a decrease could occur if the phytoplankton species produced are not suitable for the herbivores, possibly because of their size, or because of other chemical factors inhibiting grazing.

Zooplankton patchiness

These examples imply that patchiness will develop only in the absence of zooplankton grazers. Theoretically, this can be explained from the formula on page 107. In this a was the growth rate of phytoplankton and the critical diameter l_c increases as a decreases. If grazers are present, we can take a to represent the net growth expressed as the difference between actual production and loss due to grazing. Then, in any circumstance where grazing is not negligible a will be smaller and so l_c larger than the value deduced from phytoplankton only. If, again theoretically, we think of summer conditions as approximating to a steady state, then on average $a = 0$ and no patchiness would be possible. Yet, as illustrated by the observations in the North Sea (Figs. 5.2 and 5.3), patches occur when large zooplankton populations are present. For these reasons, although simple combinations of diffusion rates and phytoplankton growth can illuminate some types of patchiness, they do not fully explain the more general features.

One question concerns this idea of a 'steady state'. Simple models of the interactions of phytoplankton and zooplankton (e.g. Steele 1958) treat the latter as 'biomass' with a growth rate, rather like the phytoplankton growth rate. Such 'pictures of reality' produce those steady states after the spring outburst. Zooplankton, however, go through life cycles and may do so in

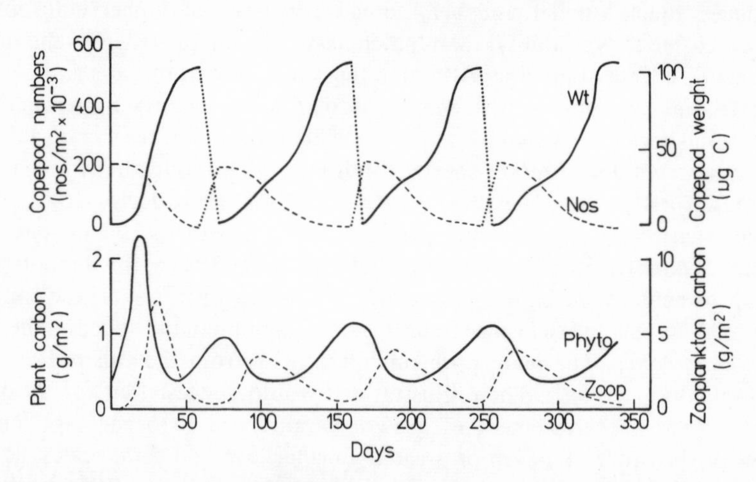

Fig. 5.9. The results of a simulation model showing the limit cycle produced by a zooplankton population consisting of a simple cohort (Steele 1974a).

cohorts. Thus, *Calanus* has about three generations in the North Sea (Marshall & Orr 1955) from spring to autumn. If this cohort structure is simulated in a model, then cycles of phytoplankton and zooplankton result, Figure 5.9 (Steele 1974a). When these cycles are compared with the observations in Figure 5.2, they show the same type of variation (Steele 1974a). The difficulty is that the theoretical cycles occur as changes with time while the observations are distributions in space at about the same time. Can such temporal fluctuations, which arise from zooplankton growth cycles, be the cause of variations in space?

Theoretically, the problem is that we are now dealing with changes in space and time of both phytoplankton (P) and zooplankton (Z). These changes are connected by grazing which depends on both P and Z, thus,

$$\text{change in } P = \text{growth } (P) - \text{grazing } (P\,Z) + \text{diffusion } P \quad (5.3)$$
$$\text{change in } Z = \text{grazing } (PZ) - \text{predation } (Z) + \text{diffusion } (Z). \quad (5.4)$$

These are similar to the Lotka–Volterra equations (Pielou 1969) with diffusion added, although the grazing term may be more complicated than the simple Lotka–Volterra form.

The addition of terms multiplying P and Z not merely makes the equation more complicated than the simple form used for phytoplankton only, it alters the whole character of the solutions. Patchiness can be thought of as the superposition of perturbations at different wavelengths corresponding to the typical length scales, l, of the earlier discussion of single patches. For the simple form (Kierstead & Slobodkin 1953), any perturbation at one time with a wavelength, l, would remain at that wavelength and decrease or increase in amplitude depending on whether $l < l_c$ or $l > l_c$. The so-called 'non-linearity' of the equations for P and Z, means that a perturbation at one wavelength can propagate changes at other wavelengths. In particular, perturbations at short wavelengths, less than l_c, can alter the mean values of P and Z (Steele 1974b). Thus, small-scale variations of P and Z which, considered separately, would be damped out by diffusion, in conjunction could lead to changes at larger scales beyond the critical value of l_c. There are, mathematically, certain conditions on these perturbations. It is necessary that they be correlated (Steele 1974c) but this is a probable feature of such fluctuations, as shown by the model of cohort structure.

An *analogy* with physical turbulence may be helpful. In the description of turbulence given earlier, it was stated that, in general, larger-scale flows (i.e. currents) were gradually broken down and dissipated by the random motions of turbulence. Under certain conditions, however, for example atmospheric jets or the Gulf Stream (Webster 1961), energy is taken from the surrounding turbulence to feed the currents and so increase their momentum. This apparent reversal is explained mathematically by the non-linearity of the equations describing these situations. In the same (mathematical) way, small spatial fluctuations in phytoplankton and zooplankton may affect the distributions observed on a larger scale. These concepts are obviously speculative but they offer an explanation of patches observed on a larger scale without requiring 'behaviour' of plankton at this scale. One possible source of perturbation could be the inherent changes in grazing rates resulting from growth in cohorts of the zooplankton.

Population stability

Questions about diffusion and patchiness are related to questions about the factors determining the stability of populations of phytoplankton and zooplankton. The term stability, as used here, means the ability of plankton populations to absorb fluctuations imposed by 'external' factors. These could be either the physical environment or predators such as shoals of herring moving through an area. Nearly all theoretical studies of this problem consider the distributions or organisms to be uniform in space and deal with changes in time (May 1973). For plankton this approach (Steele 1974a) leads to the idea that the zooplankton must be prudent grazers and have low feeding rates when phytoplankton is below some threshold concentration. It would seem possible that the dispersing effects of diffusion might provide an alternative method of eliminating any perturbations.

Considering the simple relation between growth and diffusion, the critical wavelength l_c for patchiness would also be critical for stability since any perturbation on a smaller scale would be dispersed. On the other hand, perturbations on a larger scale could be unstable. Such large-scale perturbations can occur, typically, when the seasonal thermocline is formed in the spring over large areas of the North Sea or North Atlantic and the spring outburst of phytoplankton begins, followed by the growth of zooplankton.

If this remained as a smooth process with, at any one time, the same concentrations of P and Z everywhere, then the need for 'clever' zooplankton would remain. If, at the other extreme, instability tended to appear as local perturbations then, using the simple concepts, diffusion would tend to smooth these out. Further, as the zooplankton populations developed, the net rate of increase of phytoplankton, taken as the balance of growth and grazing, would become much lower and the critical patch size much larger leading again to a relatively 'smooth' world. If, however, larger-scale patchiness can be continually produced by smaller-scale interactions, then these perturbations at all wavelengths up to the average values (i.e. zero frequency) of P and Z would produce instabilities unless counteracted by some general behavioural adaptations such as threshold feeding levels. In other words, some of the conditions presumed by theories which ignore spatial heterogeneity would still be necessary in a patchy world.

Patchiness and fish

As I mentioned at the beginning, many pelagic fish species may migrate long distances to areas of generally high food concentration. Yet within these general areas variations in food concentration may be equally important. For adult fish, the patches may provide a source of food which can be consumed at a rate much greater than would be the case if only the average concentration were available. Of course, in turn, this predation may be another source of patchiness in the plankton and illustrates the simplifications of the previous discussion where plankton was considered as a relatively closed system.

For larval fish, patchiness may be even more important. Jones (1973) has shown that larval fish may require relatively high concentrations of the smaller stages of zooplankton if they are not to starve to death during their early stages. Also, they need these above-average concentrations for periods of one or two weeks. Such conditions would occur in large patches since the larger the patch the slower is exchange of water through its circumference.

On this basis, for fish or fish larvae, patchiness is not merely a random variation in their environment, but an essential requirement for obtaining adequate concentrations of food. From this viewpoint, the variance of plankton distributions (i.e. the maximum possible food concentrations) may be more important than the average. If this is so, then a relatively constant variance is needed or, in other words, patchiness must be a regular and normal feature of planktonic environment. Further, any differences in variance that did occur from year to year at the time of larval development could be of greater significance to larval survival than differences in average zooplankton populations.

Summary

The evidence from observations of spatial heterogeneity is rather scant but suggests that, although variability occurs at all scales, there may be patches with, typically, dimensions of 10–100 km. Many of these features can be explained by a combination of accumulation due to phytoplankton growth and dispersion due to turbulent diffusion. However, combined phytoplankton and zooplankton patches are less easy to explain. It is possible that small-scale perturbations resulting from cohort structure or predation can generate large-scale patches, but these perturbing effects are balanced by functional responses of the zooplankton to changes in their food. This balance could lead to a relatively regular structure of patchiness which is utilized by higher trophic levels.

Chapter 6. Vertical Migration

A. R. Longhurst

Introduction
Patterns of migration
Diel migration
Longer-term migrations
Regional differences
Control of migration
Light as the stimulus for migration
Other constraining factors
Biological significance of migrations
Avoidance of predation
Horizontal dispersion and transport
Bioenergetic advantages

Introduction

The small-scale distribution and migration patterns of the animals of the oceanic plankton and nekton are primarily determined by the simple, but often overlooked, fact that their environment changes far more rapidly in the vertical than in the horizontal plane. With the important exception of chlorinity, the environmental variables that are generally accepted to be important in determining the occurrence of planktonic organisms follow this generalization. Thus, within the upper kilometre of an idealized oceanic water column one passes from warm, lighted, food-rich surface waters to dark, food-poor waters as much as 10° or 20°C colder; for many variables these trends continue to the bottom at slower rates of change, which may yet be in excess of rates in the horizontal plane, in which it may be necessary to pass from polar to equatorial regions to achieve similar effects.

One might expect, *a priori*, that planktonic animals and their nektonic predators would, in the idealized oceanic situation, have simple causal relationships with their environment which determined their vertical patterns of distribution. However, the actual situation proves to be otherwise, and is dominated by an apparent anomaly: almost globally at least some of the zooplankton and nekton, and often a majority of them, are found in different 'vertical life-zones' in daytime and at night. Almost throughout the oceans and seas, and also in fresh water, as night approaches very great quantities of animals migrate actively upwards from the deeper, dark waters towards the surface layers, where they spend the night until at dawn they descend again to their daytime residence depths, where direct observation of fish, euphausiids and sergestids suggests that these spend the day lying inert in the water. It is with an analysis of this migration that we are concerned in this chapter, and in particular to attempt to draw some general conclusions concerning vertical migration and the advantages it must bring to those organisms which daily perform such long 'walks', frequently between 10,000 and 50,000 body lengths in each direction. It is also intended to examine vertical migrations of similar, or the same, organisms over similar vertical intervals but on much longer time scales, as between summer and winter conditions.

We have to be careful, in drawing such generalizations, that we are not confounded by basing our analysis on investigations that have been performed in situations where the distribution and migration of organisms are distorted by a depth restriction, as in continental shelf seas, by very active upwelling or divergence, by absence (over the depths investigated) of significant variation of environmental variables in the vertical plane, or by the day/night differences in some of the same variables being slight or absent. Rather, we should hope to use the anomalies found under such conditions as natural experiments to investigate the general phenomenon.

Vertical migration has been known to occur in some planktonic animals for the last 150 years, but there is still no widely accepted explanation of the precise advantages it confers; Cushing (1951) has reviewed the early history of investigations from the earliest notes, such as those of Cuvier (1817) on freshwater cladocera. Towards the end of the nineteenth century, it was already understood that a variety of types of organisms were involved, and that some migrated daily over distances in excess of 500 m. It also became clear that daily migrations occurred in all oceans and at both high and low latitudes. Some species, however, were found not to migrate and others, but very few, appeared to migrate in a reverse sense, usually under anomalous conditions.

Cushing's review (*op. cit.*) demonstrates the progression of methods of investigation used to examine vertical migration. From the earliest nighttime observations of animals at the surface (from which they were absent in daylight), progress was rapidly made towards comparisons between day and night integrating hauls with non-closing nets hauled from standard depths up to the surface, such as those of the *Challenger* expeditions (Murray 1885). In the first decades of this century, closing nets were first used (e.g. Michael 1911) to establish the sub-surface distribution of planktonic organisms, thus establishing the basis for the classical investigations by Russell (1925, 1927) in the western English Channel, by Hardy and Gunther (1935) and Frazer (1936) in the Antarctic, and by Esterly (1911) off California, to name but a few. Opening–closing nets of increasing sophistication (e.g. Bé 1962) have been used until the present time, notably by the Soviet oceanic plankton investigations of the 1950s and 1960s; in these expeditions, reviewed extensively by Vinogradov (1968), samples were routinely taken with Juday and other nets of this type, in 10–12 horizons down to the ocean floor at 8000–10,000 m.

More recent developments in sampling equipment include the use of fast flow-rate pumping systems, which are mostly useful for detailed investigations of the smaller plankton in the upper 250 m (e.g. Beers & Stewart 1969), and the multiple serial plankton sampler, or LHPR (Longhurst–Hardy Plankton Recorder; Longhurst *et al.* 1966), which is hauled obliquely, usually from 1000 m, and which collects about 50 samples of macrozooplankton during its ascent, recording by electronic means for each sample the depth, temperature and volume of water filtered. The LHPR has been used to investigate zooplankton migration in a subsurface O_2 minimum off California (Longhurst 1968), the zooplankton component of a DSL (Kinzer 1969, 1970) and

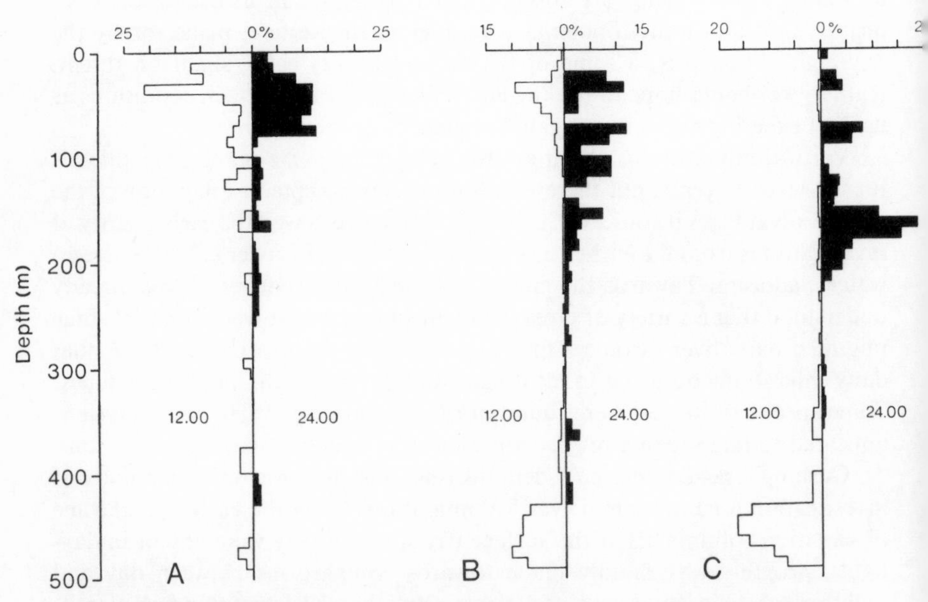

Fig. 6.1. Three LHPR profiles of zooplankton species at Ocean Weather Station 'J' (52°N 20°W), 13 May 1970, to show three different patterns of diel migration. On each, the left profile is midday, the right profile midnight. (a)—*Calanus finmarchicus*, adults only, a non-migrating taxon in this situation with only some withdrawal from the surface at night; (b)—*Metridia lucens*, adults only, to show newly moulted individuals remaining near the surface, older individuals migrating over a range of about 350 m; (c)—*Pleuromamma robusta*, adults only, to show the whole population migrating strongly. (Figure based on unpublished data of R. Williams.)

seasonal vertical migrations of copepods at weather ship *India* (Williams 1973), amongst other projects (Fig. 6.1). None of these devices, including the more sophisticated recent equipment, has escaped the general sampling problems associated with nets or pumps and around which an extensive literature has been developed: extrusion (being pressed through the meshes), avoidance (of the mouth of the net) and plankton patchiness are among the more important effects which have had to be assessed.

However, sonic methods, having a different set of problems depending on the acoustic properties of sea water, have been used to good effect in investigations of vertical migration by plankton; the development of acoustic depth indicators and acoustic submarine-hunting devices (sonar) in the early 1940s quickly showed that a 'deep scattering layer' (DSL) was almost globally present at some depth from 100 to 400 m, or even deeper, and was frequently mistaken for a bottom echo; it was also soon observed that at least part of the DSL appeared to rise towards the surface at dusk and to descend again at dawn. In the late 1940s a paper by Hersey and Moore (1948), among others, established that the DSL was indeed the reflection of sound scattered by layers of siphonophores, copepods, euphausiids, cephalopods and small bathypelagic fishes, amongst other organisms. Direct visual evidence was obtained, especially by Barham (e.g. 1963) by the use of research submersibles during the late 1950s and early 1960s.

In addition to confirming the occurrence of vertical migration of some plankton and nekton, DSL records have also been very important in demonstrating the fundamental fact that such organisms are often distributed in lamellar sheets, whose vertical depth may be less than ten metres and usually less than several tens of metres, but whose horizontal continuity may be traced for thousands of kilometres; it is within these sheets that the zooplankton patchiness and aggregation, which has been so much studied in recent years, occurs.

Patterns of migration

The term 'migration' applied to the vertical movement in the water column of plankton organisms is misleading, for in this context it has little in common with the more familiar animal movements to which the term is usually applied; the migrations of birds, fishes, insects and mammals, primarily in the horizontal plane, generally maintain the organisms in favourable conditions of food supply and temperature, as the loci for such conditions alter seasonally. Vertical movements of plankton, on the other hand, usually place the organisms under alternating conditions of temperature and food supply, and only in the case of light to maintain them within a roughly constant regime.

Diel migration

It is with the daily movements that most workers have been concerned; the literature is so large, and the situations in which vertical migration has been described are so diverse, that it is not easy to identify general principles. The topic has been reviewed several times, notably by Russell (1927), Cushing (1951), Bainbridge (1961), Banse (1964) and Vinogradov (1968). Of these, Cushing's review gives the most useful general statement on the topic as it stood in 1950, Bainbridge's the most useful list of references to studies of migration in more than 150 individual crustacean species, while Vinogradov's is particularly helpful in its analysis of vertical distribution and migration, region by region, on a global basis.

What is now clear is that diel vertical movements occur in all planktonic phyla, and in most of the smaller taxonomic groups, and that there are few generalities to be drawn on the basis of size, form, feeding habits or taxonomic relationships between those that do, and those that do not migrate. Predators as different as siphonophores and chaetognaths, herbivores as different as copepods and euphausiids, and animals ranging in size from tintinnids to large oceanic squid are all now known to migrate vertically. Amongst each of these groups there are, of course, specialists in migration and in non-migration; among copepods, for example, most species of *Pleuromamma* specialize in migrations to near the surface from rather deep daytime residence levels, often exceeding 300–400 m in low latitudes, while, on the other hand, all species of *Pontella* remain in the upper mixed layer throughout the 24 hours. There may, however, be a size limitation on the extent to which a species migrates; epiplankters which show only weak migration are relatively smaller than their deeper living relatives, from among which the strong migrants come. As Hure and Scotto di Carlo (1971) have recently shown for Adriatic copepods having daytime residence depths below 250 metres, the migrants are large species of *Pleuromamma* and *Euchaeta*, while smaller species (of *Spinocalanus*, *Temoropia* and *Oncaea*) at the same depths are weak migrants which may even apparently sink slightly at night.

Apparent nighttime sinking of this nature may be a similar phenomenon to the dispersal, perhaps partly by passive sinking, seen commonly in strongly migrating forms after the dusk ascent to the surface waters has been completed, and the migrating organisms find themselves under lower ambient light levels than those at which they occur in the daytime and during their ascent. With the first traces of dawn, the dispersed animals re-aggregate themselves closer to the surface, before their downward migration begins. Cushing (1951) reviews many cases of this phenomenon, among them the classical investigations over a 2-year period off La Jolla by Esterly (1911); the exception, that on bright moonlit nights this dispersal and sinking is apparently reduced (Moore, 1949), appears to prove the rule.

We have been concerned so far only with migrations of species having their daytime residence depths within the zone to which some daylight penetrates; for present purposes, following Clarke and Denton (1962), it is assumed that organisms cannot perceive diel changes in light intensity below 750–1000 m, though of course this parameter, dependent on water clarity, is highly variable from place to place in the oceans. Below this depth, in any case, the intensity of biological light exceeds that of the remaining downwelling sunlight.

The now widely-quoted 'ladder of migrations' hypothesis of Vinogradov (1953, 1955) supposes that organic material is actively transported downwards to an important degree by coprophagy and predation between overlapping patterns of vertical migration extending to at least 4000–5000 m below the surface. If this model could be validated it would have important consequences in food-chain biology, and since it stands or falls on a demonstration that diel vertical migration occurs far below the lighted zone, we should examine the evidence on which it is based rather closely.

Vinogradov (1955, 1968), as direct evidence for such migration, quotes Banse's (1964) reference to upward migration at sunset of a DSL at 4000 m which, if it were correct, would be almost an order of magnitude greater depth than that to which vertical migration of DSL's have otherwise been observed; however, examination of the original evidence in a paper by Koczy (1954) on another topic, shows that Banse misinterpreted an echogram and that the DSL in question was only at some hundreds of metres below the surface. However, Waterman and Berry (1971) recently showed that 25 of 207 species taken in opening–closing hauls centred at 600, 1000, 1400 and 1700 m showed a nocturnal rise. Species of *Eukrohnia*, *Eucopia*, *Acanthephyra* and *Gennadas* rose from 1400 m to 1000 m at dusk and two species of chaetognaths descended from 1700 m at dawn. This appears to be the greatest depth for which we have good evidence of diel migration.

Vinogradov also cited Chindonova (1959) who suggested that some abyssal organisms from 4000–5000 m must make regular, short-term migrations to the lower photic zone to feed on phytoplankton, her evidence being based on the presence of radiolarians in the guts, and the occurrence of mandibles adopted for microphagous diets, among some abyssal mysids (*Boreomysis*, *Hyperamblyops*), isopods (*Eurycope*), euphausiids (*Bentheuphausia*) and copepods (*Gaetanus*). However, this evidence is slender on which to base such an important principle, for Vinogradov (1968, p. 124) says that, in the same NW Pacific region as Chindonova worked, aulacanthid radiolarians are an important constituent of the 5000–6000 m plankton, sinking from the surface layers aggregated into clumps. He also writes of crustacea which 'at maximum depths had their stomachs filled with a greenish mass', but Chindonova, to whom he refers, writes that diatoms are only insignificantly found in her material from great depths. Vinogradov (1962) also mentions abyssal pteropoda 'which seem to rise to the surface layers of the ocean' and contain a greenish mass with diatoms, radiolarians and foraminifera. The original tabulated data of Chindonova (1959) to which he refers, however, contains data on only six individuals of pteropods, none of which need have been taken much deeper than about 500 m.

The literature on deep vertical migration abounds in such confusion as this and on balance, Vinogradov's formal 'Ladder of migrations' seems unsubstantiated, but it seems probable that diel vertical migration will be shown to occur, in some measure, at depths below the photic zone by some of the larger filter feeding organisms moving up to the lower fringes of the zone of abundant sinking phytoplankton at depths corresponding to pycnocline; however, it also seems probable that most abyssal zooplankton will be found to subsist on sinking material, such as faecal pellets and clumped radiolaria referred to above, as well as on microorganisms such as the olive-green cells of uncertain affinity (Hentschel 1936, Fournier 1971) which seem to be ubiquitous at great depths and of which Chindonova (*op. cit.*) wrote 'the intestines of abyssal species (of copepods) would quite often reveal certain greenish, oval cells of unknown origin, measuring some 10–14 μm'.

Longer-term migrations

Vertical displacements are also known to occur over longer periods than daily, and these have been separated in the past as 'seasonal', in which a species occurs at different depths at different times of the year, or as 'ontogenetic', in which a species occurs at different depths at different stages in its life history. Since most, though not all, plankton species which show these sorts of migrations have several generations each year it is not useful to separate ontogenetic from seasonal migrations, since so frequently they are one and the same thing. Thus, in many species of calanoid copepods, the population passes the winter (or other non-reproductive period) in the pre-adult copepodite V stage, rather deep and often below the photic zone. The classical investigations of *Calanus finmarchicus* (Gunner) in the north-east Atlantic show this very clearly: a population of larvae, dominated by copepodite Vs, overwinters in deep water generally at around 1000 m, moults there to the adult form in the spring, and as the phytoplankton bloom develops, begin to rise into the surface water and, except in very high latitudes, begin diel migrations (Marshall & Orr, 1955; Ostvedt, 1955). Elsewhere *Calanus plumchrus* (Marukawa) in the North Pacific overwinters as a dense layer of copepodite stage V at less than 100 m (Barraclough *et al.*, 1969); off Ghana *Calanoides carinatus* (Kröyer) does the same between upwelling periods (Bainbridge 1972) as does *Calanus pacificus* (Brodskii) off California (Longhurst 1967a), and off Peru (Mullin personal communication).

Except in cases such as these, it is a general rule that younger stages tend to lie closer to the surface than older stages of the same species, thus following the familiar generalization that body size increases with depth in planktonic animals (e.g. Mauchline, 1972). While there are exceptions, the literature is full of accounts of species distributions which follow this pattern. Thus, Brinton (1962) describes many cases in Pacific euphausiids of both high and low latitudes, Vinogradov (1968) reviews many cases in Pacific copepods, and Alvarino (1964) describes several instances in chaetognaths. As Vinogradov shows for two species of *Hymenodora* the phenomenon can be traced to at least 5000 metres depth.

Regional differences

Under conditions of continuous daylight in Polar seas during the summer, diel migration may be almost entirely suppressed (e.g. Bogorov 1938, 1946), though Digby (1961) found diel migration in quieter, more stratified areas off Spitzbergen, as did Marshall and Orr (1955) in Tromsø Sound; seasonal migrations are clearly more important in high latitudes, where their mass effect determines the winter–summer differences in biomass profiles (Wiborg 1954) and, indeed, they have not been clearly described in low latitudes except in locations where marked seasonal changes occur, as in the Californian and West African coastal upwelling areas referred to above.

In the Antarctic, the species of mass occurrence concentrate at depths of 250–500 m during the winter where there is a southerly component in the mass transport, rising to the surface in the spring earlier in the north and progressively later southwards; as the summer progresses, the northerly component in the flow of the Antarctic surface water concentrates them near the Antarctic convergence, to descend again to more than 250 m in the autumn. This generalization is evident in the work on *Rhincalanus gigas* (Brady) of Hardy and Gunther (1935) and Mackintosh (1937), on *Calanoides acutus* (Brady) of Andrews (1966), on *Euphausia superba* (Dana) of Frazer (1936) and Mackintosh (1937). Diel migrations are made, especially in the early summer in the latitude of South Georgia by development stages of *E. superba*, and by species of *Calanus*, though Vinogradov (1968) says that these decrease in the highest latitudes.

Migrations in mid-latitudes have been best described in the northern hemisphere; the presence of cold intermediate water in the North Pacific has important consequences for migration, even to the extent of apparently separating stocks of *Calanus plumchrus* above and below it. In the North Pacific in winter the epiplankters are distributed throughout the deep mixed layer, to about 200 m, while such forms as *Eucalanus bungii* (Johnson), *Calanus cristatus* (Kröyer) and *C. plumchrus* overwinter below this depth, rising to surface layers in the spring, in some cases as mid-developmental stages. Diel migrations are performed during the summer, Vinogradov (1955,

1968) showing how those species which make long seasonal migrations by wintering very deep make diel migrations of small amplitude and do not penetrate the cold intermediate layer, while species such as *Metridia pacifica* (Brodskii) and *M. ochotensa* (Brodskii), which do not descend deep to over-winter make intensive diel migrations throughout the year. In the autumn, the intensity of diel migration apparently increases after the deep overwintering species make their seasonal descent, leaving the migrating plankton to be dominated by species such as *Metridia pacifica*.

In the North Atlantic, winter stratification of the water column is rather different, and mixing extends to below 500 m; a single species of mass occurrence, *Calanus finmarchicus*, dominates the biomass in mid-latitudes and generally overwinters at 500–1000 m, while the high arctic *C. hyperboreus* (Kröyer) lies deeper, at 1000–2000 m. *Pseudocalanus elongatus* (Boeck) winters deeper than *Calanus finmarchicus* but lies closer to the surface in the summer. As in the Pacific, epiplankters such as *Oithona* are spread in winter throughout the mixed layer. Summer diel migrations of the Arctic *C. hyperboreus* (as in the Pacific *C. cristatus*) are less intense (Sømme 1934) than those of *C. finmarchicus*, which appears to migrate more intensely than the Pacific counterpart, *C. plumchrus*, perhaps because of the absence of an Atlantic cold intermediate layer. *C. finmarchicus*, as does *C. plumchrus*, may at times mass at the surface, even in full daylight (Marshall & Orr 1955); the possible significance of this will be discussed later.

Differential depth distribution of developmental stages appears to be at least as widespread in low as in high latitudes; Alvarino (1964) gives several examples for chaetognaths, Brinton (1962) many for euphausiids, Angel (1969) several for ostracods. *Calanoides carinatus* is a very abundant and widespread copepod of low latitudes in the Atlantic, and apart from the 'overwintering' of stage V copepodites in the West African intermittent upwelling regions already mentioned, several studies, especially those of Vervoort (1963), have suggested that in the open ocean it performs seasonal and ontogenetic migrations.

Diel migration, of course, is well developed in low latitudes differing, as elsewhere, in its intensity and in its effects on plankton mass distribution between different oceanographic regions (e.g. King & Hida 1954, Legand 1958, Bogorov & Vinogradov 1960). Essentially, the greatest intensity of diel migration occurs in central gyre waters with high-water clarity and a deep-mixed layer, while the lowest intensities occur in the equatorial divergence regions with low clarity and a very shallow mixed layer. The migrating organisms perform either minor vertical movements within the mixed layer (epiplankters such as *Undinula*, *Scolecithrix*, *Candacia*, many *Sagitta* spp.) or major movements frequently observable as a DSL rising from 300–500 m in daytime to the region of 0–50 m at night: species of *Pleuromamma*, *Rhincalanus*, *Euchaeta*, and myctophid fishes dominate in this fauna.

Control of migration

While it is now generally agreed that the proximate stimulus for vertical migration is the level of ambient illumination, there still remains some confusion between the way in which light levels may act as stimuli for migration, and the possible biological advantages to be gained by thus responding to photic stimuli. This confusion was important in the early literature reviewed by Cushing (1951); much of the work on vertical migration in the first half of this century was directed towards an analysis of the mechanisms of 'tropisms', 'taxes' and 'kineses' as responses of organisms to stimuli of light and gravity. 'Geotropism', meaning a directional response to light from which deviations are corrected by gravity, was coined by Parker (1902), and finally led Esterley (1911, 1912) to propose an explanation of vertical migration by two tropisms: negative phototropism and negative geotropism, with temperature change as the mechanism to trigger one or the other tropism. Such ideas were discussed by a number of authors around this time, but achieved no great success.

Light as the stimulus for migration

Michael (1911) and later Rose (1924) explored the simpler idea that organisms followed an optimal light intensity by means of 'kineses' whose signs might be frequently and continuously changed by changes in ambient illumination, so that migration could become a continuous process following an isolume rather than a 'go–no go' mechanism occurring only at dawn and dusk. These proposals were taken up, and extended, by Russell (1926) who discussed the concept of a limited range of light intensities to which organisms were adapted, and though he wrote 'by what mechanism this is brought about will not concern us', his was one of the first clear appreciations of the difference between light as the controlling factor of, and light as the ecological reason for, vertical migration.

Russell showed how this line of reasoning led to an explanation for observed descents away from the surface in the middle of the night, suggesting that those forms which came by this means to be more or less evenly distributed found themselves at light levels even lower than their minimal intensity and so dispersed because of lack of orientation; at dawn, the return of the isolume corresponding to their minimal light level again allows them to orient, to disperse themselves within their light range and to descend with it as daylight comes.

Clarke (1934) used very similar concepts in his studies in the western Atlantic, and since this time the groundwork of our understanding of the control of vertical migration has changed very little; the main problem not now clearly resolved seems to be the existence and importance of endogenous circadian rhythms which might in some way modify the simple control of migration by light. Harris (1963) demonstrated with *Calanus helgolandicus* and *Daphnia magna* (Strauss), that these organisms had a 24-hour cycle of locomotor activity in small containers, with increased activity in the hours of darkness; Harris used this confirmation of much earlier work by Esterly (1912) on *Acartia* as an explanation of the 'dawn rise' of dispersed organisms which seemed to occur too early to be cued by lightening of the sky. However, Herman (1963) was unable to demonstrate endogenous rhythms in the diel migrant *Neomysis americana* (Smith) and Enright and Hamner (1967) obtained very mixed results with several species. One has to conclude that physiological condition, physiological light adaptation and endogenous rhythms may all interact in a complex fashion, with light as a controlling factor.

A combination of laboratory experimentation and field observation has now brought us to a reasonable model of the mechanism by which radiance levels regulate the vertical movements of both vertebrate and invertebrate animals within the photic zone; this model depends very largely on our understanding of the physics of underwater illumination.

The photo-environment in the open ocean is extremely complex, but the use of electronic multi-spectral photometers in recent decades has gone far to describe the changing spectral quality of light with increasing depth in ocean waters of differing turbidities; the geometric distribution of underwater light has, by the use of the same instrumentation, been adequately described in recent years. The fauna of the photic zone reacts extremely sensitively to small changes in the photic environment from day to day, induced by daytime cloud cover (Blaxter & Currie, 1967), sea-state (Lendenvald 1901), solar eclipses (Backus *et al.* 1965), lunar phase (Worthington 1931), artificial lights (Blaxter & Currie 1967). Kampa (1970) has demonstrated that some of the physicists' assumptions of the distribution of underwater light are insufficiently precise for photo-biological purposes. Thus, Jerlov (1963) assumes that the attenuation of daylight below 100 m is essentially uniform in all tropical ocean regions, falling between $K = 0{\cdot}03$ and $0{\cdot}04$ where K is the extinction coefficient; actual measurements off San Diego and Tenerife increase this range to $K = 0{\cdot}06$ and $0{\cdot}02$ respectively—calculations based on Jerlov's assumptions would result in imprecisions in the estimates of the photo-environment at daytime residence depths greater than the photosensitivity of the organisms.

Below the surface, the nature of the light field is such that light can be detected from every direction, though not uniformly, so that even at considerable depths, towards the extremes of light penetration, the difference in zenith and nadir radiation remains significant in terms of the light sensitivity of organisms.

Selective absorption of the red and infra-red regions of the spectrum occurs below the surface and, in the first metre or so, about 50 per cent of the total incident solar radiation has been absorbed, and remaining radiation is confined to 300–700 nm in wavelength. Light of about 360–370 nm penetrates deepest in clear oceanic water and the visual pigments of twilight-adapted, vertically migrating animals may have maximum sensitivities in this range; Kampa (1955) found *Euphausia pacifica* (Hansen), for instance, to have maximum sensitivity at 436–468 nm.

The polarization of sky light is lost in transmission through the sea surface, but the irradiance distribution pattern at depth imposes a plane polarization pattern whose net effect is a residue of polarized light at right angles to the main axis of the irradiance distribution pattern. Thus, between zenith/nadir illuminance ratios and plane polarization there appears to be an adequate signal in the photo-environment to support an hypothesis that light alone sufficiently explains the control of vertical migration of plankton within the lighted zone.

The investigation in the laboratory of the responses of planktonic organisms to changes in their photo-environment is exceedingly difficult; the design of the experimental container is critical to success yet many reports in the literature have been based on observations in simple culture jars or beakers, in which the geometric distribution of irradiance and the spectral quality of light had little in common with that in the ocean, and in which the photo-environment contained stimuli caused by refraction, reflection and point-source effects, quite unlike anything encountered by the organisms in nature.

However, it is possible to deduce some principles from the experimentation so far performed, much of which has been on freshwater cladocerans. As Baylor and Smith (1957) showed for *Daphnia magna*, the crustacean eye is sensitive to plane polarized light, and under experimental conditions *Daphnia* can orientate itself at right angles to a beam of plane polarized light; the mechanism for this appears to be a comparison between radiance levels reaching lateral cones and those directed in the sagittal plane of the organism. Using the same organism, Ringelberg (1964), showed that it was unable to maintain a normal orientation in homogeneously diffuse angular light fields, but swam normally when an angular light gradient was introduced.

Electronic instrumentation has enabled us to perform both experiments and observations at sea on the reactions of plankton to light that would have been impossible until very recently. A good example is the equipment used by Blaxter (1973), based on matched thermistor pairs in a vertical perspex tube with a very precisely controlled light environment which he used to demonstrate the response of herring larvae to the light field. Such instrumentation holds great promise for investigation of a wide range of planktonic organisms in the future.

Using a surface irradiance meter on deck, a submarine irradiance meter matched to it and a 10 kHz precision depth recorder, Boden and Kampa

(1967) beautifully demonstrated the accuracy with which sound-scattering layers of migrating organisms followed the vertical movement of the 5×10^{-4} $\mu W/cm^2$ isolume at a wavelength of 474 mμ, which was the isolume associated with the top of the layer at its daytime residence depth. Figure 6.2 shows how,

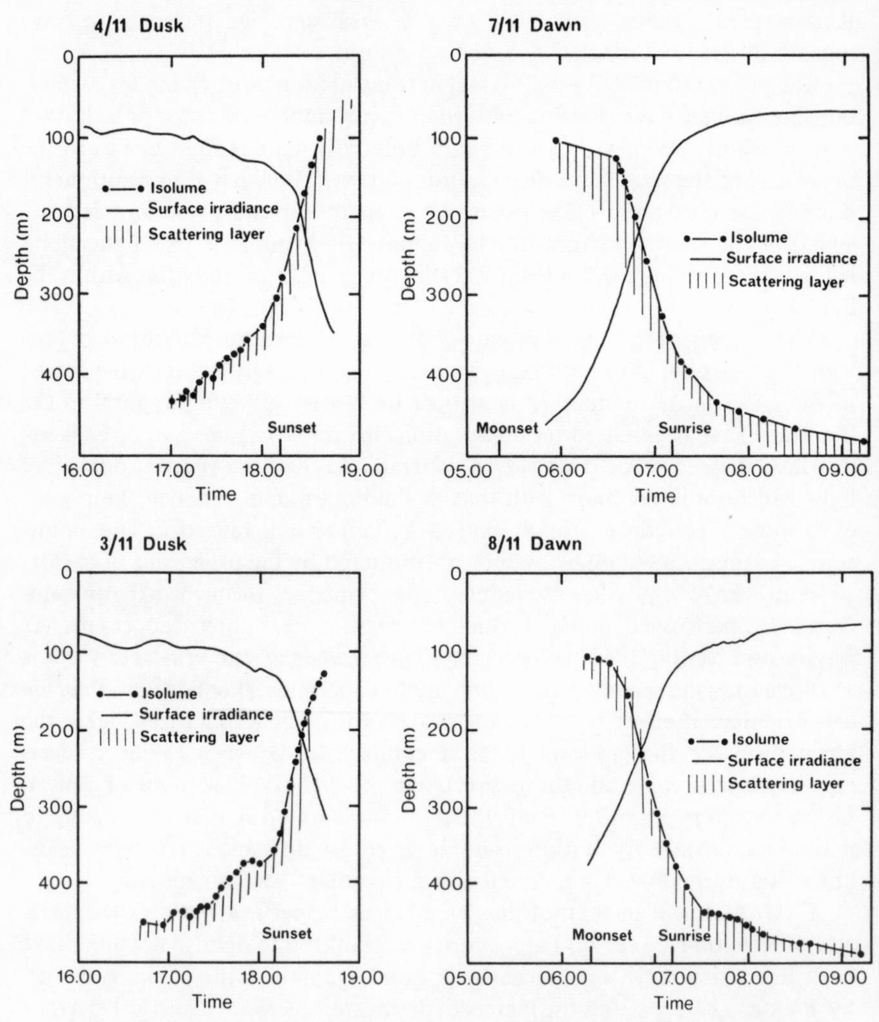

Fig. 6.2. Two dawn and two dusk examples of the correspondence between an isolume, surface irradiance and a scattering layer near the Canary Islands. Note the lag in zooplankton response to the descent of the isolume on 8/11, perhaps because of the 'light-adapting' effect of the full moon, setting only an hour before sunrise. Note also the lag in response at sunset on 3/11 to a sudden decrease in irradiance and related sudden shoaling of the isolume as a cloud covered the setting sun briefly. (Redrawn from Boden & Kampa, 1967.)

at two dawn and two dusk migrations, the layer adjusted its depth to follow the isolume. The effect of a sudden alteration of irradiance levels by the passage of clouds over the sun was reflected at both dawn and dusk by temporary imprecision in the tracking of the isolume by the scattering layer, and even the effect which bright moonlight in the hour before dawn on one morning had on the physiological dark-adaptation of the optical sensors of the scattering layer organisms could be deduced; being adapted to a brighter light, reaction to the downward movement of the key isolume was initially slightly delayed, the animals returning to their normal relationship with the layer only subsequently.

Ringelberg (1964) suggested that relative rate changes in irradiance cue organisms to move vertically, and that this occurs over rather wide ranges of absolute levels of irradiance to which the organisms may have become physiologically adapted; this complication of the simple hypothesis has some merit, and appears to be supported by the observations of Bary (1967).

Other constraining factors

As we discussed earlier, the ocean characteristically lacks homogeneity in the vertical plane and it is not to be expected that light, or any other controlling factor of vertical migration should operate entirely independently of the changing physical, chemical and biological factors encountered by a migrating organism; one is forced to postulate that there must be situations in which an organism is constrained from further migration, even though impelled to do so by its photic environment, either because it is physically impossible for it to move in the vertical plane, or because some local attractive factor overrides the stimulus of light.

The most obvious constraints on vertical movement are, of course, the sea surface and the ocean floor. There is ample evidence of accumulation of biota close under the sea surface but it is not clear whether this can be regarded as the effect of a constraint or an attraction; it seems very likely that the accumulation of nocturnal migrants close under the surface at night may simply be a feeding aggregation in relation to the very abundant microneuston. There is also evidence (e.g. Isaacs & Schwartzlose 1965) of horizontal strata of sound scatterers colliding with the sea bed during the course of dawn descents which would otherwise carry them to greater depths. The best evidence for this effect comes from echo sounder records of the interactions between DSL traces and the continental edge, or the summit of a seamount. The ecological implications are complex and vary from situation to situation, but must usually result in increased mortality of the migrants. This phenomenon may well account for the richness of pelagic predators over isolated oceanic banks and, at least in part, for the increased abundance of pelagic predators right along the edge of the continental shelf. This phenomenon must, on a wider scale, be responsible for the inability of organisms which normally migrate each day or seasonally, to depths greater than 200 m, to maintain populations in large continental shelf regions where the requisite depth of water is not available to them.

Such effects are not, of course, restricted to the organisms of deep scattering layers. The diel vertical migrations of other organisms may similarly be constrained by water depth, and such constraint may explain the well-known aggregation of pelagic fish on the bottom in daytime in continental shelf areas. Thus, mature oceanic herring *Clupea harengus* (Linnaeus) spend daylight hours between 200 and 400 m deep, rising to upper mixed layer depths at night (Ryzhenko, 1961), while an inability to achieve such depths in the North Sea may explain, in part at least, the difficulty Blaxter and Parrish (1965) had in finding a reasonably consistent relationship between daytime depth and irradiance at that depth.

It is to be expected that both salinity and temperature, but especially the latter, will constrain the vertical movement of organisms under some circumstances. The most obvious case is the temperature differential between a mixed layer and the sub-thermocline water, for differentials of as much as 10°C above and below a thermocline are commonplace, and in low latitudes may exceed 15°C. The number of examples known in which diel migrants cross a thermoclinic temperature barrier of these magnitudes is very great, and it is clear that the normal situation in mid and low latitudes is for this to occur; for example, Vinogradov's review (*op. cit.*) discusses many examples in all oceans of species of both mass and sparse occurrence. Thus, *Calanus finmarchicus* has been reported as being thus restricted in its seasonal migrations in the Black Sea (Nikitine 1929) and in its diel migrations in the North Sea (Savage 1926). In general, it seems to be in high latitudes that thermal barriers to vertical migration have been demonstrated, though in not every case has a sufficient distinction been made between organisms whose vertical migration is constrained by temperature and those whose distributional range is normally limited either above or below, by a thermocline.

The studies of Angel (1968) on the vertical migrations of ostracods off Morocco demonstrate dissimilar specific responses within a single group of animals to a thermocline as a barrier. In this case, the temperature discontinuity occurred at about 80 m. The ostracods fell into three categories in this respect: seven species moved freely and abundantly across the thermocine each day; five species penetrated up through the thermocline only in low numbers, most individuals remaining below it; and for four species the thermocline was the upper ceiling for vertical migration.

Hansen and Dunbar (1970) give one of the few satisfactory accounts of a constraint due to a salinity discontinuity; below Arctic pack ice, brackish surface water lies over more saline intermediate water with a sharp boundary at about 50 m at which a layer, only a metre or so in thickness, of *Spiratella helicina* (Phipps) occurs at night, separating itself from the general mass of the DSL as this ascends at dusk and passes through the interface. Experimentally derived salinity tolerance of *Spiratella helicina* supports the hypothesis that in this case the constraint is salinity.

Layers of reduced oxygen tension occur in the ocean, either below sill depths in enclosed basins, or at depths of minimal horizontal motion (Wyrtki 1962), and these layers may be accompanied by the production of hydrogen sulphide, as in the Black Sea (Caspers 1957). Oxygen minimal layers in mid-depth, with or without the presence of H_2S, occur in the open ocean especially off regions of strong coastal upwelling, as Peru, California and the Arabian Sea. Below about 0·2 ml/l oxygen, or in the presence of H_2S, there may be exclusion of plankton (Ivanenkov & Rozanov 1961, Longhurst, 1967a, Mullin personal communication) though exclusion is seldom complete and at oxygen levels of 0·2–0·5 ml/l, diel migrants and seasonal 'hibernators' may establish their day or longer-term residence depths (Longhurst, 1967a). The codlet *Bregmaceros nectabanus* (Whitley) migrates daily over a vertical range of more than 800 m in the area of the Cariaco Trench, and has its daytime residence depth at 850 m below the surface (Baird, Wilson & Milliken, 1973); this is below sill depth, where the water is anoxic and smells strongly of H_2S from about 400 m downward. These codlets form a single-species DSL which resides for 10–11 h daily within the H_2S zone at depths far greater than would be necessary to attain to avoid predators. Whatever biological advantage is gained by this behaviour must be very powerful, to offset the difficulties of the environment.

Probably the most important localized attractant which would constrain vertical migration, especially diel migration, is the presence of discrete layers of food-rich water. The vertical distribution of living phytoplankton frequently includes sharply discontinuous maxima within the mixed layer or within the thermocline; these maxima, for various reasons, frequently coincide with density discontinuities. Aggregation of migrating zooplankters at such pycnoclines during the night may reasonably be interpreted as resulting from such layers being the destinations to which vertical migration has evolved to carry herbivorous zooplankters and, by extension, the organisms predatory upon them.

Hobson and Lorenzen (1972) found that phytoplankton maxima tended to occur at pycnoclines at many places in the Atlantic and Caribbean, and suggested that this was because nutrients are increased by diffusion from below at such sites, so that plant growth is not dependent only on remineralization, as in the mixed layer; they also observed, in some locations, a clear correlation of these phytoplankton layers with zooplankton maxima.

Anderson, Frost and Peterson (1972) show that nighttime aggregations of zooplankton off Oregon correspond very closely with the seasonal subsurface chlorophyll maxima at 55–65 m, and that the aggregated zooplankton was primarily herbivorous, compared with the non-aggregated animals. From a similar study off California on *Calanus pacificus*, Mullin and Brooks (1972) concluded that such aggregation was active, rather than passive or coincidental. Such relationships are inherently more probable explanations of observations than the purely physical explanations suggested by, for instance, the experiments of Harder (1968) on aggregation at water stratification in small experimental containers. Studies such as those of Venrick, McGowan and Mantyla (1973) and Love (1970) show how widespread is the subsurface phytoplankton maximum, and hence the possibility of this relationship.

Cooper (1967), however, postulated a social aggregating mechanism, which again might constrain vertical migration, at density layers well below thermocline depths; he suggests that pheromones, released at and diffusing along such layers, would physiologically trap receptive individuals whose migrations might otherwise have carried them through this layer. However, there is no experimental evidence to support this hypothesis.

Biological significance of migrations

The diel vertical migrations of planktonic organisms certainly result in far greater mobilization of biological material than any other animal migrations: using an average figure for plankton biomass of 25 g/m^2 to a depth of 100 m as a reasonable estimate over all oceans and assuming that diel migration adds 10 per cent to this at night as a minimal figure from an average daytime residence depth 250 m deeper, then one arrives at vertical translocation of 25 $tons/km^2/d$ over a distance of 250 m. Summed, for all ocean areas, this becomes a minimal figure of $0{\cdot}9 \times 10^9$ tons/d.

Such calculations are not to be taken very seriously, but give point to a search for unifying theories of the biological advantages accruing to the migrants, or the ecosystem of which they form a part, from such migrations. It is no longer possible to take seriously the opinions of those who maintain that the migrations are simple reactions to external stimuli, are sufficiently explained by analysis of tropisms, or even (Rudjakov 1970) that 'diel migrations have not been developed as a result of the evolution of planktonic organisms, but were inherent to the ancestors of recent plankton'.

Admittedly, the diversity in the general pattern of vertical migration is very great, and exceptional organisms and situations have both attracted more attention than they warrant and have frequently been used (e.g. Rudjakov 1970) to illustrate the supposed fallacy of general interpretations. The literature also contains numerous examples of simplistic, single-benefit explanations of what has to be recognized as a very complex phenomenon, or set of phenomena, occurring on various time and space scales, each subset conferring different biological benefits.

The following discussion of the more probable hypotheses follows the pattern set previously by, for example, Vinogradov (*op. cit.*) but attempts to lay emphasis upon the diversity of benefits, rather than upon a more simple search for a single unifying hypothesis concerning the reasons for the evolution of the phenomenon.

Avoidance of predation

It is not clear to whom this widely-discussed hypothesis should be attributed, but it has early roots in the literature of vertical migration, and has recently been reviewed by Vinogradov (*op. cit.*) and McLaren (1963). There are many reasons to suppose that this is neither a universal nor a complete explanation, but it remains true that many organisms are, in fact, less vulnerable to potential upper photic zone predators (such as epipelagic fish and oceanic birds) by virtue of their daytime residence depths—not perhaps because it is too dark there for such predators to locate them, but rather because they only rise to depths at which such predators can function when light levels are at their lowest. It is not possible to investigate this directly, but it seems likely that if, for instance, euphausiids and myctophid fish regularly occurred in daylight in the upper tens of centimetres of the water column then they would be far more easily taken by gulls, petrels, auks and shearwaters than they are at night by some of these birds in reality. The apparent fallacies in the predator-avoidance hypothesis are well known and well reviewed, so that it is only necessary to notice them here. Firstly, at daytime residence depths of many migrants there remains sufficient light for location by predators (Clarke 1936). Secondly, other migrants appear to go deeper than required by this hypothesis (Moore 1958, Baird *et al.* 1973). Thirdly, many migrants bioluminesce actively at night (Hardy 1953). Fourthly, many transparent, jellylike forms (ctenophores, siphonophores, chaetognaths and salps) including predators, are also strongly migratory.

Contradictory though the evidence is, one must agree with Vinogradov that 'high concentration of zooplankton in the surface layers facilitates the predatory activity of fishes . . . by sinking and dispersing the zooplankton largely escapes annihilation'.

Horizontal dispersion and transport

In a number of cases, both on diel and seasonal time-scales, it has been demonstrated that by migrating vertically between two water masses, organisms might be able to utilize the current differential in order to maintain themselves in one location or to transport themselves horizontally to extend or to maintain a species-range.

Diel migration might, it has been suggested, serve by these means to enable zooplankton to avoid the supposed toxic effects of dense phytoplankton patches (Hardy & Gunther 1935) and though this hypothesis is no longer taken seriously it is likely that, in a complex way, diel migration and horizontal patchiness of phyto- and zooplankton are related, in view of the complex nature of such spatial relationships (e.g. Beklemishev 1957).

However, on the seasonal scale, the use of differential current systems in a water column seems fairly clearly demonstrated in numerous cases, and Hardy (1953), in fact, considered this to be the main reason for the evolution of vertical migration. We have already discussed the maintenance of Antarctic zooplankton within limited zonal boundaries by this mechanism, and Kashkin (1962) suggests that *Calanus finmarchicus* uses a similar mechanism in the North Atlantic. In the tropical regions Longhurst (1967b, 1968) suggested that larvae of the planktonic galatheid crab *Pleuroncodes planipes*, which are expatriated by the California Current several thousand miles to the south-west of the range of the adults, have a potential transport mechanism back again in an opposing under-current; careful depth-discrete sampling with an LHPR revealed that the vertical, seasonal distribution of larvae and post-larvae was in accordance with that demanded by the hypothesis. As Vinogradov points out, Frasseto, Backus and Hays (1962) have shown how DSL organisms maintain themselves in the current differentials of the Straits of Gibraltar, and Grindley (1964) how plankton may maintain themselves in an estuary by directed vertical migrations.

Bioenergetic advantages

Apart from the seasonal displacements which are clearly widespread, but differ in detail in each case, the nearest approaches to unifying theories are those postulated by McLaren (1963) and McAllister (1969). McAllister (1969), apparently independently of a similar suggestion by Petipa and Makarova (1969), suggested that the effect of discontinuous nocturnal grazing of copepods on phytoplankton would be to increase the production of phytoplankton cells in comparison with a population continuously grazed day and night. This appeared to be substantiated by computer simulation of a situation in the north-east Pacific and might be based on 'better utilisation of the growth potential of the phytoplankton with less of the primary product being lost to phytoplankton respiration, as well as permitting unimpeded growth by the plants during the daylight hours', thus echoing the speculations of Wimpenny (1938) on the implications for oceanic production of nocturnally feeding copepods and diel variations in diatom cell division rates. However, McAllister (1971) as a result of further simulation modelling and of laboratory grazing experiments, modified his earlier conclusions concerning the advantages which might accrue to the individual zooplankton organisms, though the later work appears to uphold the general hypothesis of increased primary production resulting from pulsed, rather than continuous grazing by herbivores.

The hypothesis stated by McLaren (1963) rests upon the partition of physiological activities, daily or seasonally, between the warm surface and the cold subsurface environment in order to obtain an energetic advantage from the temperature differential. Any such hypothesis rests fundamentally on a demonstration that the energy gained by this stratagem is greater than that expended during migration; such a demonstration has not been made and, in fact, the consensus until very recently was that very considerable amounts of energy must be expended in migration which, as noted above, may commonly involve distances of 10,000 to 50,000 body lengths twice daily. Translated into human terms this seemed to mean a walk to breakfast of perhaps 25 miles, and this led Hardy (1956) to comment that 'it is an extraordinary thing that vertical migrants will expend so much energy to climb up to the surface, only to sink or swim down again in the daytime'.

This opinion was apparently confirmed by the observations of Petipa (1964) that diel migrant adult *Calanus helgolandicus* in the Black Sea showed a remarkable daily variation in gut content and in the apparent amount of oil droplets around the gut (though these were at the limits of optical resolution) compared with weakly migrating younger stages. Using these data, she later (Petipa 1966) calculated the expenditure of energy in diel migration, assuming that the lipid apparently observed to accumulate rapidly during nocturnal feeding was equally rapidly utilized to fuel the vertical migration. It has recently been shown (Lee *et al.* 1971), Nevenzel 1970) that zooplankton, including *Calanus*, are unusual in that wax esters may comprise up to 50 per cent of their total lipid reserves, and that this percentage is high in deep-living and migrating copepods but low in nonmigrating epiplankters. The presence of wax esters has variously been attributed to a structural function, to a flotation function, to chemical effects of temperature and pressure, and to an energy reserve; during starvation, triglycerides are utilized as fuel preferentially over wax esters (Lee *et al.* 1971, though at rates of several orders of magnitude slower than those apparently demonstrated by Petipa; Gatten and Sargent (1973) found some diel variation in total lipid in weakly migrating *Calanus finmarchicus*, but a very constant ratio of total lipid to wax ester, perhaps suggesting that factors other than energy requirements and hence utilization of triglycerides were the cause of the lipid fluctuations. If this is correct or if Petipa's observations were mistaken, then it may explain the results of Vlymen (1970) who determined the drag on *Labidocera*, by ingenious experimentation, to be about six orders of magnitude less than that required for *Calanus* by Petipa's theory; Vlymen's calculations of the energy expenditure of copepods at swimming speeds required for migration may then be accepted as reasonable; he finds that this is only 0·3 per cent above basic metabolic demands. Vlymen's determination of copepod drag coefficients then confirms Hutchinson's (1967) calculation, ignoring drag, that the energy required to migrate vertically by an approximately neutrally buoyant plankter placed very small demands upon its metabolism. One can therefore accept, in principle, that the energetic advantage accruing from partitioning physiological functions between warm and cold environments need not be very great in order to make diel or seasonal migrations profitable. McLaren's hypothesis (*op. cit.*) suggests that an organism might receive an energy bonus through feeding at night in warm water where the efficiency of food uptake will be high, and by the more efficient directing of energy to growth at lower temperatures, deeper down in the daytime. Clearly, adult size, and hence absolute fecundity are positive functions of temperature (Mauchline 1972), yet any bonus derived from maximizing these by growth at minimal temperatures must be offset by the fact that growth itself, and hence generation time, is a negative function of temperature.

Except in cases where generation time is constrained by some physical event to be of long duration, the size of the energy bonus or deficit derived from vertical migration must depend on the balance of these functions. Starting from the von Bertalanffy model of catabolism being a mass-dependent rate and anabolism a surface-dependent rate, McLaren (*op. cit.*) argues that for any given temperature differential over the migrational range (and hence between loci of anabolism and catabolism), there will be a tendency for an energy deficit to accrue at low anabolic temperatures and for a bonus to accrue when these are higher.

Such a deduction accords with the general observation that at low latitudes under stable water-column conditions vertical migrants are especially abundant and of wide occurrence; a suggestion is also made by McLaren that the nonmigratory, small epiplankters may require to spend more hours per day filtering their small-particle food than do the larger migrants and are thus constrained from obtaining the energy bonus accruable from vertical migration. McLaren points out that in very high latitudes there is a high frequency of situations in which temperature differences through the water column are small, and shallow temperatures (and hence those of the loci of anabolism) are low; in accordance with the hypothesis, it is here also that diel vertical migration seems most frequently to be suppressed, and it is here that the frequently quoted anomalous mass occurrences at or near the surface in daytime of migrating species have been observed, mostly in spring when food is abundant near the surface, and surface temperatures low to moderate.

Of course, anomalous daytime mass occurrences at the surface may also occur in low latitudes, but these can usually be ascribed to instability in the water column. I have seen abundant daytime populations of the normally strongly migratory *Euphausia eximia* (Hansen) at the surface in the equatorial divergence, in situations in which surfacing of the thermocline had reduced both thermal stratification and surface temperatures.

The same arguments that have been applied to diel migrations may be involved, at least in part, in analysis of seasonal or ontogenetic migrations. It seems to be generally accepted that overwintering deep populations of, for instance, *Calanus* either in high latitudes (e.g. Marshall & Orr 1955) or in low latitude intermittent upwelling regions (Longhurst 1967a) descend at the end of the feeding season as diatoms become scarce and pass the winter in cool water using up their accumulated lipid reserves for maximal adult size and fecundity. Though in certain situations (Firth of Clyde, Marshall & Orry 1958) nitrogen metabolism remains active enough during winter to require continuous feeding at a reduced rate, it seems probable that the general statement will be validated. There are sufficient observations of the relative immobility of migrants at daytime residence depths—for instance, of myctophid fish hanging motionless and head down in the water, and of sergestids and euphausiids hanging totally motionless—to suggest that this is a general habit and that metabolism is minimized under these conditions; Lee, Hirota and Barnett (1971) found that *Gaussia princeps* (T. Scott) was able to maintain itself at 5°C for 35 days without total depletion of its lipid reserves and one may reasonably assume that under laboratory conditions an organism would be unlikely to be able to run down its metabolic rate as completely as a 'hibernating' animal in the dark, still depths of the ocean. It may be, therefore, that the suppression of vertical migration in high latitude populations of vertical migrants after primary production runs down in late summer is due to the fact that after a critical point it is no longer possible to achieve an energy bonus by vertical migration, so that the organism is faced with the choice of remaining near the surface—with perhaps more food but certainly a greater predation pressure, or remaining at depth—with less food but less predation. Observation suggests that it chooses the second course in most cases.

If we fully recognize the complexity of the fascinating problems posed by diel and seasonal vertical migrations, and the patchiness of the data and naïveté of the models with which we seek to understand the processes, we soon realize that we have not progressed so far as we may sometimes wish to think in the ninety years since Fuchs (1882) wrote, presciently and simply, that planktonic animals preferred darkness, and that vertical migration was a feeding migration because the plant plankton lies much nearer to the surface.

PART III. FUNCTIONS IN THE MARINE ECOSYSTEM

In this section, the productive engine of the sea is described, but our account of the degradation of energy remains incomplete. Transfer from the upper layers to the sea bed is not well understood; the range of food sizes on which the benthos depends is well established but the generation of that distribution is not. This range comprises the triturated grains dissipated by herbivores, their faecal pellets, the pieces spilled by predators and the rotting remnants of mortality; to compile a budget of such quantities is obviously hard. Similarly, a description of the role of bacteria in water or in the sediments has been omitted because there is much controversy over how their numbers should be estimated. Marine microbiologists such as Zobell, Kriss, Wood, Sieburth, Morita, Watson, and Caldwell have already described what is known about marine decomposers.

Although the complete cycle of energy or material in the sea can adequately be described in vague pictorial terms, such an account largely ignores detritus food chains, and only the autotrophic-based movement of energy and transfer through the trophic levels has been described quantitatively in this section. This, however, will not be our sole concern here, for study of the productive engine gives estimates of the potential yield at any trophic level. The special yield to man is discussed in the next section, but more generally the yields from any one trophic level to the next raise the question of how an ecosystem works. In such a way, analysis of the productive engine allows us to begin to understand the structure of life in the sea.

The prime force in the engine is the production of material or energy by the algae in the euphotic layer. This is restrained by the grazing activity of the herbivores and by nutrient lack (at rather low levels of nutrient). Material is transferred to herbivores and subsequently to pelagic fish or eventually to benthic animals and thence to demersal fish. Nutrient lack is analysed in full detail in Chapter 7, because it is the system's fail–safe mechanism and so is very important, while the text of Chapter 8 describes autotrophic production, but ignores heterotrophy because it is probably a low level or subsistence process. In temperate waters or in upwelling areas most of the material has been transferred to herbivore flesh by the time primary production is restrained by nutrient lack. In the quasi-continuous production cycle of the deep tropical ocean, however, the condition of nutrient restraint may be permanent unless there is a diurnal input from vertically migrating animals that balances the daily loss by uptake.

The production of herbivores (Chapter 9) or of the benthos (Chapter 10) is controlled by their growth and death rates, about which rather little is known. In contrast, the growth of fishes is well known and is described in Chapter 11 in some detail, particularly in terms of the energy used in swimming. The growth of other animals in the sea is almost ignored because it is not as well known as that of fishes. Indeed, until quite recently the growth of crustacea in the sea had not been well described. The death rates of most animals are poorly documented, at least in workable form.

Just as the structure of an animal follows from its functions, so the fluxes through an ecosystem determine its structure. The most difficult problem relating to the structure of a single population is to understand how it stabilizes itself in the face of environmental variability. The mechanisms of stabilization are the means whereby single populations combat competition yet support those parts of ecosystems needed for their survival. This central and important subject (the stock/recruitment problem of fisheries biologists) is discussed in a preliminary way in this text because no full solution is available. The problem has been simulated in reasonable constructions, but such models have not yet been absorbed into the general scientific framework. The stability of ecosystem is an allied and more complicated problem, the mathematics of which is being actively explored at the present time. But the two lines of progress, the stock recruitment problem of the fisheries biologist and the stability of the multiple predator/prey system should converge because they are both at the centre of ecological science today.

Chapter 7. Nutrient Cycles

R. C. Dugdale

Introduction

Distribution of nutrients

Nutrient uptake
Nitrogen
Factors influencing the uptake of ammonia and nitrate
- Effect of concentration
- Effect of light
- Nitrogen interactions
- Nitrite uptake
- Nitrate and ammonia uptake in the water column
- Organic nitrogen

Factors influencing the uptake of silicate
- Concentration

Nutrient intake *continued*
- Light
- Temperature
- Uptake in the water column

Factors influencing the uptake of phosphate
- Concentration
- Light
- Temperature

Nutrient uptake and growth

Nutrient circulation
Nitrogen productivity measurements
Sources of regenerated nitrogen
Nitrogen fixation
Sources of regenerated phosphate and silicate

Introduction

The primary nutrients—phosphorus, nitrogen and, for some organisms, silicon—are present in inorganic form in sea water mostly as phosphate, nitrate and silicate ions, respectively. The major exogenous source of phosphorus and silicon is land drainage with the reservoir represented by the vast volume of sea water being very large in comparison as shown in Table 7.1, a summary of the marine nutrient budget made by Emery *et al.* (1955). Nitrogen compounds also are introduced into the sea by land runoff, but the larger proportion comes from the atmosphere and is a result of evaporation of ammonia from the land surface and *in situ* fixation. Nutrients are removed from sea water by the phytoplankton in the euphotic zone, resulting in a thin

Contribution No. 836 from the Department of Oceanography, University of Washington, Seattle, Washington 98195, U.S.A. and No. 76002 from the Bigelow Laboratory for Ocean Sciences, W. Boothbay Harbor, Maine 04575, U.S.A.

nutrient-depleted layer at the surface over large regions of the world ocean. The surface nutrient content is replenished by mixing from the nutrient-bearing deeper water and by local regeneration processes. Representative vertical distributions of phosphate, nitrate, and silicate for the three major oceans are shown in Figure 7.1. Considerable variation occurs between oceans although nitrate and phosphate tend to vary together; silicate is the most variable, depending heavily on river inputs as suggested in Table 7.1. Surface depletion resulting from uptake of nutrients by phytoplankton is a universal feature of these profiles.

Table 7.1. Nutrient budget of world's oceans (from Emery *et al.* 1955)

	Millions of metric tons		
	Nitrogen	Phosphorus	Silicon
Reserve in ocean	920,000	120,000	4,000,000
Annual use by phytoplankton	9,600	1,300	—
Annual contribution by rivers	19	14	4,300
Dissolved	19	2	150
Suspended	0	12	4,150
Annual contribution by rain	59	0	0
Annual loss to sediments	9	13	3,800

Phosphate and nitrate occur in relatively constant proportion in deep water. Deep water usually is formed in the high latitudes during winter and contains nutrients in both inorganic and organic form before it sinks. The nutrients of inorganic origin are 'preformed' and the quantity is increased with time by oxidation of organic matter. The slope of a plot of pairs of nitrate and phosphate values through a water column usually has an atomic ratio of about 15:1 (Redfield *et al.* 1963), close to the proportion in which these compounds are found in phytoplankton cultured with sufficient amounts of both nutrients. Dissolved silicate does not show a constant relationship to nitrate or phosphate, but tends to show an increase relative to both compounds with increasing depth. The uptake of silicate and of phosphate in diatoms occur in approximately the same ratios (Redfield *et al.* 1963). These authors also show that the amount of carbon contained in plankton tends to be in constant proportion to nitrogen and phosphorus; the ratio is given commonly to be $C:N:P = 106:16:1$ by atoms.

The concept of a constant proportionality of uptake between carbon, nitrogen and phosphorus has been used widely in subsequent studies of the cycles of these nutrients in the sea. The oxygen required to remineralize phytoplankton phosphorus and nitrogen occurring in this ratio is calculated, on the basis of the final oxidation products, to be $-276{:}16{:}1 = \Delta O_2{:}\Delta NO_3{:}\Delta PO_4$ since phosphorus is present in the oxidized state in the cell and only nitrogen and carbon are reacted with oxygen. The $\Delta O_2{:}\Delta NO_3{:}\Delta PO_4$ ratio is useful in evaluating the amount of regenerated nitrate and phosphate present. The difference between the saturation concentration of oxygen and the measured concentration is the apparent oxygen utilization value, AOU, and the oxidative ratios are applied to give an estimate of the nitrate and phosphate of oxidative origin. The differences between measured and oxidative phosphate and nitrate values provide in turn an estimate of the preformed nutrients.

Although the proportions of nitrate and phosphate occurring in deep

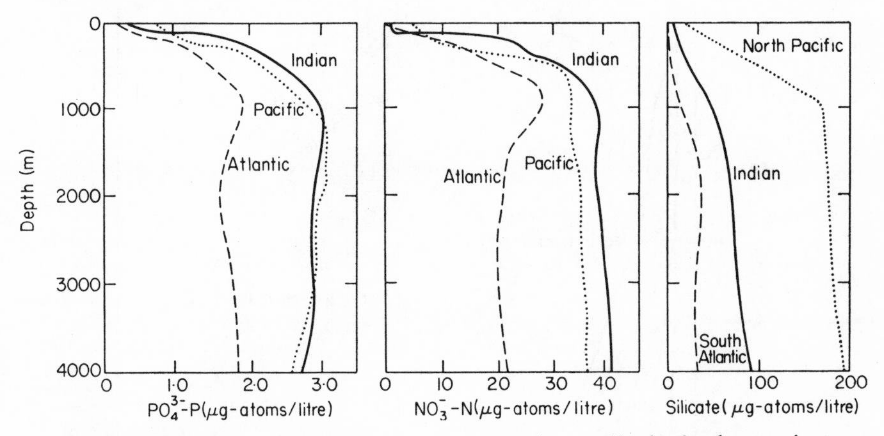

Fig. 7.1. Phosphate, nitrate and silicate concentration profiles in the three major oceans (from Richards 1968).

water appear to be controlled largely by the ratios of nitrogen and phosphorous in organic matter of phytoplankton origin at the time of sinking, the ratio of uptake by phytoplankton is known to vary widely. Ketchum (1939) showed that in cultures of *Phaeodactylum tricornutum* (Bohlin) the ratio varied systematically and only at certain concentrations of phosphate and nitrate did the ratio correspond to the 'normal' value of 15:1. Conway (1974) found the N:P ratio reduced to 6:1 for *Skeletonema costatum* (Greville) grown under ammonia limitation in chemostat culture and many other examples of variation in this ratio can be found (e.g., Parsons & Takahashi 1973). Nitrogen fixation occurs at significant rates in some parts of the ocean tending to restore the normal abundance ratio to nitrogen-deficient surface areas of the ocean, but denitrification has an opposite effect. The balance between these processes will be discussed later; however, it is clear that the quantities of phosphate and nitrate found in sea water result from a series of complex interrelated processes.

Distribution of nutrients

An adequate supply of nutrients to the euphotic zone is necessary for high rates of primary production to be sustained. The rate of supply is a function

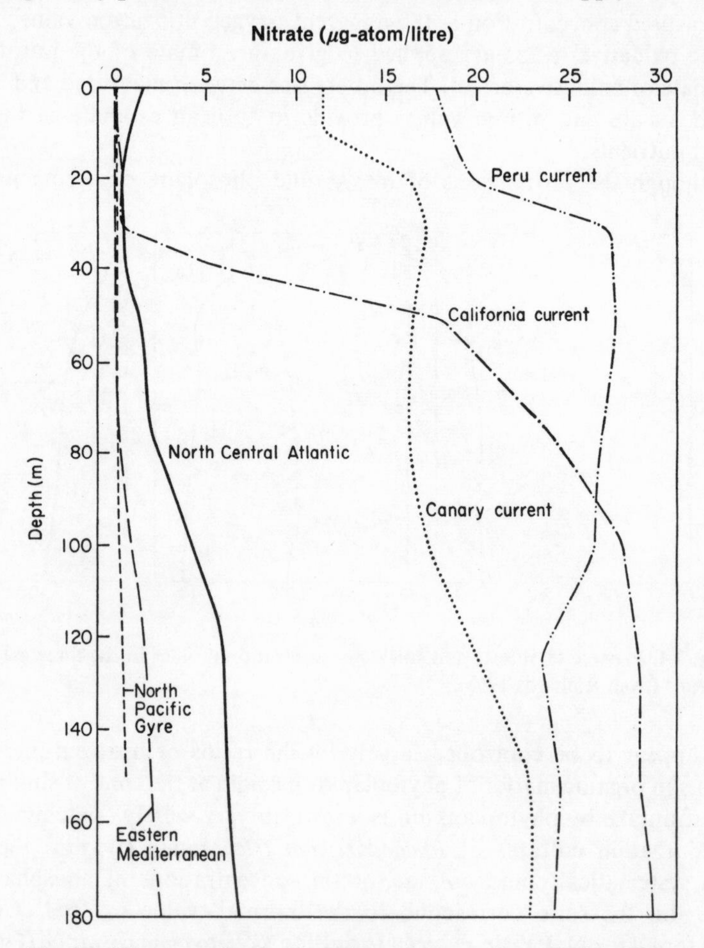

Fig. 7.2. Nitrate profiles in six regions of the sea (from Walsh 1974).

of advection, eddy diffusion, and the vertical gradients of nutrient. In Figure 7.2 nitrate profiles for the first 180 metres of depth are shown for six regions. The high nitrate levels exhibited at 50 metres by the California current, Peru current, and Canary current profiles typify conditions in regions of divergences, where nutrients are transferred into the lighted zone at high rates, resulting in eutrophic conditions. The profiles for the north

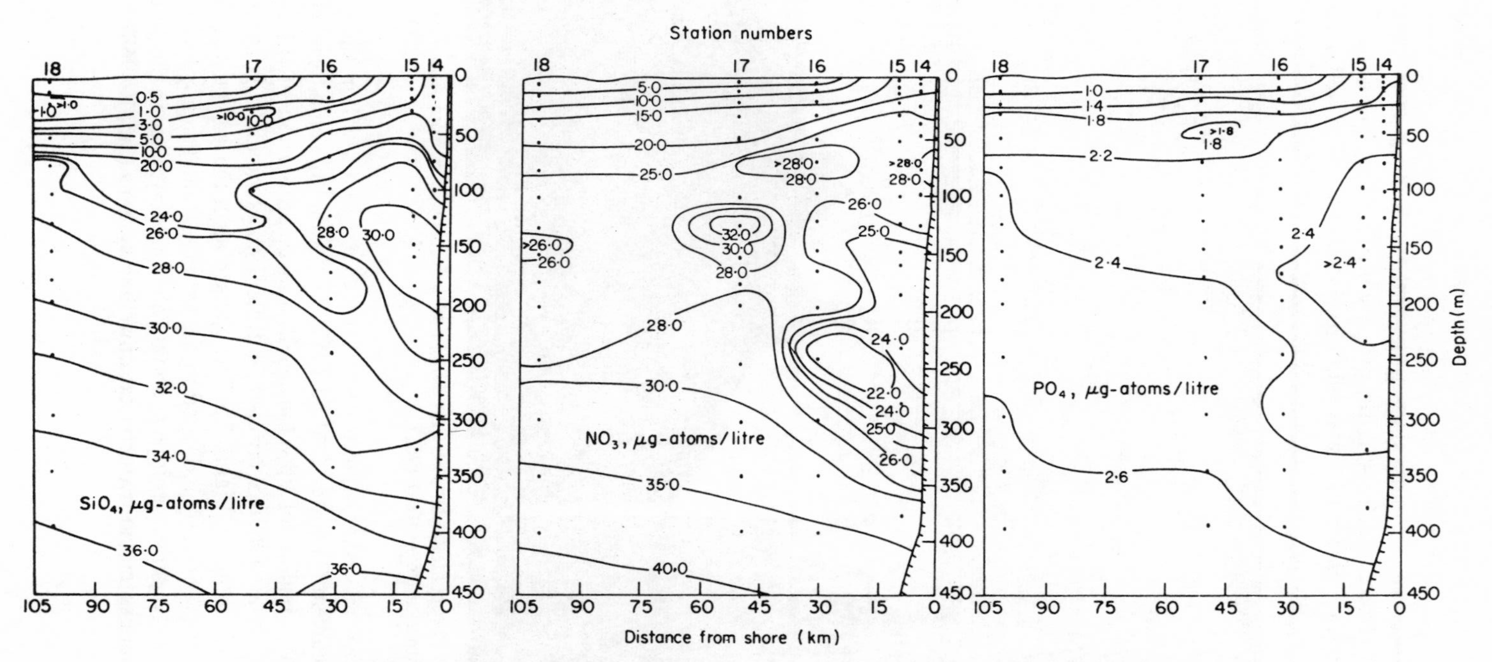

Fig. 7.3. Silicate, nitrate and phosphate sections in the Peru upwelling region (from Dugdale 1972).

Pacific gyre and the north central Atlantic are typical of convergences, with low nutrient concentrations prevailing to greater depths; these regions are referred to as oligotrophic since little nutrient is physically transported toward the surface, resulting in low primary productivity. The eastern Mediterranean nitrate profile indicates oligotrophic conditions as well, a result of the negative estuarine circulation of that body of water (Redfield *et al.* 1963).

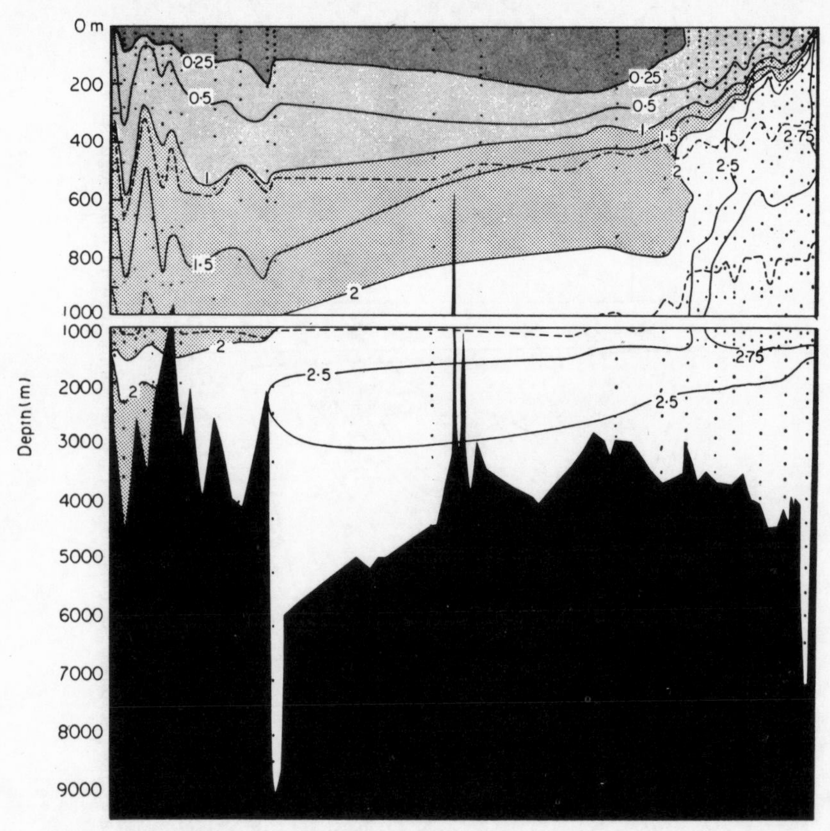

Fig. 7.4. Phosphate (μgat/l) section along approximately 27°S from Australia to South America (from Reid 1965).

The most important regions of divergence are found within the eastern boundary current (Wooster & Reid 1963), and within the equatorial current system. These regions of upwelling yield a large proportion of the world's fish catch (Ryther 1969). Sections of nitrate, phosphate and silicate for the Peru coast are shown in Figure 7.3. The upwarping of the isopleths of each of these properties is clearly indicated in the upper 100 m. The high concentrations occurring against the slope at about 250 m are contained in the poleward undercurrent associated with coastal upwelling systems (Smith 1968).

The regions of potential eutrophy and oligotrophy can be seen in oceanographic atlases. The phosphate section shown in Figure 7.4 (Reid 1965) delineates large oligotrophic regions corresponding to the South Pacific convergent gyre and the eutrophic upwelling region associated with the Peruvian eastern boundary current system. Figure 7.5, also taken from Reid (1965), is a phosphate section from Antarctica to Alaska. Here, the equatorial divergence is shown by the upwarping of phosphate isopleths from about 10°S to 10°N latitude. The convergent gyres located just north and south of

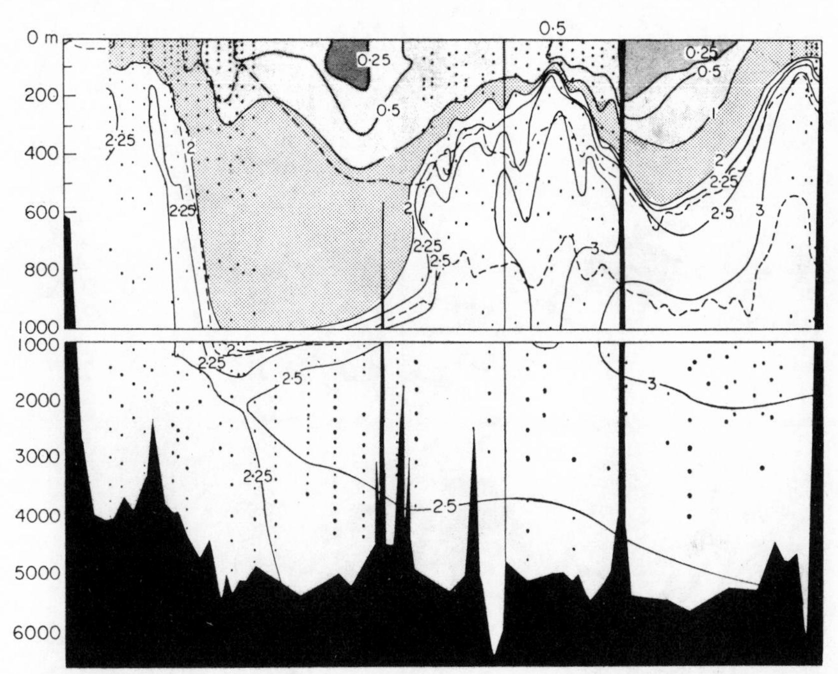

Fig. 7.5. Phosphate (μgat/l) section along approximately 160°W from Antarctica to Alaska (from Reid 1965).

the equatorial current system are indicated by downwarping of the phosphate isopleths. The divergent gyres lying to the north and south of the convergent gyres are indicated by the high phosphate concentrations occurring near the surface. The distribution of primary production shown in Figure 7.6 (Koblentz-Mishke *et al.* 1970) reflects these patterns of nutrient distribution and transport.

The patterns of nutrient uptake and regeneration in an upwelling region are shown in Figure 7.7. This representation is somewhat oversimplified. For example, bacterial regeneration within the water column is ignored for silicate and phosphate and no nitrite pool is shown. Also omitted are the pools of

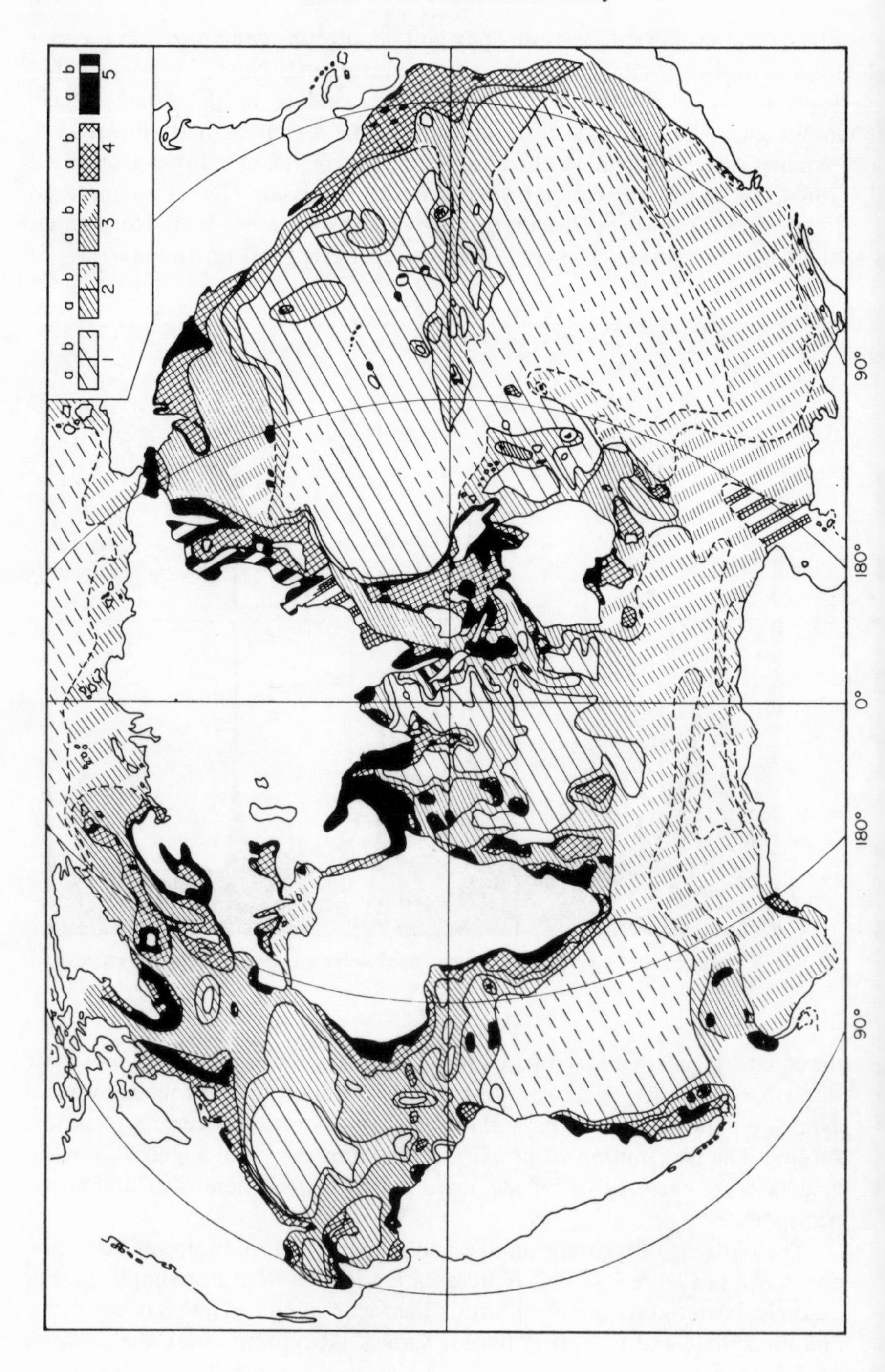
a b
1
2
3
4
5
90°
180°
0°
180°
90°

dissolved organic nitrogen and phosphorus. The nitrite pool can arise either by oxidation of ammonia or by reduction of nitrate (Vaccaro 1965). However, the diagrams shown are essentially correct in presenting the dominant pathways and they apply equally to open ocean or non-upwelling coastal regions as well. The magnitude of the various rates may change with depth, however. For example, over the deep ocean the regeneration at the sediment-water interface becomes much less important and water column bacterial and zooplankton grazing regeneration become the dominant processes. General discussions of the phosphorus, nitrogen and silicon cycles may be found in

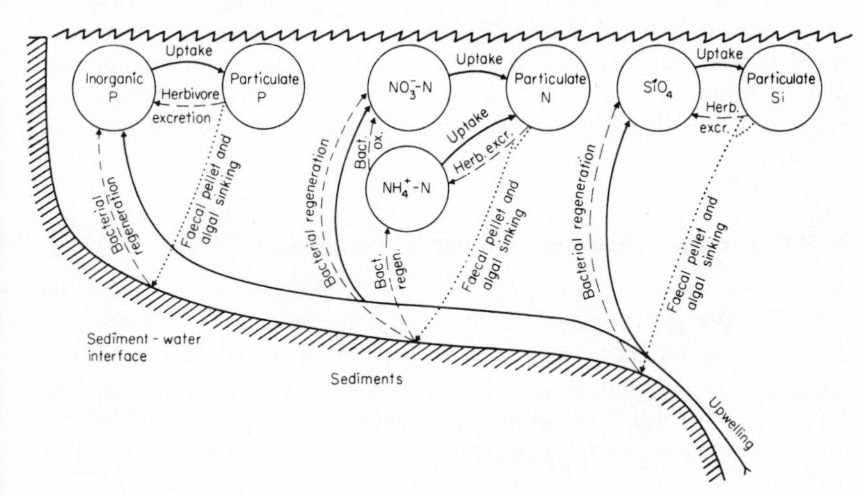

Fig. 7.7. Simplified diagram of the circulation of phosphorus, nitrogen and silicon in an upwelling ecosystem (from Dugdale 1972).

Armstrong (1965a), Vaccaro (1965), and Armstrong (1965b), respectively. The remainder of this chapter will deal with uptake and regeneration processes in the upper regions of the sea. For these discussions it will be necessary to remember that phosphate and silicate are regenerated in the euphotic zone into the same ionic form as they are present in deep water, while nitrate is transformed into ammonia, urea, and a number of other forms.

Nutrient uptake

Nitrogen

Measurements of the rates at which nutrients are removed from sea water by phytoplankton could be made, under favourable circumstances, by observing the rate of decline in concentration in a sample of sea water placed in some kind of vessel and incubated in the light for a suitable period of time. In practice the technique usually is unsuitable, since the phytoplankton populations are low and nutrient methods are relatively insensitive. Long periods of time are required to observe significant changes in nutrient concentration, and bacterial populations tend to develop rapidly on the surface of glass incubation containers, modifying the *in situ* processes. The solution has been to develop tracer techniques analogous to the ^{14}C method for photosynthesis. Although there is a radioisotope suitable for use as a tracer for nitrogen, the stable isotope ^{15}N has provided the means for a useful and practical method for measuring the uptake of nitrogen compounds using incubation techniques similar to those of the ^{14}C method (Neess *et al.* 1962). Requiring a mass spectrometer and relatively time-consuming sample preparation techniques, the method has been used only by a few laboratories. However, the entire procedure may be carried out at sea and its use is becoming more widespread.

Factors influencing the uptake of ammonia and nitrate

Although the uptake of inorganic nitrogen is influenced by all factors that act to set the growth rate of a phytoplankton population, the uptake of nitrate and ammonia by phytoplankton is influenced strongly by the concentration of the compound in the surrounding sea water and by light. Interaction also occurs; for example, the uptake of nitrate is strongly affected by the ambient concentration of ammonia.

EFFECT OF CONCENTRATION

The transport of certain ions from the external medium into the interior of the cell is performed by specialized enzymes, the permeases, embedded in the cell membrane. These enzymes cannot be investigated directly with the methods of normal enzyme research since their action is dependent upon structures that are destroyed in any attempt to isolate or purify them. Nevertheless, it appears from indirect evidence that the uptake of many compounds is controlled in a manner that is compatible with the concept of permeases. A specific permease functions both to control the rate of entry of the compound or ion into the cell and to increase the relatively low concentration of ions found outside the cell to sufficiently high levels inside the cell for the efficient functioning of the succeeding enzyme systems.

Monod (1950) found that the following expression described the uptake of limiting nutrient in continuous cultures of bacteria:

$$V = \frac{V_{max} \cdot S}{K_s + S} \qquad (7.1)$$

where V = uptake rate of limiting nutrient
V_{max} = maximal uptake rate of limiting nutrient
S = concentration of limiting nutrient
K_s = concentration of limiting nutrient at which $V = 1/2 \cdot V_{max}$.

The equation is often referred to as the Monod expression; however, it is identical to a limiting form of the Michaelis–Menten expression for enzyme kinetics and K_s is usually called the Michaelis constant. The equation describes a rectangular hyperbola, as shown in Figure 7.8.

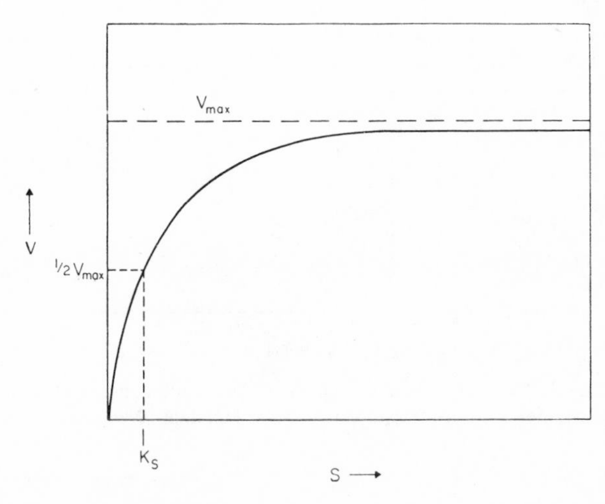

Fig. 7.8. The Michaelis–Menten hyperbola described by the expression $V = V_{max}\dfrac{S}{K_s + S}$.

The application of the Michaelis–Menten formulation to nutrient limitation in marine ecosystems was suggested in a theoretical paper (Dugdale 1967). Previously, the nutrient uptake term in models of phytoplankton production was represented as a straight line levelling off abruptly at some saturating level (Riley *et al.* 1949, Steele & Menzel 1962). Confirmation of the applicability of the Michaelis–Menten expression to the uptake of inorganic nitrogen by marine phytoplankton has been obtained for cultures in laboratory experiments and for natural populations through shipboard experimentation. The data for the kinetic constants, V_{max} and K_s, of natural populations have been obtained primarily with the tracer ^{15}N (MacIsaac & Dugdale 1969, 1972). Under appropriate conditions successive additions of labelled ammonia or nitrate followed by a short period of incubation in natural light result in uptake data points that fit a Michaelis–Menten hyperbola. Figures 7.9 and 7.10 show ammonia and nitrate uptake hyperbolae, respectively, both obtained under conditions where healthy phytoplankton occurred in sea water only recently depleted in ammonia or nitrate, i.e. from eutrophic water. In oligotrophic regions of the sea showing chronically low or undetectable nutrient levels, it is usually not possible to reproduce the upper,

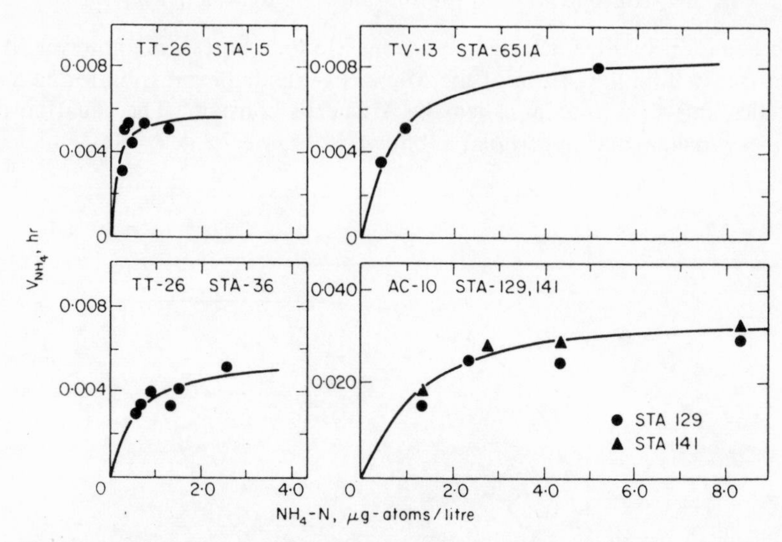

Fig. 7.9. Uptake of ammonia as a function of ammonia concentration, following Michaelis–Menten kinetics (from MacIsaac & Dugdale 1969).

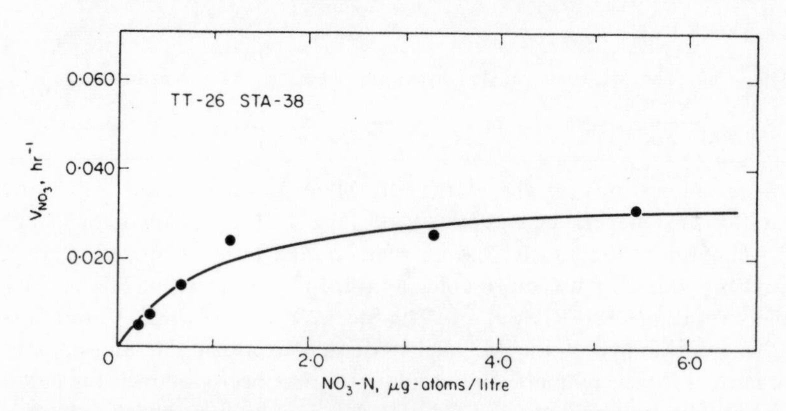

Fig. 7.10. Uptake of nitrate as a function of nitrate concentration, following Michaelis–Menten kinetics (from MacIsaac & Dugdale 1969).

saturating nutrient portion of the curve and a truncated hyperbola results. Possible explanations may be found in limitation by non-nitrogen nutrients or in the response of cells to continuously low nitrogen concentrations.

The constant, K_s, is a measure of the affinity of the permeases for the substrate and thus an important indicator of the ability of a phytoplankton species to compete for limiting nutrient. A summary of the K_s values for nitrate and ammonia uptake obtained for natural populations with ^{15}N are listed in Table 7.2. Measurements of V_{max} are also shown. However, the values obtained by the ^{15}N method using natural populations are underestimated

Table 7.2. A summary of nitrate and ammonia uptake obtained with ^{15}N for oligotrophic and eutrophic populations in various areas

Source	Area	K_s-NO_3 μg at/l	K_s-NH_4 μg at/l	V_{max}-NO_3 h^{-1}	V_{max} NH_4 h^{-1}
		Oligotrophic			
MacIsaac & Dugdale 1969	Tropical Pacific	0·07(6)	0·42(3)	0·0026(6)	0·0067(3)
MacIsaac & Dugdale 1972, Figure 8	Mediterranean	0·1 – to 0·3(4)	<0·1 (2)	0·0014– 0·0020(4) x̄ (0·0017)	0·0030– 0·0053(2) x̄ (0·0041)
Eppley *et al.* 1973, Figure 8	North Pacific central gyre		0·15(1)		
MacIsaac & Dugdale 1972, Table 5	Pacific central gyre			0·0029(8)	0·0049(8)
		Eutrophic			
MacIsaac & Dugdale 1969	Tropical Pacific	0·98(1)	1·30(1)	0·0361(1)	0·0362(1)
MacIsaac & Dugdale 1972, text, Table 5	Peru Coast	—	1·11(4)	0·0240(27)	0·0115(6)
MacIsaac & Dugdale 1969	Subarctic Pacific	4·21(1)	1·30(1)	0·0163(1)	0·0362(1)

Note: The values of V_{max} are low by about 30 per cent due to an instrument error described by Pavlou *et al.* (1974) and may be underestimated further depending on the detrital portion (Dugdale & Goering 1967). These errors have little or no effect on the value of K_s.

according to the amount of detritus present (Dugdale & Goering 1967) and the value of V_{max} can be affected by many factors and is not so easily interpreted. Although the number of measurements is small, it is clear that the K_s

and V_{max} values for oligotrophic populations are an order of magnitude less than for eutrophic populations. These conclusions are supported by the results of measurements made in various ways on single species of marine phytoplankton (Eppley *et al.* 1969). The nutrient concentrations occurring over large areas of the sea match these values of K_s for ammonia and nitrate very well. When phytoplankton production is viewed as a continuous process (Dugdale 1967) it is clear that it is the phytoplankton that are responsible in large part for controlling the euphotic zone ammonia and nitrate concentrations to the approximate levels of K_s.

EFFECT OF LIGHT

The energy required for the functioning of permeases must come directly or indirectly from that captured by chlorophyll. High energy phosphate bonds

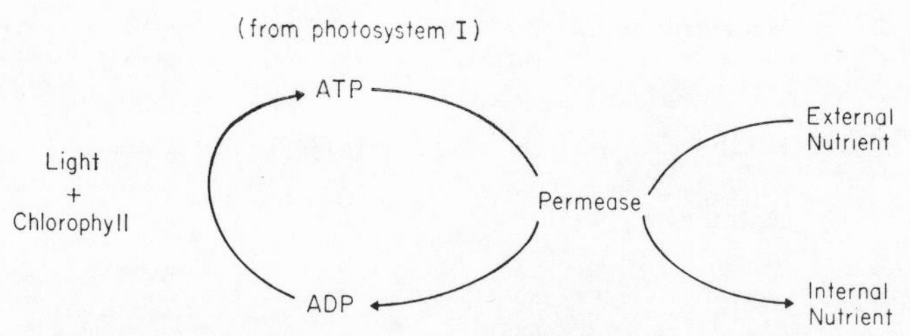

Fig. 7.11. Relationship between light and nutrient uptake.

in the form of adenosine triphosphate (ATP) are generated and the energy extracted producing adenosine diphosphate (ADP); the cycle is shown in Figure 7.11.

Uptake of ammonia and nitrate is strongly dependent upon light. Experiments with ^{15}N in which natural populations of phytoplankton are exposed to a gradient in light result in data that usually can be fitted accurately to Michaelis–Menten hyperbolae (MacIsaac & Dugdale 1972). These experiments are made by taking samples from a selected light depth, adding appropriate amounts of ^{15}N, and incubating in a series of bottles with neutral density screens to remove known fractions of the incident light. The final nutrient level must be approximately saturating, so that the effect of nutrient concentration is removed. K_{lt} values obtained in such experiments fall in the range of 1–10 per cent surface light intensity depending in part upon the depth from which the experimental population is obtained and the extent of vertical mixing. Deeper populations apparently adapt under stable conditions to lower ambient light levels. Typical light-nutrient uptake hyperbolae are shown in Figure 7.12.

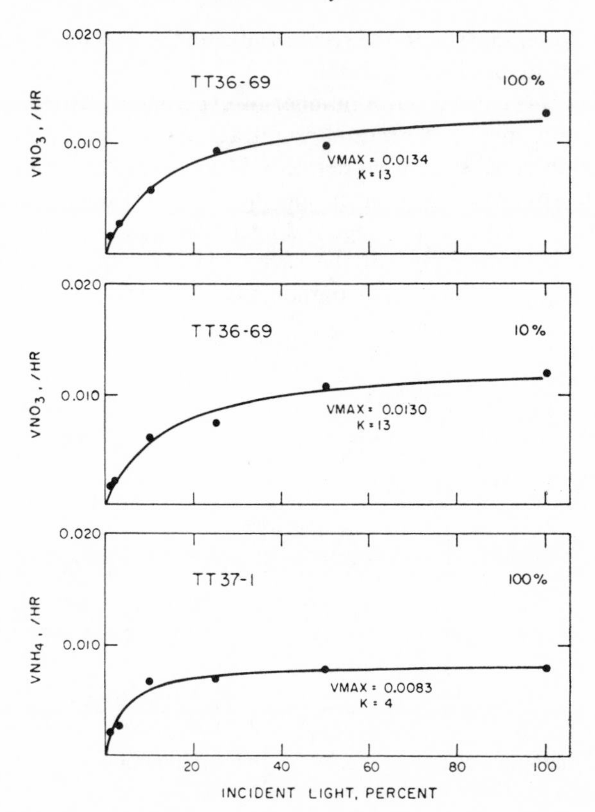

Fig. 7.12. Uptake of nitrate and ammonia as a function of light, following Michaelis–Menten kinetics (from MacIsaac & Dugdale 1972).

NITROGEN INTERACTIONS

Ammonia is taken up and incorporated into amino acids without a change in oxidation state. Nitrate, however, must be reduced by the enzymes nitrate reductase and nitrite reductase (Syrett 1962). Energy is conserved by phytoplankton by suppressing the uptake of nitrite and nitrate in the presence of ammonia, the degree of suppression depending upon the concentration of ammonia present. To some extent the suppression effect probably is also time and species dependent.

In natural populations in a bay in Greece receiving the sewage of Athens, the ratio of saturated nitrate uptake to saturated ammonia uptake appears inversely related to the ambient ammonia concentration, Figure 7.13. Similar results were obtained from the Peru upwelling region (Dugdale & MacIsaac 1971). In these experiments the populations used had been exposed to relatively high ambient ammonia concentrations for some period of time, in the range of one to ten days, and the results indicate the nature of the long-term response of nitrate uptake to ammonia.

Conway (1974) carried out a number of experiments using ^{15}N to study the effect of ammonia concentration on nitrate uptake. In experiments made on a population consisting of 90 per cent *Skeletonema costatum* and 10 per cent *Amphora* sp. grown in a chemostat, he showed a linear suppression of V_{max} for nitrate uptake in the range 1–3 μgat/l ammonia. Ambient ammonia appears to control the shape of the hyperbola for nitrate uptake through reduction of V_{max} with increasing ammonia concentration to some minimum

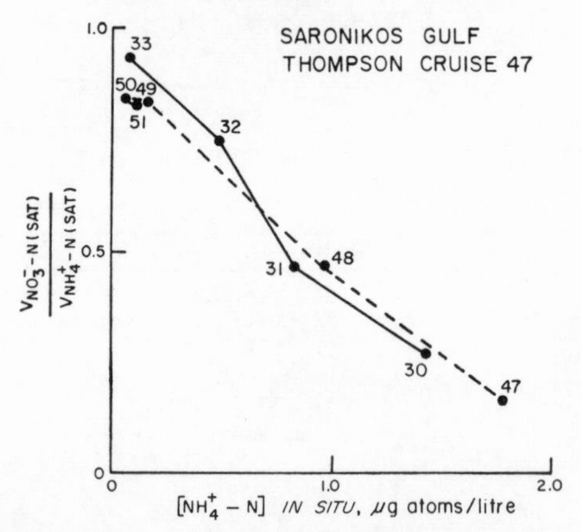

Fig. 7.13. Inhibition of nitrate uptake by ambient ammonia in the Saronikos Gulf, Aegean Sea (from MacIsaac & Dugdale 1971).

level of V_{max}-NO_3. This relationship between V_{max}-NO_3 and ammonia concentration was confirmed by Conway (1974) for *Skeletonema costatum* and he found the time lag for the inhibitory effect of ammonia on nitrate uptake to be in the range of 0·1–1 h and the time lag for recovery of nitrate uptake from ammonia suppression apparently to be in the same range.

NITRITE UPTAKE

Conway (1974) studied nitrite uptake by *Skeletonema costatum* by the ^{15}N technique and found the kinetics of nitrate and nitrite uptake to be identical. The same author also found that in a mixed culture of *Skeletonema costatum* and *Amphora* sp. the inhibition of nitrite uptake by ammonia was identical to the inhibition of nitrate.

NITRATE AND AMMONIA UPTAKE IN THE WATER COLUMN

The maximum rate of uptake of nitrate and ammonia observed at a specific depth is influenced primarily by the factors discussed previously, i.e. by the average light intensity experienced by the phytoplankton during the previous day or days, by the concentration of nitrate and ammonia prevailing during the same period of time and by any other factors that may act to reduce the maximum growth potential of the population. The short-term or instantaneous rates of uptake which are measured in ^{15}N uptake experiments are reduced from the maximum rates by contemporary light, nitrate and ammonia levels. Disregarding pre-conditioning effects for the moment, the following examples will serve to illustrate the effects and interactions of light, nitrate and ammonia in controlling the uptake of these inorganic nitrogen ions.

Eutrophic profiles

Figure 7.14 illustrates the simple condition of light control of nitrate uptake at one station in the Peru upwelling region (Dugdale & MacIsaac 1971). The

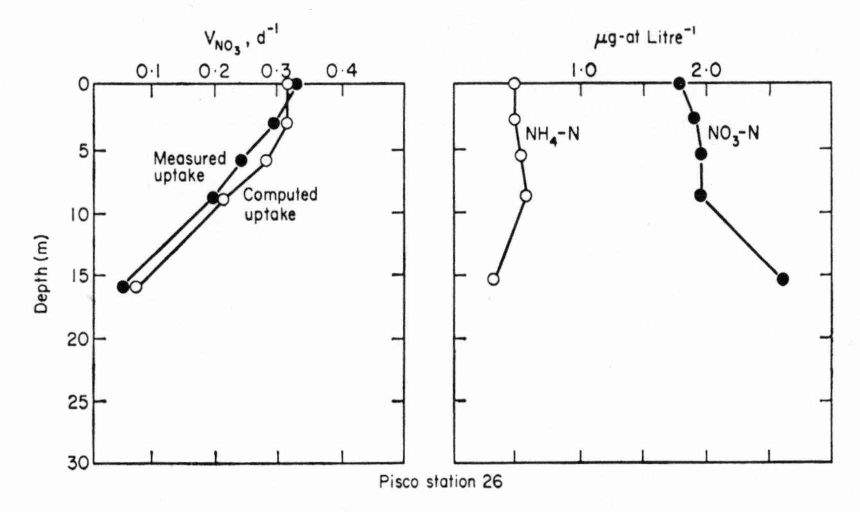

Fig. 7.14. Agreement between computed and observed nitrate uptake rates in the euphotic zone of the Peru upwelling region with nearly constant ammonia levels (from Dugdale & MacIsaac 1971).

nitrate concentration was greater than 2 μgat/l at all depths, thus approximately saturating, and the ammonia concentration nearly constant at 0·5 μgat/l. The measured points are plotted along with those computed from a model requiring a knowledge of only K_s-NO_3, K_{lt}-NO_3, V_{max}-NO_3 and a term for the inhibition of nitrate uptake by ammonia. The values of these parameters were obtained from experiments carried out in the Peru upwelling region. In the model, for any given light and nutrient condition, a value of

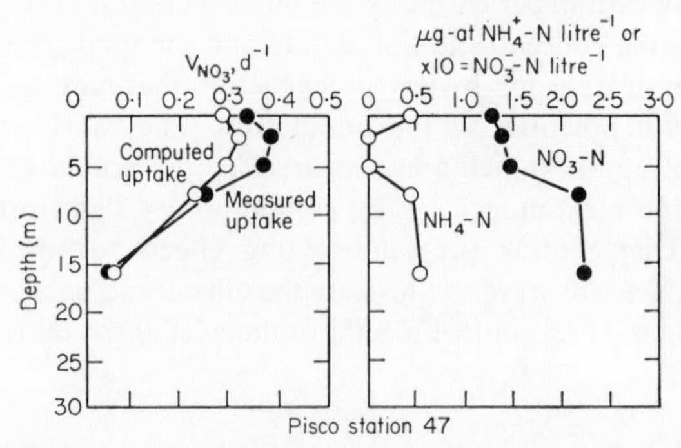

Fig. 7.15. Computed and measured nitrate uptake profiles illustrating effect of subsurface minimum in ammonia (from Dugdale & MacIsaac 1971).

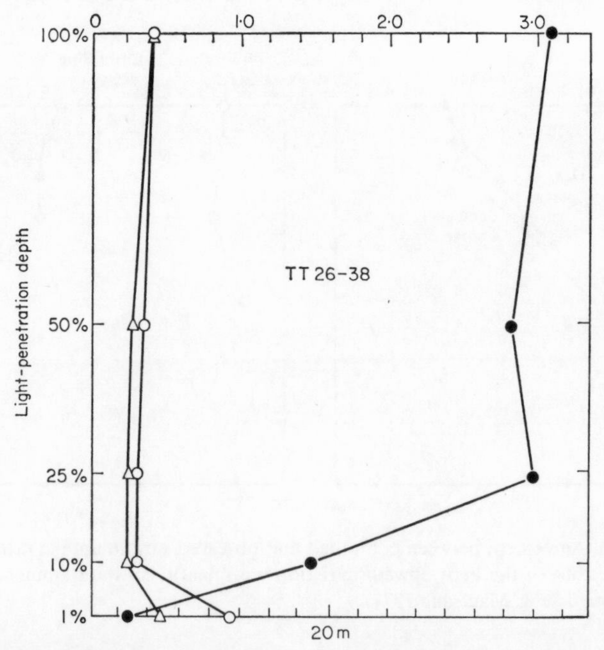

Fig. 7.16. Example of the agreement between calculated and measured values of approximate *in situ* rates of nitrate in a euphotic zone profile. The ambient nitrate concentration at this station was 0·10 μg-atom/l at the upper four depths and 0·39 μg-atoms/l at the bottom depth. The solid circles show values of enhanced uptake rates measured with an enrichment of 5·31 μg-atoms NO_3^-–N/l. The open circles show values of *in situ* uptake rates calculated from the measured-enhanced rates; the open triangles are measured *in situ* rates, with an enrichment of only 0·05 μg-atom NO_3^-–N/l. 20 m was the 1 per cent light-penetration depth (from MacIsaac & Dugdale 1972).

V_{NO_3}, the specific uptake rate of nitrate, is computed from the Michaelis–Menten equation first using the K_s-NO_3 and second the K_{lt}-NO_3; the lower of the resulting two values of V_{NO_3} is selected. Agreement between measured and computed values of uptake is good in this example and suggests that in this case there are no significant unknown factors influencing uptake.

The effect on nitrate uptake of ammonia concentration changing with depth is illustrated in Figure 7.15. The observed subsurface maximum in nitrate uptake is a feature of both the computation model and the observed profile. In the former it is a result of the reduced ammonia concentration and calculated reduced inhibitory effect at the second and third depths, and it may be assumed that the shape of the observed uptake profile was determined by this effect of ammonia concentration.

The effect of reduced nitrate levels on nitrate uptake is illustrated in Figure 7.16. The profile on the right was measured by adding saturating concentrations of $^{15}NO_3$ to one set of bottles, and it is equivalent to the profile of Figure 7.14 where naturally saturating nitrate conditions prevailed. One of the two profiles on the left is the result of making low $^{15}NO_3$ additions to the same set of samples to avoid enhancing the uptake rates; the other is a profile resulting from computing uptake rates for each depth from the usual kinetic constants and the light and nitrate values for each depth. In addition to demonstrating the large differences that may exist between maximum uptake rates and *in situ* rates, the *in situ* profiles on the left show the uptake maximum to be at the 1 per cent light-penetration depth, a result of the increase of nitrate at that depth. Although station TT26–38 was eutrophic as indicated by the high maximal nitrate uptake rates, a subsurface maximum in the *in situ* uptake rate of nitrate has been observed often in oligotrophic regions of the sea as a result of increasing nitrate concentrations at those depths.

The shape of the ammonia-uptake versus depth curve is determined in much the same way as for nitrate except that no effect of nitrate concentration on ammonia uptake has been observed. The K_{lt} for nitrate is often about twice the value of K_{lt} for ammonia in eutrophic regions, a condition that may lead to light limitation of nitrate uptake at a shallower depth than for ammonia uptake. The degree of light dependence for uptake of these two compounds may be expressed in another way. The result of many comparisons in several parts of the ocean shows that hourly rates of nitrate uptake may be converted approximately to daily rates by multiplying by 12 h, as for carbon work; the factor for ammonia, however, is 18 h (Dugdale & Goering 1967).

Oligotrophic profiles

The two nitrate uptake curves on the left in Figure 7.17 represent measured and calculated values and may be considered typical for oligotrophic regions.

The low values of uptake in the upper region were the result of nearly undetectable nitrate concentrations; the maximum at the 10 per cent light-penetration depth was the result of an increase in nitrate concentration to 3·5 μgat/l (Table 4, p. 228, MacIsaac & Dugdale 1972). The decrease at the 1 per cent light-penetration depth occurred from light limitation although the nitrate concentration increased further to 32·2 μgat/l. The ammonia uptake curves on the right of Figure 7.17 reflect probable inhibition by surface light and light limitation below the 50 per cent light-penetration level.

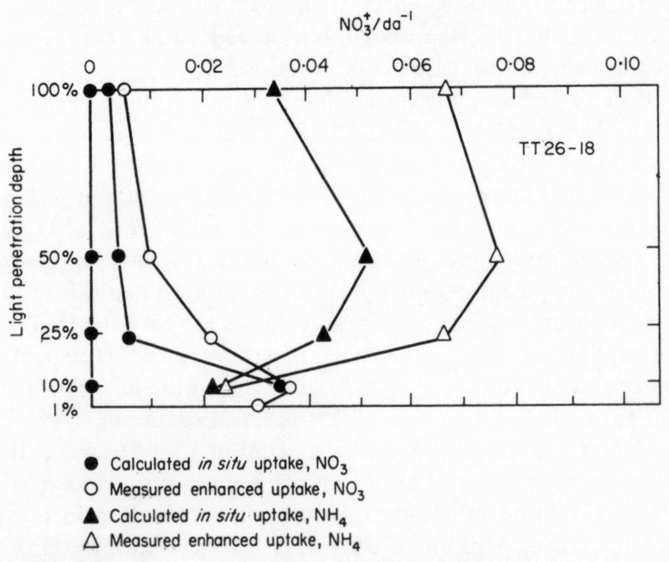

Fig. 7.17. Computed *in situ* (solid symbols) and measured enhanced (open symbols) nitrate (circles) and ammonia (triangles) uptake profiles in an oligotrophic area. The 1 per cent light depth was 78 m (from MacIsaac & Dugdale 1972).

ORGANIC NITROGEN

The potential role of urea as an important source of regenerated nitrogen for marine phytoplankton has been established by recent studies. Carpenter *et al.* (1972) showed that 5 diatoms and a flagellate were able to grow on urea at the same rates as on nitrate or ammonia and the uptake of urea by *Skeletonema costatum* was found to follow Michaelis–Menten kinetics with a $K_m = 8{\cdot}5$ μgat urea-N/l. From urea concentrations of about 2·5 μgat/l as seen by Remsen (1971) in waters between Panama and Peru, a doubling time of about 2·2 days can be computed for phytoplankton using only urea. McCarthy (1972) using ^{15}N, measured the uptake of urea, ammonia and nitrate at various depths at a number of stations along the California coast. The contribution of urea to the sum of nitrogen uptake of the three sources varied from 1–60 per cent with an average of 28 per cent. Both urea concentration and urea uptake tended to decrease with depth in this set of stations.

Marine phytoplankton are able to take up and grow on a variety of amino acids. According to Wheeler *et al.* (1974) 25 species of unicellular marine algae were able to use several out of 9 common amino acids. Uptake rates were enhanced by nitrogen deprivation but were quite low. Doubling rates were calculated to be about 10 days at substrate concentrations of 1 μgat urea-N/l. Some specific uptake rates as high as 0·04 h^{-1} were measured, however. In another recent study Schell (1974) used ^{15}N to study the uptake and regeneration of glycine and glutamic acid in south-east Alaskan marine waters. He observed the levels of uptake to be low in agreement with the results of Wheeler *et al.* (1974). From these results, it seems that amino acids are unlikely to play a significant role in the nitrogen economy of marine phytoplankton under most normal circumstances.

Factors influencing the uptake of silicate

Quantitative studies of silicate uptake by natural populations of marine phytoplankton have begun only recently and as a result much less is known about the process in comparison with the uptake of inorganic nitrogen. A recently developed method (Goering *et al.* 1973) using the stable isotope, ^{29}Si, should contribute significantly to silicate work, especially in the field.

CONCENTRATION

Michaelis–Menten kinetics characterize the uptake of silicate by marine diatoms according to a number of recent investigations. Harrison (1973) collected the published values of K_s for silicate; four authors using three different methods reported K_s values in the range 0·7 to 3·4 μgat Si/l. Paasche (1973) obtained values of K_s for 5 species of diatoms using batch culture techniques; Davis (1973) and Harrison (1973) used chemostats and Goering *et al.* (1973) used ^{29}Si tracer techniques with natural populations. Guillard *et al.* (1973) studied two clones of *Thalasiosira pseudonana* (Hasle & Heimdal) and reported K_s values of 0·98 and 0·19. As with nitrate and ammonia the range of K_s matches well the lower values of silicate concentration found in the euphotic zone of much of the ocean.

LIGHT

The uptake of silicate does not appear to be so closely related to incident light as does the uptake of nitrate and ammonia in preliminary results obtained by Goering (personal communication) using ^{29}Si and natural populations. Davis

(1973) studying the reactions of *Skeletonema costatum* to differing light levels found the V_{max} for silicate uptake to be a function of the light levels provided to his chemostats. The uptake of silicate showed no cyclic pattern under either a 16 h light–8 h dark regime or under continuous light.

TEMPERATURE

Temperature appears to affect the V_{max} for silicate uptake approximately in accordance with a Q_{10} of about 1·7, from two experiments of Harrison (1973) using chemostat cultures of *Skeletonema costatum*. This value is close to the Q_{10} of 1·88 suggested by Eppley (1972) for phytoplankton growth. When the temperature was increased from 12°C to 18°C in Harrison's work there appeared to be virtually no lag period before increased uptake was observed.

UPTAKE IN THE WATER COLUMN

The few simulated *in situ* measurements of silicate uptake made by Goering (personal communication) using ^{29}Si tend to support his earlier conclusion that silicate uptake is not strongly and immediately coupled to light. His measurements were made in two upwelling regions and showed little change from the top of the euphotic zone to the bottom. He also reports quite high uptake rates at depths of up to 150 m when populations of diatoms are present.

Factors influencing the uptake of phosphate

Studies of phosphate in the marine environment have been concerned largely with its circulation and partition between various fractions (Watt & Hayes 1963). Use of ^{32}P as a tracer has resulted in the general conclusion that phosphorus turns over quickly, with the dissolved inorganic fraction being highly mobile and available as a source for the requirements of phytoplankton (Pomeroy 1960).

CONCENTRATION

The uptake of phosphate follows Michaelis–Menten kinetics. The few values of K_s obtained so far for phosphate uptake by marine phytoplankton are remarkably similar to those for nitrate, ammonium and silicate. Data obtained by Harvey (1963) for oxygen evolution by *Phaedactylum tricornutum* as a function of added phosphorus show clear Michaelis–Menten kinetics, with a K_s of 0·33 μgat/l. Working with the same organism, Ketchum (1939) showed dependence of nitrate uptake on concentration and an interaction with phosphate. Fuhs *et al.* (1972) report values of K_s = 0·58 μgat/l for

Thalassiosira pseudonana and 1·72 μgat/l for *Thalassiosira fluviatilis* (Hustedt); nutrient-limited cultures were grown for these experiments.

LIGHT

Phosphate uptake in relation to light has not been investigated extensively. Fuhs *et al.* (1972) observed that phosphate-limited cultures grown in continuous light were able to take up phosphate at high rates in the dark.

TEMPERATURE

The relationship between phosphate uptake and temperature is shown in Figure 7.18 (Fuhs *et al.* 1972). A Q_{10} of about 2 can be inferred, in agreement

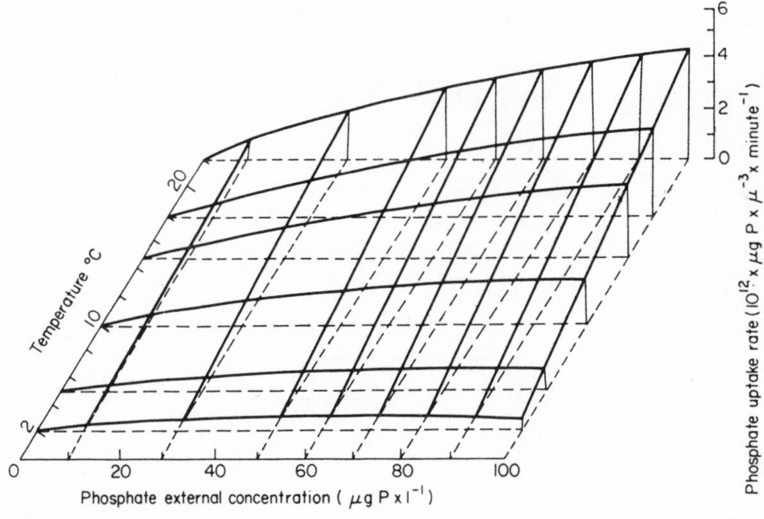

Fig. 7.18. Relationship between temperature and phosphate uptake for *Thalassiosira fluviatilis* (from Fuhs *et al.* 1972).

with the scanty evidence already mentioned for silicate uptake in relation to temperature.

Nutrient uptake and growth

Nutrient uptake and growth are separate processes that are coupled through various mechanisms including feedback control of uptake. Under steady state conditions specific uptake and specific growth rates must be equal; during transient phases these rates may differ. With nutrient-limited growth, the specific growth rate is controlled by the uptake of the limiting nutrient through the permease system. Under internally or non-nutrient controlled growth, the uptake is controlled to the level required for cell synthesis. The Michaelis kinetics describing uptake of limiting nutrient have been discussed previously.

Marine phytoplankton react to nutrient deficiencies by reducing the amount of limiting nutrient in each cell, a finding first reported from chemostat studies by Droop (1968) for vitamin B_{12} and confirmed for nitrogen by Caperon and Meyer (1972a), Eppley and Renger (1974), and Conway (1974); for silicate by Paasche (1973), Davis (1973), and Harrison (1973); for phosphate by Fuhs *et al.* (1972). The chemostat is a continuous culture device in

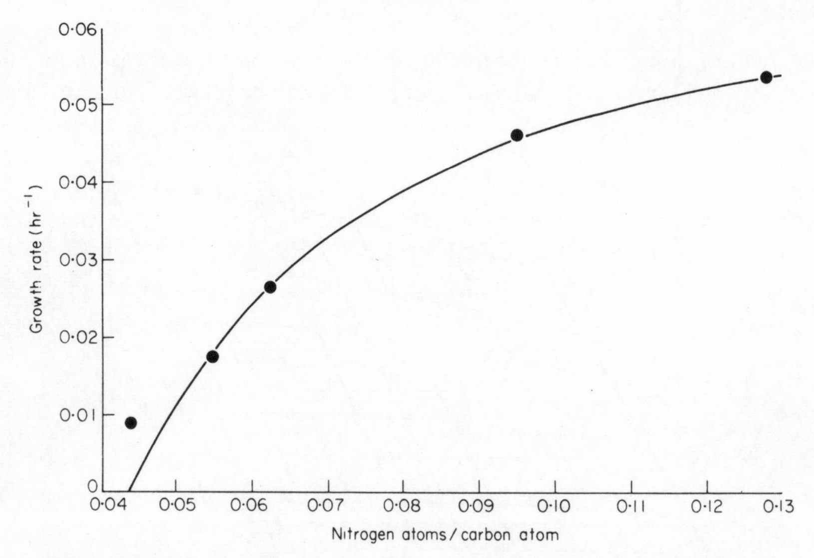

7.19. Relationship between q, cell quota as nitrogen/carbon by atoms, and growth rate or dilution rate in *Dunaliella tertiolecta* (from Caperon & Meyer 1972a).

which the growth rate of a population can be controlled by the rate of turnover of medium in the reactor, the dilution rate; the medium is designed to assure that one nutrient will become limiting (Herbert 1959).

A summary of the current understanding of nutrient uptake and growth may be found in Droop (1973). All of these investigators have found that a saturation curve describes the relationship between growth rate and q, cell quota; the latter is the amount of limiting nutrient contained in each cell, as in Figure 7.19.

The curve is described by:

$$\mu = \frac{\mu_m(q-q_0)}{K+q-q_0} \tag{7.2}$$

where q_0 is the concentration of q at which $\mu = 0$.

Since the flow of limiting nutrient into a cell is necessarily a function of the uptake rate, models incorporating the uptake rate as the limiting step and cell quota as the growth rate controlling step can be made. The approximate hyperbolic expression:

$$\mu = \mu_m \frac{[S - q_0 k_s/b]}{S} \tag{7.3}$$

is simplified slightly from Caperon and Meyer (1972b). From it, $\mu = \frac{1}{2}\mu_m$ when $S = 2q_0K_s/b$ giving an estimate of the K for growth, K_g. Since $b = V_m/\mu$

$$K_g = 2q_0\, k_s \frac{\mu}{V_m} \tag{7.4}$$

With one set of data Caperon and Meyer computed a value of $K_s = 0{\cdot}32$ μgat N/l and $K_g = 0{\cdot}058$ μgat N/l illustrating that the K for growth is about an order of magnitude less than the K for uptake.

The studies of control of growth by internal concentrations of limiting nutrient discussed above were made with chemostats. Most of these studies were made within a relatively restricted range of growth rates, however, and Harrison (1973) and Conway (1974) operated chemostats to grow *Skeletonema costatum* within the mid-range and also at low and high dilution rates. From their results, four operating regions can be recognized as shown in Figure 7.20. In Region 2, the cell quota is substantially lower than maximal values and the kinetics discussed above apply, i.e., $K_g \ll K_s$. This region is characterized by the small change in concentration of limiting nutrient in the reactor with changing dilution rate. Cells growing in such a regime are nutrient deficient as a result of the reduced cell quota. At the lower dilution rates in Region 1, cell division is inhibited, resulting, under silicate limitation, in large chloroplasts; silicate concentration may increase slightly in the reactor and the population may be said to be nutrient starved. In Region 3 at moderately high dilution rates, the cell quota approaches its maximum value and the chemostat begins to show larger changes in limiting nutrient concentration with changes in dilution rate; the organisms are growing near μ_{max} and $K_g \simeq K_s$. If the dilution rate is increased slowly and carefully, the culture can be grown at rates exceeding the pre-existing μ_{max} suggesting that these Region 4 cells have been selected for the ability to grow at these very high rates or that their physiology has become adapted for rapid growth. In the bacteriological literature, as cells proceed through these various regions of nutrient sufficiency, they are said to 'shift-up' their physiological state (Mandelstam & McQuillen 1968).

Since the sea is a continuous culture system, and primary production is limited by nitrogen over large areas (Ryther & Dunstan 1971) it may be possible eventually to recognize regions or regimes in the sea that correspond to those just described for the chemostat culture of *Skeletonema costatum*. For example, the southern Sargasso Sea or the eastern Mediterranean may correspond to Region 1, the northern Sargasso Sea and the western Mediterranean to Region 2, while Regions 3 and 4 might be represented in upwelling and coastal areas.

Nutrient-based growth rates can be measured directly with ^{15}N and ^{29}Si if the assumption can be made that uptake and growth are in steady state at

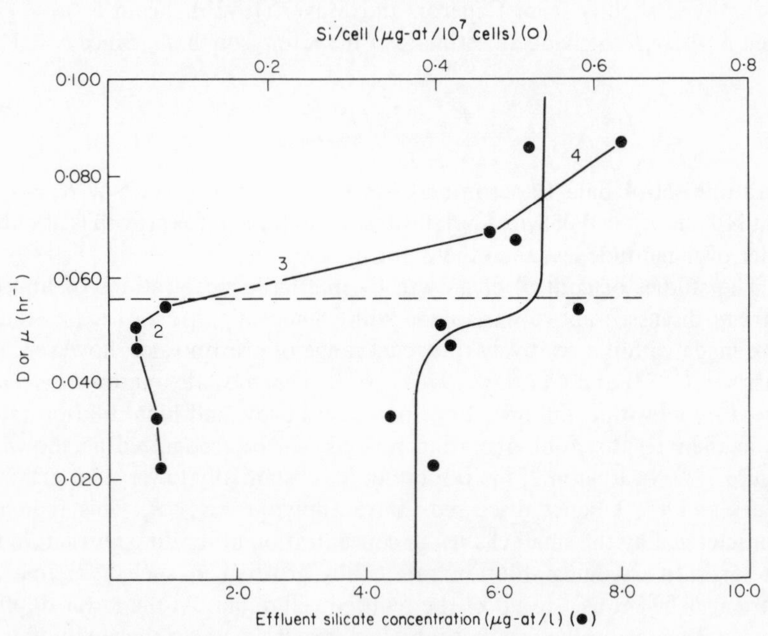

Fig. 7.20. Characteristic operating regions for *Skeletonema costatum* grown under silicate limitation in chemostat culture (after Harrison 1973).

the time the tracer uptake measurement is made (Neess *et al.* 1962), since the outcome of the measurement with the mass spectrometer is normalized to the element being studied. For example, the specific nitrate uptake rate, V_{NO_3}, equals the amount of nitrate–nitrogen taken up per unit particulate nitrogen per unit time and is obtained from the expression (Dugdale & Goering 1967):

$$V_{NO3} = \frac{\rho_{NO3}}{\text{PN}} = \frac{da_1/dt}{a_2 - a_1} \tag{7.5}$$

where ρ_{NO_3} is the rate of transport of nitrate into the cells, PN is the particulate nitrogen concentration, a_2 is the atom per cent ^{15}N in the liquid fraction, and a_1 is the atom per cent ^{15}N in the particulate fraction at the end of the incubation. In natural populations, V_{NO_3} values obtained in this way are reduced according to the amount of detritus present and care must be taken in interpreting such data.

Nutrient circulation

Nitrogen productivity measurements

The availability of a variety of ^{15}N labelled compounds makes it possible to measure rates along some of the specific pathways of nitrogen in the euphotic

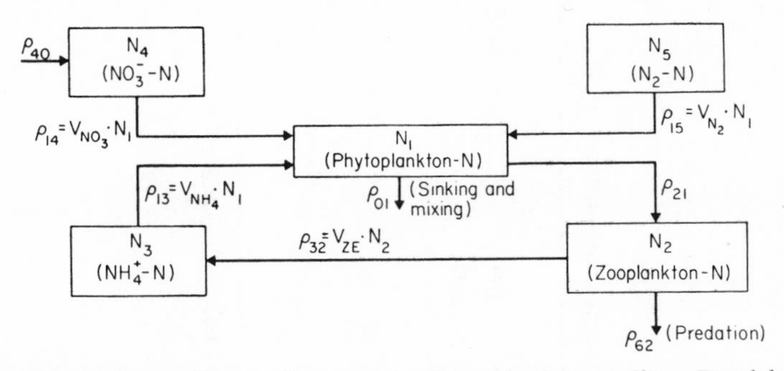

Fig. 7.21. Major pathways of nitrogen in the euphotic zone (from Dugdale & Goering 1967).

zone as indicated in Figure 7.21. Especially important is the ability to distinguish between the uptake of new nitrogen in the form of nitrate, and regenerated nitrogen-ammonia, urea, and other organic nitrogen compounds. The biomass that may be removed from a marine ecosystem as yield cannot exceed the amount of new nutrient arriving in the euphotic zone if a steady state is to be maintained. The fact that distinct species of nitrogen compounds are formed during regeneration provides a powerful method for separating the new and regenerated components of primary productivity.

As discussed previously, the amount of new nutrient injected into the euphotic zone usually controls the productivity status of an ocean area. Values of nitrate uptake measured with ^{15}N are given in Table 7.3 arranged in order of increasing uptake. The most extreme cases, a central ocean gyre and the Peru upwelling region, are included, giving a range of about 0·2–30 mgat/m^2/d for the uptake of nitrate integrated through the water column. The ratio of nitrate uptake to nitrate-plus-ammonia uptake, primary productivity, and the ratio of primary productivity to nitrate uptake are also included in the table and these values also can be assumed to represent the range of extreme oligotrophy to extreme eutrophy. The proportion of nitrate

Table 7.3. Value of nitrate uptake measured with ^{15}N

	1% light depth m	ρNO_3 mgat/m²/d.	$\frac{\rho NO_3 \times 100}{\rho NO_3 + \rho NH_4}$	^{14}C mgat/ m²/d.	$\frac{^{14}C}{\rho NO_3}$	Source
North Pacific central gyre	~130	0·18	6	18·92	105	Eppley *et al.* 1973
Eastern Mediterranean	76	0·45	23	16·42	36·49	MacIsaac & Dugdale 1972
Eastern tropical Pacific	79	1·14*	57	57·67	50·59	MacIsaac & Dugdale 1972
Costa Rica dome	35	8·12*	47	337·04	56·64	MacIsaac & Dugdale 1972
Peru upwelling	22	29·96	66	438·89	14·65	MacIsaac & Dugdale 1972

* This value is low by about 30% due to an instrument error described by Pavlou *et al.* (1974).

to nitrate-plus-ammonia uptake in the oligotrophic Sargasso Sea near Bermuda was 8·3 per cent with a range of 0·5–18·9 per cent in one study (Dugdale & Goering 1967). The high values occurred during the period of winter mixing that brings nitrate into the euphotic zone. In Table 7.3 it appears that when nitrate uptake reaches about 1·5 mgat/m^2/d the proportion of nitrate uptake increases to about 50 per cent, a value that may be considered typical for eutrophic areas as opposed to the lower values of oligotrophic areas where recycled nutrient forms the basis for nitrogen productivity. The ratio of ^{14}C production to nitrate uptake is high in oligotrophic areas compared to eutrophic areas, also reflecting the capacity of oligotrophic systems to conserve nutrient through rapid recycling. A compound not included in the table is urea, shown recently by McCarthy (1972) to be about equally as important as ammonium to the regenerated nitrogen pool. The true proportion of nitrate to total regenerated nitrogen uptake thus is smaller than the values given and the maximum proportion probably is in the neighbourhood of 30 per cent for eutrophic regions. Urea uptake was measured in the North Pacific central gyre study cited in Table 7.3; when included, the proportion of nitrate to the sum of nitrate, ammonium and urea uptake is reduced from 6 to 3 per cent (Eppley *et al.* 1973).

Sources of regenerated nitrogen

In the Peru upwelling region, the nitrogen budget (Dugdale & Goering 1970) suggests that the regeneration in the euphotic zone is carried out primarily by the anchoveta population through grazing on the phytoplankton. Regeneration of nitrogen through grazing by zooplankton was estimated to be much less, about one-fourth of that of the anchoveta. Measurements of anchoveta and zooplankton excretion rates (Whitledge 1972, Whitledge & Packard 1971) for ammonia, urea and creatine when combined with fish distribution data for Peru indicate that anchoveta account easily for most of the regenerated nitrogen requirements of the phytoplankton; the contribution of ammonia by zooplankton amounted to only 10 per cent of the total contributed by both populations. However, these estimates of regeneration were probably high as a result of the shipboard experimental methods used. Subsequent laboratory experiments with *Engraulis mordax* (Girard) showed that ammonia comprised 82·7 per cent, urea 16 per cent and creatine 1·3 per cent of the excreted nitrogen. The total nitrogen excretion in these experiments on a dry weight basis was about half of that reported for the shipboard experiment with *Engraulis ringens* (Jenyn), or about 0·7 μgat N/mg d.w./d (McCarthy & Whitledge 1972).

In the North Pacific central gyre, where the population of fish is negligible in comparison to the zooplankton, the regeneration of nitrogen by zooplankton accounted for about 50 per cent of the regenerated nitrogen requirements of the phytoplankton (Eppley *et al.* 1973). Urea and ammonia were excreted in approximately equal amounts. Regeneration by bacteria in the water column presumably accounts for the remainder of the regenerated nitrogen. Other estimates of the contribution of zooplankton excretion to phytoplankton nitrogen requirements in oligotrophic regions were 10 per cent for the Sargasso Sea near Bermuda (Dugdale & Goering, 1967) and 7·8 per cent and 1·6 per cent for two stations near the Costa Rica Dome (Whitledge & Packard, 1971). These lower values may result from lack of urea and other organic nitrogen measurements. The excretion rates reported are minimal estimates since the biomass of zooplankton smaller than 187 μ were not sampled and the relative activity of these smaller sizes is often very high.

Nitrogen fixation

The use of molecular nitrogen for growth is common in the freshwater environment, but it appears to be rare in the open sea. The blue-green alga, *Trichodesmium* sp. is capable of fixing nitrogen and large blooms of this organism are found in tropical oceanic regions (Dugdale *et al.* 1964). Measurements of the absolute rates of N_2 uptake were made in a study of the nitrogen economy of the Sargasso Sea (Goering *et al.* 1966). Rates as high as 0·28 μgat N/1/d were measured. The organism is a difficult one to study and never has been cultured successfully. Mague *et al.* (1974) and Carpenter (1972) recently have applied an acetylene reduction technique to studies of nitrogen fixation in the central Pacific and central Atlantic gyres. Bacteria that can fix nitrogen also are found in the open sea but direct measurements of their nitrogen-fixing activity have not been made.

Sources of regenerated phosphate and silicate

Because phosphate and silicate are not regenerated into distinctly different ionic species as occurs with nitrogen, assessment of the cycling rates of these elements is more difficult. The Peruvian anchoveta studied by Whitledge (1972) excreted nitrogen, phosphate and silicate in the ratios of 3·92:1:0·23. Nitrogen is retained relative to phosphate, presumably for growth, and silicate is poorly regenerated. In the Peru region this pattern of regeneration results in silicate decreasing to zero at the outside edge of phytoplankton plumes, nitrogen decreasing at a slower rate and phosphate showing relatively little change as shown in Figure 7.22. Making a number of assumptions about the size of anchoveta schools and their distributions, Whitledge (1972) estimated that in the Peru upwelling area, the zooplankton and nekton contributed 22 per cent of the ammonia-plus-nitrate fraction, 37 per cent of the phosphorus and only 4 per cent of the silica used by the phytoplankton.

The silicate concentration of about 2 μgat/l was surprisingly high in Pacific central gyre waters which are almost devoid of phosphates and nitrogen (Eppley *et al.* 1973). The cause of this condition which occurs also in the Mediterranean Sea is not clear, but may reflect the inability of diatoms to

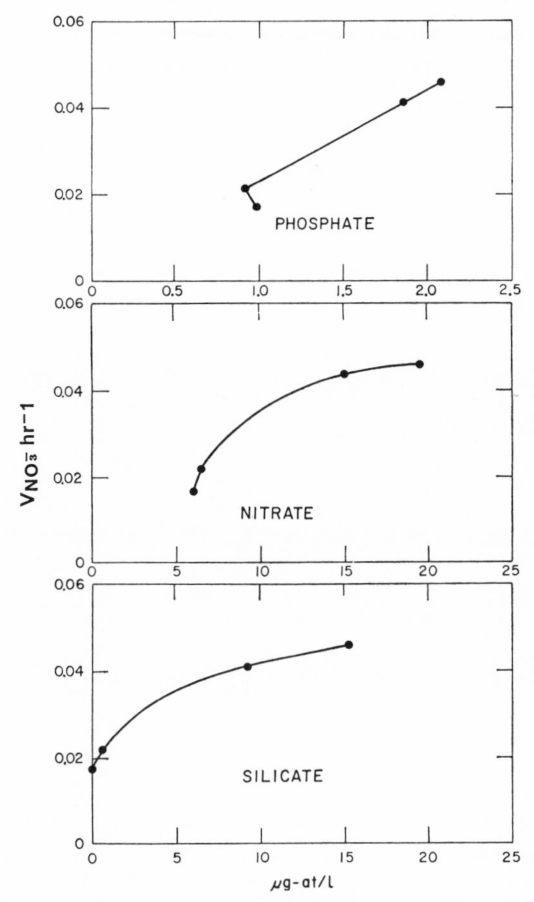

Fig. 7.22. Nitrate uptake rate as a function of nitrate, silicate and phosphate concentration in the Peru upwelling region (from Dugdale & Goering 1970).

use silicate normally at low growth rates, as seen in continuous studies (Harrison 1973).

Zooplankton excretion rates were compared with nitrogen and phosphorus utilization rates by phytoplankton by Eppley *et al.* (1973) in the central Pacific gyre. The mean rate of total excreted nitrogen to total excreted phosphorus was 8·7. The ratio of excretion to assimilation was about 0·5 for ammonia and urea, and a little over 1 for phosphorus, suggesting that phosphorus should appear in the water column in excess as in Peru. However, phosphate levels were about 0·05 μgat/l throughout most of the water column. The discrepancy probably results from assuming the usual 1:15 P:N uptake ratio, instead of a likely lower value, to compute phosphate uptake from ^{15}N measured rates of nitrogen uptake.

Chapter 8. Primary Production in the Sea

C. J. Lorenzen

Introduction

Algal growth in the oceans
Light
Nutrients

Algal growth in the oceans *continued*
Fates of primary fixed carbon
Distribution of production

Introduction

Plants, in the presence of adequate light and nutrients, will grow. The relationship between growth and the various environmental factors affecting growth have been and are now of considerable interest to biological oceanographers. The actual mechanisms, on a subcellular level, of the incorporation of inorganic carbon into organic carbon have been intensively studied but need not be discussed here. Broadly speaking, the photosynthetic process can be described as follows:

$$CO_2 + H_2O \underset{\text{chl}}{\overset{\lambda}{\rightarrow}} CH_2O + O_2 \quad (8.1)$$

Carbon dioxide and water combine in the presence of light and mediated by chlorophyll to produce free oxygen and a carbohydrate product. It has been determined that the free oxygen is derived from the water. Energy is also incorporated in the plant cell via photosynthetic phosphorylation. As far as can be determined, photosynthesis in marine plants is not uniquely different in any aspect from the same process that occurs in all other chlorophyll bearing plants. On the other hand, many of the physical, chemical and biological factors in the marine environment impose certain constraints and stresses which give rise to situations which are uniquely different from terrestrial environments. These differences are not always obvious, but it is uncertain if they are more apparent than real or are the result of and reflected in the dynamics of the overall system. Thus, we find some real differences

Contribution No. 848 from the Department of Oceanography, University of Washington.

between terrestrial and open-ocean environments. For example, on land we find a large standing crop of carbon relative to a unit area's daily production rate. We also find a large reservoir of primary fixed carbon relative to the biomass of herbivores. In the ocean it appears that most of the primary fixed carbon enters the food web through grazers rather than detrital feeders as it must on land (Steele 1974). These are some of the fundamental differences in the structure and functioning of food webs in terrestrial and marine systems which will be enlarged upon elsewhere in this volume.

During the last decade or so a number of reviews have appeared dealing with both the conceptual notions and technical practicalities involving primary production in the sea. For example, broad general coverage can be found in Raymont (1963), Ryther (1963), and Goldman (1966). A more recent coverage of the state of the art is provided by Strickland (1965, 1972), and in Parsons and Takahashi (1973). More practical considerations involving techniques and summaries of methodological problems can be found in Strickland (1960), Strickland and Parsons (1968), and in parts of Vollenweider (1969). Useful discussions of nutrients can be found in Redfield *et al.* (1963) and of the light field in the ocean in Holmes (1957) and Jerlov (1968). Older treatises exist which are almost classical in Biological Oceanography. A few that may be mentioned are Gaarder and Gran (1927), Gran and Braarud (1935), Harvey (1955), and Riley, Stommel and Bumpus (1949). Many others exist and are recorded in the references of the above-mentioned papers. In view of the availability of background information, a historical development will not be attempted.

Algal growth in the oceans

The basic photosynthetic process is the same in all chlorophyll-bearing plants but much of the information in hand has been derived from the study of *Chlorella* or other algal species and is the subject of numerous treatises and monographs (Vernon & Seely 1966, Goodwin 1966, 1967). In the ocean this process is strongly affected by extracellular processes and properties of the marine environment.

Light

One of the major factors controlling the rate of photosynthesis is the light field to which the algal cells are exposed. Sea water itself attenuates light energy so that the light field found at increasing depths in the water column changes in both quantity and quality. Superimposed on this is the selective absorption and scattering of naturally occurring suspended and dissolved organic and inorganic materials. Consequently, we find that the quantity and quality of light penetrating to any fixed depth is a property of the type of water under discussion (Jerlov 1968). Different oceanic waters have been characterized on the basis of their optical properties and we find that open ocean water is more transparent to visible light than coastal waters. In addition, open ocean waters are most transparent in the blue part of the spectrum ($\sim$480 nm) while the region of maximum transmission is shifted to green ($\sim$550 nm) for the more turbid coastal waters, primarily due to increased attenuation by dissolved and suspended materials. It is not very clear what significance these spectral shifts might have on marine primary production since the major groups of marine phytoplankton appear to be able to utilize different portions of visible light equally efficiently in photosynthesis, because of the transfer of energy absorbed at shorter wavelengths to chlorophyll α. As a result, energy absorbed by the accessory pigments can contribute to overall cell photosynthesis (Haxo 1960). Energy outside the visible range (400–700 nm) is normally not considered important since it does not appear to be utilized in photosynthesis and is attenuated by water much more rapidly than visible light. While this may be true, the effects of energy outside of the visible may have important consequences for primary production in the sea. During the course of a study of the effects of ultraviolet light (290–320 nm) on phytoplankton in this laboratory the earlier suspicions of Steemann Nielsen (1964) were confirmed. In *in situ* and simulated *in situ* productivity experiments in which ultraviolet transmitting material, either quartz or vycor glassware, was used, a significant inhibition of ^{14}C incorporation was noted when compared to parallel samples from which ultraviolet light was excluded by a thin film of mylar. Normal laboratory glassware and plastics used in the construction of simulated *in situ* incubators are opaque to ultraviolet. The effect is quite large, up to 50–80 per cent reductions at the surface, and extends through a considerable portion of the euphotic zone. In turbid coastal waters the effect is measurable down to a level encompassing the upper third of the euphotic zone. This depth corresponds to an ultraviolet intensity of approximately 0·01 per cent of its irradiance level at the surface. If one extrapolates Jerlov's (1968) extinction coefficients at 310 nm for clear ocean waters (Type I), ultraviolet should affect phytoplankton in the upper half of the euphotic zone. Roughly, one would estimate that primary productivity measurements taken with ultraviolet opaque materials overestimate water column carbon fixation by approximately 25 per cent.

In practice, the vertical extent of the euphotic zone, which can be defined as that layer lying above the point in the water column where daily algal photosynthesis is equal to daily algal respiration, is delineated by the level at which one finds 1 per cent of the surface light intensity. This point in the water column is measured by a variety of techniques ranging from the Secchi Disc to submarine photometers, both filtered and unfiltered, to spectral radiometers and quantum meters. The techniques and the physics involved are discussed in a variety of places (Jerlov 1968, Jerlov & Steemann Nielsen 1974). The general equation used to describe light attenuation is:

$$I_z = I_o e^{-KZ} \tag{8.2}$$

where I_z and I_o are light intensities at depths z and o, respectively, and K is the average extinction coefficient. While the 1 per cent level is a handy reference and on the whole appears to be adequate in many situations, especially in temperate regions, there are certain fallacies in its use. For example, it is possible to define a euphotic zone at high latitudes throughout the year. In the dead of winter the sunlit portion of the day may only be an hour or so. Contrast this with the light conditions at the same geographical location on the

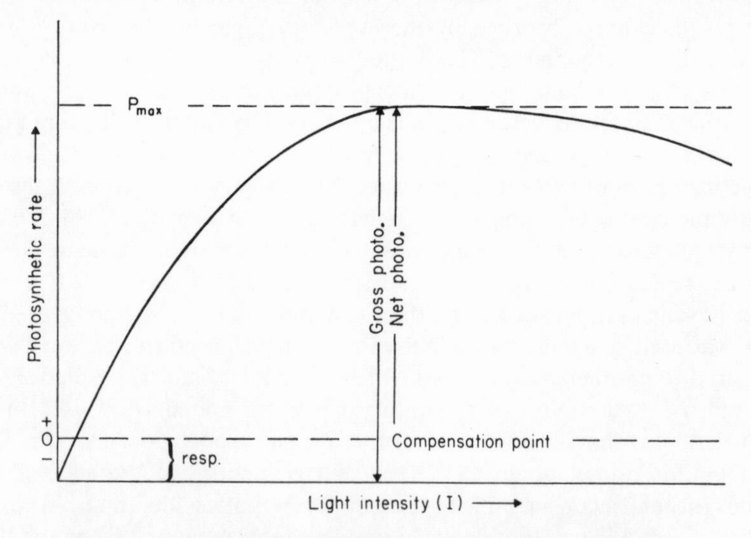

Fig. 8.1. Generalized photosynthesis v. light intensity relationship.

longest day of the year when the sunlit portion of the day may be 20 h or more. The biological consequences to an algal cell at the point where the light intensity is equal to 1 per cent I_o is quite different in these two extremes. Obviously, the critical factor is the total amount of energy available for photosynthesis per unit time.

The relationship between photosynthesis and light intensity is an area of considerable interest and its importance to photosynthesis in the ocean is quite obvious. At low to moderate levels of light, photosynthesis, P, is proportional to intensity, I. This relationship holds until a plateau is reached. At still higher intensities photosynthesis decreases. The shape and nature of this relationship, usually called a P/I curve when depicted by a graph, show important parameters when discussing marine primary production (Fig. 8.1).

Yentsch and Lee (1966) and Steemann Nielsen and Jorgensen (1968) present general discussions of *P/I* curves. The lower limit of the euphotic zone is found at the compensation point which is defined as the point where photosynthesis is equal to respiration when measurement is made over a daily cycle. The daily average light intensity at this point is the compensation intensity. At shallower depths, i.e. higher light intensities, the photosynthetic rate will reach a maximum, P_{max}. If incident radiation on the sea surface is higher than P_{max}, supraoptimal light intensities are reached and one observes a depression in the photosynthetic rate. As a result, the specific photosynthetic rates, expressed as unit C fixed per unit biomass on a volume basis (usually $mg\ C \cdot mg\ Chl^{-1} \cdot m^{-3}$) will vary throughout the euphotic zone and frequently a graphic display of the data for a station will look similar to a *P/I* plot tipped on its axis.

There is at least one other interesting consequence of *P/I* and primary productivity and that is the relationship between incident radiation and the total amount of carbon fixed under a square metre of sea surface ($mg\ C \cdot m^{-2}$). At moderate to low incident radiation levels the photosynthetic rate is directly proportional to light intensity. Changes in incident radiation are reflected in changes in the thickness of the euphotic zone since the compensation point will move vertically in response to changes in light intensity. Under the 1 per cent I_o rule of thumb mentioned earlier, the thickness of the euphotic layer is independent of incident radiation. Consequently, $mg\ C \cdot m^{-2}$ is directly proportional to incident radiation until P_{max} is achieved at the surface. Further increases in incident radiation will further deepen the compensation level thereby thickening the euphotic layer and increasing the standing stock of phytoplankton exposed to light, but further increases in $mg\ C \cdot m^{-2}$ are offset by inhibition of photosynthesis at the surface. Thus, we find the situation where $mg\ C \cdot m^{-2}$ appears to be independent of incident radiation when incident radiation is above a moderate level (Strickland 1970).

Nutrients

In addition to radiant energy, plants need an adequate supply of nutrients if they are to grow and the relationship between phytoplankton growth and nutrient supply has been the subject of study for an extended period of time. Many of the earliest studies and theories were direct transfers from the field of plant physiology which in turn was driven by interests in agriculture. Marine algae have been shown to have absolute requirements, for certain elements and organic compounds in common with terrestrial plants. Some of these materials, such as C, H, O, N, Si, P, Mg, K, and Ca are needed in relatively large quantities and are incorporated into nonstructural components of the cell such as enzymes, etc. (Hewitt 1957). Unlike terrestrial plants, which may experience growth limitation because of the short supply of any of the essential nutrients, marine algae are normally exposed to an adequate supply of most of the essential nutrients. Much of the early literature concerning nutrient relationships in the sea deals with phosphate and nitrate not because the importance of the other nutrients was not appreciated, but because of the poorly developed analytical techniques. Thus, we find extensive literature dealing with nitrate and phosphate limitation of phytoplankton growth extending back to the 1930s. As more reliable analytical techniques became available, interest spread to include other chemical species. As a result, nowadays we find in programmes that study phytoplankton ecology the desire and capabilities to measure an entire suite of nutrient salts. Unfortunately, these measurements are standing stock estimates and while they have been used to infer certain conditions of the phytoplankton in the water column, there is obviously some degree of uncertainty in this approach. For example, the total quantity of phosphorus or nitrogen in the euphotic zone surely limits the absolute magnitude of the algal standing crop that could be realized, but it is not at all certain if the same could be said of the quantity of carbon that could be fixed. The latter is more closely related to the rate of supply rather than the instantaneous standing crop.

Furthermore, the rate of supply is also related to subsequent events further along in the food web. In the open ocean for example, nutrients are frequently present at extremely low levels, but a ^{14}C experiment will indicate that the plants are photosynthesizing. Observations over a number of days will suggest that the phytoplankton biomass and nutrient content of the euphotic zone is constant and it can be hypothesized that each daily increment of plant growth is grazed down. Additional nutrients for the following day's plant growth are derived primarily from regenerative processes within the euphotic zone since stratification minimizes the input of new nutrients from the nutrient-rich deeper waters. One can view the open 'blue' ocean as a system containing rather low-standing crops of nutrients, plants, and animals turning over at some moderate rate. It is almost a closed system. At the other end of the spectrum we find a rather different set of circumstances in the coastal regions of the world's oceans. At intermediate latitudes at least, we find markedly higher levels of nutrients, plants and animals. At the same time we also find larger fluctuations in these properties than in the central regions of the oceans. While we may generally describe the seasonal cycle in the 'classical' framework of spring bloom, low summer levels, autumn bloom and winter conditions, we notice many shorter-term variations. The major difference between these regions and the open ocean is the mechanisms which supply nutrients necessary for plant growth to the euphotic zone. While regenerative processes may be important, especially during periods of stability, the major processes supplying nutrients for plant growth are physical mixing and advection. The extreme of this situation is found in the upwelling regions along the eastern boundaries of the oceans. These regions encompass only a small portion of the oceans' surface, approximately 0·1 per cent, but maintain high annual productivity levels on an areal basis (Table 8.1). Ryther (1969) has suggested that the high potential fish yield in upwelling regions in comparison to oceanic regions is due to a combination of both the higher unit area productivity and shorter food chain. I would suggest in addition that this is perhaps a consequence of nutrient supply for phytoplankton growth. In the open ocean, total nutrients, including nutrients bound up in plants and animals in addition to the dissolved and particulate non-living fractions, are not very abundant. Much of this annual primary productivity

Table 8.1. Comparison of different marine regimes in respect to area, and primary production (after Ryther 1969).

	% of ocean surface	Total primary production (tons C . yr^{-1})
Open ocean	90	$16{\cdot}3 \times 10^9$
Coastal zone	9·9	$3{\cdot}6 \times 10^9$
Upwelling areas	0·1	$0{\cdot}1 \times 10^9$

utilizes nutrients produced as a result of and destined to participate in regenerative processes. Thus, a significant fraction of the annual production in oceanic regions is re-cycled near the primary stage in the food chain and is not passed along to the terminal predator. Perhaps the effective annual primary productivity, that which may find its way into a terminal predator, is an order of magnitude less than that suggested by Ryther (1969).

Fates of primary fixed carbon

The fate of organic carbon produced by the photosynthetic activity of algae is variable. Obviously, some fraction is incorporated into new cellular material. This fraction, which the ^{14}C technique most closely approximates, can be called growth in the broadest sense (Antia *et al.* 1963). Some of the newly incorporated organic carbon is destined to be lost from the cell either in the form of respiration or excreted as dissolved organic carbon. Neither of these entities are accounted for in routine ^{14}C productivity measurements. This could be considered the difference between gross and net productivity. The magnitude of this difference is uncertain but appears to be 10–40 per cent of total photosynthetic activity.

The fate of the particulate fraction, the algal cells themselves, is of interest primarily in the food web context. The factors affecting the distribution and loss of algal cells from the water column can be attributed to three basic processes: sinking, grazing activity of zooplankton and turbulence. These three factors can be viewed as being competitive depending on geographical and seasonal considerations (Sverdrup 1953, Riley 1946). If turbulence is excessive, as it is in temperate latitudes from late autumn to early spring, cells are vigorously moved throughout the surface layers. Plant growth is diminished or essentially non-existent since, on average, the plants do not receive sufficient radiation to offset respiration losses. Obviously, a 'bloom' can only occur if vertical turbulence is restricted to a layer approximating the euphotic layer. During summer in temperate latitudes, or at low latitudes throughout the year, outside areas of divergence, vertical turbulence is restricted. Floristic studies consistently demonstrate that certain species groups are found at depths in the water column which can be related to average light conditions. In addition, the persistence of other biological features such as chlorophyll maxima would suggest that turbulence as a dissipating force is quite variable in its effect. As pointed out in earlier sections, turbulence is effective in bringing nutrients conducive to plant growth to the surface, but on the other hand, the plants entrained in the water must remain near the surface in order to receive sufficient light for photosynthesis.

The two other factors suggested as being important in the dispersal of primary fixed carbon, sinking and grazing, can be viewed as being competitive or mutually exclusive in their significance. In the oceanic regions of the North Pacific the 'spring bloom' is observed not as an increase in phytoplankton biomass, but rather as an increase in photosynthetic activity and zooplankton biomass. In coastal regions the other extreme is found. In both the Long Island Sound and Gulf of Maine regions, the 'spring bloom' is observed as a large increase in phytoplankton standing crop which is not coupled to increases in zooplankton abundance and the blooms are terminated by an exhaustion of the nutrient supply. One spring bloom in the coastal waters of the North Sea was coupled to zooplankton production and was terminated by grazing at high nutrient levels (Cushing *et al.* 1963). In addition, sporadic blooms occur throughout the summer. These are associated with transient weather conditions and the breakdown of vertical stability which inject nutrients into the euphotic zone. In both cases, significant portions of the plant biomass finds its way down onto the sea floor (Heinrich 1962).

It is unclear if sinking of algal cells beneath the upper layers of the water column constitutes a loss from the plankton economy in the open ocean. Laboratory studies have consistently measured sinking rates on the order of 1–10 m d^{-1} for intact living cells (Smayda 1970). If algal cells do sink out of the euphotic zone they clearly penetrate to only a few hundred metres. Chlorophyll concentrations at 1·3–3 times euphotic zone depths are exceedingly low and the vertical distribution of particulate organic carbon would also suggest that surface-produced particulates do not penetrate to depths greater than a few hundred metres.

The significance of grazing as a controlling mechanism of plant growth has been alluded to from studies of the distribution of chlorophyll and chlorophyll-like pigments (Lorenzen 1967). These chlorophyll-like pigments, the 'phaeopigments', which are degraded forms of chlorophyll can be estimated by either spectrophotometric or fluorometric techniques if readings are taken before and after acidification. The phaeopigments are composed primarily of phaeophorbide α with traces of phaeophytin α. Both of these pigments are identical spectrophotometrically but are readily distinguished by chromatography. They are structurally identical to chlorophyll α except that the central magnesium atom is replaced by hydrogen in the case of phaeophytin. Phaeophorbide is further modified by having the phytol side chain removed. The phaeopigments are associated with particles in the water column but it is not known if they are bonded to protein as is the case of chlorophyll in the intact algal cell. Zooplankton faecal pellets are rich in phaeophorbide and it has been recently noted that this conversion of chlorophyll α to phaeophorbide α by passage through a copepod's gut is 100 per cent efficient on a molar basis. On a weight basis, 1 mg chlorophyll ingested appears as 0·66 mg phaeophorbide in faecal pellets (Shuman & Lorenzen, 1975).

Statistical relationships have been found between the abundance of small zooplankton, phaeopigments, and the level of primary production (Lorenzen 1967). The interpretation was that the data reflected a closely coupled system approximately in equilibrium with the phaeopigments being an index of grazing activity and *in situ* regeneration of nutrients.

Preliminary studies on the stability of phaeopigments contained in faecal pellets have shown them to be quite light labile. On a moderately clear sunny day in the summer in the temperate latitudes, with incident radiation of approximately 700 langley d^{-1}, one would expect to find significant loss of phaeopigments as a result of photo-oxidation in the upper half of the euphotic layer. At the surface the reduction should be on the order of 99 per cent while at the 10 per cent I_0 level, reduction would be approximately 30 per cent. Effects should be measurable further down in the euphotic zone also. Although the porphyrin ring, the central portion of the phaeophorbide molecule, is stable within geological time scales, phaeophorbide in the surface layers of the ocean is clearly much less durable. Phaeopigments measured in routine pigment samples probably represent an integrated value of grazing and photo-oxidation on a time-scale of a week or less.

Distribution of production

The level of primary production in all the world's oceans has been investigated since the advent of the ^{14}C technique (Steemann-Nielsen 1952). From this activity has emerged a broad-scale picture of regional distribution of

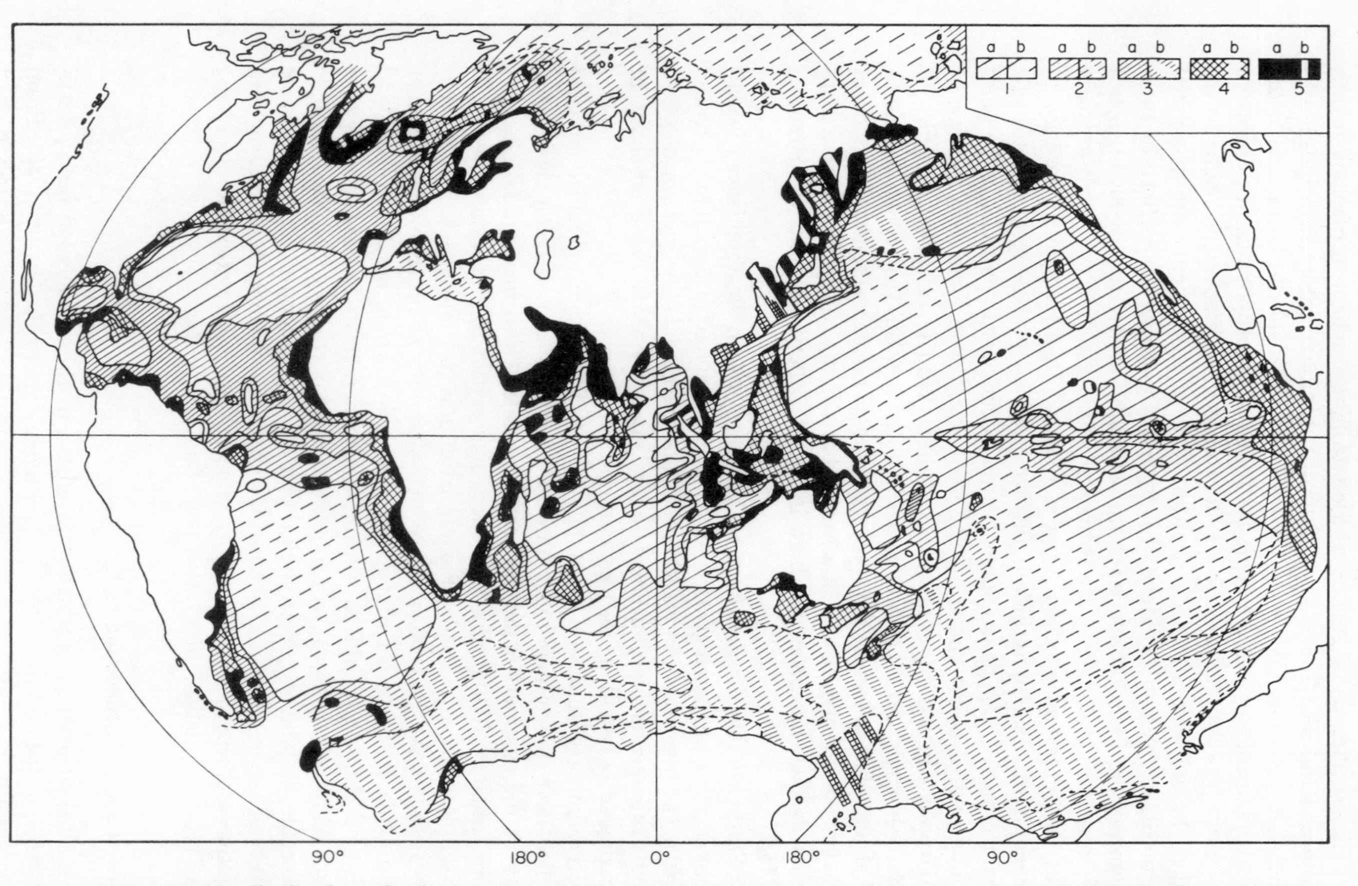

Fig. 8.2. Geographic distribution of primary productivity in the world's oceans in units of mg of C/m²/d. (1) Less than 100; (2) 100–500; (3) 150–250; (4) 250–500. a = data from direct ^{14}C measurements; b = data from phytoplankton biomass, hydrogen,

primary production (Fig. 8.2). Coastal areas exhibit higher levels of primary production than the central regions of the oceans. Certain coastal areas, the upwelling regions, consistently show the highest production rates. As pointed out earlier, the relative ranking is proportional to the magnitude of vertical mixing.

Distribution of productivity and the standing crop of phytoplankton throughout the water column in these different regions show systematic differences. Water columns exhibiting high levels of primary production

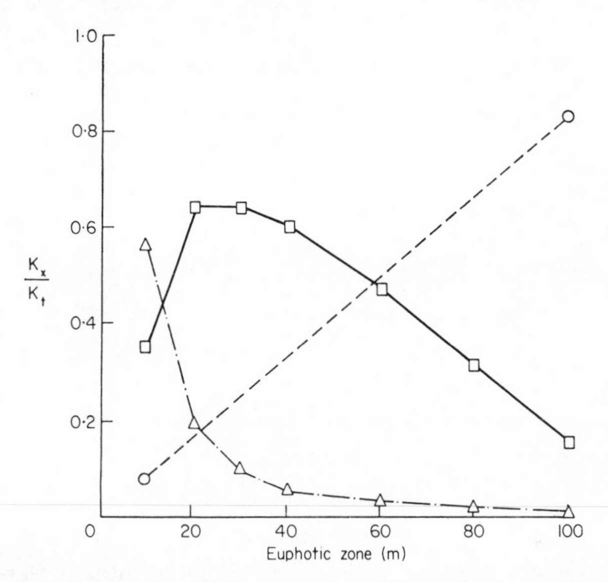

Fig. 8.3. Partitioning of total light attenuation within euphotic zone into fractions attributable to water (circles), plants (triangles), and 'other materials (squares) as a function of total euphotic zone depth'.

usually have shallower euphotic zones, 10–30 m, production rates of greater than 0·5 gC m^{-2} d^{-1}, and a tendency towards uniform vertical distribution of plant biomass. The other extreme, the central portions of the oceans, have euphotic zone depths of greater than 50 m, production rates of less than 0·5 gC m^{-2} d^{-1}, and a vertical distribution of plant biomass characterized by a chlorophyll maximum near the base of the euphotic zone. Productivity levels usually show a maximum not at the surface but rather at some intermediate depth (Talling 1966). In the more productive regions, this productivity maximum is found relatively higher up in the euphotic zone than in the less productive regions. On the average, the productivity maximum could be found at 2·5 m in a 10 m euphotic zone, and at 40–50 m in a euphotic zone of 100 m.

The above would suggest that there are some substantial differences between the conditions of the phytoplankton in the two extreme situations cited. Surely 'green water', the 10 m euphotic zone, is more productive than 'blue water', the 100 m euphotic zone. It is also implied that there is some fundamental difference in the condition of the phytoplankton inhabiting these two different regions since nutrients, primarily nitrogen, is frequently undetectable at oceanic stations. A close examination may suggest this not to be the case.

An analysis of data in hand, all collected by the author, was carried out to investigate the relationship between euphotic zone depth and the chlorophyll content and the primary production of the euphotic zone. A similar

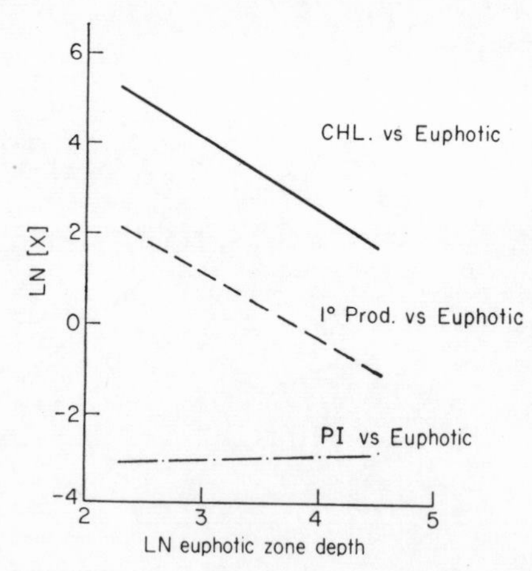

Fig. 8.4. Relationship between euphotic zone chlorophyll (mg . m^{-2}), euphotic zone primary production (mg C . m^{-2} . d^{-1}), and Productivity Index (mg C . mg chl . d^{-1}) as a function of total euphotic zone depth.

analysis has been previously published. This further analysis slightly modified the previous results (Lorenzen 1972, p. 264, Fig. 1) but does not change the conclusions. Thus, we find that the amount of radiant energy that could be absorbed by the phytoplankton in the euphotic zone for the two cases mentioned in the previous paragraph is rather different (Fig. 8.3). In the case of the 10 m euphotic zone, approximately 56 per cent of the energy in the light field within the euphotic zone could be absorbed by the algae. In a 100 m euphotic zone, only about 1 per cent of the energy could be absorbed. This is in direct proportion to the standing crop of chlorophyll. The situation for primary production is similar. The same group of data, but including the calculated productivity index, *PI* (mg C mg Chl^{-1} d^{-1}), are shown in Figure 8.4. An analysis of variance is shown in Table 8.2. The data clearly show the following features: 1 Chlorophyll content decreases logarithmically with increasing euphotic zone depth. 2 Total water column productivity decreases with increasing euphotic zone depth, and 3 The productivity index does not measurably change with increasing euphotic zone depth. The first two points are not surprising and have been demonstrated in the literature, but the third point has not. The latter point, if correct, has some serious implications. Increasing euphotic zone depths are associated with water columns with decreasing nutrient content. The nutrient analyses for this set of data are far

Table 8.2. Statistical relationship between water column chlorophyll content (mg . m^{-2}), primary production (mg . m^{-2}), productivity index (mg C . mg chl^{-1} . d^{-1}) and euphotic zone depth. Asterisk (*) denotes significance at the 99 per cent level, and ns denotes not significant.

	df	r
ln chlorophyll content = 8·85−1·57 ln euphotic zone depth	90	0·75*
ln primary production = 5·48−1·47 ln euphotic zone depth	86	0·74*
ln productivity index = 3·39+0·14 ln euphotic zone depth	86	0·06 ns

from complete, but they do not indicate anything unusual. This being the case one could infer that either the Productivity Index is insensitive to nutritional status of the phytoplankton, or that measurements of the nutrient status in the water column contains little useful information relevant to phytoplankton productivity. Both of the above could be modified by considerations of the chlorophyll content of the cells, but it is only fair to point out that the Productivity Index does estimate in a rough way the cells' ability to utilize absorbed energy in the complete photosynthetic process.

Chapter 9. Herbivore Production

D. J. Tranter

Introduction

Life cycles and development

Methods for estimating herbivore production
The population dynamics approach
The carbon-budget approach
Food intake (C)
Assimilation efficiency (A/C)
Growth efficiencies
The biomass turnover approach

Information on transfer rates and efficiencies
Birth rates
Death rates
Population growth rate
Food consumption
Gross growth efficiency
Assimilation efficiency
Respiration rate
Net growth efficiency
Biomass turnover rate (production: biomass)

Estimates of secondary production

Discussion
Methodology
Principles
Notation

Introduction

The herbivore is a living machine transforming plant material into animal tissue, the total mass of herbivore tissue formed within a given time being the herbivore production. If, on the first of two successive occasions, the total biomass of the herbivore stock is B_1 and, on the second, it is B_2, then the production (P) during the intervening interval is given by the equation

$$P = (B_2 - B_1) + M \tag{9.1}$$

where M is mortality (see notation p. 223).

The herbivore is also a living organism, a member of a species whose continuity from one generation to the next is by way of reproduction within specific populations. If N is the number of individuals in the population and $\bar{w}$ is the mean weight of an individual, then production is given by the equation

$$P = (N_2 \cdot \bar{w}_2 - N_1 \cdot \bar{w}_1) + M \tag{9.2}$$

Production can be considered either at the level of the individual organism or at the level of the population. It may be visualized as a process taking place within a pool of biomass, a reaction vessel into which flow newly-born individuals and the plant material that they eat, and out of which flow dying individuals and wastes of various sorts (Fig. 9.1).

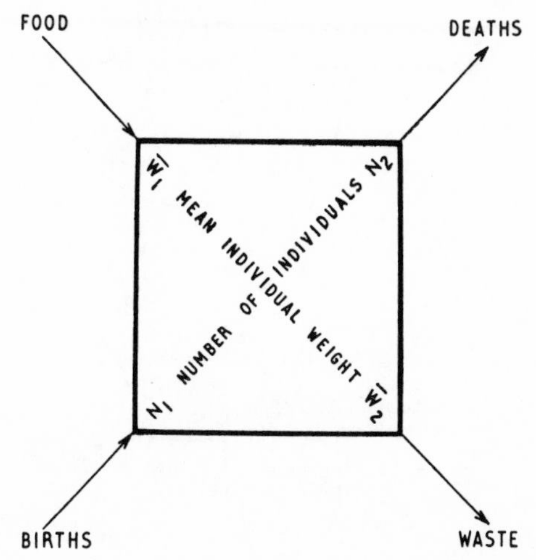

Fig. 9.1. Diagrammatic representation of secondary production.

When production and mortality are considered separately, it may be seen that

$$P \approx \frac{(N_1+N_2)}{2} \times (\bar{w}_2 - \bar{w}_1) \tag{9.3}$$

and

$$M \approx (N_1-N_2) \times \frac{(\bar{w}_1+\bar{w}_2)}{2} \tag{9.4}$$

or, more accurately,

$$P = \int_{\bar{w}_1}^{\bar{w}_2} N_t \, d\, \bar{w}_t \tag{9.5}$$

and,

$$M = \int_{N_1}^{N_2} \bar{w}_t \, d\, N_t \tag{9.6}$$

From the analogy of the stream comes the concept of turnover time, viz. the time taken for the stream to flush out the contents of the pool, in the present case the time taken for the biomass of the population to be replaced by fresh production at the rate P', i.e. B/P'. The reciprocal (P'/B) is the turnover rate.

In shallow seas, macrophytes (benthic macro-algae and sea-grasses) are often the dominant plants, and a large part of primary production enters a

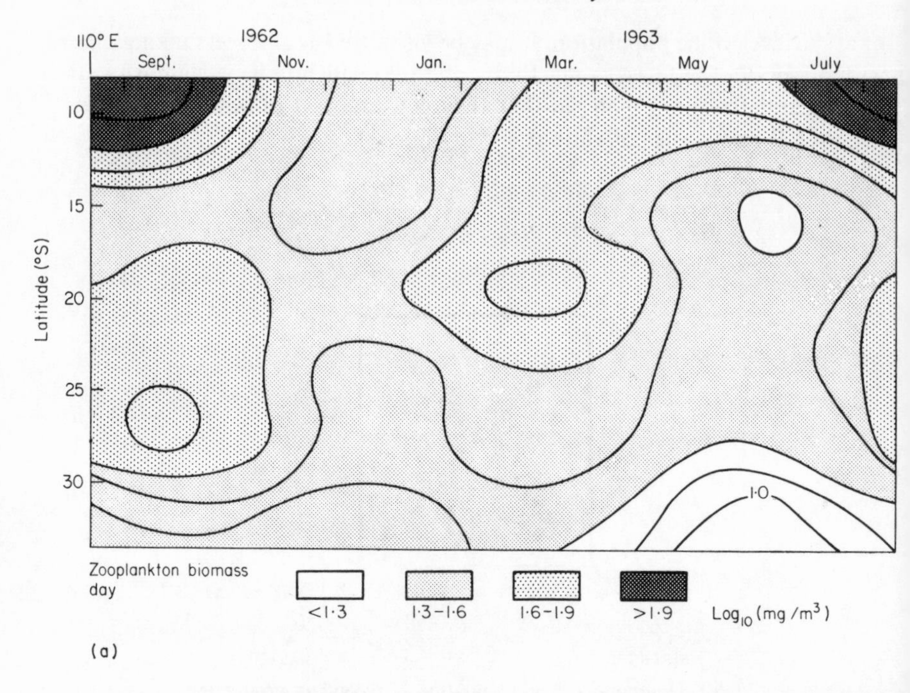

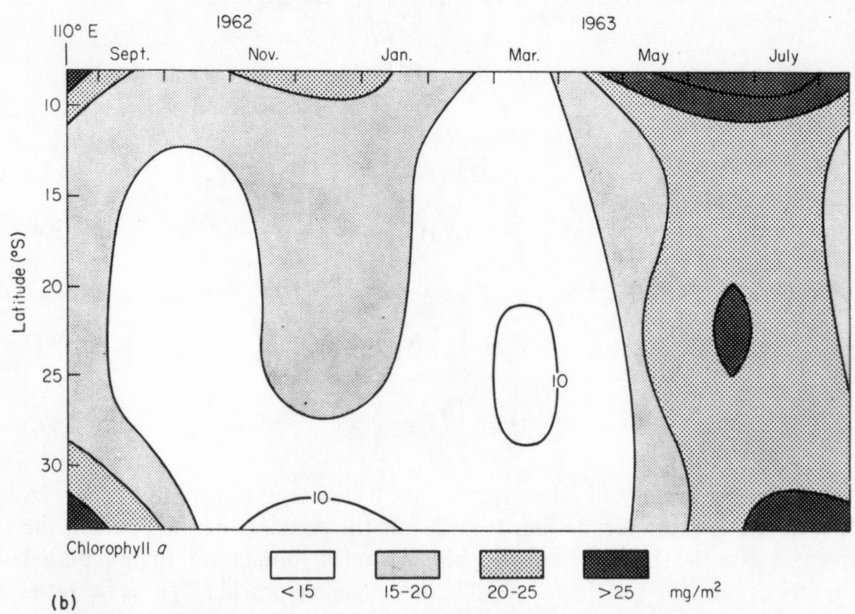

Fig. 9.2. Zooplankton and Phytoplankton stocks in the eastern Indian Ocean, and their seasonal variation. (a) Wet weight of zooplankton (b) Chlorophyll *a* concentration, an index of phytoplankton stock. (After Tranter 1973.)

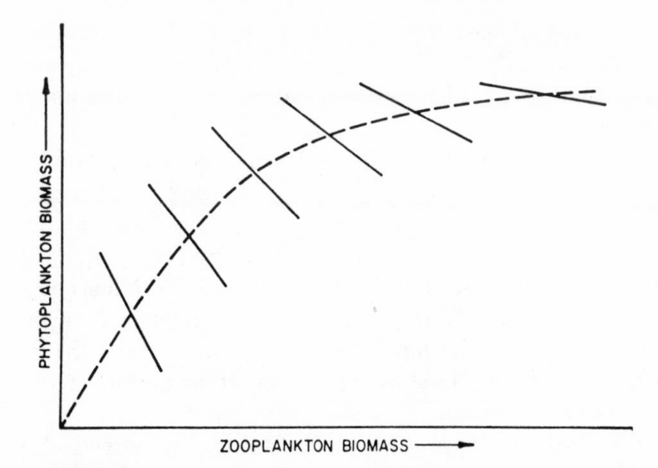

Fig. 9.3. Theoretical relationships between phytoplankton and zooplankton biomass within particular systems (solid lines) and between systems (dashed line). (After Brocksen 1970.)

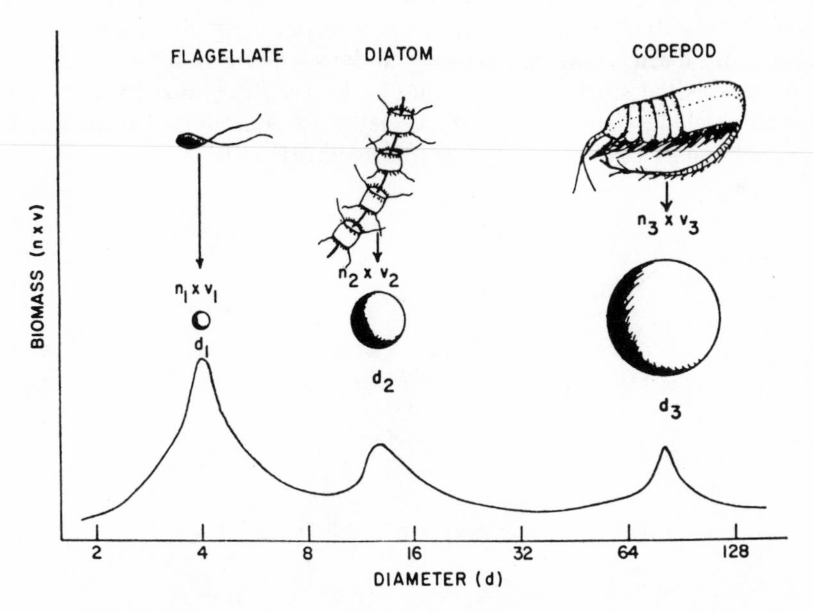

Fig. 9.4. The size of a typical zooplankton herbivore in relation to that of the plants on which it feeds as indicated by the diameter of a sphere of equivalent volume. The biomass curve (below) is the product of the numerical frequency of each class of organism (*n*) and its average size expressed in volume (*v*). (After Parsons and LeBrasseur 1970.)

detrital (decomposer) phase before being taken up in secondary production. Detrital production is dealt with in Chapter 10. Here we are concerned with the plankton food chain where grazing pressure is directed mainly on the living autotrophs.

In contrast with the regime on land, herbivores in the plankton seem to be more often limited by food than predators (Steele 1972). Where the phytoplankton stock fluctuates with the seasons so too does the zooplankton stock (Fig. 9.2), the time lag between the two generating a family of negative correlations (Brocksen 1970), shown in Figure 9.3 as a series of lines of decreasing slope. These represent a sequence of production regimes in which grazing pressure is continually outstripped by primary production, the broken line representing the path traced put by the zooplankton–phytoplankton relationship in reaching each new steady state.

The nature of herbivore production in plankton ecosystems is determined largely by the constraints involved in harvesting food of microscopic size (Fig. 9.4). The planktonic autotrophs are small unicellular algae, their high surface: volume ratio endowing them both with the buoyancy needed to keep them near the light and also the capacity to take up scarce nutrients from the surrounding medium. The harvesting of these micro-algae by planktonic animals is usually done by filtration. There are a number of structural solutions to the filter-feeding problem (Fig. 9.5), each usually associated with locomotory mechanisms; the capacity to draw water through the filter is as important as the structure of the filter itself. The filters that have been developed include mucus nets (salps), meshwork secretions (copelates—see Fig. 9.5b) and basketworks of setae (crustacea) (Fig. 9.6).

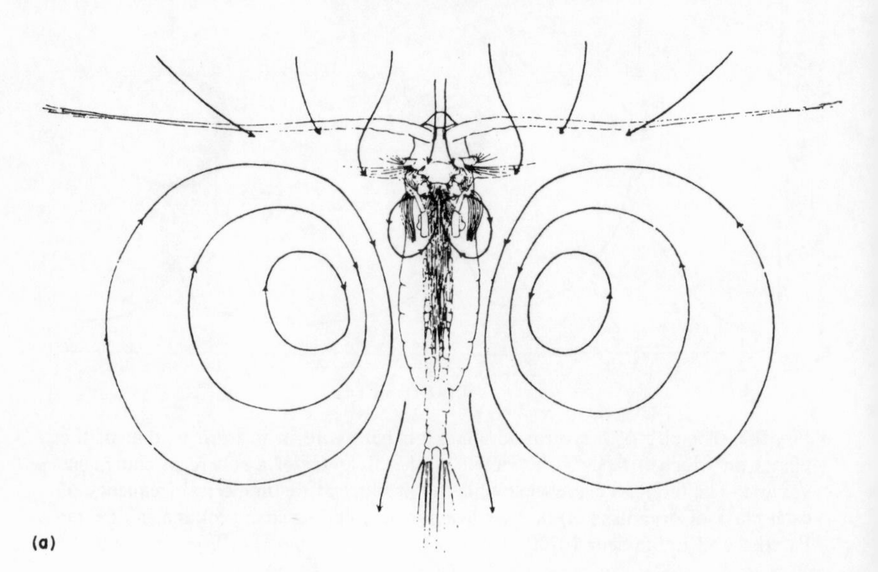

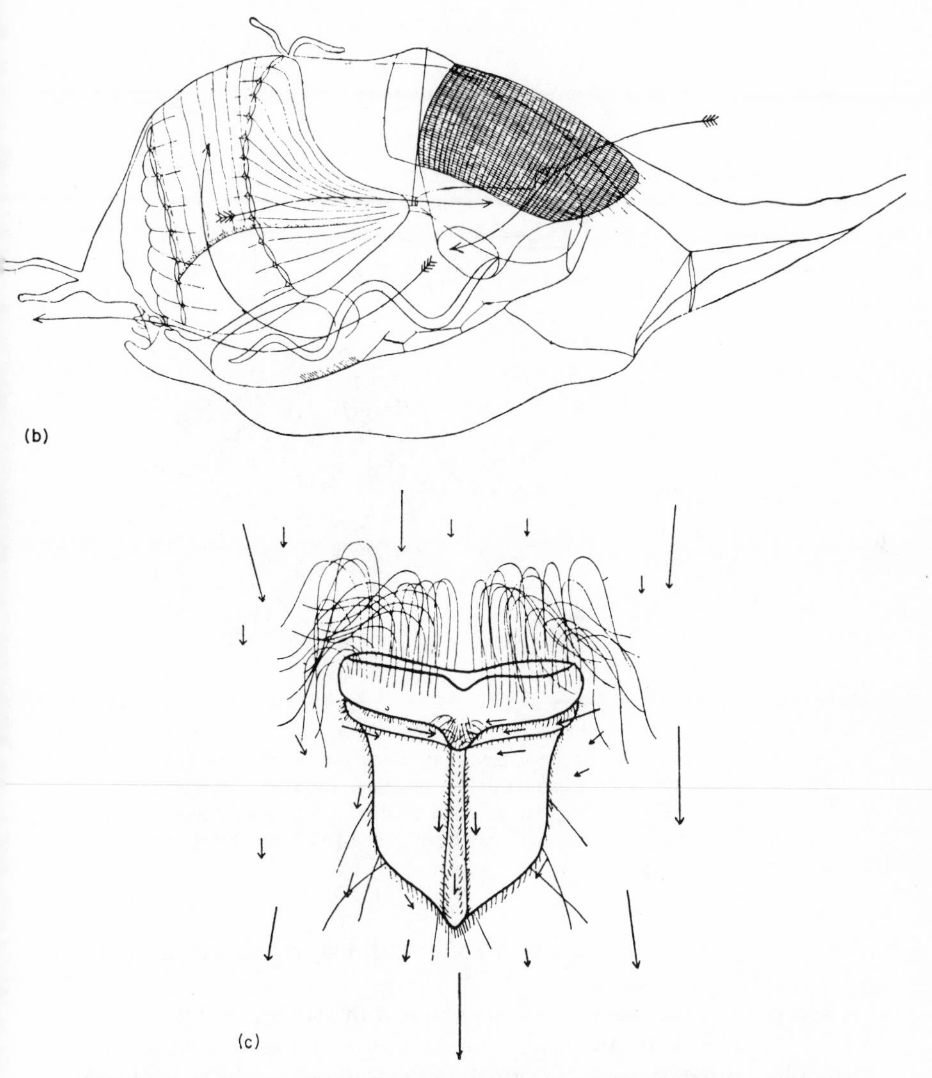

Fig. 9.5. Three types of zooplankton herbivores showing the water currents generated by the animal in locomotion and filter-feeding. (a) the copepod *Calanus finmarchicus*. (After Marshall & Orr 1955a.) (b) The copelate *Oikopleura albicans*. (After Lohmann 1933.) (c) The nudibranch veliger *Archidoris pseudoargus*. (After Jorgensen 1966.)

The fact that such highly specialized filters have been developed does not mean that grazing is an indiscriminate process. Herbivorous copepods, at least, actively select particular types of food (Conover 1966a, Marshall 1973) and can discriminate in favour of larger particles (Fig. 9.12). However, they

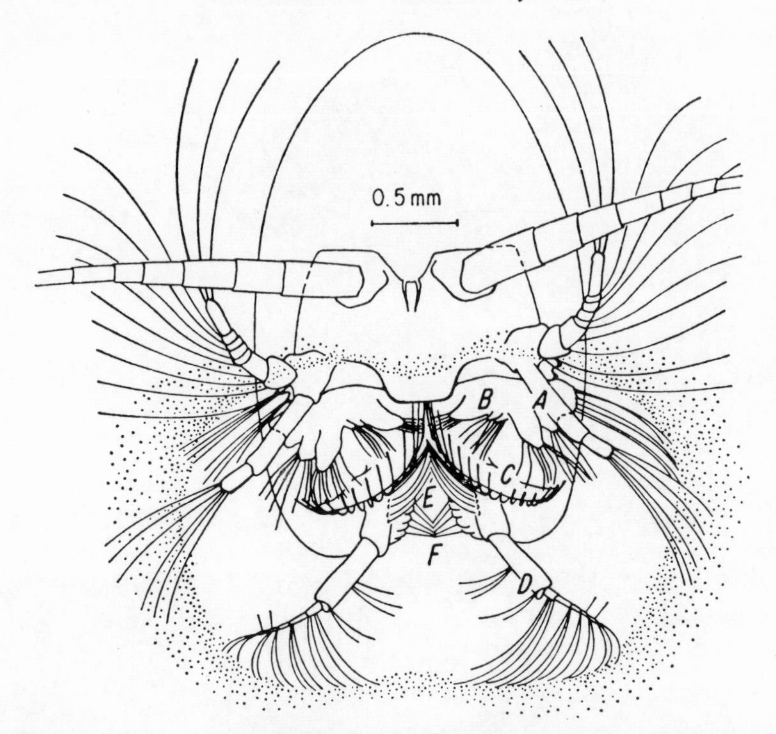

Fig. 9.6. Anterior and slightly ventral view of a *Calanus hyperboreus* female showing limbs in feeding position. Stippled margin surrounds the area within which contact with a large particle can result in a successful capture. *A*, second antenna; *B*, first maxilla; *C*, second maxilla; *D*, maxilliped; *E*, interdigitated setae originating from the second maxillae and maxillipeds; *F*, tips of swimming legs. (After Conover 1966a.)

do not do so all the time, and their preferences may change, even from plant to animal food.

Recent evidence suggests that plankton with soft permeable outer membranes (e.g. pteropods and bivalve larvae) can also take significant amounts of dissolved organic matter directly from the water (Sorokin & Wyshkwarzev 1973). Strictly speaking, 'herbivore production' excludes such inputs, but in practice it is difficult to make the necessary distinctions. In this account herbivore production is taken to mean the production of species which feed principally, but perhaps not exclusively, on particles of plant origin.

A special characteristic of plankton herbivores is the diversity of different species and developmental stages apparently sharing the same food resources as, for example, in low latitudes. In addition to the permanently planktonic forms ('holoplankton') there is a miscellany of planktotrophic larvae of marine bottom invertebrates ('meroplankton') particularly in shallow waters.

In such a diverse ecosystem, the production of a single cohort of herbivores, or even a single population, is of less concern than the total production of the whole herbivore assemblage. In high latitudes, however, and in some upwelling areas, production may be restricted to few species and even a small number of cohorts.

Life cycles and development

Larvae with large yolk reserves are termed 'lecithotrophic' and those poor in yolk, which start to graze at an early age, are termed 'planktotrophic' (Fig. 9.7). The larvae of plankton herbivores are generally planktotrophic (Thorson

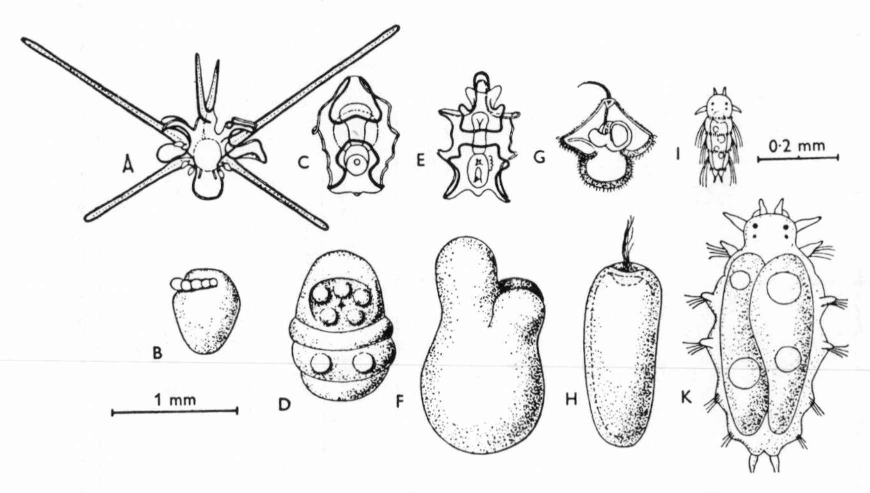

Fig. 9.7. Planktotrophic larval types (above) and lecithotrophic larval types (below) of each of the groups Echinoidea, Holothuroidea, Asteroidea, Nemertini, and Polychaeta. (After Thorson 1949.)

1949) the need for wide dispersal apparently being more important to the species than the need for a high frequency of survivorship to sexual maturity. The marine herbivore begins to grow of its own accord when the yolk reserves which it has inherited from its mother are exhausted and the larva starts to graze. During the planktotrophic phase, the young are exposed to continual predation, and mortality is high. Reproductive success depends upon high fecundity synchronized with phytoplankton blooms (Thorson 1949).

Secondary plankton production is dominated by the crustacea. These have a common pattern of larval development, despite their wide diversity. There is usually metamorphosis from one phase to another and, within each phase, progression from stage to stage by a series of moults. The basic phase-

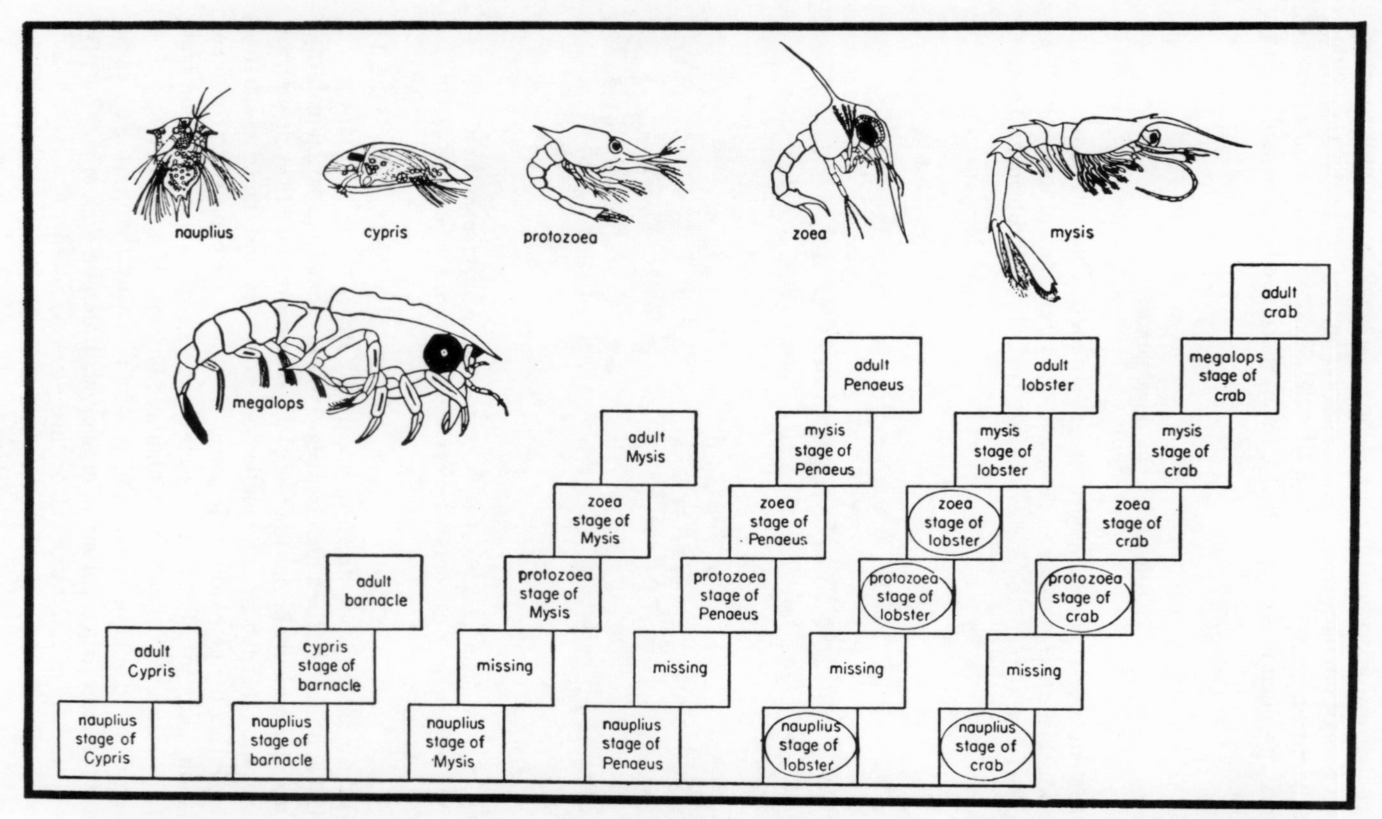

Fig. 9.8. General developmental history of marine crustacea, together with some variations from the basic pattern. (After MacGinitie & MacGinitie, 1949.)

types in the crustacea are the nauplius, zoea, and megalopa (Williamson 1969) distinguished one from the other by a changing pattern of locomotion—first by antennae, then thoracic appendages, then abdominal appendages. From this basic pattern, there are many variations (Fig. 9.8): in the barnacles and some lower crustacean groups there is a post-naupliar 'cypris' stage; the 'protozoea' is an intermediate (pre-zoeal) stage characteristic of prawns, euphausiids ('calyptopis'), and copepods ('copepodite'); the 'mysis' is an intermediate (post-zoeal) stage characteristic of prawns and lobsters; and so on.

Since the production of a herbivore species depends, amongst other factors, on brood survival, the duration of the planktotrophic period, when larvae are exposed to continual predation, is a crucial issue. Within a given species, the rate of development is generally accelerated by higher temperatures, and the duration of development is shortened (Thorson 1949, McLaren 1963, Winberg 1971); in other words, duration is an inverse power function of temperature. This relationship is shown for the specific case of the copepod

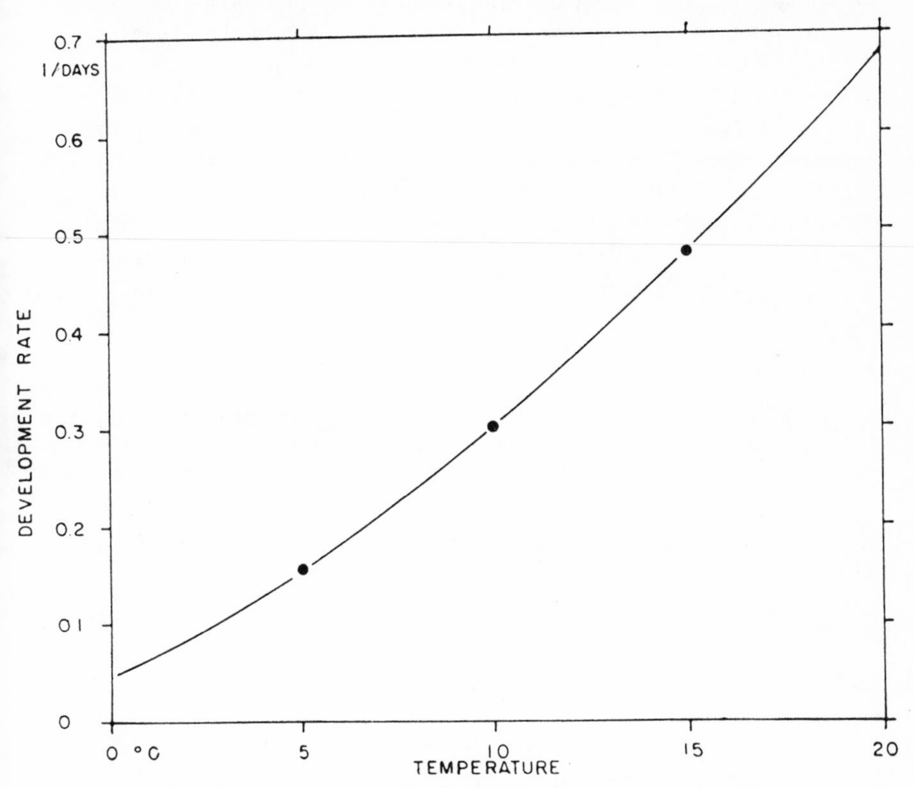

Fig. 9.9. Development rate of eggs of *Pseudocalanus minutus* from Loch Striven. (After Ray 1966.)

Pseudocalanus minutus in Figure 9.9. Taken on a global scale, however, there is no clear relationship between larval duration and temperature—the larvae of tropical species take just as long to develop as those of their temperate relatives (Thorson 1949).

Methods for estimating herbivore production

Plankton was one of the first biocoenoses to be given a special name (Hensen 1887) and interest at the level of the biocoenosis is inherent in the present question of how to measure plankton production. Unfortunately, suitable methods have yet to be developed and we are forced to make indirect estimates at the species level. The possibilities of a direct approach are dealt with in the discussion. The whole study of secondary plankton production is in its infancy. Heinle (1966) drew attention to the fact that only one of the 42 papers presented at the 1961 symposium on zooplankton production (Yablonskaya 1962) came up with actual quantitative estimates of production. In more recent years there have been several significant studies, at the species level.

There are several recent reviews on the subject (Mullin 1969, Mann 1969, Allen 1971, Edmondson & Winberg 1971, Winberg 1971, Marshall 1973, Parsons & Takahashi 1973) from which this summary is mainly drawn. There are three main approaches: the population dynamics approach, focused on the population, with births as input and deaths as output; the carbon-budget approach, focused on the organism, with 'food' as input and 'wastes' as output; and the turnover approach, focused on the biomass of the stock rather than the population or the individual organism as such.

The population dynamics approach

The population dynamics approach provides a means of determining the mortality component M in the production equation

$$P = (B_2 - B_1) + M$$

where B_1 is the biomass of the population at the beginning, and B_2 is the biomass of the population at the end of a short interval t_2–t_1.

The numerical component of the production of a population may be represented by the equation

$$r = b - d \tag{9.7}$$

where r is the instantaneous rate of numerical increase;
b is the instantaneous birth rate;
d is the instantaneous death rate.

Given r and b, d may be derived, thence $d \cdot \bar{w}$ which is the mortality flux M.

Measurements of the number of individuals present at t_1 and t_2 allow r to be derived as shown below

$$r = \frac{ln\ N_2 - ln\ N_1}{t_2 - t_1} \tag{9.8}$$

It remains, now, only to determine b.

Andrewartha and Birch (1954) and more recently Caswell (1972) have shown that, in populations with a stable age distribution, the instantaneous birth rate, b, may be derived from the finite birth rate β, by the equation

$$b = \frac{r\beta}{(e^r - 1)} \tag{9.9}$$

In those herbivores in which the eggs are so clearly visible that they may readily be counted (e.g. rotifers), the finite birth rate β may be determined from the number of eggs per female (E) and the duration of the embryonic state (D) (Edmondson 1960), the equation being

$$\beta = E/D \tag{9.10}$$

Where births are periodic, cohorts of individuals of the same age are generated, and the estimation of production is somewhat easier. Taking the life of the cohort, from birth to death,

$$B_1 = B_2 = 0 \tag{9.11}$$

Consequently,

$$P = M \tag{9.12}$$

that is, cohort production is equivalent to the mortality flux from the stock. Expressed in population dynamics terms, there are no more births once the cohort is established, only deaths, and equation 9.7 reduces to the simple form

$$r = -d \tag{9.13}$$

Given a satisfactory measure of the death rate d, the accuracy of the population dynamics approach to measuring mortality depends upon how closely the mean weight $\bar{w}$ of the study population represents the weight of an individual at death ($M = d \cdot \bar{w}$). As we have already seen, plankton life cycles are characterized by a succession of developmental stages each with a characteristic mean weight and death rate. For maximum accuracy in measuring the overall production, separate estimates need to be made for each developmental stage.

One of the first to follow this approach was Buchanan-Wollaston (1923) in estimating the production of plaice eggs in the North Sea. He monitored the abundance of eggs through some 25 developmental stages and derived a more or less continuous mortality function. His work, in turn, was based on that of Hensen and Apstein (1897).

Heinle (1966) adopted a similar practice to estimate the death rates of nauplii (d_n) and copepodites (d_c) of the copepod *Acartia tonsa* in the Patuxent River estuary, using the equations:

$$-d_n = (ln\ C_{t_n} - ln\ N)/t_n \tag{9.14}$$

$$-d_c = (ln\ A_{t_c} - ln\ C)/t_c \tag{9.15}$$

where t_n and t_c are, respectively, the times required to grow from pivotal (median) age nauplius to pivotal age copepodid, and from pivotal age copepodid to pivotal age adult; N is the nauplius density and C_{t_n} the copepodid density at the beginning and end, respectively, of the nauplius-copepodid interval t_n; and C is the copepodid density and A_{t_c} the adult density at the beginning and end, respectively, of the copepodid-adult interval t_c. [Strictly speaking, Heinle's values are not death rates as he stated (K.R. Allen. personal communication) the decline in numbers from the pivotal point of one stage to the pivotal point of the next being accounted for by deaths in each stage, not only in the first.]

The biomass of the population being at a generally steady state during the period of his observations, Heinle assumed, apparently, that mortality was balanced by production. The accuracy of such methods depends, among other things, on the estimates of developmental rate which, we have already seen, are governed by the prevailing temperature. This means that development needs to be observed under the same conditions as the field observations.

It is possible to obtain death rates from the one sample taken at a single point in time (Mullin & Brooks 1970), by the relative frequencies of consecutive developmental stages. However, this approach assumes a constant input flux at birth and a constant output flux (mortality) after birth, assumptions which severely limit the generality and accuracy of the method (Fager 1973). Harding and Talbot (1973) have extended the method of Buchanan–Wollaston. The numbers of each developmental stage are sampled successively in time and each is divided by the stage duration at the appropriate temperature to estimate production and the result is a distribution of production of that stage in time. The ratio of such integrated production for one stage to that of its predecessor is a good estimate of mortality.

The carbon-budget approach

The carbon-budget approach to herbivore production is focused on the individual herbivore rather than the herbivore population. It is based on the simple stream analogy model of production (Fig. 9.1) with food as input and wastes as output. The equation developed by Ricker (1968) and applied to plankton organisms by Mann (1969) is

$$P = C-(F+R+U) \tag{9.16}$$

where P is Production;
C is Food Consumed;
F is Food Waste (Faeces);
R is Metabolic Waste (Respiration);
U is Excretory Waste (Urine).

It may be argued that food which passes through the gut undigested is a direct flux from algae to detritus and that 'assimilation' is a better basis for production estimates than ingestion. In this case,

$$P = A-R \tag{9.17}$$

where A is the food Assimilated $(= C-F-U)$
and A/C is the Assimilation Efficiency.

It is, of course, impracticable to measure all components of the carbon-budget equation in routine estimates of production. The goal is to determine coefficients for growth efficiency which can be applied either to ingestion ('gross growth efficiency') or to assimilation ('net growth efficiency'), to estimate production, as follows:

$$P = K_1\, C \tag{9.18}$$

and

$$P = K_2\, A \tag{9.19}$$

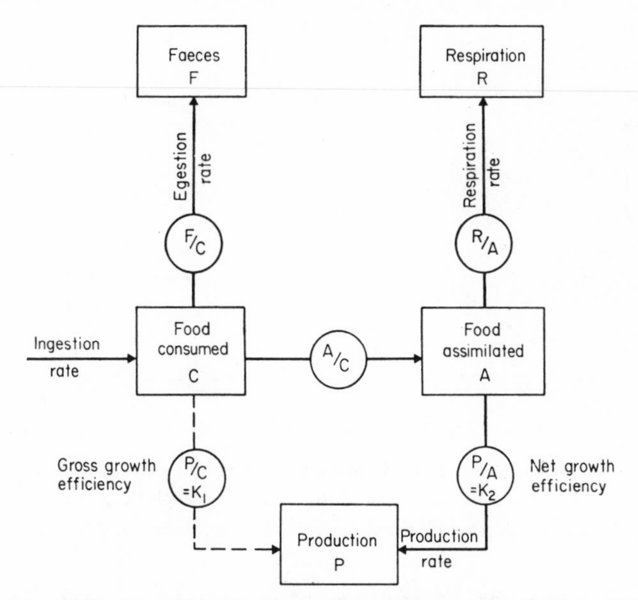

Fig. 9.10. Compartmentalized budget for the production of a plankton herbivore. The arrows are the fluxes; the boxes are the 'stocks'; and the circles are the efficiencies. (Modified from Edmondson & Winberg 1971.)

where K_1 is Gross Growth Efficiency
and K_2 is Net Growth Efficiency

In practice, estimates of plankton production using the carbon-budget approach are based mainly on Ingestion (C), assimilation efficiency (A/C), and growth efficiency (K_1 or K_2) (Fig. 9.10).

FOOD INTAKE (C)

Food intake is a function of the filtration rate of the grazing organism and the food concentration in the water from whence its food is drawn (Marshall & Orr 1955b). Given the biomass of the grazer, it would be simple to determine the food intake if the grazing rate per unit biomass were relatively constant. Even so, there are technical problems in obtaining accurate measures of grazing rate, due to sedimentation, concentration, and reproduction of the algal cells in the experimental containers, and a curious behavioural reaction of plankton herbivores to container volume (Anraku 1964). However, there are limits, determined mainly by the quality and quantity of food available, beyond which the grazing rate is no longer constant. There is a lower threshold, below which plankton herbivores do not feed (Fig. 9.11), and an upper

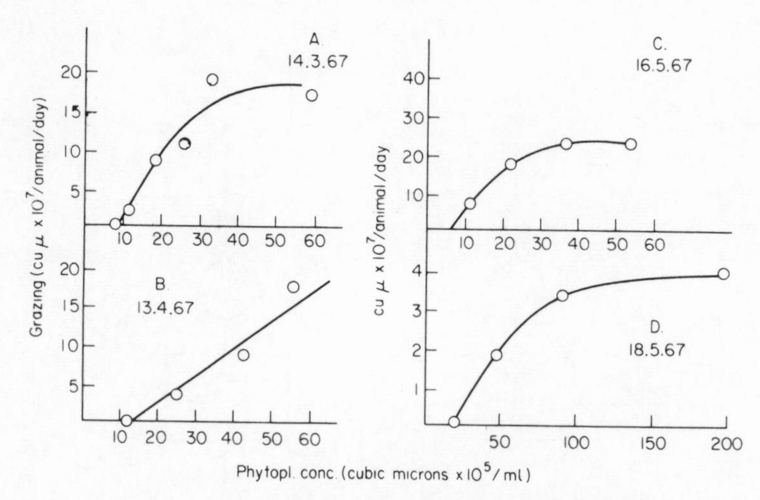

Fig. 9.11. Grazing rate of *Calanus* at varying algal concentrations showing the threshold at which the herbivore starts to feed and the plateau of food saturation. (After Parsons *et al.* 1969.)

limit, beyond which their capacity to filter is no longer matched by their ability to ingest and digest the food that their filtration activities have made available (Parsons *et al.* 1969, Suschenya 1970). Parsons and Takahashi

(1973) have developed the Ivlev equation for expressing the food consumption of a plankton organism in terms of the food available. The equation is (using my notation):

$$C = C_{max}\ (1 - e^{k(p_o - p)}) \tag{9.20}$$

where C_{max} is the upper limit to the rate of food consumption, p_o is the threshold food concentration below which the animal does not feed, and k is a proportionality constant.

There is also evidence that plankton herbivores can adapt their grazing rate to the type of food available. For example, copepods fed with algae of different sizes tend to select the larger (Gauld 1964, Marshall 1973), perhaps by using their second maxillae and maxillipeds as an active 'scoop-net' rather than as a passive filter—for example as shown in Figure 9.6. The consequences of such selectivity are shown in Figure 9.12.*

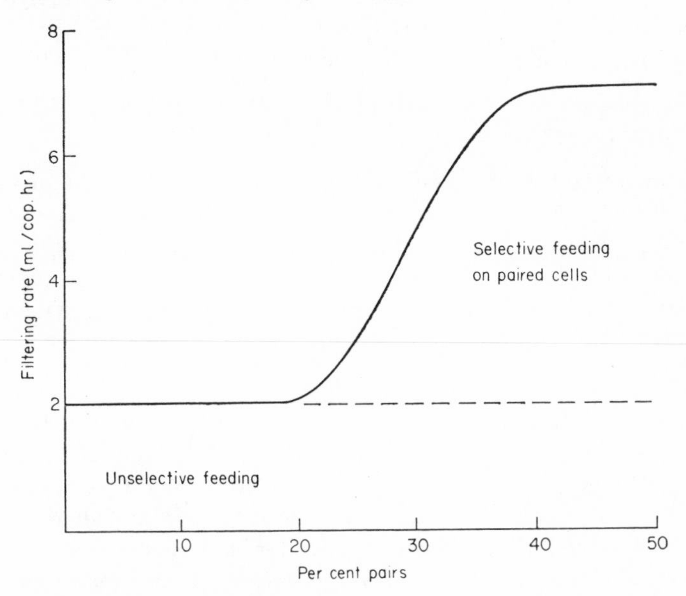

Fig. 9.12. Filtering rate of *Calanus* on algal suspensions containing varying proportions of paired (dividing) cells which are selected preferentially as food. (From Steele 1972, after Richman & Rogers 1972.)

According to Suschenya (1970), the food intake of crustacea is a function of their body weight, the relationship being of the form

$$C = k\ W^{\phi} \tag{9.21}$$

* Recent work by Wilson (1973) and Poulet (1973) shows that certain filter-feeding copepods can discriminate between particles of slightly different size, perhaps by prior 'scanning' the available size spectrum, and can select those whose total biomass is greatest per unit volume of sea water filtered, thus favouring maximum food intake for minimum effort.

where k defines general level of food intake in response to factors such as food concentration; ϕ defines rate of change of food consumption with body weight.

ASSIMILATION EFFICIENCY (A/C)

Assimilation efficiency may be determined from the ratio of total weight to ash-free dry weight in food and faeces (Conover 1966b). It is assumed that only the organic fraction of the food is assimilated. Alternatively, labelled algae may be used, but special care must be taken to measure the material excreted which may be a relatively important factor in short-term studies, particularly if radioactive phosphorus is used (Marshall 1973).

It has frequently been proposed that 'superfluous feeding' takes place among plankton herbivores when there is abundant food, as in phytoplankton blooms; this has been contested by Conover (1966c). The history of this controversial concept has been reviewed by Corner and Davies (1971).

GROWTH EFFICIENCIES

Once assimilation has been determined, the most important factor yet to be accounted for to estimate production is the respiration—that is, the cost of running the herbivore machine. There is a general relationship between respiration and body weight which may be used in production estimates (Suschenya 1970). It is of the same form as that relating food consumption to body weight, viz.

$$R = \alpha W^{\gamma} \tag{9.22}$$

The factor α describes the general metabolic level, and the factor γ the general type of metabolism. The latter is related to ϕ, the factor describing the rate of change of food consumption with body weight (eqn. 9.21).

The respiration rate may vary with the season (Marshall & Orr 1958, Conover 1959, 1962, Conover and Corner 1968), particularly in response to seasonal availability of food. *Calanus hyperboreus* stores food as fat reserves in times of plenty and uses this reserve in times of need, starved animals respiring less than those that are well fed. Herbivores have a lower respiration rate than carnivores (Conover 1960).

There is evidence that the specific respiration rate (respiration rate per unit weight or 'metabolic rate') is higher in small plankton organisms than in large, and higher in tropical than in boreal waters (Parsons & Takahashi 1973, Ikeda 1973).

The biomass turnover approach

The 'stream' analogy of production (Fig. 9.1) suggests that it would be useful to know the time that it would take for the stream to flush out the contents of the pool ('dilution time'), i.e. for the herbivore biomass to be replaced by fresh production ('turnover time'). This interval is given by the ratio of biomass (of the stock) to production rate (biomass increment per day). Its reciprocal (production rate/biomass) is the 'turnover rate'.

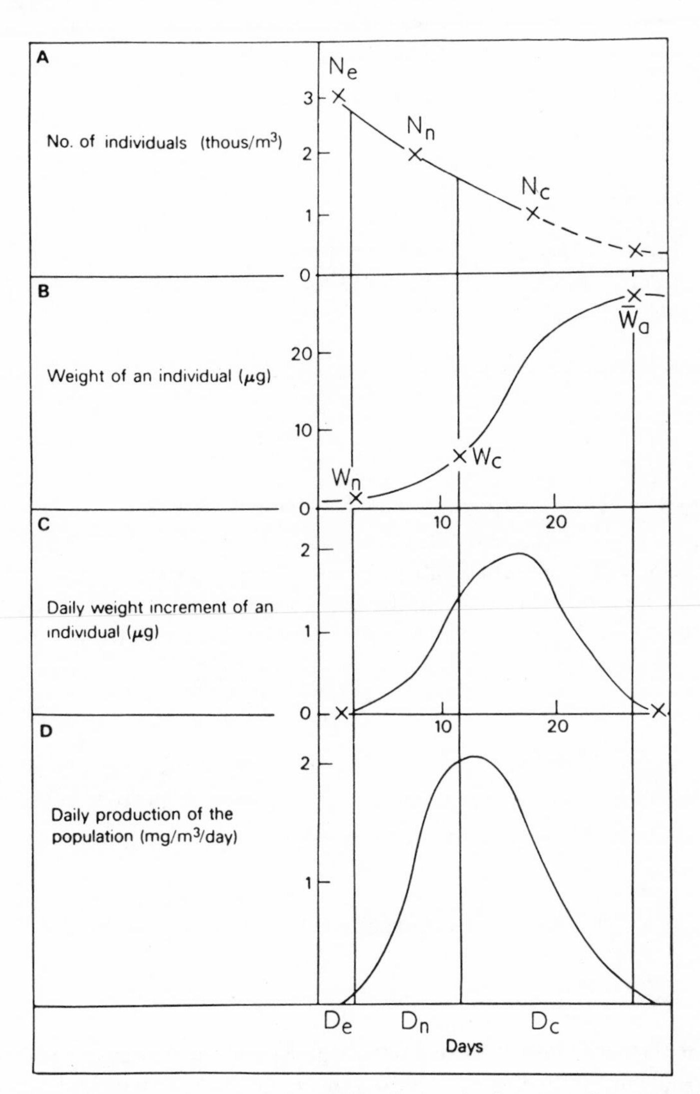

Fig. 9.13. Graphical method of estimating the production rate of a steady-state copepod population N = no. of individuals; e = egg; n = nauplius; c = copepodite; a = adult; D = development time; W = initial weight; w = mean weight. (From Edmondson & Winberg 1971.)

If the turnover rate were known, then it would be possible to calculate production directly from the biomass of the stock. This proposition was first introduced by Juday (1940) and Lindeman (1941), but their results were in error because they did not distinguish between life span and turnover (K.R. Allen personal communication). Mean age (Eqn. 9.23) is a better approximation to the biomass: production ratio than is the mean life span (Eqn. 9.24) (Allen 1971).

$$\text{Mean age} = \frac{1}{\int N_t\, dt}\int t\, N_t\, dt \qquad (9.23)$$

$$\text{Mean life span} = \frac{1}{N_0}\int t\, dN \qquad (9.24)$$

The turnover method estimates the time taken for a quantity of biomass to be produced equivalent to the standing stock. A more accurate approach is to calculate the growth increment over shorter periods of time, using the equation

$$P' = \frac{N\Delta\bar{w}}{t} \qquad (9.25)$$

where N is the number of individuals in the population, and $\Delta\bar{w}$ is their mean individual growth over time interval T (Edmondson & Winberg 1971). Since herbivores grow at different rates throughout their life, and since steady state populations are made up of all ages, growth increments need to be determined for a number of points in the life history of the species. For example, the production of a copepod population may be calculated from the formula:

$$P' = \frac{N_e\Delta\bar{w}_e}{D_e} + \frac{N_n\Delta\bar{w}_n}{D_n} + \frac{N_c\Delta\bar{w}_c}{D_c} \qquad (9.26)$$

where e is eggs, n is nauplii, and c is copepodites, and D_e, D_n, and D_c are the duration of the egg, nauplius, and copepodite stage respectively. The graphical representation of this calculation is shown in Figure 9.13.

Information on transfer rates and efficiencies

Birth rates

Most of the estimates of birth rates in plankton herbivores are of finite (β) rather than instantaneous (b) rates. For example, Marshall and Orr (1952) determined the finite birth rate of the copepod *Calanus finmarchicus* under laboratory conditions by keeping individuals under continuous observation, making daily counts of eggs laid, and transferring the females to fresh containers. Starved of food, the copepods laid less than 25 eggs per day, but with abundant food the egg-laying rate increased by an order of magnitude (Fig. 9.14).

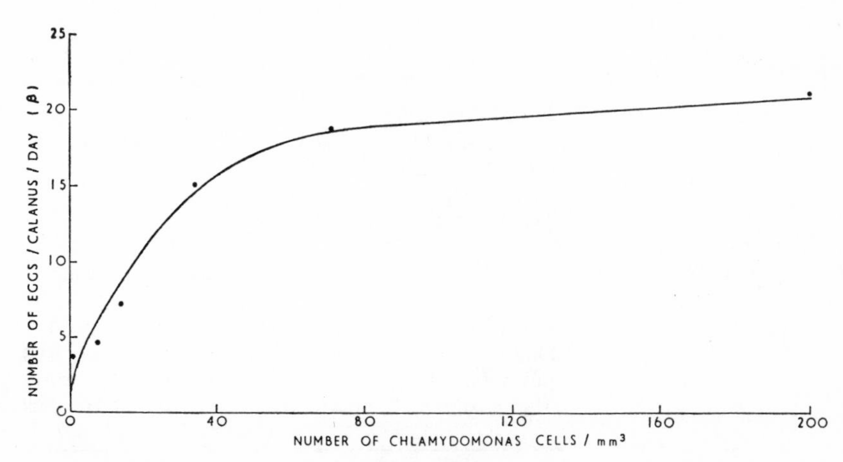

Fig. 9.14. Finite birth rate (β) of *Calanus finmarchicus* fed on Chlamydomonas at a range of concentrations. (After Marshall & Orr 1952.)

There is also evidence that fecundity is influenced by temperature, countered to some extent by an opposite trend in survival.

Death rates

Few estimates have ever been made of death rates in plankton herbivores some notable exceptions being the work of Heinle (1966) on *Acartia tonsa* in the Patuxent River estuary and Mullin and Brooks (1970) on *Calanus helgolandicus* off the Californian coast. Heinle's estimates of death rate at the copepodid stage (Table 9.1) are of the order of 0·6–0·7, mortality in the nauplius stages being somewhat higher (up to 0·9) and the turnover time somewhat shorter (1·8 days). In *C. helgolandicus*, mortality is much heavier at the nauplius stage (up to 0·65) than at the copepodid stage, and heavier at late copepodid than at early copepodid stages.

Population growth rate

One of the few estimates that have been made of instantaneous rate of population increase in a plankton herbivore is that of Heron (1972) for the salp *Thalia democratica* (Forskål). The high innate capacity for increase of this species (r = 0·4–0·91 per day) allows populations to double in numbers in a

Table 9.1. Estimates of death rate (*d*)* made by Heinle (1966) for *Acartia tonsa* and by Mullin and Brooks (1970) for *Calanus helgolandicus*.

Author	Date of observations	Death rate (*d*) Nauplii		Copepodids			
Heinle 1966 (*Acartia tonsa*)	July 16–July 22	0·80		0·47			
	July 22–July 29	0·74		0·71			
	July 29–Aug. 6	0·89		0·74			
	Aug. 6–Aug. 13	0·80		0·53			
	Aug. 13–Aug. 20	0·89		0·64			
	Aug. 20–Sep. 9	0·77		0·47			
	Sep. 9–Sep. 17	0·83		0·44			
Mullin & Brooks 1970 (*Calanus helgolandicus*)				Early		Late	
		Stn. 2	3	2	3	2	3
	Apr. 18–May 19	0·33	0·20	0	0	0·02	0
	May 20–June 30	0·25	0·35	0	0	0	0·03
	July 1–July 31	0·45	0·65	0	0	0	0·12
	Aug. 1–Aug. 18	0·45	0·65	0	0	0·04	0·08

* Based on abundances of successive developmental stages in serial plankton samples.

day, generating large 'swarms' immediately after phytoplankton blooms, so permitting this transient resource to be exploited to the maximum (Fig. 9.15).

Food consumption

Measurements of the filtration rate of plankton herbivores range from mean values of the order of 100 ml per day for the copepod *Calanus finmarchicus* (Marshall 1973) to 600 ml per day for the euphausiid *Euphausia pacifica* (Table 9.2). These rates are equivalent to approximately 40 ml/mg wet wt/d (Parsons and Takahashi 1973). There is evidence that this rate of filtration declines with increasing algal concentration (Fig. 9.16) and age of algal culture.

The estimates of Riley (1947) for zooplankton grazing rates on Georges Bank per unit biomass of grazers ranged from 5 per cent per day in winter to 30 per cent in early summer. On the other hand, Petipa *et al.* (1970) have estimated the daily rations of epiplanktonic crustacea in the Black Sea to be 129 per cent for nauplii and 194 per cent for early copepodites, while the estimates of Cushing (1962, 1964) for *Calanus* range as high as 390 per cent. The highest values recorded by Parsons *et al.* (1969) for *Calanus plumchrus* were 60 per cent (Table 9.3), and Marshall (1973) considers Petipa's values to be overestimates. Cushing's values were calculated from a plot of algal mortality on weight of herbivores.

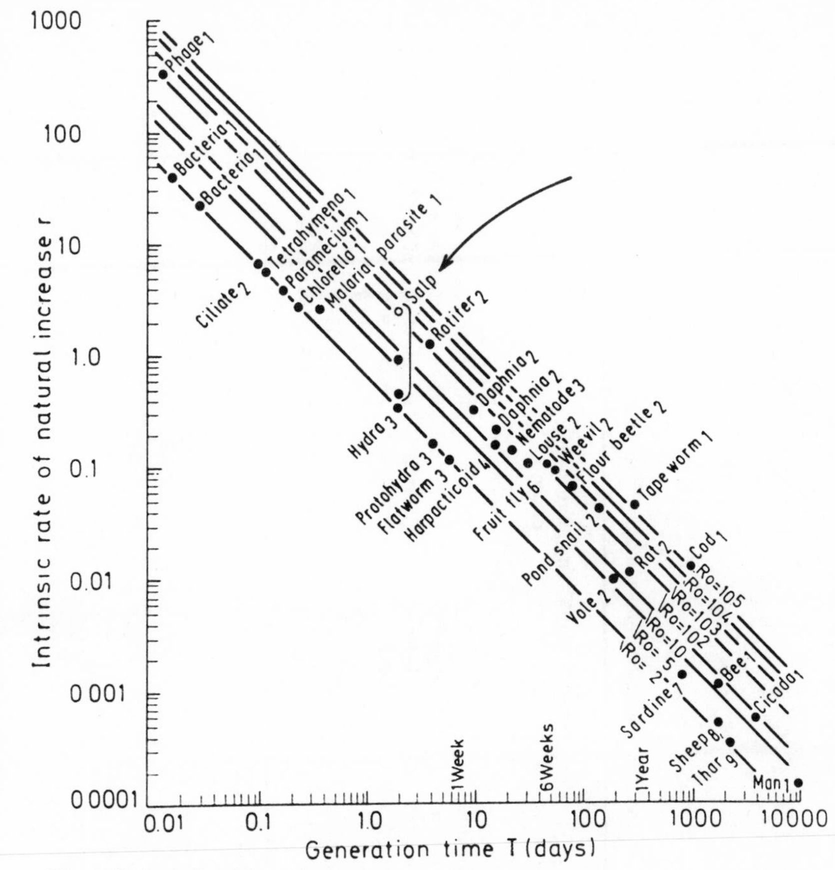

Fig. 9.15. Relationship of intrinsic rate of population increase r (per day) and generation time T (in days) for a variety of animals including the salp *Thalia democratica*. (After Heron 1972.)

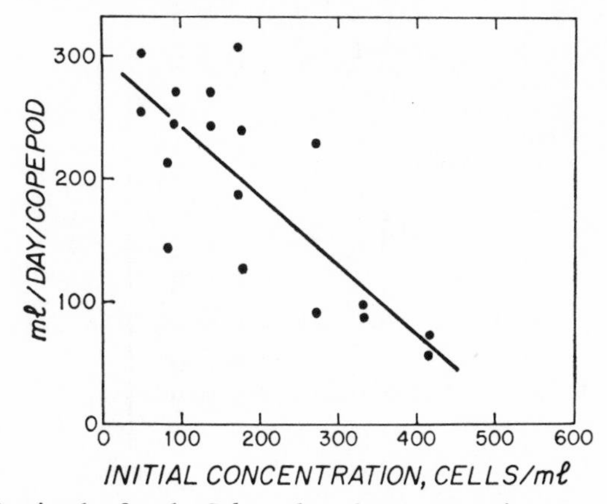

Fig. 9.16. Grazing by female *Calanus hyperboreus* on various concentrations of *Ditylum brightwellii* (three experiments). (After Edmondson 1966.)

Table 9.2. Filtering rates of euphausiids. (After Mauchline & Fisher 1969).

Species of euphausiid	Food organism	Filtering rates (ml/animal/h) range	mean	Authorities
Meganyctiphanes norvegica	*Skeletonema costatum*		2·3	Raymont & Conover 1961
Meganyctiphanes norvegica	*Thalassiosira fluviatilis*		1·1	Raymont & Conover 1961
Thysanoëssa sp.	*Skeletonema costatum*	1·8–2·6	2·2	Raymont & Conover 1961
Thysanoëssa sp.	*Thalassiosira fluviatilis*	1·1–1·6	1·3	Raymont & Conover 1961
Euphausia pacifica	*Platymonas subcordiformis*	12·0–38·0	25·0	Lasker 1966
Euphausia pacifica	*Gonyaulax polyedra*	6·7–20·0	14·6	Lasker 1966
Euphausia pacifica	*Thalassiosira fluviatilis*	2·3–7·1	4·1	Lasker 1966
Euphausia pacifica	*Dunaliella tertiolecta*	1·3–24·9	6·3	Lasker 1966

Table 9.3. Estimates of daily rations taken by herbivorous copepods in British Columbian Waters (After Parsons *et al.* 1969).

Predominant zooplankton	Ration (% body wt/d)
Pseudocalanus minutus Kroyer	4·0
Calanus pacificus Brodsky	16·8
Calanus pacificus	18·4
Calanus pacificus and *Calanus plumchrus* IV	20·2
Calanus plumchrus III and IV	5·7
Calanus plumchrus III and IV	60
Calanus plumchrus V	14·8
Pseudocalanus minutus and *Oithona* sp.	45

Gross growth efficiency

Marshall (1973) has summarized a number of recent estimates of gross growth efficiency in herbivorous copepods (Table 9.4). The values range, in general, from 10 to 50 per cent. Conover's estimates for *Calanus hyperboreus* (Table 9.5) also fall within this range. Efficiency is greatest at the first copepodite stage, naupliar stages being inefficient feeders and later copepodite stages losing considerable energy in vertical migration.

Assimilation efficiency

Estimates of assimilation efficiency range from 50 to 95 per cent (Marshall & Orr 1955b, Lasker 1960, Corner 1961, Berner 1962, Conover 1968). Conover (1964) provides good evidence that, for a given species, assimilation efficiencies are more or less constant over a wide range of food concentrations (Table 9.5), his estimates for *Calanus hyperboreus* averaging 55 per cent. Assimilation efficiency appears to be also independent of temperature (Conover 1966c).

Respiration rate

Raymont and Gauld (1951) and Conover (1959, 1968) showed that the factor γ in the respiration equation (9.22) was of the order of 0·8 (Fig. 9.17).

The (log) relationship between specific respiration rate (respiration rate per unit weight) and body size is shown in Figure 9.18. There are differences between (a) boreal, (b) temperate, and (c) tropical waters as described by the following equations:

$$\text{(a)}\ R' = -0{\cdot}\,054W^{-0{\cdot}169} \qquad (9.27)$$

Table 9.4. Gross growth efficiencies (K_1) of *Calanus*, *Rhincalanus* and *Acartia* on Different Foods. (After Marshall 1973.)

Source:	Mullin and Brooks, 1970						Paffenhöfer		Paffenhöfer, 1971		Petipa, 1967		
Species:	*Calanus pacificus*		*Rhincalanus nasutus*				*Calanus pacificus*		*C. pacificus*		*C. helgolandicus*	*Acartia clausi*	
Food:	*Thalassiosira*		*Ditylum*		*Thalassiosira*		*Lauderia*	*Gymnodinium*	*Lauderia*	*Gymnodinium*	Natural sea water		
μg C/l	226	177	200	148	352	196	101	95	101	95			
T°C:	10°	15°	10°	15°	10°	15°	15°	15°	15°	15°			
							Newly moulted body wts		Medium body wts				
Stage N_I													Stage N_I
N_{II}													N_{II}
N_{III}	21	18	39	39	22	21							N_{III}
N_{IV}									7·6	9·8			N_{IV}
N_V							17·3	20·1	14·7	14·1	34	14	N_V
N_{VI}									29·8	36·7			N_{VI}
C_I									22·0	22·0	50	17	C_I
C_{II}	72	26	31	32	55	32	20·9	29·6	17·6	21·2	39	16	C_{II}
C_{III}							18·6	34·7	22·4	27·2	28	23	C_{III}
C_{IV}									15·7	25·3	21	16	C_{IV}
C_V	30	37	34	48	25	40							C_V
♀								27·6	19·6	22·2	5	11	♀
Total N_I–C_{VI}	35	34	34	45	30	37							

N = nauplius; C = copepodite.

Table 9.5. Assimilation efficiency (A/C), gross growth efficiency (K_1), and net growth efficiency (K_2) of *Calanus hyperboreus*. (After Conover 1964.)

Stage of develop-ment	Exp. temp. (°C)	Species	Food conc. (mg ash free dry wt/l)	A/C	Growth efficiency: Based on weight (K_1)	(K_2)	Based on calories (K_1)	(K_2)
IV	2	*Thalassiosira fluviatilis*	6·4	44·0	3·7	8·5	5	12
V	2	*Thalassiosira fluviatilis*	6·4	47·6	17·3	36·4	24	50
IV	5	*Thalassiosira fluviatilis*	6·7	52·7	13·0	24·1	18	34
V	5	*Thalassiosira fluviatilis*	6·7	50·9	14·6	28·6	20	40
V	5	*Thalassiosira nordenskioldii*	2·6	39·6	13·9	32·4	—	—
V	2	*Thalassiosira fluviatilis*	1·7	71·1	28·4	39·4	39	55
V	5	*Thalassiosira fluviatilis*	1·7	64·1	18·6	27·6	26	38
V	2	*Ditylum Brightwelli*	0·6	53·0	32·3	60·6	46	86
V	2	*Rhizosolenia seticera*	1·7	65·4	29·0	44·2	41	62
V	5	*Rhizosolenia seticera*	1·4	63·1	30·4	48·4	43	68
V	4	*Thalassiosira fluviatilis*	0·3	57·2	13·3	23·3	18	32
V	4	*Thalassiosira fluviatilis*	1·8	56·6	36·4	64·0	50	89

$$\text{(b)}\ R' = -2{\cdot}275\ W^{-0{\cdot}309} \tag{9.28}$$

$$\text{(c)}\ R' = -7{\cdot}482\ W^{-0{\cdot}464} \tag{9.29}$$

where R' is the specific respiration rate and W is the (dry) body weight (Parsons & Takahashi 1973).

Relative to other losses, a great deal of the food assimilated by plankton herbivores is used in respiration. According to Suschenya (1970) and Parsons and Takahashi (1973), this is of the order of 40–85 per cent. The daily respiratory loss of epiplanktonic herbivores in the Black Sea is of the order of 28 per cent of their body weight. By comparison, the mean daily loss due to moulting is only 1 per cent (Corner *et al.* 1967, Petipa *et al.* 1970); the 'Ecdysal Waste' factor may therefore usually be ignored in growth efficiency considerations.

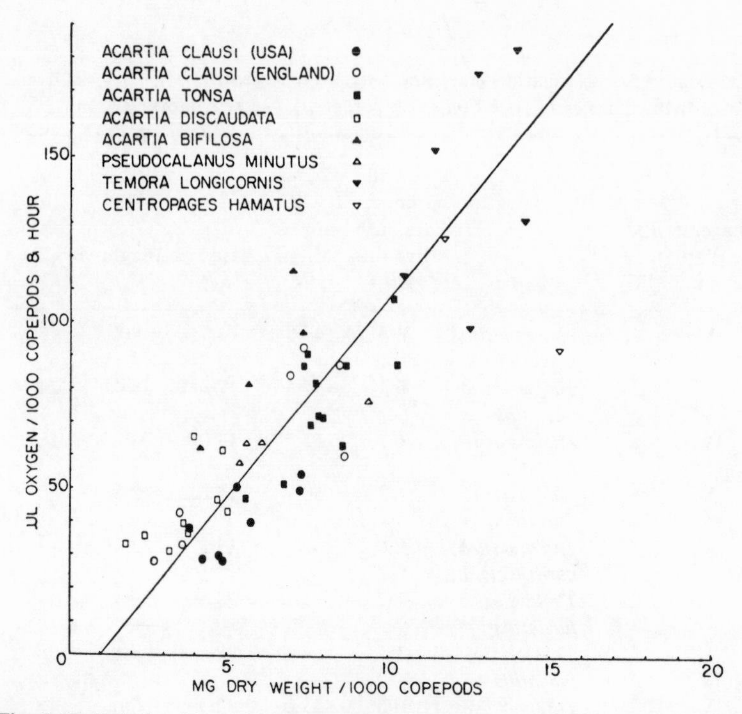

Fig. 9.17. Relationship between respiration rate and herbivore biomass. (After Conover 1968.)

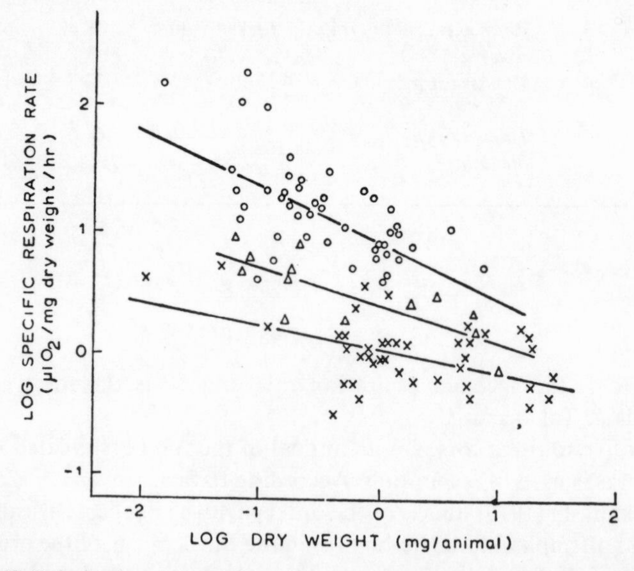

Fig. 9.18. Relation between specific respiration rate and dry weight (log values) in planktonic animals from boreal (X), temperate ($\triangle$), and tropical ($\circ$) waters. (From Parsons & Takahashi 1973, after Ikeda 1970.)

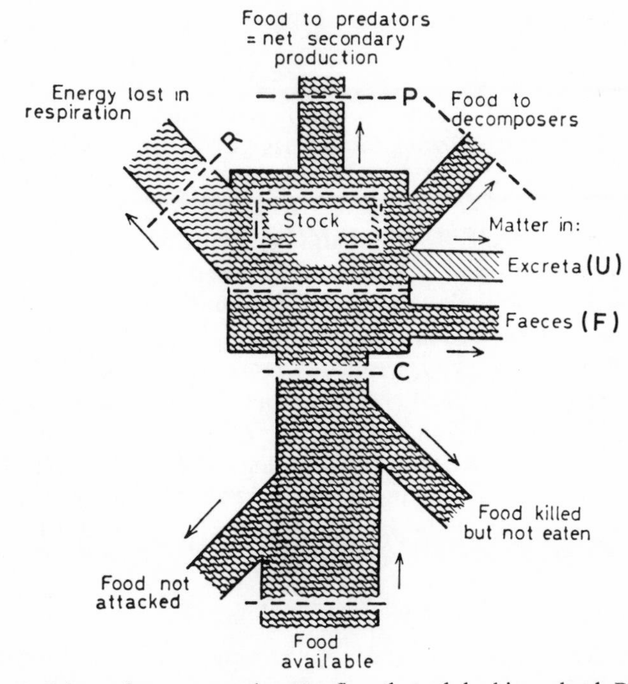

Fig. 9.19. Scheme for energy and matter flow through herbivore level. Ratios of flows at points marked are 'efficiencies': P/C = Gross growth efficiency. (After Macfadyen 1964.)

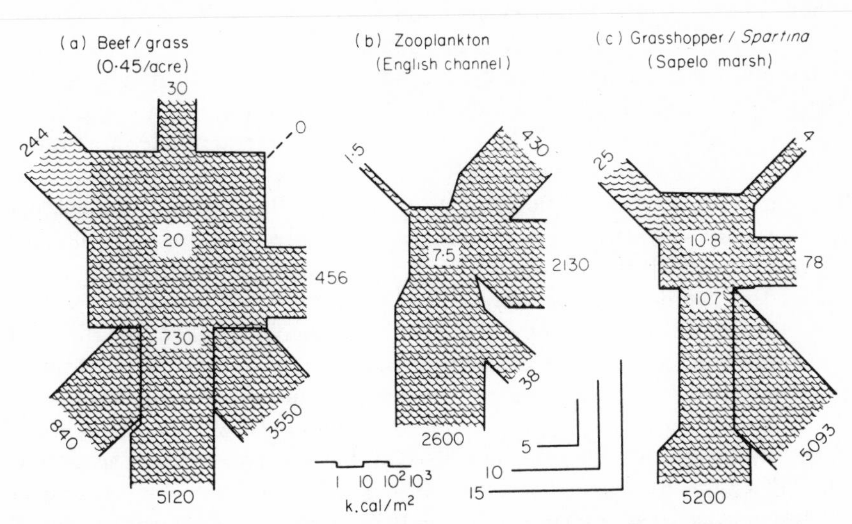

Fig. 9.20. Scaled flow paths for three herbivore systems. Conventions as in Fig. 9.22. (In *b* and *c*, the productions cropped by predators and by decomposers have been combined.) (After Macfadyen 1964.)

Net growth efficiency

Net growth efficiencies lie usually between 20 and 60 per cent (Conover 1968, Parsons & Takahashi 1973), but may be higher for actively growing young, and for animals storing fat. Values for *Calanus hyperboreus* average 36 per cent but vary with temperature, food, and stage of development (Table 9.5).

To sum up, the available information on plankton energetics is relatively sparse compared with what is known of other herbivore systems. In approximate terms, plankton herbivores assimilate about half of their daily rations, the remainder being voided in the form of faeces. Of the assimilated, about half is used in metabolic activity, frequently more than is available for growth. Secondary plankton production differs from that for other grazing systems such as beef production on land or grasshopper production in *Spartina* marsh (Macfadyen 1964, Steele 1972) in that most of the available food is ingested, the subsequent flux to detritus is relatively high, and the stock of herbivores is small relative to their production (Figs 9.19, 9.20).

Biomass turnover rate (production: biomass)

The most useful measurements of turnover rates are those of Greze and Baldina (1964) and Greze (1970) for the plankton of the Black Sea. Their estimates are based on time-series observations on the relative frequencies of developmental stages in the plankton, the turnover time being taken as the observed interval between one generation and the next. Table 9.6 lists some of their P'/B ratios for a variety of zooplankton herbivores. The values are of the order of 0·1 to 0·3, corresponding to turnover, times (B/P') of 10 days and 3·3 days, the latter being for the nanoplankton feeder *Oikopleura*. By comparison, the turnover rates for phytoplankton are an order of magnitude more rapid.

Greze has also determined turnover rates on a seasonal basis (Table 9.7a) and on a depth basis (Table 9.7b). In Sevastopol Bay and in the Black Sea off Sevastopol, winter rates were about half those in spring and summer. Turnover rates in the water column did not vary much above 200 m, but below this depth fell appreciably.

For comparison with the estimates of turnover rate made by Greze, we have estimates of the seasonal cycle on the Georges Bank (Fig. 9.21), made by Riley (1947). The seasonal range is much greater than in the Black Sea.

Table 9.8 shows P'/B ratios compiled by Mullin (1969) together with his later estimates for *Calanus helgolandicus* (Mullin & Brooks 1970). These values were derived in a variety of ways and vary in their accuracy. They range from 0·002 (turnover time, 500 days) for the Barents Sea to 0·98 (turnover time, 1 day) for the Gulf of Panama. Boreal species such as *Calanus cristatus* and *Calanus plumchrus* gave ratios of the order of 0·01 (turnover time, 100 days),

Table 9.6. Turnover rate (P'/B) and turnover time (B/P') for species in the Neritic Zone off Sevastopol during summer, by trophic level. (After Greze 1970.)

Trophic levels and their components	Dry matter (%)	Biomass* (mg/m³)	Daily production* (mg/m³)	P'/B co-efficient	B/P' (days)
I Phytoplankton	8	36·00	28·80	0·80	1·3
II Phytophages		8·82	1·37		
Cladocera (*Penilia, Evadne*)	15	1·81	0·35	0·19	5·3
Paracalanus parvus	12	1·41	0·19	0·09	11·1
Pseudocalanus elongatus	13	2·30	0·32	0·14	7·1
Centropages ponticus	13	0·13	0·01	0·09	11·1
Acartia clausi	12	1·46	0·18	0·12	8·3
Oithona similis	14	0·07	0·01	0·08	12·5
O. minuta	14	0·50	0·12	0·11	9·1
Oikopleura dioica	10	0·15	0·04	0·30	3·3
Mollusca larvae	15	0·57	0·09	0·15	6·7
Polychaeta larvae	12	0·42	0·06	0·15	6·7
III Carnivores		4·07	0·67		
Acartia clausi	12	0·24	0·02	0·12	8·3
Oithona similis	14	0·29	0·01	0·08	12·5
O. minuta	14	1·21	0·12	0·11	9·1
Sagitta setosa	8	1·78	0·35	0·20	5·0
Pisces larvae	15	0·55	0·17	0·30	3·3
IV Detritus feeders					
Noctiluca miliaris	4	8·52	10·21	1·20	0·8

* dry weight

Table 9.7. Variation in turnover rates with (*a*) season and (*b*) depth. (After Greze 1970.)

	(a) Season			
Area	Spring	Summer	Autumn	Winter
Sevastopol Bay	0·15	0·15	0·10	0·06
Black Sea	0·17	0·15	0·12	0·08

	(b) Depth		
Area	0–50 m	50–200 m	200–500 m
Atlantic	0·15	0·17	0·10
Ionian Sea	0·18	0·15	0·09

whereas the more temperate species *Calanus helgolandicus* yielded a ratio of the order of 0·1 (turnover time, 10 days). Similarly, the highest values obtained for the cool-temperate neritic copepod *Acartia clausi* were of the order of 0·2

Table 9.8. Summary of data available on secondary plankton production. (After Mullin 1969.)

Reference	Organism or group	Area and period	Production (P') (mg C/m²/d)	P'/B (per day)	B/P' (days)
Kamshilov 1958	*Calanus finmarchicus*	E. Barents Sea; year	7·8	0·002	500
Yablonskaya 1962	*Diaptomus salinus*	Aral Sea; year	0·66	0·007	143
Mednikov 1960	*Calanus cristatus*	N.W. Pacific: summer	5·6	0·012	83
	Calanus plumchrus	N.W. Pacific; summer	4·6	0·010	100
	Eucalanus bungii	N.W. Pacific; summer	3·5	0·014	71
Riley 1947	zooplankton	Georges Bank; year	200	0·03	33
Greze & Baldina 1964	*Acartia clausi*	Black Sea; year	0·38	0·035	29
Steele 1958	zooplankton	N. North Sea; Apr.–Sept.	180	0·048	21
Cushing 1959	herbivorous copepods	North Sea; Jan.–June	4·9	0·08	33
Harvey 1950	zooplankton	English Channel; year	75	0·10	10

Conover 1956	zooplankton	Long Island Sound; year	166	0·17	6
Mullin & Brooks 1970	*Calanus helgolandicus*	La Jolla; spring	104	0·12	8
		summer	26	0·15	7
Petipa 1967	*Acartia clausi*	Black Sea bay; June	15	0·17	6
	Acartia clausi	Black Sea, open sea; June	6·6	0·23	4
	Calanus helgolandicus	Black Sea, open sea; June	28	0·15	7
Heinle 1966	*Acartia tonsa*	Chesapeake Bay estuary; summer	77	0·50	2
Smayda 1966	zooplankton	Gulf of Panama; Jan.–April	70 or 234	0·29 or 0·98	3·4 or 1·0

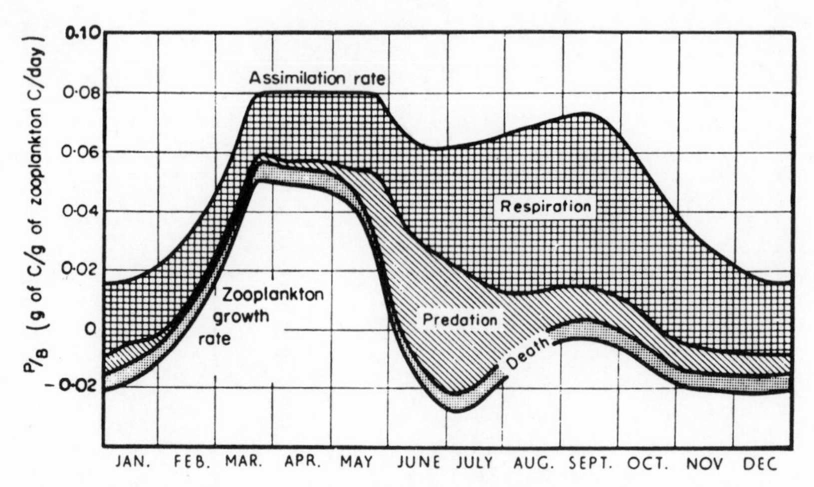

Fig. 9.21. Zooplankton production (net and gross) on Georges Bank relative to zooplankton biomass. (After Riley 1947.)

(turnover time, 5 days) whereas the estimate for its warm-temperate counterpart was 0·5 (*Acartia tonsa*) (turnover time, 2 days).

It seems, therefore, that the miscellany of methods used to estimate the P'/B ratio fail to hide the general trend towards slower turnover rates in high latitudes and faster turnover rates in low latitudes. This is consistent with the relationship between development rate and water temperature on the one hand (Fig. 9.9), and between specific respiration rate and latitude on the other (Fig. 9.18). It emphasizes the inherent dangers in equating large stocks of zooplankton with high rates of production, and vice versa.

Estimates of secondary production

In contrast with the frequency of primary production estimates in the sea (Chapter 8), there are few good values for secondary production, and those that are available are rarely comparable with one another. Table 9.8 extracted from Mullin (1969) summarizes the current state of knowledge, mostly on single species rather than on the whole complex of herbivores in the biocoenosis.

Production values for single species (all copepods) range from less than 10 to more than 100 mg C/m^2/d. Values for total zooplankton range to 200 mg/m^2/d.

Perhaps the most reliable measurements that have been made to date are those of Heinle (1966) for the copepod *Acartia tonsa* in the Patuxent River estuary of Chesapeake Bay (Table 9.9). It is interesting to note that the greater

Table 9.9. Dry weight production of *A. tonsa* (mg/m³ h). (From Heinle 1966.)

Interval	Biomass (mg/m³)			T_n (turnover time of nauplii) (hours)	T_c (turnover time of copepodids and adults) (hours)	Production (P')		
	Nauplii	Copepodids	Adults			Nauplii	Copepodids and adults	Total
1	80·030	34·567	22·071	43·4	64·1	1·84	0·88	2·72
2	93·151	44·003	17·995	45·8	47·3	2·03	1·31	3·34
3	124·467	38·237	6·304	40·8	45·8	3·05	0·97	4·02
4	80·570	24·021	11·340	43·4	58·3	1·86	0·61	2·47
5	92·774	36·399	8·730	40·8	50·6	2·27	0·89	3·16
6	51·557	24·555	17·717	44·6	64·1	1·15	0·66	1·81
7	66·541	16·274	3·079	43·0	67·4	1·55	0·29	1·84
Average production								2·77
Mean population	83·965	30·628	12·886	46·8	60·5	1·79	0·72	2·51

part of the production took place in the early (nauplius) stages of development.

Discussion

Methodology

Each of the methods used to measure secondary plankton production has severe limitations. The result has been a paucity of reliable estimates, and an undue emphasis on single species studies, an approach more relevant to fish production than to plankton production. The production of a single algal species, for example, is an inadequate basis for estimating primary production in the sea.

No matter how difficult it may be, further attempts must be made to measure the secondary production of the plankton biocoenosis *in situ*. According to Mullin (1969) there are no reliable ways to make such measurements, even approximately. However, the errors involved in extrapolating from species to biocoenosis are so great that it is worthwhile to take a fresh look at possible alternatives. Two approaches would seem to merit further investigation: the size-fractionation approach, and the labelled algae approach.

Although continuity of reproduction in plankton biocoenoses rarely allows the classical cohort approach to be used effectively, it is possible to fractionate in space, if not in time, by means of a battery of graded sieves. The result would be a spectrum of spatially separate size-groups containing mainly, but by no means exclusively, organisms of the same approximate age. The fraction containing the smallest organisms would, in general, be the youngest 'cohort', and the fraction containing the largest organisms would be the oldest 'cohort'.

This approximation might be sufficiently satisfactory to permit the total production of the plankton biocoenosis to be estimated more accurately than can be done at present by existing methods. The one strategy permits production to be estimated in two different ways: by the incremental growth per organism in each size group over short intervals of time (e.g. 1 day); and also by the mortality in the captive populations. The major assumptions would be, first, that the effect of captivity could be ignored over the observational period; and, second, that the summation of production in each spatially-separate size-group is an adequate representation of production as a whole.

One of the greatest obstacles in deriving reliable measurements of secondary plankton production is the difficulty of distinguishing herbivores from omnivores and carnivores so that herbivorous production may be isolated for independent measurement. If one aims to avoid the errors inherent in equating zooplankton production with herbivore production, one has recourse to few techniques other than that of using radioactive labelled algae.

However, it is deceptively easy to obtain wrong information using tracer isotopes to measure the transfer of materials in aquatic food chains, so much so that Conover and Francis (1973) advise potential users to first consider the alternatives available. The problem arises in the inherent complexity of the grazing system. When a tracer is introduced, say in the form of labelled algae, it may take any one of a number of paths, of which the direct path to the grazer, by way of grazing, is only one (Fig. 9.22a); for example, before the isotope can be incorporated in the tissues of the herbivore as secondary production it may be remineralized via excretion and used again in photosynthesis (Fig. 9.22b).

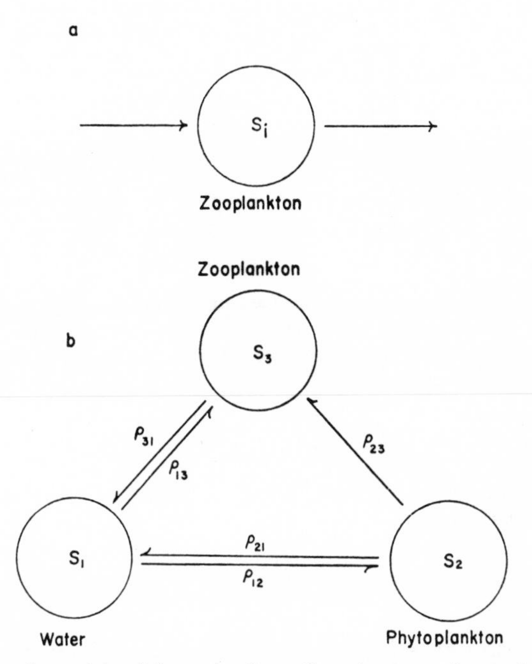

Fig. 9.22. Simple models of the path of a radioactive tracer in plankton grazing system. (a) Single component open model in steady state; (b) Three-component, closed, steady-state model, with recycling. (After Conover & Francis 1973.)

It takes some time for a grazing system incorporating radioactive tracer to reach a steady state where meaningful observations may be made (Fig. 9.23). During this time the tracer needs to be monitored through all the major 'pools' in which it may appear. All the available methods for measuring secondary plankton production require time-series observations on one variable or another, so the demands of the tracer technique are not unique. It would

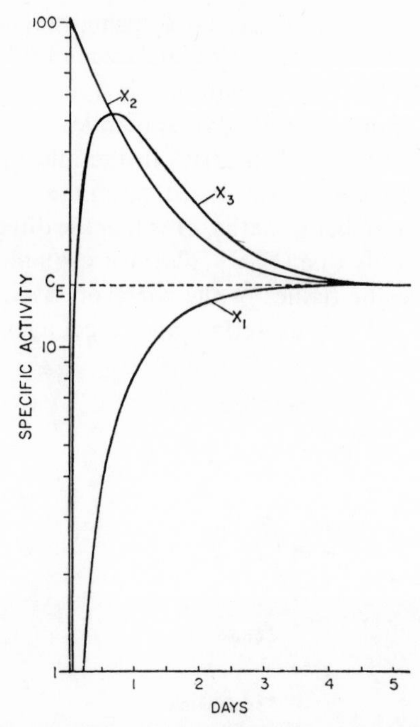

Fig. 9.23. Theoretical curves showing the time course of specific radioactivity in water (X_1), phytoplankton (X_2), and zooplankton (X_3), interacting as a 3-compartment closed system in steady state. C_E is the specific activity at equilibrium. (After Conover & Francis 1973.)

seem that more attention could be given to the use of this method for *in situ* estimates of production.

Principles

From this account it may be seen that secondary production in the plankton has much in common with secondary production in other grazing systems, and much that is quite different. These essential similarities and differences need to be keenly appreciated if we are to understand the role of plankton grazing in the marine ecosystem as a whole. For example, the advantage that phytoplankton derive from their buoyancy in exploiting the energy of sun-light is provided at considerable cost on land by the structure of vegetation. Perhaps, for this reason, the predator–prey interface may be a more important controlling mechanism on land than it is in the sea (Steele 1972). On the other hand the grazing interface in the sea is a control mechanism that has few parallels on land, the net effect being to maximize production—one consequence of grazing is to rapidly remineralize plant material and replenish the nutrient pool; another is to keep the water column 'window' open to the sunlight.

Paloheimo and Dickie (1970) have considered the implications of the possibility that production at the grazing interface in the sea may be substantially different from production at the predator–prey interface. They have also drawn attention to the fact that there are large differences from one species to another in the proportion of production that is channelled into somatic growth and reproductive growth. In this context it is interesting that Murphy (personal communication) should enquire whether man's exploitation of fish stocks that are low in the food chain (e.g. pilchards and sardines) may have qualitatively different consequences from exploitation of fish stocks higher in the food chain. Steele (1972) has pointed out that community structure appears to be affected more severely by eutrophication than by overfishing, and enquires whether the consequences of nutrient enrichment on marine grazing systems may be different from those on land.

There is a close interaction between production in the water column and on the substratum below. By controlling the concentration of phytoplankton in the water, grazing influences the amount of light available for photosynthesis by benthic plants. By feeding to excess, grazers transfer organic matter to the bottom in the form of faecal detritus. Whether the greater part of the detritus in the water column is formed from plant material prior or subsequent to grazing, there is here a large pool of organic matter available for secondary production. This detrital phase of secondary production may prove to be more important than is often thought and merits closer study.

Notation

General

t = time
N = number of individual herbivores
$\bar{w}$ = mean weight of an individual
B = biomass of the stock ($= N \cdot \bar{w}$)
P = production (units: mass)
P' = production rate (units: mass/time)
p = prey concentration (phytoplankton)

Population dynamics

RATES

b = instantaneous birth rate

d = instantaneous death rate
r = instantaneous rate of numerical increase ($= b-d$)
β = finite birth rate
δ = finite death rate
λ = finite rate of numerical increase ($= 1+\beta-\delta$) $= e^r$
M = mortality ($= d \cdot \bar{w}$)

'CONSTANTS'

D = duration of development
T = generation time (birth of parents to birth of offspring)
E = number of eggs per female

Carbon budget

RATES

A = food assimilated
C = food consumed
F = faeces
R = respiration
R' = respiration per unit biomass
U = excretion (urine)

'CONSTANTS'

K_1 = gross efficiency of growth
K_2 = net efficiency of growth
ϕ = food intake: body-weight function
γ = respiration: body-weight function

Chapter 10. Production on the Bottom of the Sea

Kenneth H. Mann

Introduction
Types of benthic plants and animals
Benthic plants
Benthic animals
Supply of food to the benthos
Distribution of the benthos
Distribution of benthic animals according to their feeding habits
Life histories of benthic animals
Methods of estimating production
Production to biomass ratios P/B
Commercially important shellfish stocks: population dynamics
Population dynamics of lobsters
Population dynamics of scallops
Conclusions
Acknowledgements

Introduction

By far the greater part of the sea floor is below the euphotic zone where the light intensity is too attenuated to support photosynthesis. The organisms that live there are therefore forced to rely on a supply of food from above, sinking through the water column and settling out on the interface between water and sediments, or on food supplied by horizontal currents which move sediments after initial deposition. Animals on the sea bed have evolved three main methods of feeding: they may filter material from the water, feed on the bottom deposits, or prey on other animals.

The quantity of food reaching the bottom varies greatly from place to place. At the very edge of the sea the bottom may be within the euphotic zone, so that a variety of macroscopic and microscopic plants, particularly algae, grow there and supply an abundance of organic material to consumer organisms directly through grazing, or indirectly as detritus. In slightly deeper water there may be an input of plant detritus from the rich coastal areas, and a good proportion of the primary production from the euphotic zone reaches the sea bed. In deeper waters, beyond the continental shelves, the sinking plant material has to make a long, slow journey to the sea floor, and most of it is consumed by organisms living in the water column. Hence, the food supply to the sea floor in the main ocean basins is very restricted, and production of the benthos is correspondingly reduced. Most of the commercially important species are found on the continental shelves, or on the slope at the edge of it. The three main feeding types of invertebrates: filter feeders, deposit feeders and carnivores, form the food of many commercially important species of fish. Some of the invertebrates, notably the crustaceans and the molluscs, are harvested directly and are of commercial importance.

The aim of this chapter is to trace the production process, from its origins in the plankton and macrophytes of surface waters to its realization as harvestable protein in the bodies of bottom living fish, molluscs and crustaceans.

Types of benthic plants and animals

Benthic plants

Production of planktonic algae has been treated in an earlier chapter (see Lorenzen, Chapter 8) but there are various kinds of plants attached to the sea bed which require separate mention. They are all found in shallow, coastal waters where the light intensity at the bottom is sufficient to support photosynthesis. First, there are the large marine algae, of which the most important groups are the kelps (Laminariales) and rockweeds (Fucales). The kelps are found below low-tide level on almost all temperate shorelines (Fig. 10.1). They are attached to rocky substrates by a holdfast which serves only for attachment, not as a root system. There is a stem, or stipe, to which is attached a large (1–10 m) blade or lamina. The action of wind and waves washes these plants with a virtually inexhaustible supply of nutrients, making possible very high rates of production, of the order of 1000–2000 gC m^{-2} yr^{-1} (Mann 1973). Rockweeds have approximately the same geographical distribution as kelps, but occur chiefly between high- and low-tide levels. Their productivity is usually about half that of the kelps.

Another important group of marine macrophytes are the sea-grasses and marsh-grasses. They are flowering plants which also have world-wide distribution, but colonize the more sheltered, sedimented areas of the coastline. Their productivity is of the order of 200–1000 gC m^{-2} yr^{-1}. In the tropics, large tracts of shoreline are occupied by mangrove swamps which contribute organic matter to the sea through falling leaves and twigs. All of these shoreline plant communities generate large amounts of organic detritus (Mann 1972) which is carried considerable distances and accounts in part for the high level of productivity of food chains in coastal waters.

It has been estimated that on a global scale marine macrophytes account for at least one-tenth of the primary production in the sea (Ryther 1969). Since this production is concentrated at the land/water interface it is an

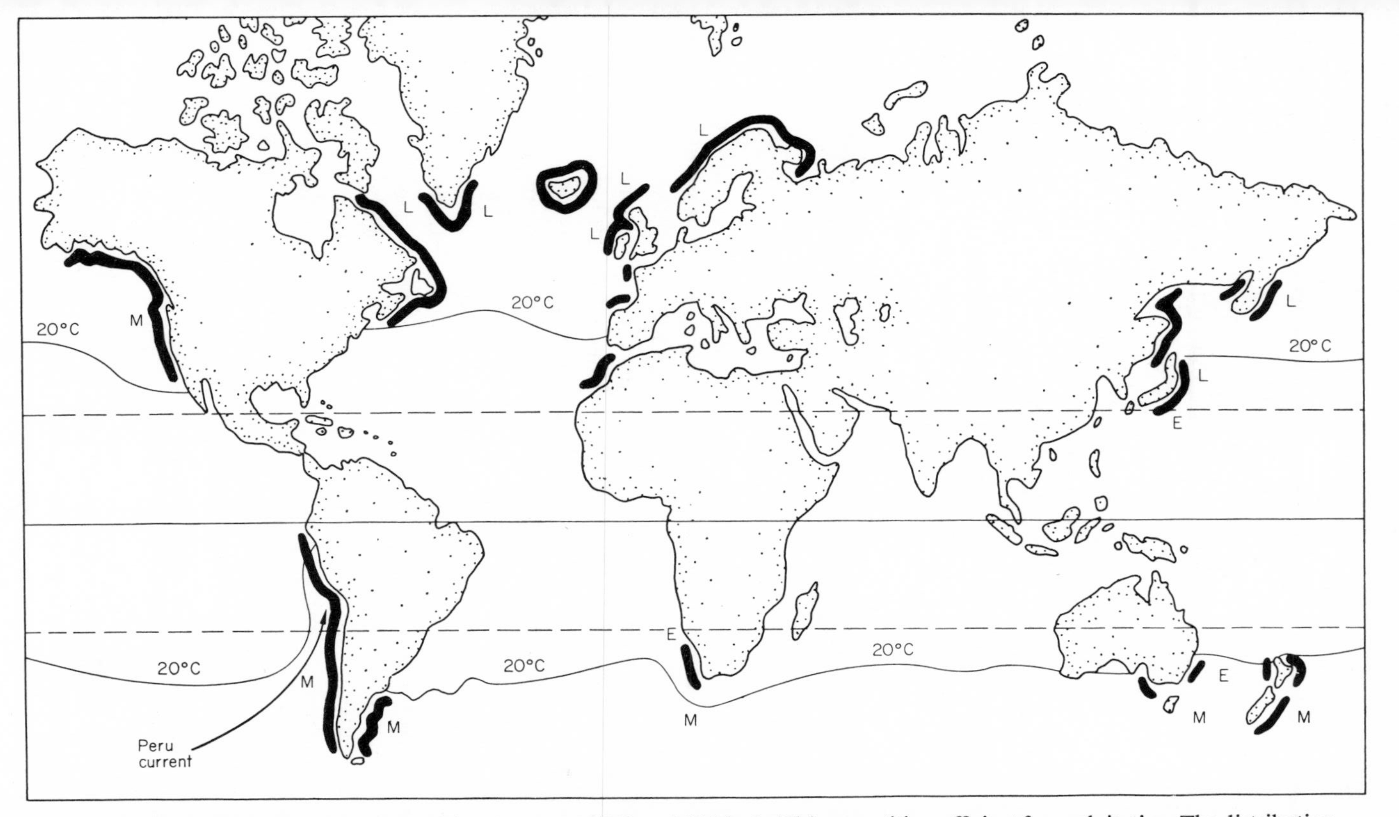

Fig. 10.1. The distribution of *Laminaria* (*L*), *Macrocystis* (*M*), and *Ecklonia* (*E*) in quantities sufficient for exploitation. The distribution of rockweeds (*Fucus* and *Ascophyllum*) is approximately the same as that of the laminariales. The 20°C isotherms are for summer in northern and southern hemispheres, respectively. (From Mann 1972.)

important contributor to food chains leading to species of commercial importance in the coastal zone.

A less obvious, but none the less important source of primary production on the sea bed in shallow waters, are the algae which live in the interstices of sand deposits, and on the surface of mud. For example, Leach (1970) found that in a Scottish estuary there was an epibenthic algal community dominated by motile, pennate diatoms. He recorded uptake of ^{14}C down to 4 cm in the mud, and estimated annual primary production at 31 gC m^{-2}. In an area of much greater solar radiation, Pomeroy (1959) estimated primary production by algae on the surface of a Georgia salt marsh to be about 200 gC m^{-2} yr^{-1}. Fenchel and Straarup (1971) found that on a sandy beach in Denmark there were between 1 and 3×10^7 cells per cm^2, predominantly diatoms. Munro and Brock (1968) found viable diatoms as deep as 15 cm in the sand of a Scottish inlet, yet light penetration was only 1 per cent at 3 mm. Studies of the mixing of the sand suggested that only the top 5 cm regularly moved under wave action, so it seems that the algae can remain buried and viable for long periods.

Benthic animals

The fauna of the sea floor is very diverse: almost all major groups of invertebrates are represented, as well as the fish. There are several ways of grouping these organisms: by size, by mode of life, or by feeding habits. A rough and ready division often used by benthic ecologists is: (1) infauna, organisms that live within the bottom deposits, and (2) epifauna, those that live on or above the sea floor. Polychaete worms, burrowing crustaceans and burrowing clams are important components of the infauna, while scallops, starfish, sea urchins, and mussels are examples of the epifauna. The infauna is often divided again on a size basis into macrobenthos, meiobenthos and microbenthos, with the dividing lines set at about 1 mm and 0·1 mm respectively.

The macrobenthos is the component about which most is known. It is easier to sample and sort than the smaller forms, and is of importance both to man directly and as the food of fish. However, before going on to discuss the macrobenthos, some attention should be given to smaller forms. The meiobenthos comprises two components: the permanent meiofauna in which all stages of the life history including the adults are small, and the temporary meiofauna which comprises the younger stages of large animals such as worms and molluscs. The permanent meiofauna is dominated by nematode worms and copepod and ostracod crustaceans. These organisms are found in large numbers in the interstices of sand and mud deposits. They are most abundant just below the surface, to a depth of 3–4 cm, where the sediments are reasonably aerobic, and where food is most plentiful (McIntyre 1969, Coull 1973).

The microfauna is comprised mainly of protozoa and bacteria. Their food source is fine particulate and dissolved organic matter in the sediments. They in turn are preyed upon by the meiofauna. Although it is almost certain that worms and other forms that ingest the sediments will consume meiofauna in the process, there is little evidence of there being predators adapted to feeding selectively on the meiofauna. Hence, it has been suggested by McIntyre (1969) and Fenchel (1969) that the meiofauna is relatively unimportant as a link in the food chain from dead organic matter to large invertebrates and fish. The main ecological role of the meiofauna appears to be one of mineralizing organic matter and releasing nutrients to the overlying water. This conclusion, with qualifications, was also reached by McIntyre *et al.* (1970) and by Coull (1973).

The macrofauna is most conveniently classified, according to its feeding habits, into predators, filter feeders and deposit feeders. The deposit feeders can be further subdivided into those that selectively feed on the surface deposits and those that ingest surface and subsurface deposits indiscriminately. Conspicuous among the predators of the macrobenthos are demersal fish, crustaceans such as crabs and lobsters and various carnivorous gastropods. Since these predators depend on the non-predators for food, it is natural that the numbers and biomass of carnivores are normally less than the numbers and biomass of deposit feeders and filter feeders. Many of the commercially important species taken from the sea floor are carnivores.

Examples of filter feeders in the benthos are mussels, scallops, clams, sponges, ascidians, barnacles and fan worms. It is beyond the scope of this volume to review the mechanics of filter feeding, suffice it to say that it is one of the more remarkable ecological processes, since it is the means by which microscopic organic particles diffusely scattered in suspension in the water are converted to packages of protein of the size of scallops and clams. An excellent review of filter feeding is found in Jørgensen (1966).

Animals feeding from the surface deposits may live in, above or below the sediment surface. Many crustaceans, such as mysid and crangonid shrimps or some amphipods, spend much of their time swimming a few centimetres above the bottom deposits, making short excursions down to collect food. They may be collected by nets hauled along the surface of the deposits, and are collectively referred to as hyperbenthos.

Examples of those living on or in the surface deposits are isopod and some amphipod crustaceans, gastropod molluscs, and some ophiuroids (brittlestars). Many of these feed by ingesting the organic detritus which surrounds them. Several kinds of burrowing organisms, which might be expected to feed unselectively on bottom deposits are adapted to taking surface detritus. For example, the polychaete *Amphitrite* spreads ciliated tentacles on the sediment surface, and several burrowing clams direct their siphons so as to suck up the surface deposits (Fig. 10.2). Finally, polychaete worms such as *Arenicola* are

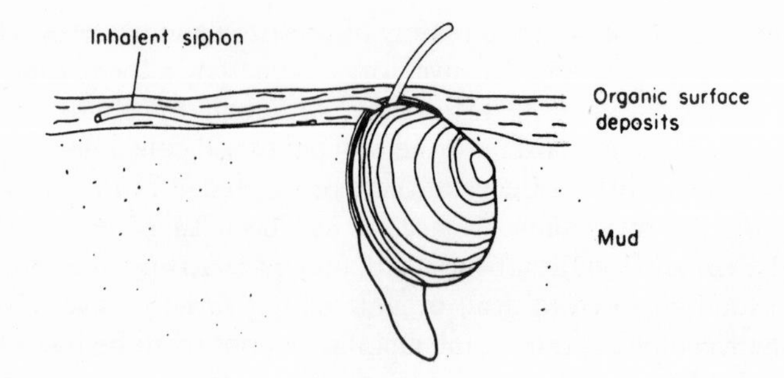

Fig. 10.2. Bivalve mollusc, *Scrobicularia plana* (da Costa), showing mode of feeding on surface deposits. (After Hesse, Allee & Schmidt, *Ecological Animal Geography*.)

examples of benthic animals which obtain their food by ingesting sub-surface deposits, and egesting copious faeces from their burrows at the surface. Such organisms often penetrate the anaerobic layers 10–20 cm or more below the surface, and have to have mechanisms for bringing water currents down in their burrows. These organisms probably have a major influence on sediment/water exchange of nutrients.

Supply of food to the benthos

Since most of the world's benthos is living below the photic zone it is dependent on a supply of organic matter from above. Until recently, attention was focused on the phytoplankton and zooplankton, their dead bodies, faeces and derived organic detritus, which forms a gentle rain of small particles sinking slowly and being decomposed as it goes down. However, evidence from baited cameras (Isaacs 1969) as well as accumulated evidence from deep-water trawling has shown that there are large numbers of fish in the deep benthos which are equipped only for feeding on relatively large particles. It now seems probable that a significant amount of food reaches the deep ocean floor as a result of the rapid sinking of large particles. These may range in size from the bodies of whales to fragments resulting from the attack of a pelagic carnivore on its prey. In the middle range would be the dead bodies of various kinds of fish.

As an example of the process involving sinking of planktonic organisms, we may take Riley's (1970) calculation for the Sargasso Sea. He found that primary carbon fixation in the upper 100 m of the water column amounted to, on average, 320–80 mgC m^{-2} d^{-1}. He estimated that 75–80 per cent of this was consumed by zooplankton, protista and bacteria in the surface and middle depths. Of the material sinking to deep-water layers, he found that about 80 per cent was consumed in the water column. Hence, the material reaching the bottom contains only about 3 per cent of the carbon fixed at the surface.

In the northern temperate and sub-arctic parts of the Pacific, it is estimated that about 10 per cent of the organic matter produced in the euphotic zone reaches depths below 3000 m, and in south temperate and tropical latitudes this figure is less than 5 per cent (Bogdanov 1965).

The absolute amount of organic matter reaching the sea floor also depends on the level of primary production in the surface waters. As has been shown in other chapters, primary production is high in coastal waters, low in the mid-ocean gyres. It is also higher to the north and to the south of the main ocean gyres, i.e. in high latitudes, in an equatorial belt and also along the west sides of continents, especially in areas of upwelling. Some examples of levels of primary production are: kelp beds, 3–6 gC m^{-2} d^{-1} (Mann 1973), eutrophic coastal upwelling areas 1–10 gC m^{-2} d^{-1} and oligotrophic central gyres of the ocean 0·1 to 0·5 gC m^{-2} d^{-1} (Walsh 1974). This difference in primary productivity and in the amount reaching the sediments is reflected in sediment analyses (Bogdanov *et al.* 1971), with very small quantities of organic matter and correspondingly small biomass of micro-organisms occurring in the sediments of mid-ocean basins (Sorokin 1970).

Measurements of oxygen uptake by sediment cores also provide an index of the supply of oxidizable organic matter received by bottom communities. (Pamatmat 1971). In areas not receiving terrestrial debris such measures are directly correlated with primary production but inversely related to water column depth, indicating increased mineralization during longer periods of sedimentation in deep water (Hargrave 1973).

In shallow, coastal waters the amount of organic matter reaching the bottom is high for three reasons: (1) phytoplankton production is higher in coastal waters than further offshore; (2) the distance through which material has to sink, and hence the opportunity for it to be consumed in the water column is less; and (3) there is lateral transport from the areas of intensely high primary productivity in seaweed beds and coastal marshes. Riley (1956) estimated that in Long Island Sound 30–40 per cent of the primary production amounting to 60–80 gC m^{-2} yr^{-1} was consumed by the benthos. Some attempts have been made to verify this figure by placing sediment traps in positions where they would catch the rain of detritus but would not, it was hoped, catch material resuspended from the bottom by water movements. Steele and Baird (1972), working in a Scottish inlet, collected about 30 gC m^{-2} yr^{-1}, mainly in the form of zooplankton faeces. Zeitschel (1965) collected the equivalent of 40 gC m^{-2} yr^{-1} in the Baltic, and Stephens *et al.* (1967) collected about 200 gC m^{-2} yr^{-1} in a bay on the west coast of Canada. About half of this was identified as material of terrestrial origin, carried by river runoff. In St Margaret's Bay, Nova Scotia, Webster *et al.* (1975) estimated that organic material reaching the bottom amounted to 134 gC m^{-2}

yr^{-1}, and that this was derived from the sinking of phytoplankton and the lateral transport of detritus derived from seaweeds and eelgrass. There was no evidence of a substantial contribution from the land.

Hence, there is evidence to suggest that the annual input of organic matter to the benthos may be as high as 200 gC m^{-2} close inshore, dropping to about 30 gC m^{-2} a little further offshore, and to only 1 or 2 gC m^{-2} in the main ocean basins. Values of the same order of magnitude have been calculated from measures of sediment oxygen uptake in different areas (Hargrave 1973).

Distribution of the benthos

It is inevitable that the biomass of benthic organisms supported by the food source described above will decrease as one moves from close inshore to mid-ocean depths. There are numerous examples of this in the literature. In the northern Pacific Ocean, off the coast of the U.S.A., at depths of 400 to 1200 m, the biomass of the benthos is in the range 20 to 200 g m^{-2} (fresh weight). (Zenkevitch & Filatova 1960, Filatova & Levenstein 1961). Beyond the 2000 m contour it drops to an average of 1·4 g m^{-2} and beyond 4000 m to 0·2 g m^{-2} (Vinogradova 1962). Average biomass figures for the Pacific Ocean at various depths are given by Filatova (1970) as follows: 2001–3000 m, 7·08 g; 3001–4000 m, 4·00 g; 4001–5000 m 0·63 g, and 5001–6000 m, 1·25 g. Average figures for the Indian sector of the Antarctic are as follows: (Belyaev & Uschakov 1957)

100– 200 m	1347 g m^{-2}
200– 500 m	239 g m^{-2}
500–1000 m	43 g m^{-2}
1000–3200 m	13 g m^{-2}

Clearly, the absolute values in the Pacific and in the Antarctic are quite different, but the trend is the same.

After assembling numerous sets of figures of this sort, Zenkevitch, *et al.* (1960) made a tentative estimate of the total biomass of benthos in the world oceans (Table 10.1). It showed that more than 80 per cent is found on the continental shelves at depths of less than 200 m, while less than 1 per cent occurs in the abyssal depths. The authors pointed out that they had not taken into account the populations of the intertidal zones of the world, so that the proportion occurring in the shallow zone is probably greater than their estimate shows.

There have been a number of attempts to distinguish the effect of surface productivity (basic food supply) from the effects of depth (utilization of the food supply in the water column), in determining the biomass of benthos. Rowe (1971) showed that when the log of benthic biomass is plotted against

Table 10.1. Biomass of bottom fauna in the world ocean. (After Zenkevitch *et al.* 1960.)

	Area		Mean biomass	Total biomass	
Depth (m)	$(km)^2 \times 10^6$	%	$g\ (m)^{-2}$ or $t\ (km)^{-2}$	$t \times 10^6$	%
0–200	27·5	7·6	200	5500	82·6
200–3000	55·2	15·3	20	1104	16·6
> 3000	278·3	77·1	0·2	56	0·8
Whole ocean	361	100	18·5	6660	100

depth on various transects from shore to deep ocean, there is always a statistically significant least-squares linear regression, but the slope of the line varies, and reflects the way in which surface productivity changes with distance from shore. For example, the effect of euphotic zone productivity is very marked in a transect from the coast of New England (U.S.A.) where coastal productivity is high, but much less marked in the Gulf of Mexico, where coastal productivity is lower.

A rather clear indication of the effect of surface productivity is given by the regression (Fig. 10.3) of benthic macrofauna biomass (dry weight) on average March–May chlorophyll concentration (Hargrave & Peer unpublished data). The data are for Eastern Canadian sites, in the Gulf of St Lawrence and adjacent areas, and for published data from Long Island Sound (near New York), the North Sea, and a Scottish inlet.

The distribution of benthic animals according to feeding habits

The relationship between environmental conditions and feeding habits of the benthos is seen most clearly in the deeper parts of the great ocean basins (Sokolova 1959, 1972). Since, as we have seen, a very small amount of organic matter reaches these sediments, their organic carbon content is low (<0·25 per cent), the upper oxidized layer is very thick (1–10 m) and commonly consists of red clays or of globigerine or radiolarian ooze. Clearly, a mode of life which involved ingesting these sediments would be extremely unrewarding. The most effective feeding method is to filter the water just above the sediments, or scoop the detritus from the sediment surface. On the other hand, in the shallower waters at the fringes of the great ocean basins, and in the equatorial Pacific Ocean there is a more rapid settlement of organic matter, resulting in organic carbon concentrations in the surface sediments of 0·5–1·5 per cent and a thin oxidized layer, with reducing conditions beneath. It is then possible for animals to obtain sufficient nutrition by ingesting the subsurface

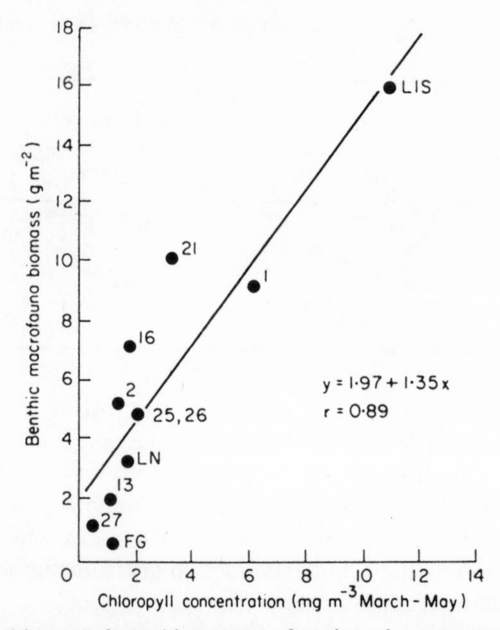

Fig. 10.3. Benthic macrofauna biomass as a function of average total chlorophyll concentration in the photic zone during March to May, in various areas. (From B.T. Hargrave & D.L. Peer, unpublished data.) 1 = Bedford Basin, 2 = St Margaret's Bay, both on the Atlantic coast of Nova Scotia. 13 = Magdalen Shallows; 16 = Gaspé Passage; 21, 25, 26, Central Area, all in the Gulf of St Lawrence. 27 = Ellerslie, Prince Edward Island. *FG* = Fladen Ground, North Sea; *LN* = Loch Nevis, Scotland; *LIS* = Long Island Sound, New England.

deposits. In depressions or on extensive flat surfaces there is a good accumulation of organic detritus and deposit feeders tend to be concentrated there, while on ridges and escarpments movements of bottom water tend to keep the organic matter in suspension, so that filter-feeders are locally abundant. The net result is a mixture of feeding types in the areas of higher organic sedimentation.

Sokolova (1972) labelled the two types of bottom area as oligotrophic and eutrophic (Fig. 10.4) and found a good correlation between bottom type and the fauna collected in a trawl. She found that deposit-feeding animals such as starfish of the family Porcellanasteridae, holothurians of the family Molpadidae, and various types of sea urchins, were more or less confined to the eutrophic, peripheral and equatorial areas of the ocean floor. In contrast, suspension-feeders such as sponges of the orders Hexactinellida and Tetraxonida, serpulid polychaetes, and cirripede barnacles of the family Scalpellidae were widely distributed in both eutrophic and oligotrophic regions of the oceans. The distribution of predators, such as starfish of the families Brisingidae, Pterasteridae, Astropectenidae, and so on, brittle stars of the family

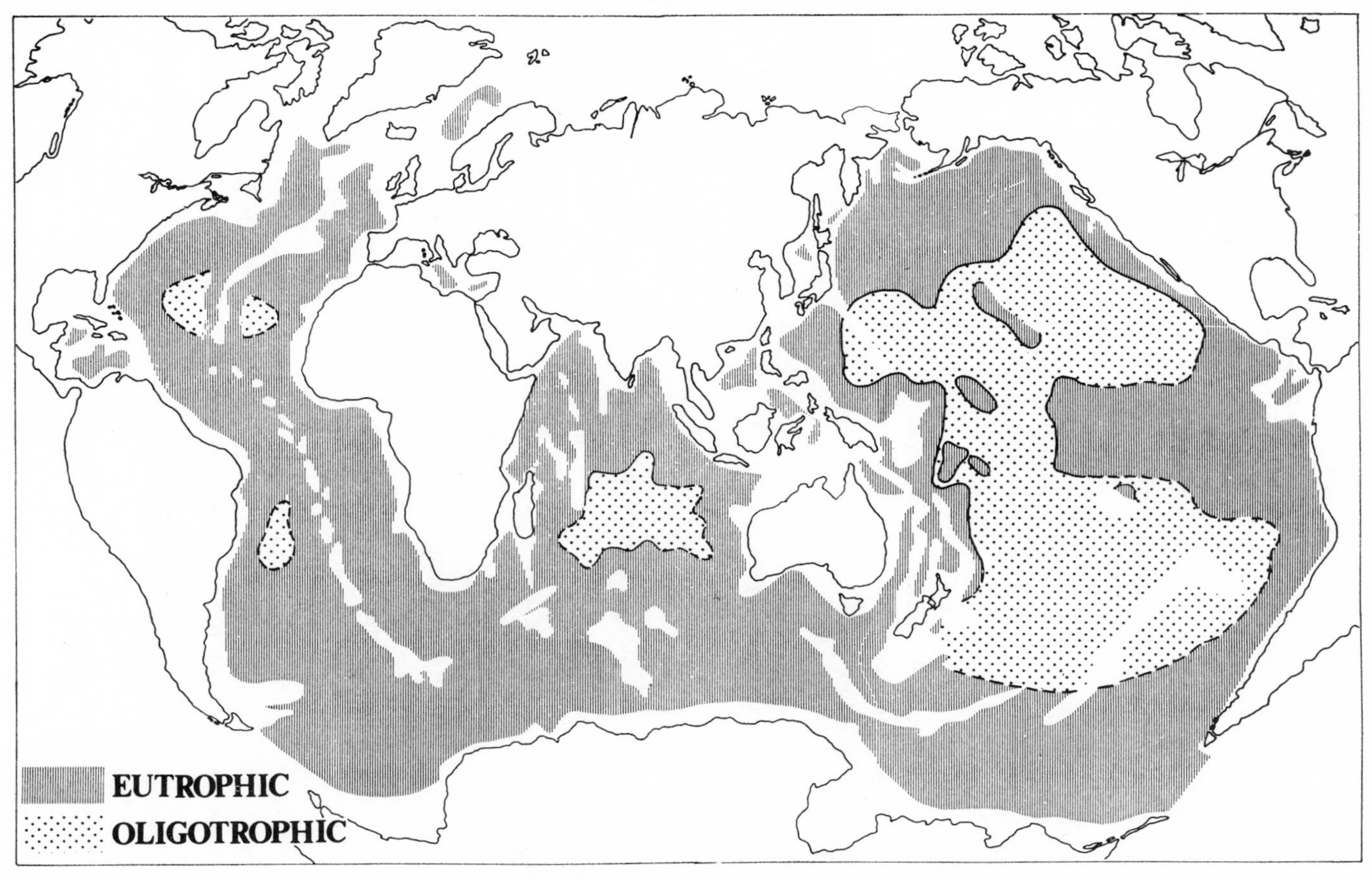

Fig. 10.4. Division of the ocean floor into oligotrophic and eutrophic areas. (After Sokolova 1972.)

Ophiuroidae, and sea anemones, was seen to be related to the areas of higher benthic production—the eutrophic zones.

Hessler and Jumars (1974) obtained evidence which conflicts with Sokolova's hypothesis. When sampling an oligotrophic bottom under the North Pacific Central Water Mass, using a box corer, they found that deposit feeders constituted the overwhelming majority in terms of numbers of animals. Of the macrofauna, polychaetes comprised over half the animals taken, and almost all of these were deposit feeders. Overall, animals which might have been suspension feeders comprised no more than 7 per cent of the numbers in the total fauna.

In comparing their results with those of Sokolova, Hessler and Jumars (1974) pointed out that many of the differences could be explained by differences in collecting methods used. Sokolova's hypothesis concerns the distribution of large animals caught in trawls (plus small animals encrusting larger objects caught in trawls). These animals are too sparsely distributed to be sampled adequately by grabs or corers, yet they contribute a significant biomass on account of their large body size.

An integrated account of the distribution of deep-sea benthic animals requires the sampling of representative areas by an array of gear which will enable us to build a picture of the relative contributions of animals of different sizes and distributions.

Life histories of benthic animals

Most benthic invertebrates live sedentary or slow-moving lives on or near the sea floor. By comparison with planktonic or nektonic animals in the water column, they are at a considerable disadvantage. They lack the ability to readily colonize new areas for their species, or to recolonize areas from which the species has been eliminated. They are not able to exchange individuals readily with neighbouring populations, thus spreading favourable genetic combinations or maintaining the integrity of the species. Furthermore, they are condemned to life in areas of relatively sparse food supply and low temperatures.

It is not surprising, therefore, that a large percentage of benthic invertebrates have a stage in their development when benthic life is abandoned temporarily in favour of life in the plankton. About 70 per cent of all marine species (Thorson 1950) have planktonic larvae which are able to take advantage of the higher temperatures and more abundant food in the surface waters, while drifting some distance from their parent populations. Planktonic larvae are found in the life histories of more than 80 per cent of the species on the continental shelves of temperate and tropical waters, but in fewer species in arctic waters and in the deep ocean.

Two main types of pelagic larvae are found: (1) those which are produced in very large numbers, but have very little food reserve. Their survival depends on their being able to obtain a food supply soon after hatching. By virtue of their large numbers, they are very susceptible to predation in the plankton. (2) Those which are given a generous supply of food reserves (yolk) and are therefore produced by the parent in relatively small numbers. These larvae can exist independently of an external food supply for considerable periods. The first type is far the most common among marine invertebrates. Since survival to the end of development is dependent on achieving just the right combination of environmental factors, it is usual to find that recruitment of

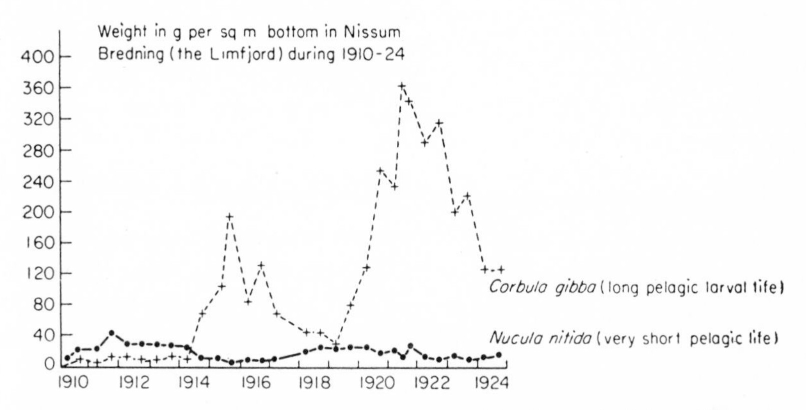

Fig. 10.5. An example of a benthic population with a long pelagic larval life and large fluctuations in population density, contrasted with one with a very short pelagic larval life and relatively constant population density. (From Thorson 1950.)

young animals to the adult population, and hence the adult population size, varies greatly from year to year. In contrast, recruitment in species which have yolky eggs tends to be reasonably constant, and the population in a given area tends to be much more constant in time (Fig. 10.5).

Opinions differ as to the geographic range over which genetic exchange takes place as a result of larval drift. Mileikovsky (1968) maintains that less than 10 per cent of the larval population of the shelf waters is carried away from the general area of the parental populations, and that those that do move over the deeper areas settle out to form populations which are incapable of reproducing themselves, and may be called 'pseudo-populations'. Scheltema (1971) on the other hand, drew attention to the evidence of larvae surviving year round in oceanic currents and serving to mediate gene flow between populations as far apart as the east and west shores of the Atlantic Ocean. This seems to be the case for a number of mollusc species, as well as for certain coelenterates, annelids and crustaceans.

Methods of estimating production

The biomass of the benthos has been determined at hundreds of sites in many parts of the world's oceans, but for an understanding of the ecological role of the benthos in an area it is necessary to know the rate at which new material is being produced—the productivity or rate of production. This is a function of the growth rate of each organism present, and varies according to species and according to the age of each animal within a species. Determination of productivity is a daunting task, yet only a knowledge of productivity gives an understanding of rates of transfer of energy and materials through food chains and ultimately to man.

Details will be given below of the methods used to estimate the productivity of various species. Fortunately, we are beginning to see evidence of predictability in the ratio between production and biomass, P/B. There is reasonable hope that before too long it will be possible to infer production from a knowledge of biomass and a few environmental parameters. Meantime, it is necessary to painstakingly collect data for accurate assessment of production in particular situations.

In any given population, the production of new tissue by reproduction and growth is balanced by losses associated with mortality, if we assume that immigration and emigration are in balance. The various ways of calculating production may be illustrated by reference to an imaginary situation in which a particular age class of animals is sampled and found, at time t_1, to have a population density N_1 and a mean weight of individual, W_1. Later, at time t_2, it is found to have a population density N_2 and a mean weight W_2. The biomass change, ΔB is given by

$$\Delta B = N_1 W_1 - N_2 W_2 \tag{10.1}$$

The production is given by the average number present multiplied by the mean weight change of an individual

$$P = 1/2\,(N_1 + N_2) \cdot (W_2 - W_1) \tag{10.2}$$

The elimination, or loss from the population, is given by the change in numbers multiplied by the average weight of a specimen

$$E = (N_1 - N_2) \cdot 1/2\,(W_1 + W_2) \tag{10.3}$$

It is readily shown that $\Delta B = P - E$, i.e. ΔB is the resultant of the two processes of increase by production and loss by elimination. Measurement of any two of these parameters will enable the third to be determined by difference.

In populations in which breeding occurs at a well-defined season, so that distinct age classes can be recognized, it is possible to collect data on population density and numbers present for each age class, and to calculate production by each age class during the interval between samples. For example,

Birkett (1959) sampled a population of the bivalve *Mactra* in the North Sea on 5 occasions in a period of 616 days. He was able to project his data backwards to the known date of spatfall, and calculate production, elimination and biomass change in each of the five sampling intervals. During the first part of the study elimination exceeded production and biomass decreased, but in the later period production greatly exceeded elimination and biomass

Table 10.2. Production data from a population of *Mactra* in the North Sea. (Recalculated from Birkett 1959.)

t (days)	N_t (number m^{-2})	Mean wt (mg)	Biomass (mg m^{-2})	Production mg m^{-2} d^{-1}	Elimination mg m^{-2} d^{-1}
0	7045	1·416	9,976		
				317·38	410·53
50	990	5·364	5,310		
				17·78	26·71
225	378	9·910	3,746		
				66·28	13·94
398	289	44·286	12,799		
				35·90	11·62
616	246	73·542	18,091		

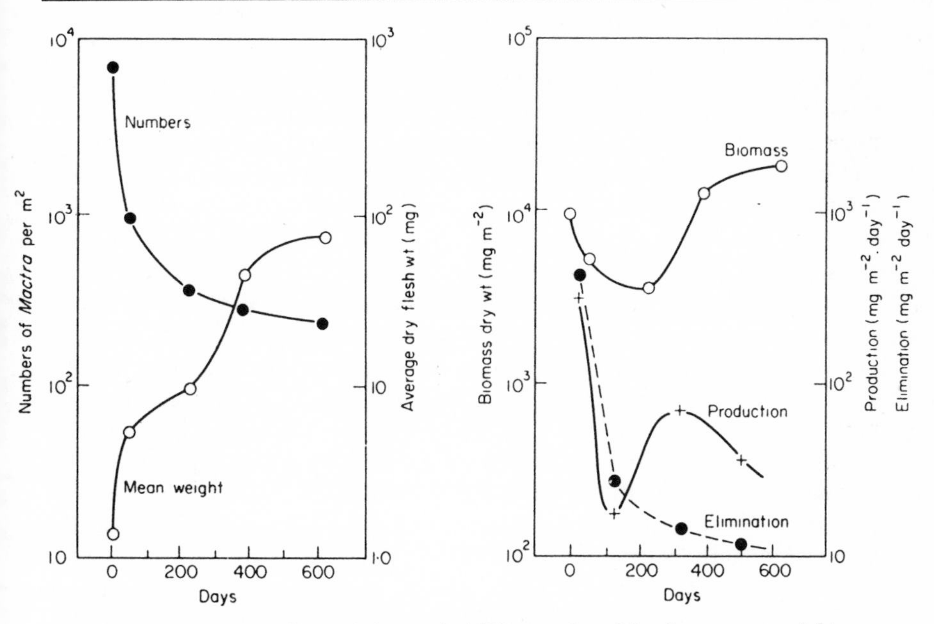

Fig. 10.6. *Mactra stultorum* (Linnaeus), 1957 year class. Numbers, mean weight, biomass, production and elimination of a year class of this bivalve mollusc from the North Sea. (From Birkett 1959.)

increased again (Table 10.2 and Fig. 10.6). The information in Figure 10.6 can also be represented as a single line, by plotting population density of a cohort against mean weight (Fig. 10.7). It has been shown that production between the first and last sample dates is given by the area under the curve (Allen 1951, Nees & Dugdale 1959, Mann 1969). Such a plot is referred to as an Allen curve, after K. R. Allen who used it to calculate the productivity of a trout population.

If the procedure described above is repeated for each age class in a population, total population production can be calculated. Sanders (1956) calculated the productivity of two polychaetes, two molluscs and an amphipod in Long Island Sound. Richards and Riley (1967) made similar calculations for other species in the same general area. Peer (1970) calculated production, elimination and biomass change for the polychaete *Pectinaria* in St Margaret's Bay, Nova Scotia, and Miller and Mann (1973) calculated production and other components of an energy budget for the sea urchin *Strongylocentrotus*. Burke and Mann (1974) worked with various molluscs of a shallow, sedimented inlet of Nova Scotia, and calculated production parameters of *Littorina*, *Mya* and *Macoma*.

Grave difficulty is encountered in attempting to calculate the productivity of species which do not have well-defined short breeding periods, and hence have populations consisting of numerous overlapping age classes. The problem is most frequently encountered in the plankton, but is applicable to many meiofauna species and probably to abyssal macrobenthos. The solution to the problem lies in determining the typical rate of growth of each size class, or instar, at a range of environmental temperatures. Each sample of the population is then analysed into its component size classes regardless of age and production is calculated on the basis of (numbers present × growth rate) for each class.

Production to biomass ratios (P/B)

Production has been measured in benthic populations for two main reasons. The first is so that the flux of energy and materials from the benthos to its predators can be quantified in particular situations. The second is in the hope of finding generalizations about P/B ratios which might be applicable under a wide range of conditions. In studying the functioning of ecosystems in temperate climates, one of the natural units of time is the year. Consequently, a number of authors have calculated the ratio of annual production to annual mean biomass and come up with answers that are in fair agreement. In Sanders' (1956) study in Long Island Sound the annual P/B ratios for two species of polychaete and two molluscs, all having a life history of over one year duration, lay between 1·94 and 2·28. Burke and Mann (1974) found that

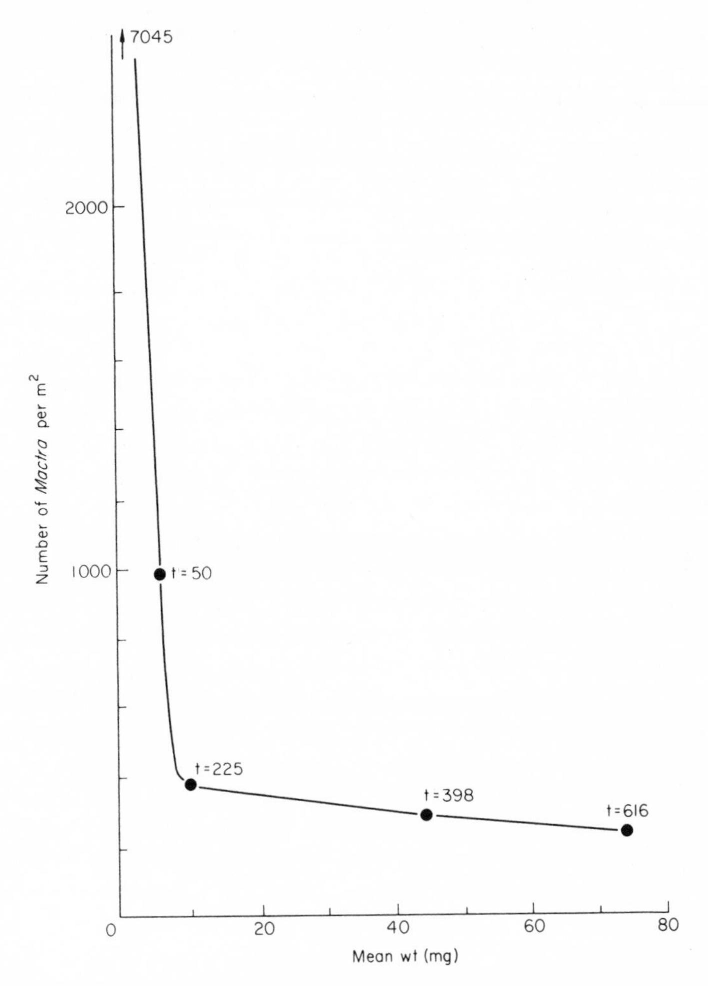

Fig. 10.7. Data from Fig. 10.6 plotted as an Allen Curve. For explanation, see text.

the annual *P/B* ratio for a *Macoma* population in Nova Scotia was 1·53 and for a *Mya* population it was 2·54. For animals completing their life history in one year or less, the annual *P/B* ratio is somewhat higher. Sanders (1956) found that the amphipod *Ampelisca*, a species which completes its life history in less than a year, had a *P/B* ratio of about 5, and Burke and Mann (1974) found that *Littorina saxatilis* (Olivi) which lives for one year, had a *P/B* ratio of 4.11.

Waters (1969) constructed a theoretical model based on an Allen Curve. For a series of hypothetical populations having different mortality and growth patterns, yielding Allen Curves of a wide range of possible shapes, he showed that most organisms have a P/B ratio in the range 3 to 6, *within the life span of an age class*. Hence, it follows that organisms requiring n years to complete their life history will have an annual P/B ratio of $1/n$ times the figure predicted for P/B in the life span, and animals having n generations in a year will have an annual P/B ratio n times the figure predicted by Waters.

It follows that P/B ratios are also a function of body size, since smaller animals have shorter life histories and vice versa. Temperature may have an effect on rate of growth, and hence on time required to complete a life history, so that in general populations in lower temperatures will have lower P/B ratios.

In the U.S.S.R. there has been a great deal of attention devoted to calculation of P/B ratios. Zaika (1970) defined the parameter C as the production of a population per unit of biomass per unit of time. Production and biomass can be expressed in a number of ways so long as both are in the same units. The dimension of C is time^{-1}. Zaika expressed the data as 'average daily C measured over a year'. In an intensive study of data on marine molluscs it was found that C ranged from 0·0003 to 0·03. This corresponds with annual P/B ratios of 0·1 to 10·9. The dependence of the P/B ratio on length of life was clearly demonstrated in a plot of these data for a variety of marine and freshwater mollusca (Zaika 1972) (Fig. 10.8). The figure shows fair agreement with Waters' model, since a predicted annual P/B ratio of 1·5 to 3 for animals requiring 2 years to complete their life history would lie in the same range as Zaika's data.

An examination of the faunal lists of continental shelf areas suggests that the greatest part of the benthic biomass is comprised of worms, clams, echinoderms which require 1, 2, or more years to complete their life histories. On the basis of the figures given above, it seems likely that annual production by the benthos may be expected to be about twice the mean biomass, at least in temperate waters.

Commercially important shellfish stocks: population dynamics

While shellfish cannot compete with fish in terms of biomass yield, they are of major importance in terms of cash value. For example, in eastern Canada in 1961 the landed value of lobsters and scallops combined was $21 million, greater than the combined value of the two most important fish species, cod ($15·4 million) and haddock ($4·6 million) (Fig. 10.9). In 1961 the catch of lobsters (*Homarus americanus* (H. Milne-Edwards)) totalled almost 22 million kg in Canada, and over 11 million in the eastern United States (Wilder 1965, Dow, 1969).

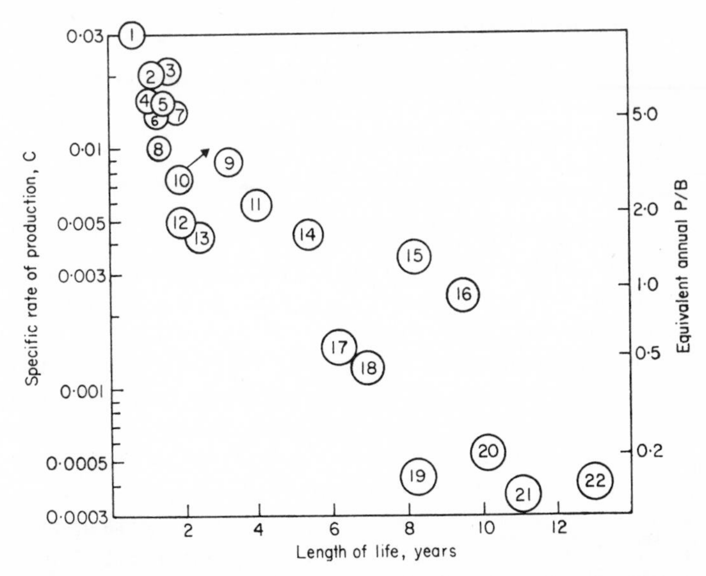

Fig. 10.8. Relation between specific daily production, *C*, and length of life, for 22 species of aquatic molluscs. (From Zaika 1972.) For comparison, annual P/B ratios, calculated as $C \times 365$ are indicated on right side. 1 = *Lacuna pallidula* (da Costa); 2 = *Spisula elliptica* (Brown); 3 = *Margarita helicina* (Fabricius); 4 = *Anisus vortex* (L); 5 = *Gyraulus albus* (Müller); 6 = *Valvata pulchella* (Studer); 7 = *Adacna vitrea* (Eichwald); 8 = *Rissoa splendida* (Eichwald); 9 = *Mytilaster lineatus* (Gmelin); 10 = *Margarita helicina*; 11 = *Abra ovata* (Sowerby); 12 = *Bithynia tentaculata* (L); 13 = *Sphaerium corneum* (L); 14 = *Cardium edule* (L); 15 = *Acmaea digitalis* (Rathke); 16 = *Mytilus galloprovincialis* (Lamarck); 17 = *Dreissena polymorpha* (Pallas); 18 = *Acmaea testudinalis* (Müller); 19 = *Modiolus demissus* (Dillwyn); 20 = *Anodonta anatina* (L); 21 = *Unio tumidus* (Philipsson); 22 = *Unio pictorum* (L).

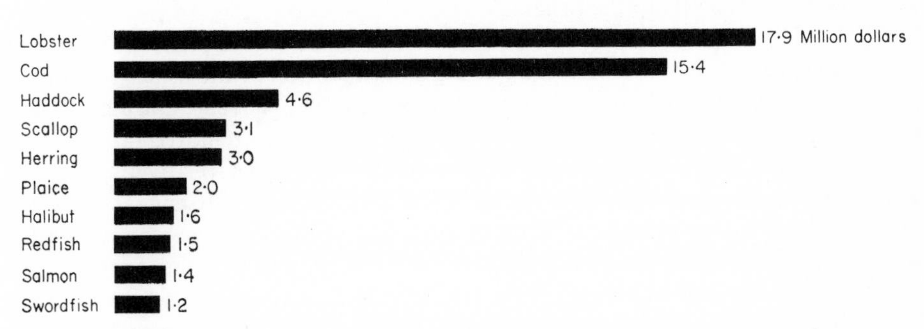

Fig. 10.9. Landed value of Canada's 10 most valuable Atlantic coast species, 1961. (From Wilder 1965.)

Population dynamics of lobsters

Female lobsters carry their eggs attached to the ventral side of the abdomen; in different parts of eastern Canada the number of eggs per female averages from 7000 to 30,000 (Wilder 1965, Squires 1970). The newly hatched larvae pass through 4 stages in the plankton, and mortality during the planktonic stage is very high, of the order of 99 per cent. Wilder (1965) showed that in one area between 1949 and 1961 the number of stage 1 larvae varied from year to year by a factor of 40, while the number of stage 4 larvae varied by a factor of only 4. He inferred that egg production was not a limiting factor in the abundance of 4th instar larvae.

In the Northumberland Strait, south of Prince Edward Island, it was found that the lobsters moult 5 to 7 times during their first growing season, passing the winter in 6th, 7th or 8th stages. In years 2 to 5 lobsters moult twice, and thereafter once per year, reaching the 20th stage at an age of 8¼ years and a length of 240 mm. Hence, a growth curve can be produced for lobsters in this area. Growth in other areas may be considerably different.

The lobsters are caught in wooden traps constructed so as to allow the escape of small lobsters and to prevent the entrance of very large specimens. The traps are laid mostly from small (i.e. *c.* 20 foot long) boats which operate close to shore. In Canada different parts of the coast have different seasons when trapping is permitted, and the effect is to spread the catching effort round most of the year, including winter in areas that are ice-free.

There is a lower size limit, of the order of 18–23 cm total length, below which it is illegal to retain lobsters in the catch. It is also illegal to retain lobsters carrying eggs. The fishing pressure is very intense, so that catches, usually high at the beginning of the season, decline sharply within two or three weeks. It has been estimated that in different areas 48–90 per cent of the lobsters which have attained legal size are removed in a season. Naturally, in areas of highest fishing pressure, most of the catch consists of lobsters which are new recruits to the fishable stock.

Historically, the Canadian catch has passed through 3 phases (Fig. 10.10). Between 1870 and 1886 there was a rapid expansion of the fishery, to a peak landing of about 100 million pounds (45 million kg) in 1886. During the next 30 years the fishery declined as accumulated stock was removed faster than it could be replenished. Since 1918 landings have been relatively constant, between 30 and 50 million pounds (14–23 million kg). It appears that the present regime of exploitation is in some kind of equilibrium with production. However, within the overall picture of moderately stable catches for the whole of eastern Canada, there are interesting fluctuations from one area to another. Dow (1969) has shown that the catch for Maine (U.S.A.) trebled during the period 1939 to 1951, and that this coincided with a warming trend in sea water temperatures, during which the annual mean at Boothbay Harbour

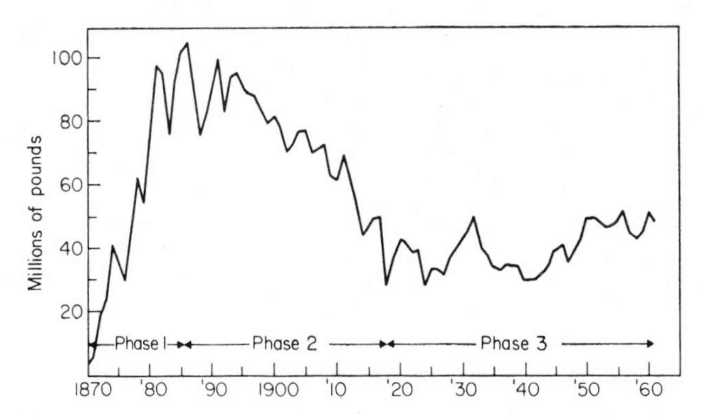

Fig. 10.10. Annual Canadian lobster landings, 1870–1961. (From Wilder 1965.)

changed from 6·5° C to 10° C. After 1953 the temperature began to fall, and so did the catch, after a lag of 4–5 years (Fig. 10.11). Even more striking is the fact that during this phase of warming, peak catches were attained first in Rhode Island (1939) then in New Hampshire (1943), followed by Nova Scotia (1951) and Newfoundland (1955), the northern limit of the lobster's range. During the subsequent cooling phase peak catches appeared in New Brunswick and Prince Edward Island (1960), Quebec (1962), Massachusetts (1965), Connecticut and New York (1967). Hence, during the warming phase record lobster catches appeared successively in a south to north direction, and vice versa during the cooling phase.

The last 20 years have seen a great development of trawling for lobsters near submarine canyons at the edge of the continental shelf, 200 to 400 kg off the coast of New England. Skud (1969) has documented the effects of exploitation, including decrease in average size, change in sex ratio and lower catch per unit effort, all of which are most marked in the canyons closest to shore. In the canyons further from shore new recruits account for only 5 per cent of the catch, which is in sharp contrast to the coastal fishery of Maine, where new recruits account for 80 to 90 per cent of the landings.

An interesting interaction is thought to occur between lobsters, sea urchins and seaweed in the coastal zone of Nova Scotia (Mann 1973). Sea urchins (*Strongylocentrotus*) eat kelp (*Laminaria*) and when their population density exceeds a critical level they eat out the seaweed beds, leaving almost bare rock where once had grown luxurious and highly productive kelp forests. Local population explosions of sea urchins occur sporadically in areas where lobster fishing has been carried on intensively. Sea urchins are high on the list of food preference for lobsters (Himmelman & Steele 1971) and the hypothesis has been advanced (Mann & Breen 1972) that there is a balanced predator–prey relationship between lobsters and sea urchins, which

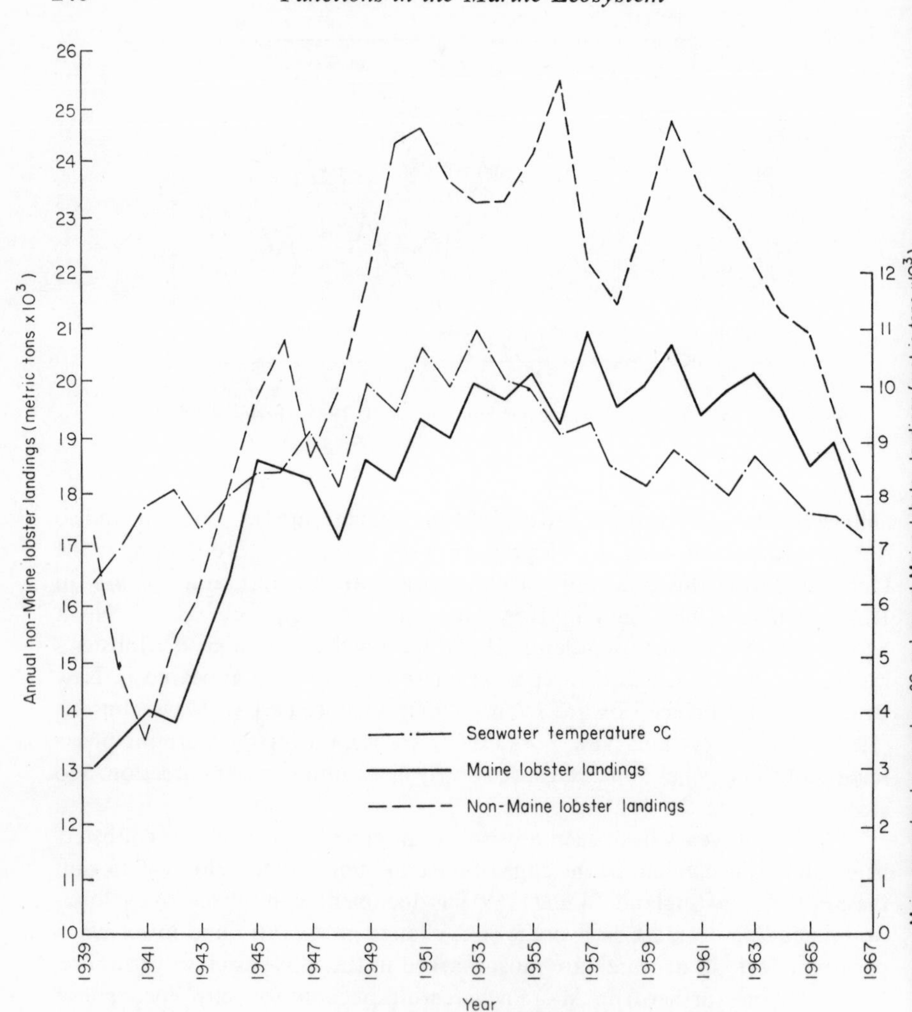

Fig. 10.11. Comparison of Maine and non-Maine lobster landings with mean sea surface temperature at Boothbay Harbour, Maine. (From Dow 1969.)

can be upset by too heavy exploitation of lobster stocks. Since, as was mentioned earlier, the productivity of seaweeds is an important factor in the high total productivity of coastal waters, devastation of seaweed beds can have a depressing effect on the level of production in the food chain leading to lobsters.

Population dynamics of scallops

For lobster populations it was shown that although the numbers of stage 4

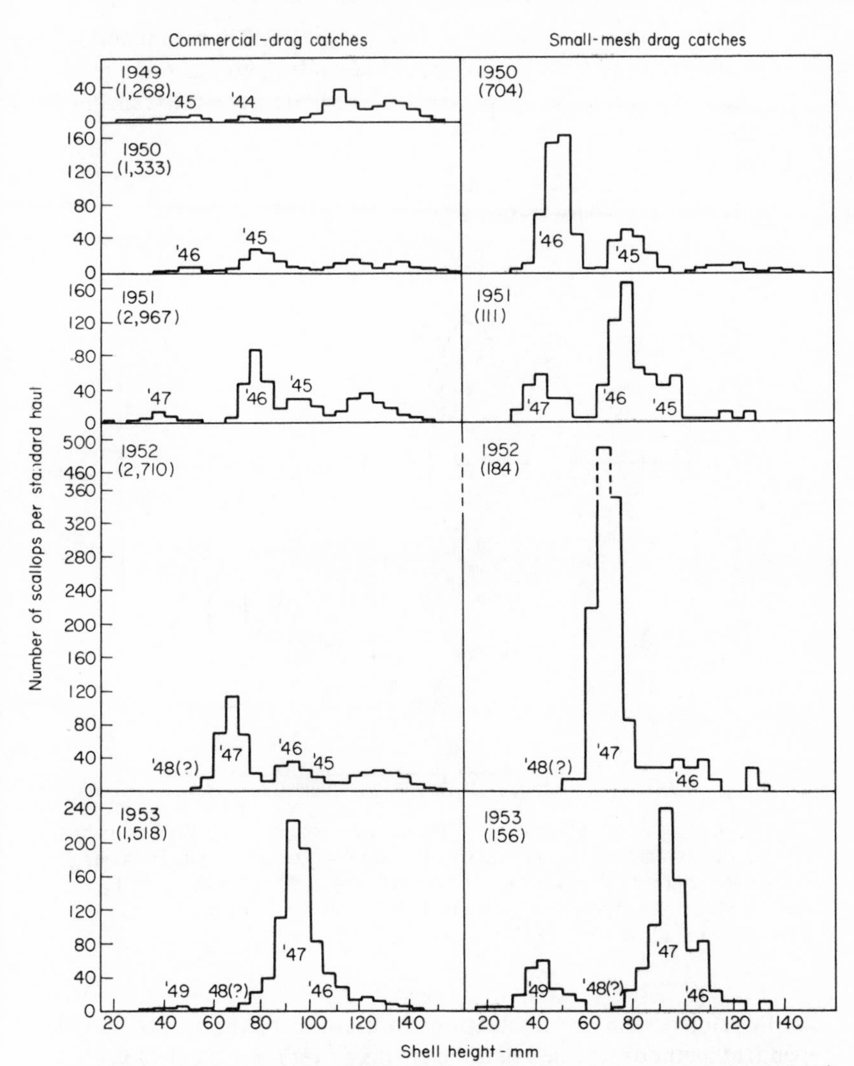

Fig. 10.12. Catches of scallop, *Placopecten magellanicus* in samples from Bay of Fundy, near Digby, Nova Scotia. (From Dickie 1955.)

planktonic lobster larvae available for recruitment to the benthic population may vary by a factor of 4 between years, there is no evidence of major failures of recruitment. Scallop populations differ in this respect, for it has been shown in a number of places that recruitment to the stocks is extremely variable from year to year, and that a 'good' year's recruitment often dominates the fishery for many years. For example, Dickie (1955) showed that for the populations of *Placopecten magellanicus* (Gmelin) off Digby, Nova Scotia, the catches of 1953 were dominated by the 1947 year class. Experimental fishing with a special small-mesh drag confirmed that while there was some representation of the 1949 year class, there were almost no representatives from 1948 or from 1946 and earlier (Fig 10.12).

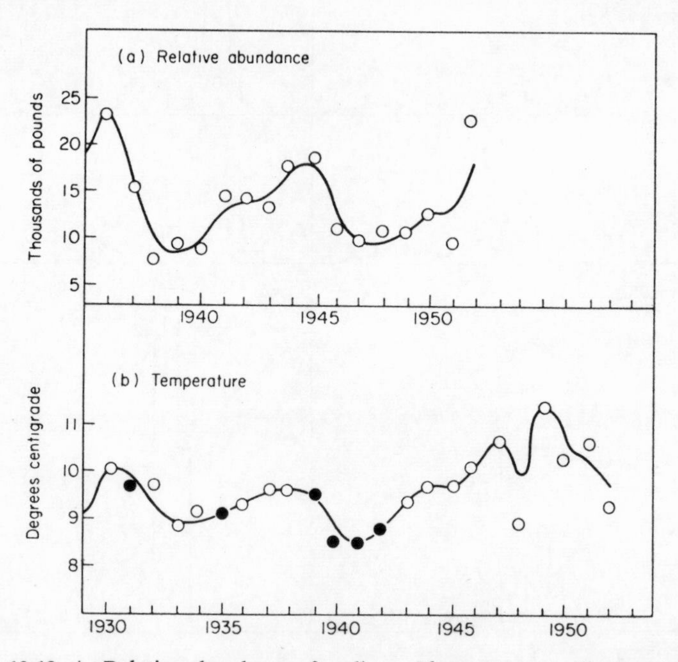

Fig. 10.13. A. Relative abundance of scallops, *Placopecten magellanicus*, in the Bay of Fundy near Digby, as shown by seasonal catch per boat. B. Average September–October–November water temperatures at 90 m depth in the Bay of Fundy. Note that temperatures are plotted against a time scale 6 years earlier than the catch data above them. (From Dickie 1955.)

An investigation of the fluctuations in biomass of the stocks by several independent methods revealed that the changes over a period of 17 years were faithfully reflected in the statistics of seasonal catch per boat (Fig 10.13A). When these were compared with water temperatures in early autumn it was found that there was a good correlation ($r = 0{\cdot}723$) between catch in a particular year and water temperature six years earlier (Fig 10.13B) Dickie interpreted this as an effect on the pelagic larvae. When temperatures are high the larvae pass through the planktonic stage and settle on the bottom fairly quickly. Also, high temperatures are associated with a relatively closed circulation in the Bay of Fundy, which holds the larvae in the vicinity of the parent beds. Conversely, years of low temperature indicate a great exchange of water in the Bay of Fundy with outside water masses. Since the larvae develop relatively slowly, there is a heavy loss of larvae from the Bay, poor settlement on the parent beds, a weak year class, and low abundance of the catchable stocks six years later.

The minimum size of scallop taken by the industry is regulated through the mesh size of the bags used to catch them. Dickie (1955) estimated that 'natural', i.e. non-fishing mortality on the Bay of Fundy stocks was about 10 per cent per annum, while direct fishing mortality was about 20 per cent per annum. Clearly, the intensity of exploitation of the stocks is not as great as for lobsters, where 90 per cent of the catch may be new recruits. Nevertheless, a parallel exists, in that the last 20 years has seen the development of U.S. and Canadian fleets equipped for exploitation of offshore populations, such as those on George's Bank, off New England, and there has been an initial period of steadily increasing catches (Bourne 1963).

Conclusions

The biomass of benthic invertebrates of the world has been estimated at 6–7 × 10^9 metric tons, of which over 80 per cent is on the continental shelves. The reason for this pattern of distribution is that primary production is higher in coastal waters than in mid-ocean, and the proportion of this fixed carbon reaching the bottom is inversely proportional to the depth of the water column. On this basis, the marine benthic fauna of the world may be divided into eutrophic and oligotrophic communities.

Most macro-invertebrate species of the coastal benthos have planktonic larvae to facilitate dispersal. These larvae are important in maintaining gene flow between otherwise isolated populations.

Estimation of annual ratio of production to biomass (P/B) enables one to estimate benthic productivity from biomass data. Larger, long-lived species have low P/B ratios, commonly less than 1·0, while smaller, short-lived animals have P/B ratios of the order of 10. It is suggested as a first approximation that the annual productivity of the world's benthic invertebrates may be of the order of twice the biomass, i.e. of the order of 13×10^9 metric tons for the world's oceans.

The economic value to man of most of this benthos is as food for the commercially important stocks of bottom feeding fish, but a proportion is exploited directly as 'shellfish'. Benthic crustaceans and molluscs are particularly vulnerable to fishing pressure, since their populations are static and readily located. The case histories quoted in this chapter illustrate the delicate balance which may exist between the fishing industry and the stocks. Species with regular recruitment may be fished so hard that 90 per cent of the recruits to the fishable stock are removed in the year of entry. Species with irregular recruitment are subject to major fluctuations in abundance. One result of this is that it is not economic in the long term to maintain a fleet of vessels capable of removing such a high proportion of the year's recruits during their first year in the fishable stocks.

Acknowledgements

I thank Dr B.T. Hargrave for the use of unpublished data (Fig. 10.3), Dr Eric Mills for valuable criticism, and Miss Leslie Linkletter for a great deal of help with the literature survey. Some of the work reported here was supported by National Research Council of Canada Grant 1200-242-037 to K.H. Mann.

Chapter 11. Growth of Fishes

R. Jones

Introduction

Description of growth
Growth in length
Growth in weight
Variations in shape
Growth equations
von Bertalanffy
Robertson
Gompertz
Parker and Larkin
Variations in growth between species
Geographical variations in growth within species
Temporal changes in growth rate
Changes in growth following transplantation
Effect of the length of the growing season on growth

Environmental effects on growth
The effect of temperature on growth
The effect of maturity on growth
Hierarchical effects on growth
The effect of genetic factors on growth
The effect of fish density on growth

Physiology of growth
Empirical relationship between food intake and growth
Relationship between the maintenance coefficient (α_1) and bodyweight (W)
Conversion efficiency
The effect of temperature on metabolism
The energy required for swimming
Metabolic growth models

Discussion

Introduction

Growth is the change of size of a living organism with age. Fish differ from the higher animals such as mammals and birds in that they appear to continue growing throughout the whole of their lives. Mammals and birds, on the other hand, tend to grow to their adult sizes relatively early in life and then to remain at approximately the same size for the greater part of their adult lives.

Description of growth

Growth in length

Growth can be recorded by making measurements of the length or of the weight of an individual and relating them to age. Figure 11.1, shows the relationship between length and age of a female *Lebistes*, and this is typical of many fish species. Growth in length never quite ceases, but tends towards an asymptotic size. At any age, the growth rate can then be estimated by the slope of the curve at that age. The growth rate in this example is most rapid when the fish is young, and diminishes progressively as it becomes older. The relationship in Figure 11.1 is an approximation since it is usual to find seasonal fluctuations in growth rate superimposed on the annual growth pattern. Fish in temperate climates grow relatively rapidly in the summer and relatively slowly in the winter, so that, a curve such as that in Figure 11.2 is required to describe age/length relationships in more detail.

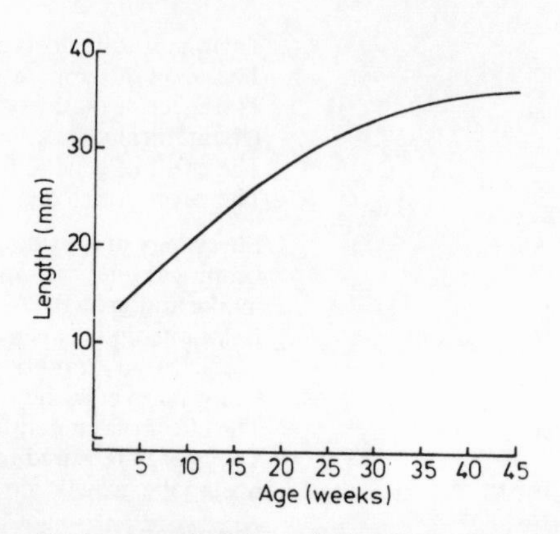

Fig. 11.1. Relationship between length and age of female *Lebistes* (Ursin 1967).

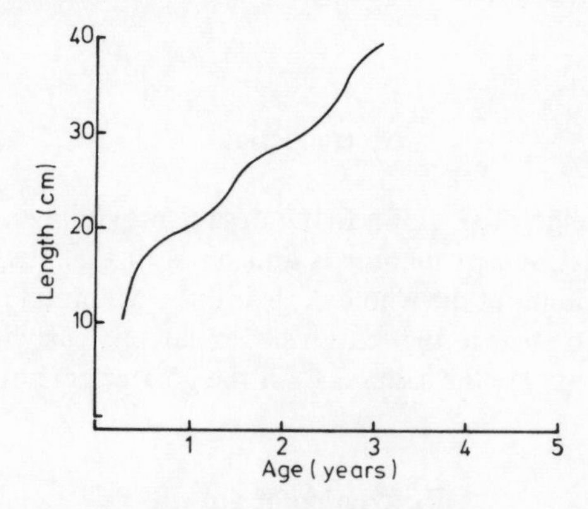

Fig. 11.2. Relationship between length and age of haddock from the east coast of Scotland (1945–54)—from Scottish research vessel data.

Growth in weight

As in length, body weight increases throughout the life of a fish and tends towards an asymptotic value (Fig. 11.3). It differs from the typical length/age relationship as shown in Figure 11.1, in that there is a point of inflexion at point A, i.e. up to the age corresponding to that point, the growth rate progressively increases with age, whereas subsequently it progressively decreases. In the length/age curve a similar inflexion may occur during the first few months of life in the larval stage, but apart from this, the growth rate in units of length declines progressively through the whole life. Length (L) and weight (W) in fishes are related in such a way that $W_0 \propto L^n$ and for many species n is close to three.

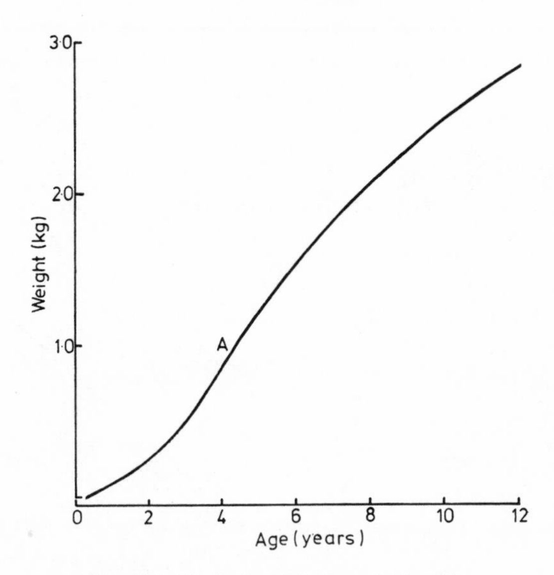

Fig. 11.3. Typical weight/age relationship.

Variations in shape

Growth is allometric. Heincke (1898) and others, for example, have described how different parts of a herring may grow at different rates at different times in its life. Particularly characteristic are differences in the relative body proportions, before and after metamorphosis.

Some relationships, such as the ratio of head length to total length may change continuously throughout life in some species. Other body proportions like the weight/length relationship vary seasonally. It varies throughout the course of the year leading to seasonal variations in the shape of the fish. A convenient way of measuring this is by means of a coefficient known as the condition factor (C). This is defined by $C = 100W/L^3$ where L is the length of the body in centimetres, and W is the body weight in grams. Seasonal variations in the condition factors of marine fish have been described by Graham (1924), Johnstone (1924), Russell (1922) and Stanek (1964). Among freshwater fish, seasonal variations in condition factors in perch have been investigated by Le Cren (1951).

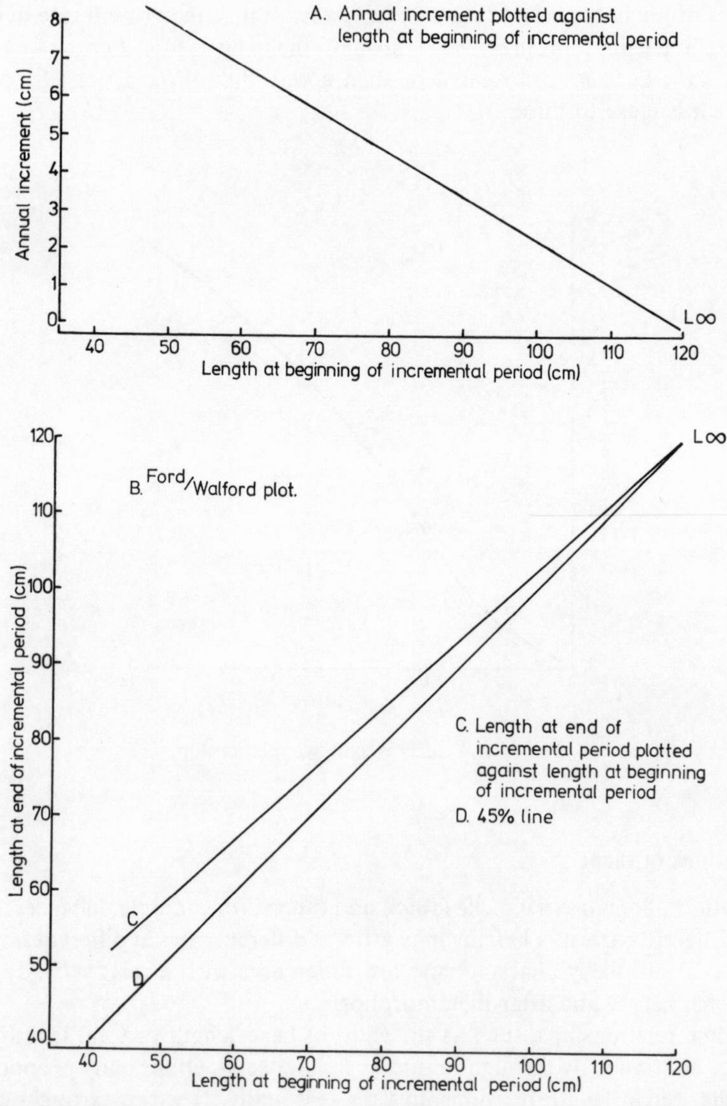

Fig. 11.4. Growth relationships of Norwegian saithe (data from Schmidt 1959).

Growth equations

VON BERTALANFFY

A number of expressions have been fitted to age/length and age/weight data. The von Bertalanffy curve (Bertalanffy 1938) is widely used, and is derived from the equation

$$L_t = L_\infty (1 - e^{-K(t - t_0)}) \quad (11.1)$$
$$[\text{or } W_t = W_\infty (1 - e^{-K(t - t_0)})^3]$$

where L_t is the length at age t, L_∞ is the asymptotic length when $t = \infty$, W_t is the weight at age t, W_∞ is the asymptotic weight when $t = \infty$, and K and t_0 are two other parameters needed to define any one curve exactly.

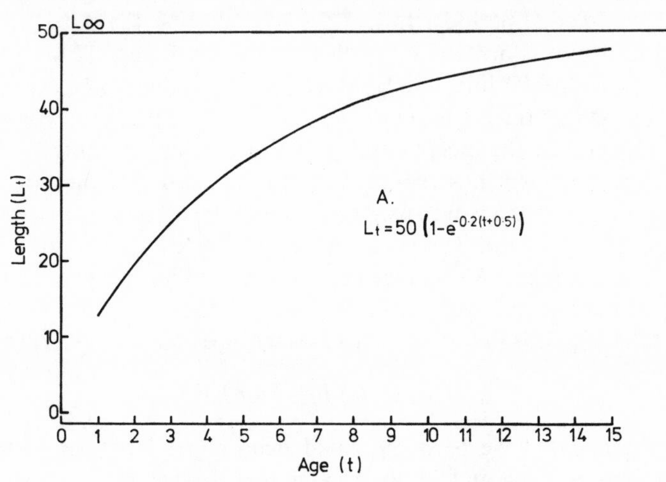

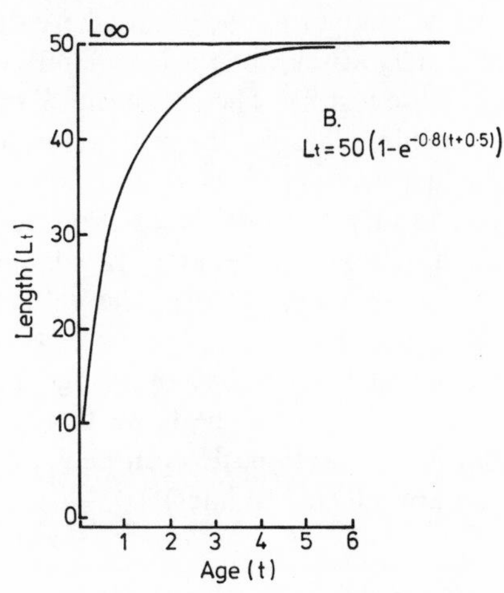

Fig. 11.5. Theoretical growth curves with different values of K.

There are various ways of deriving this equation for describing a length at age relationship. One follows from the fact that for a wide range of species, a length-at-age relationship such as that shown in Figure 11.1 can be transformed into a straight line, if it is re-plotted in certain ways. Figure 11.4a (from Schmidt 1959) shows the relationship between the annual growth increments of Norwegian saithe, *Pollachius virens* (Linnaeus), and the lengths of the fish at the beginning of the corresponding annual incremental periods. It shows that the annual increment declines linearly with increasing size of fish. Figure 11.4b with the same data shows the lengths at the end of incremental periods plotted against the lengths at the beginning of the corresponding incremental periods; this method of plotting the data is called a Ford–Walford plot (Ford 1933), Walford 1946). In Figure 11.4b, the fitted line (line C) cuts the 45° line, line D, at a length equivalent to L_∞, since at this point, the length at the beginning of the incremental period is equal to the length at the end of the incremental period and hence is equivalent to the length at which growth has ceased. In Figure 11.4a L_∞ is the point where the straight line cuts the length axis and it represents the length at which the annual increment is zero. The slopes of the two relationships are related in such a way that if $(exp-K)$ denotes the slope of the line in Figure 11.4b, then the slope of the line in Figure 11.4a is $(1-exp-K)$.

To estimate t_0, it is first necessary to re-arrange equation (11.1) to give

$$t_0 = t+(1/K)\ ln(1-Lt/L_\infty) \qquad (11.2)$$

Once L_∞ and K have been estimated from regression lines of the kind shown in Figures 11.4a and 11.4b, a value of t_0 can be obtained from each pair of values of L_t and t, using equation (11.2). The mean of a complete set of individual values of t_0 then gives an estimate of t_0, the point where the curve cuts the age axis. A non-zero value of t_0 arises when a Bertalanffy curve is fitted to date for the adult part of the life history. The coefficient K measures the degree of curvature of the relationship. Consider the two curves in Figure 11.5. These have been drawn with identical values of t_0 and L_∞. Curve A depicts a fish that grows relatively slowly towards its asymptotic length. Curve B on the other hand depicts a fish that grows more rapidly, and approaches its asymptotic size when it is relatively young. The value of K is lower for curve A than for curve B.

The Bertalanffy curve can be completely described by using only three parameters, and, experience has shown that it can be fitted to the growth patterns of a large number of fish species. Exceptions do occur (e.g. in the case of the Arctic cod and haddock, and are referred to later).

ROBERTSON

Although the von Bertalanffy growth equation has had the widest application, other equations have been used. For example, the logistic equation was used by Robertson (1923). It is based on the relationship

$$dW/dt = kW\ \ (A-W) \qquad (11.3)$$

where dW/dt is the growth rate (in weight units) and W is the body weight. The solution to this equation is:

$$W = A/(1+b\ exp-Akt) \tag{11.4}$$

where t is the age of the fish. This is a symmetrical sigmoid curve for growth in weight that rises to an asymptotic weight A when $t = \infty$.

GOMPERTZ

Another equation is the Gompertz equation and this is given by:

$$W = A\ \exp-b(\exp-ct) \tag{11.5}$$

This equation also rises to an asymptotic weight A when $t = \infty$. It has been applied by Saetersdal and Cadima (1960) to Arctic cod and haddock growth data.

PARKER AND LARKIN

A fourth equation, due to Parker and Larkin (1959), is based on the assumption that the rate of change of weight with age is proportional to the body weight to the power x, i.e.

$$dW/dt = kW^x \tag{11.6}$$

so that

$$W^{(1-x)} = (1-x)kt+W_0^{(1-x)} \tag{11.7}$$

Here W_0 is equivalent to the weight when the age t is equal to zero. This equation does not tend towards an asymptotic size as t tends to infinity. Parker and Larkin used it to describe the growth of steelhead trout, *Salmo gairdneri* (Richardson) and Chinook salmon, *Oncorhynchus tshawytscha* (Walbaum), with a different exponent in each stage of development.

There is no *a priori* reason for preferring any one of these equations, and their use is best assessed on their 'goodness of fit' to data, which means that the constants have no independent meaning as Gray (1926) pointed out.

Variations in growth between species

Growth parameters for a large number of species have been tabulated by Beverton and Holt (1959) and by Ursin (1967). For example, of the larger fish species, in blue-fin tuna, *Thynnus thynnus* (Linnaeus), $L_\infty = 270$ cm and $K = 0{\cdot}6$. Acipenser, the sturgeon, also grows to a large size (178 cm) but it grows relatively slowly for $K = 0{\cdot}05$. Of the smaller species in the male mosquito-fish, *Gambusia holbrookii* (Baird and Girard), $L_\infty = 3{\cdot}6$ cm and

$K = 1{\cdot}2$. The largest value of K is found in the female *Labidesthes* ($K = 3{\cdot}7$) which reaches its maximum size in little more than a year.

There is a negative correlation between the values of K and L_∞ in different species (see Fig. 11.6). The rate of decline in K with increasing value of L_∞ is not uniform, but is most rapid for small values of L_∞. In the length range up to 40 cm, values of K decline from as high as 3·7 to between 0·15 and 0·4. For fish with much larger values of L_∞, over 100 cm, values of K mainly

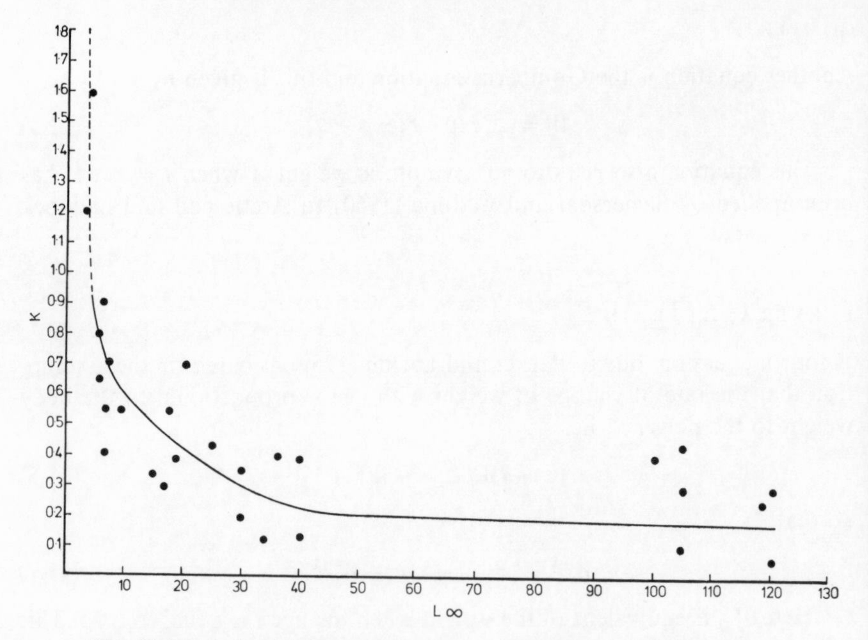

Fig. 11.6. Relationship between K and L_∞ (from Beverton & Holt 1959).

ranged from 0·05 to about 0·4 and showed no detectable trend. Also, there tends to be a positive correlation between L_∞ and the maximum ages attained by individual fish.

Geographical variations in growth within species

Bertalanffy parameters have been calculated for the growth curves of a wide variety of fish species. Each species does not necessarily have any one characteristic value. For example, the growth of various species can vary considerably in different parts of their geographical ranges.

Cod, *Gadus morhua* (Linnaeus) and haddock, *Melanogrammus aeglefinus* (Linnaeus), two species widely distributed in the North Atlantic, provide examples. Figure 11.7, shows length-at-age curves for haddock from different parts of its range. Length-at-age varies from the smallest individuals on the

Grand Banks to the largest on Georges Bank. Barents Sea haddock are unusual because their growth is more linear than that of other haddock. They are relatively small, compared with other stocks of up to 4–5 years but relatively large by the time they are 10 years old. Figure 11.8 shows similar data for cod. Generally, the smallest cod at any age are found at Labrador and the largest are found at Faroe Bank. As in the case of haddock the growth curve of cod from the Barents Sea is more linear than that of cod from elsewhere. This may be due to the fact that the larger cod and haddock in the Barents

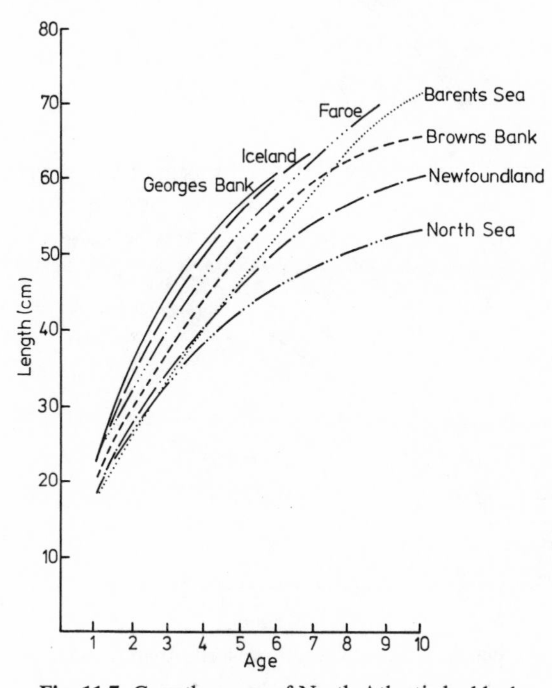

Fig. 11.7. Growth curves of North Atlantic haddock.

Sea feed on capelin. It could account for a change in growth and might be thought of as a change from one growth curve to another. For haddock, L_∞ = 47 cm for northern North Sea fish to over 80 cm for Faroe and Iceland fish. For cod L_∞ = 65 cm for Labrador cod to 120 cm for Faroe cod. For haddock $K = 0{\cdot}2$ or $0{\cdot}3$ whilst for cod $K = 0{\cdot}06$ to $0{\cdot}4$.

The herring, *Clupea harengus* (Linnaeus), also shows variation in growth throughout its range, and, there are differences between the spring and autumn spawning races of this species. The Atlantic-Scandian spring spawning race has an L_∞ of about 36 cm and $K = 0{\cdot}07$–$0{\cdot}13$. For autumn spawning North Sea herring on the other hand $L_\infty \simeq 30$ cm and higher values and $K = 0{\cdot}35$–$0{\cdot}50$.

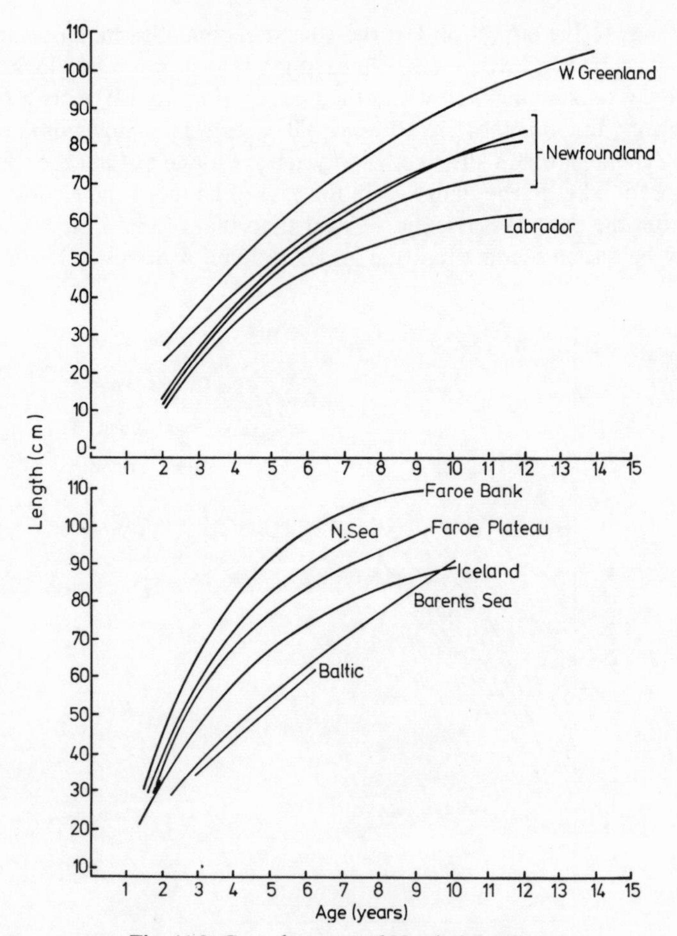

Fig. 11.8. Growth curves of North Atlantic cod.

Temporal changes in growth rate

Not only does growth vary throughout the geographical range, but it also varies with time. For example, the mean weight of 8–12-year-old Iceland cod increased considerably during the period from 1930 to 1949, decreased during the 1950s and then increased again during the period up to 1960–4 (Anon 1969). Arcto-Norwegian cod increased in length during the period from 1932 to 1936, declined during the period 1944–5, and then increased again in length until the beginning of the 1950s (Rollefsen 1953). The herring, has increased its growth rate during recent years in the North Sea, and by 1960 this species had become significantly larger age for age than it had been before 1950 (Burd & Cushing 1962).

Changes in growth following transplantation

In addition to variations in the growth of species with time and place there are instances when the growth rates of individual fish have been directly influenced under experimental conditions. For example, Burnet (1970) investigated seasonal growth in brown trout, *Salmo trutta* (Linnaeus) in two New Zealand streams, in which the fish grew at different rates. He was able to show that when fish were transferred from the stream in which they grew relatively slowly into the stream in which they grew more rapidly, the transplants proceeded to grow more rapidly. Svärdson (1949) obtained very similar results by interchanging some species of whitefish in northern Europe between lakes in which these species naturally grew at different rates. Iles (1973) mentions that certain species of *Tilapia*, when cultivated in ponds, sometimes produce populations of stunted individuals much smaller than the individuals in the natural populations from which they came. Among the marine species, Blegvad (1933) transplanted plaice, *Pleuronectes platessa* (Linnaeus), from regions where they were comparatively crowded, to areas where there was a relatively larger amount of food for them. He showed that following transplantation, the growth rates of these fish were improved.

Effect of the length of the growing season on growth

Growth is also influenced by the length of the growing season. For example, Eddy and Carlander (1940) found a correlation between the length of the growing season and the growth rate for a number of species. Gerking (1966) investigated populations of bluegill sunfish in eight Indiana lakes and found that the length of the growing season varied between the different lakes and that the most rapidly growing fish occurred in the lakes with the longer growing seasons. He noted that all populations had similar growth potentials as judged by the weight increment of the group III fish during the first month of each growing season. Hile (1936) investigated the age and growth of the Cisco, *Leucichthys artedi* (Le Seueur) in the lakes of the north-eastern Wisconsin. Ciscos were examined from four lakes and it was found that the fish from one of the lakes grew much faster than those from the other three lakes. He noted that the order of the four lakes with respect to growth rate by weight, was the same as the order with respect to the length of the Cisco's growing season and, incidentally, the reverse of their order with respect to the density of their Cisco populations.

Environmental effects on growth

The effect of temperature on growth

Investigations have shown that both of the von Bertalanffy coefficients, K and L_∞, are influenced by temperature. For example, cod growth parameters have been determined throughout their geographical range, and correlated with water temperature by Taylor (1958). Figure 11.9 shows the relationship between the von Bertalanffy coefficients K and L_∞ and temperature. When plotted on a semi-logarithmic basis

$$K \propto 10^{0\cdot065T}$$

where T is the temperature in degrees centigrade. The coefficient L_∞ is also plotted against temperature on a semi-logarithmic bases, i.e.

$$L_\infty \propto 10^{-0\cdot05T}$$

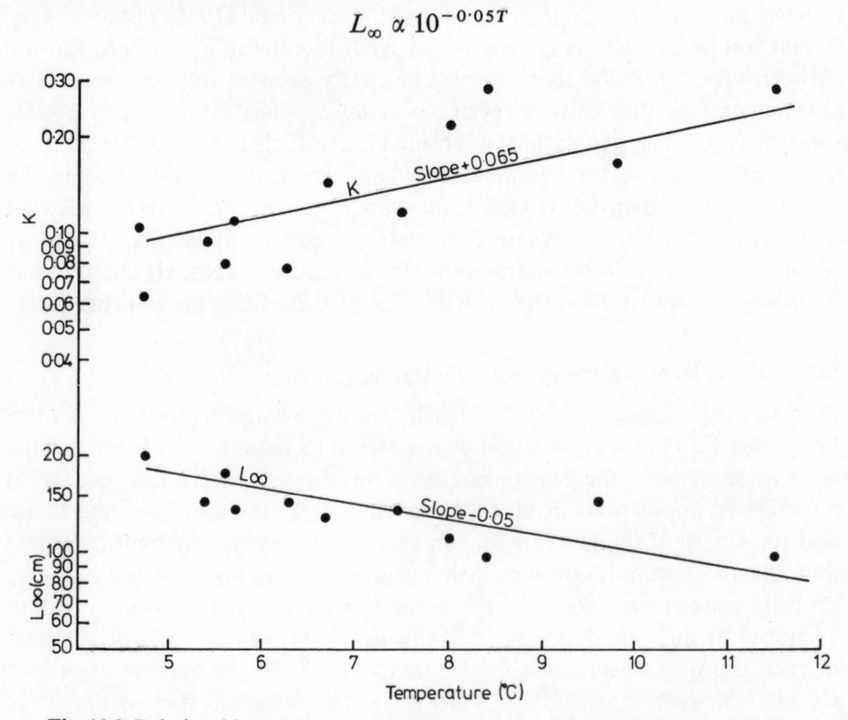

Fig. 11.9. Relationship between L_∞ and temperature and K and temperature. (Using cod data from Taylor 1958.)

The data on which Figure 11.9 has been based, come from different parts of the geographical range of the cod. Holt (1959), suggested that K ought to increase logarithmically with temperature, and that L_∞ should decrease slowly with temperature.

A more direct investigation of the effect of temperature on growth has been carried out by Kinne (1960) working with *Cyprinodon macularius* (Baird and Girard). He kept fish at temperatures of 15°, 20°, 25°, 30° and 35°C in aquaria and his results show how the length at age of individual fish varied with temperature (Fig. 11.10). The von Bertalanffy coefficient-calculated for these growth curves is linearly related to temperature when plotted on a semi-logarithmic basis, i.e. $K \alpha 10^{0 \cdot 027T}$. L_∞ was found to increase with increasing temperature over the range 15–25°C but then to decline over the range 30–35°C.

These results are consistent with the view that there is an optimum temperature for maximal growth.

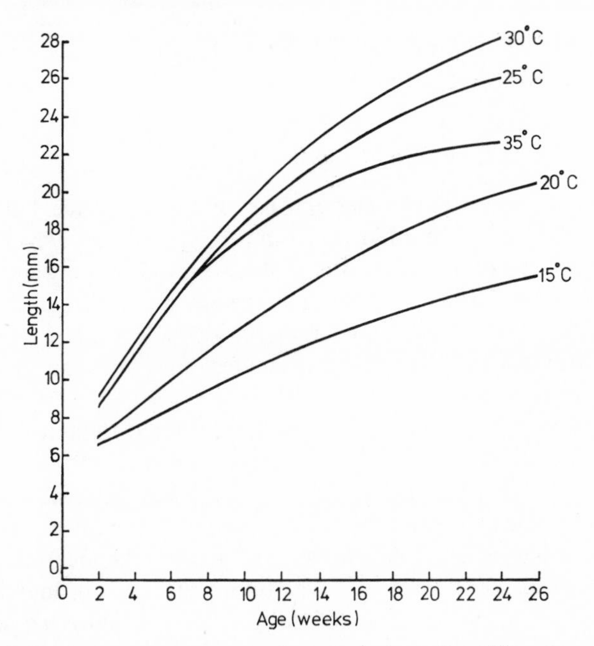

Fig. 11.10. Growth curves of *Cyprinodon macularius* at five different temperatures (from Kinne 1960).

The effect of maturity on growth

Maturity is a factor likely to influence fish growth, because, after the onset of maturity, energy that might have been used for growth will be required for developing the maturing gonad and for making spawning migrations. Thus growth can be expected to be lower after the onset of maturity than it otherwise would have been.

Alm (1939, 1949) showed, for two races of *Salmo trutta*, that the age at first spawning is inherited. In Atlantic salmon, *Salmo salar* (Linnaeus), the faster growers tend to return to spawn for the first time at a younger age than slower growers. Martin (1970) investigated the biology of the lake trout in Lake Opeongo, Ontario. He describes changes in the population during the late 1940s and early 1950s as a result of which trout became heavier and larger for their age and also matured at a larger size. Hile (1936), in his investigations of Cisco in Wisconsin lakes showed that the fish in one lake matured at the end of their second year and at a larger size, than the fish from three other lakes which mainly matured at the end of their third year. Bagenal (1969), in his experimental work with brown trout, showed that when two lots of trout were fed different amounts of food, a higher percentage of the better-fed fish matured than the others. Of the marine species, the North Sea herring, which experienced an increase in growth rate during the 1950s, now also matures earlier, fish by the end of the 1950s maturing at an age of 3 instead of 4 years (Burd & Cushing 1962).

Hierarchical effects on growth

It is well known that certain species of fish exhibit a hierarchical or dominance order such that fish may be inhibited from taking food simply by the presence of other individuals of the same species. Brown (1946), for example, describes how, when trout individuals of the same size are kept together, the difference in size between the smallest and largest individuals tends to increase with time. She attributed this to a dominance effect since the slow growers could be made to grow just as fast as the fast growers if they were removed and placed in separate tanks but not if they were offered unlimited food in the presence of other individuals. Similar results have been obtained for *Tilapia*. In the case of haddock, a hierarchical effect has been observed during aquarium growth experiments carried out at the Marine Laboratory, Aberdeen, in which individual haddock were trained to take meals at the surface from the experimenter's fingers. It was found that certain individuals were relatively slow in coming to the surface to feed when fairly large groups of nine individuals or more were kept together, but that the same individuals would come readily to the surface to feed when isolated or kept in twos or threes.

The effect of genetic factors on growth

Although growth must be influenced by genetic factors, experiments that have been done on fish are not particularly conclusive. For example, Moav and Wohlfarth (1967) reared the offspring of fast- and slow-growing carp, and found that selection for fast growth had no effect but that selection for slow growth was partly successful. It has been suggested, however, that because of the hierarchical structure of carp populations, selection for fast growth has been unsuccessful because it has really been selection for the more dominant, rather than for the genetically faster-growing individuals (Kirpichnikov 1970). Alm (1939, 1949) reared the young of lake and river trout under identical conditions. Normally, lake trout attain three to five kilograms in weight, and spawn at 5–7 years. River trout on the other hand are smaller, and spawn at an earlier age. Alm found that when reared together, the young grew at similar rates for the first three years but that the river fish matured at 3–4 years and the lake fish at 5 years as usual. He concluded that the genetic effect on growth was indirect, and due more to its effect on the age of first maturity.

The effect of fish density on growth (in freshwater)

Of the external factors that might affect the rate of growth, food intake is likely to be one of the most important. Beckman (1943) has shown how the experimental reduction in numbers of a population of rock bass led to a considerable increase in their growth. Swingle and Smith (1943) have shown, experimentally in ponds, that fish growth rate was relatively slow when the population density was relatively high. Kawajiri (1928) obtained a similar result working with rainbow trout. Eddy and Carlander (1940) compared the bluegill populations in six lakes in Ramsay county in 1938. They showed that three lakes had relatively low populations (25–49 fish per acre) but high growth rates. These fish, for example, reached a length of 12·5 cm after 2·8–3·5 years. By comparison the other three lakes had a higher population with 320–500 fish per acre, but the growth rate was much lower. It took these fish 4–5 years to reach a length of 12·5 cm. Comparisons of crappie populations showed differences, too. For example, two of the lakes showed low crappie populations with relatively high growth rates, whilst three lakes showed high crappie populations with lower growth rates. Podushko (1970) investigated the population of spawning smelt in the Amur River and showed that individuals from small year classes tended to grow faster than individuals from large year classes. Ricker (1937) obtained a very similar result for sockeye salmon yearlings and concluded that the differences were due to intraspecific competition for food.

Physiology of growth

The empirical relationship between food intake and growth

Not all of the food consumed is available for growth, and a great deal of both practical and theoretical work has been done to investigate the relationship between food intake and growth. An attempt to quantify much of this work was made by Paloheimo and Dickie (1966), who investigated the data with particular reference to the 'gross growth efficiency', i.e. the rate of growth divided by the food eaten. Paloheimo and Dickie use the symbol K for gross growth efficiency and define it by: $K = \Delta W/F\Delta t$, where $\Delta W/\Delta t$ is the growth rate and F is the food eaten.

Paloheimo and Dickie found that when fish are fed on one type of food, the logarithm of the gross growth efficiency decreases with increase in rations (Fig. 11.11) and that for a number of species and experimental situations this relationship is adequately described by a linear equation.

$$\log K = a - bF \tag{11.8}$$

Paloheimo and Dickie, and later Kerr (1971) investigated this relationship and considered its implications in some detail.

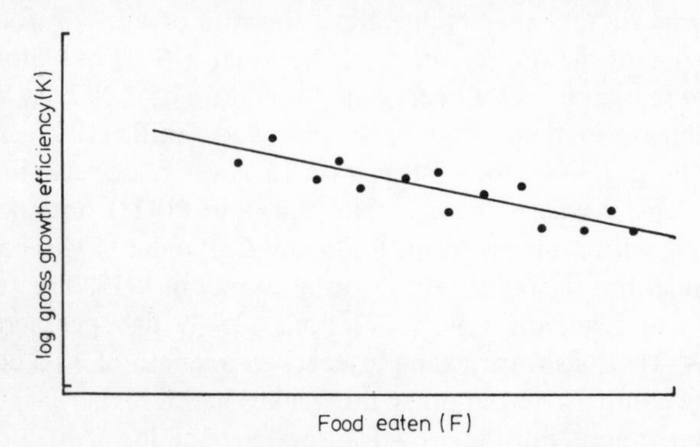

Fig. 11.11. Generalized relationship between log 'growth efficiency' and food eaten (*F*) (from Paloheimo & Dickie 1966).

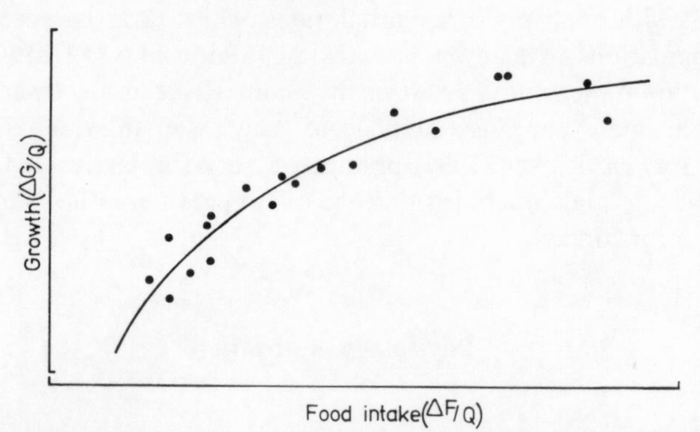

Fig. 11.12. Relationship between growth rate and rate of food intake (from Edwards *et al.* 1972).

A somewhat different approach has been used by Edwards, Steele and Trevallion (1970), working with O-group plaice, and by Edwards, Finlayson and Steele (1972) working with cod. These workers directly related food intake to growth during the same time interval as illustrated in Figure 11.12. For both plaice and cod, a curvilinear relationship was obtained. To allow for the effect of variations in the body weight on the results, the growth rates and the rates of food intake were standardized by dividing by $W^{0 \cdot 7}$ or $W^{0 \cdot 8}$. The theoretical justification for this is that basic metabolism is proportional to body weight raised to the power 0·7–0·8.

Other workers have carried out similar investigations but many have standardized their results by expressing growth and food intake in units proportional to the body weight of the experimental animals. A number of experimental results are referred to in Warren and Davis (1966) on animals with different body weights. Figure 11.13 shows the relationship between the

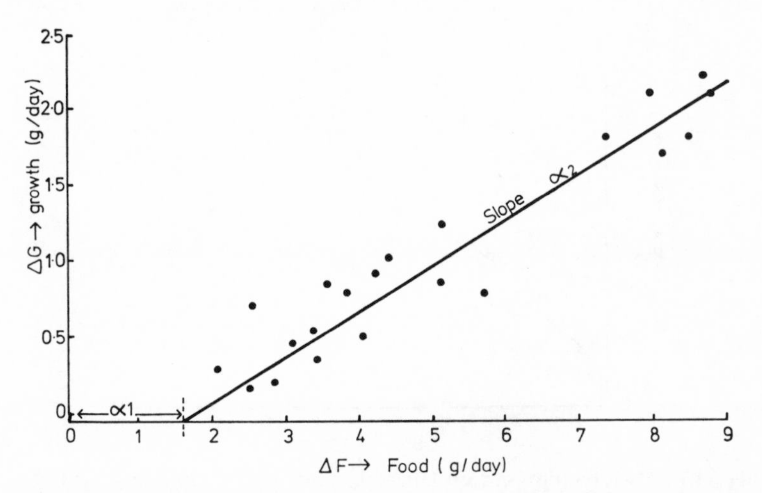

Fig. 11.13. Experimental relationship between growth rate and rate of food intake in haddock of 200–299 grams.

growth rate and the rate of feeding of cod and haddock with body weights of 200–299 g. The results suggest a linear relationship over a range of feeding rates up to a limit of the sustainable appetites of the fish.

Results like this have been obtained for groups of cod and haddock with body weights ranging from 5 to 900 g. In general, the results for each body weight were similar to those in Figure 11.13:

$$\Delta G = \alpha_2(\Delta F - \alpha_1) \tag{11.9}$$

where

$$\Delta G = \text{growth/day}$$

$$\Delta F = \text{food/day}$$

in which α_1 is the intercept of the abscissa which estimates the food requirement when growth is zero, or the maintenance ration under the experimental conditions and α_2 defines the slope of the regression and is equivalent to the net efficiency of conversion of food into growth.

The relationship between the maintenance coefficient (α_1) and body weight (W)

Much work has been done to estimate the maintenance requirements of fish of different sizes. Brown (1946) determined the maintenance requirements of fish of different body weights. Similar experiments were done by Pentelow (1939) working with trout, and by Dawes (1931) working with plaice. It was found that the maintenance requirements per unit body weight of fish decreased with increasing size of fish.

Another method that can be used for determining the energy required for maintenance is to determine the rate of loss of body weight when a fish is not

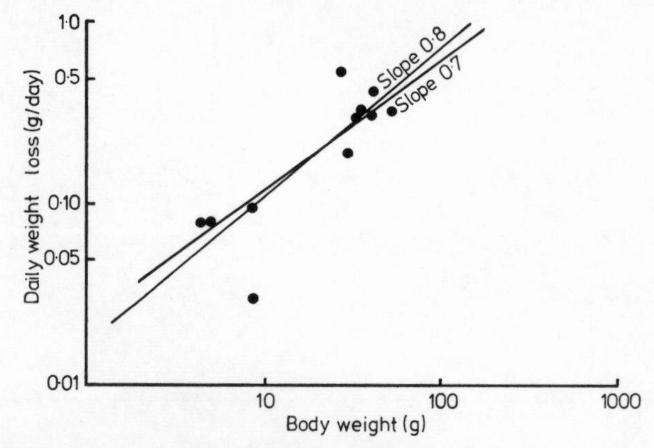

Fig. 11.14. Relationships between daily weight loss and the body weight in fasting haddock and cod (from Hislop, unpublished data).

feeding. An example is given in Figure 11.14. This shows the relationship between body weight and the daily loss in body weight in fasting haddock and cod. The data have been plotted on a double logarithmic scale and regression lines fitted to the data were found to have slopes of 0·7–0·8.

For cod and haddock direct estimates of the maintenance coefficient in equation (11.9) have been made for groups of fish with different body weights. When these are plotted against body weight on a double logarithmic scale (Fig. 11.15), the results can be fitted by a straight line with a slope of about 0·7.

Similar results have been obtained by measuring the oxygen consumption of fasting fish as an index of the rate of basic metabolism. Work of this nature has been reviewed by Winberg (1956) who arrived at the relationship

$$\text{rate of oxygen consumption} = 0{\cdot}3\ W^{0{\cdot}8}\text{ml/h at } 20°\text{C} \qquad (11.10)$$

More generally, it has been shown (Zeuthen 1953, Kleiber 1961) for a very wide range of animals including fishes, that oxygen uptake is related to body size to a power between 0·7 and 0·8.

With a suitable change of units and, where necessary, allowance for the effect of temperature, equation (11.10) can be used as a first estimate of α_1 for substitution in equation (11.9). This result it should be noted, explains why workers who expressed their results in units of the body weight of the experimental animals, observed a decline in maintenance requirement with increasing body weight; for example if rate of metabolism is proportional to $W^{0.75}$, then the metabolism per unit of body weight is proportional to $W^{-0.25}$.

In addition to this change in basic metabolic rate with change in body weight, there is evidence to show that for any one fish, body weight can be maintained over a range of feeding levels. This has been demonstrated by

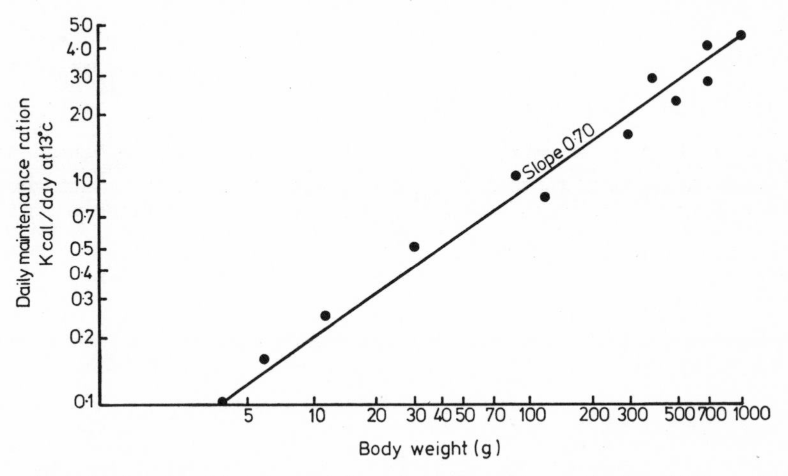

Fig. 11.15. Relationship between daily maintenance ration and body weight in cod and haddock.

Brown (1946) for trout which were able to adapt their maintenance requirements to different levels of feeding, apparently due to a capacity to lower the basic metabolic rate during starvation. It seems likely, therefore, that there is no one maintenance level for any individual fish. Because of this, equation (11.9) may not be valid when growth is negative. Also, α_1 in Figure 11.13 is more likely to represent the upper limit to maintenance than an average value.

Conversion efficiency

It is common practice to distinguish between 'gross efficiency' and 'net efficiency'. Gross efficiency has been defined above. Net efficiency is the increase in fish weight divided by the food 'available' for growth. There are various ways of calculating the food available for growth, one way being to subtract the maintenance food from the total quantity absorbed. According to this definition, α_2 in Figure 11.13 is equivalent to net efficiency. Some values of gross and net efficiency are tabulated in Table 11.1.

Estimates of the effect of body weight on net efficiency have been made

Table 11.1. Some estimates of growth efficiency.

Species	Net efficiency	Gross efficiency	Author
Carp		0·3	Ivlev 1939
Epinephelus guttatus (Linnaeus)		0·15–0·25	Menzel 1960
Cyprinodon macularius		0·27	Kinne 1960
Pike	0·44		Johnson 1966
Plaice	0·2		Beverton & Holt 1957 using data from Dawes 1931
Bluegill sunfish	0·14–0·44		Gerking 1971
Cod	0·45*		Kohler 1964
Cod and Haddock	0·3–0·65		Jones & Hislop (unpublished)

* Probably nearer to 0·225 if allowance is made for the energy content of food used.

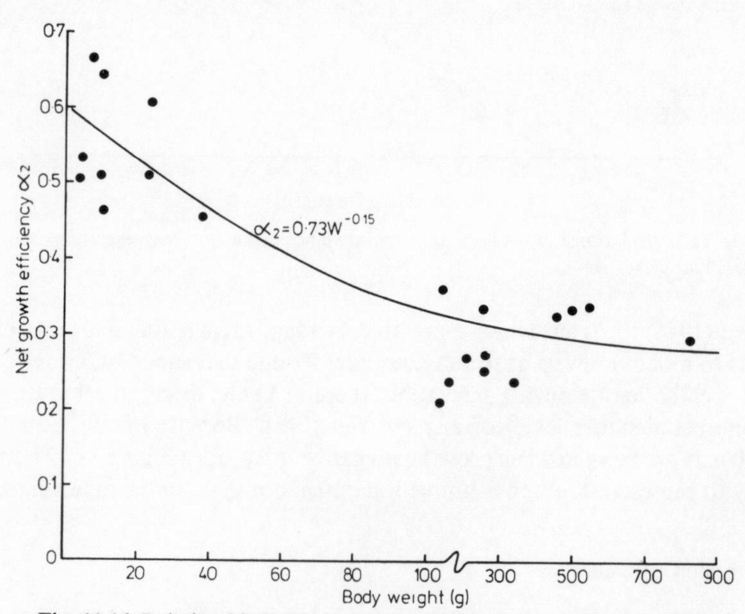

Fig. 11.16. Relationship between 'net growth efficiency' and body weight.

for gadoids and are plotted in Figure 11.16. Over a range of body weights from 100 to 900 g there was no detectable change in net efficiency, values fluctuating around a value of 0·3. Smaller fish of 5–40 g however, exhibited larger net efficiencies of 0·45–0·65. These results suggest that in gadoids, net efficiency declines with increasing body weight, but that the rate of decline is only detectable at the lower end of the weight scale. Gerking (1966) obtained similar results with bluegill sunfish, *Lepomis macrochirus* (Rafinesque). The empirical relationship $\alpha_2 = 0{\cdot}79\ W^{-0{\cdot}15}$ (11.11) was fitted to the data in Figure 11.16 and is applicable over a range of gadoid body weights from 5 to 900 g.

The effect of temperatures on Metabolism

For more general application, it is appropriate to consider the effect of temperature on metabolism with particular reference to its effect on the maintenance level of metabolism (α_1) and on the net conversion efficiency (α_2).

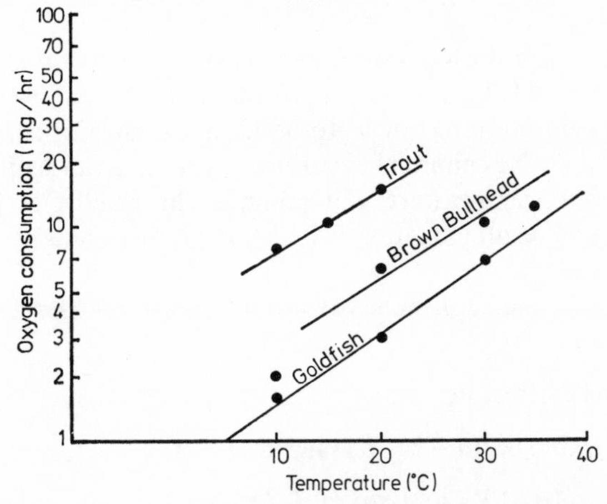

Fig. 11.17. Relationship between temperature and oxygen consumption in trout and brown bullhead (from Beamish 1964) and goldfish (from Beamish & Mookherjii 1964).

A number of experiments have been done to determine the effect of temperature on the rate of oxygen consumption (e.g. Beamish 1964; Beamish & Mookherjii 1964). Some of their data for goldfish, brown bullhead and trout are plotted in Figure 11.17, and show that the logarithm of oxygen consumption is proportional to temperature. Beamish also gives results for carp and sucker and from the regression slopes for all five species a mean slope of 0·036 has been calculated. These results show that the rate of oxygen consumption is related to temperature according to the Van't Hoff rule, which states that the velocity of a process increases geometrically with temperature. The relationship is:

$$\text{Rate of oxygen consumption} \propto 10^{0{\cdot}036T}$$

where T is the temperature in °C.

For some gadoids, Jones (in press (a)) has investigated the rate of passage of food through the stomach and obtained a linear relationship between the logarithms of the rate of digestion and the temperature. The slope of this relationship was 0·035. This is very close to the value of 0·036 referred to above and is consistent with the view that the rate of digestion is dependent on temperature.

Therefore the temperature correction $= 10^{0 \cdot 035(T_2 - T_1)}$ or exp . 0·081 $(T_2 - T_1)$ which is equivalent to a Q_{10} of $10^{0 \cdot 35} = 2 \cdot 2$.

In some studies, biologists have made use of the Arrhenius equation, which has been primarily used for describing the velocity of chemical reactions. This is:

$$k_2 = k_1 = \exp \frac{\mu}{2} (1/T_1 - 1/T_2) \tag{11.12}$$

where k_1 and k_2 are the velocity constants of the reaction at the absolute temperatures T_1 and T_2, and μ is a constant. However, over the normal biological temperature range of 0° to 35°C, the Arrhenius equation fits the data no better than the empirical equation, so that this can be used to adjust metabolic rates for temperature. For example, this result may be used, with Winberg's general result (equation (11.10)), to arrive at another equation for α_1.

That is, since metabolism involving 1 ml O_2 releases 0·021 Kjoules (Brody 1964)

energy required for maintenance

$= 0 \cdot 151\ W^{0 \cdot 8} \exp 0 \cdot 081(T - 20)$ Kjoules/d at T°C.

$= 0 \cdot 03\ W^{0 \cdot 8} \exp 0 \cdot 081T$ Kjoules/d at T°C. (11.13)

It is not known whether net conversion efficiency (α_2) is affected by temperature. Kohler (1964) gives data on the growth and feeding of Atlantic cod fed on frozen whole herring. He carried out experiments at temperatures of 2·3 and 13·6°C. Analysis of his data provides estimates of net conversion efficiency that show no significant differences with temperature.

The energy required for swimming

Data relating the rate of oxygen consumption to swimming speed have been determined for a number of species (Brett 1964, 1965, Brett & Sutherland 1965, Farmer & Beamish 1969, Alexander 1967, Tytler 1969). Tytler, working with haddock, derived the relationship:

$$\log_{10} Y = 0 \cdot 33V + 1 \cdot 77 \tag{11.14}$$

where Y is the oxygen consumption in milligrams per kilogram per hour at

10°C and V is the swimming speed in body lengths per second. Values of the first coefficient close to 0·33 have also been obtained for other species. For example, Brett (1964) gives 0·34 for young sockeye salmon at 5°C, and Brett and Sutherland (1965) give 0·31 for *Lepomis gibbosus* (Linnaeus) at 20°C. Values of the other coefficient (i.e. 1·77) are more variable and appear to be dependent on temperature.

Analysis of this result (Jones, in press (b)) leads to the following general relationship between swimming speed, body weight, temperature and energy uptake:

$$\text{energy requirement} = 0{\cdot}025\, W^{0{\cdot}8} \exp(0{\cdot}081T + 0{\cdot}76V) \text{ Kjoules/d at } T°\text{C} \tag{11.15}$$

When $V = 0$ this equation reduces to something similar to equation (11.13).

Allowing 80 per cent for the efficiency of food utilization, and converting from Kjoules to g one arrives at the following general equation for estimating the food energy needed for the basic metabolic rate (α_1) that allows for body weight, temperature and swimming speed. This is

$$\alpha_1 = 0{\cdot}008\, W^{0{\cdot}8} \exp(0{\cdot}081T + 0{\cdot}76V) \text{ at } T°\text{C} \tag{11.16}$$

therefore, this leads to the following equation for growth:

$$\Delta G = \alpha_2[\Delta F - 0{\cdot}008\, W^{0{\cdot}8} \exp(0{\cdot}081T + 0{\cdot}76V)] \text{ gm/d at } T°\text{C} \tag{11.17}$$

where ΔF is the food intake in Kjoules/d.

This equation may be used by adopting a value of α_2 of about 0·2–0·4 or, alternatively, the empirical gadoid result in equation (11.11). This alternative, for example, would give:

$$\Delta G = 0{\cdot}79\, [\Delta F\, W^{-0{\cdot}15} - 0{\cdot}008\, W^{0{\cdot}65} \exp(0{\cdot}081T + 0{\cdot}76V)] \text{ gm/d at } T°C \tag{11.18}$$

Metabolic growth models

In addition to the purely empirical growth equations, growth equations have also been derived from physiological considerations. These, however, are based on considerations not dissimilar to those involved in the empirical approach, i.e. they are based on the principle that growth can be treated as the difference between anabolism and catabolism.

Winberg (1956) considered a model very similar to the empirical growth equation (9), but with the terms re-arranged, i.e.

$$\Delta F' = \Delta M + \Delta G \tag{11.19}$$

where $\Delta F'$ is the physiologically useful food energy intake per unit time, ΔM

is the active metabolic requirement per unit time, and ΔG is the energy content of the growth per unit time.

Not all of the food energy ingested is available to the body for useful work. Energy can be lost due to incomplete absorption in the gut and due to excretion of nitrogen. To allow for these losses, Winberg assumed that only 80 per cent of the total food energy was effectively available to the body so that it is possible to write:

$$0{\cdot}8\Delta F = \Delta M + \Delta G \tag{11.20}$$

where F is the total energy content of the food per unit time.

Re-arranging terms, we have

$$\Delta G = 0{\cdot}8\Delta F - \Delta M \tag{11.21}$$

Variations of this equation have been used for calculating energy requirements (Backiel 1971, Mann 1965), and its applicability to fish has been discussed by Warren and Davis (1966)

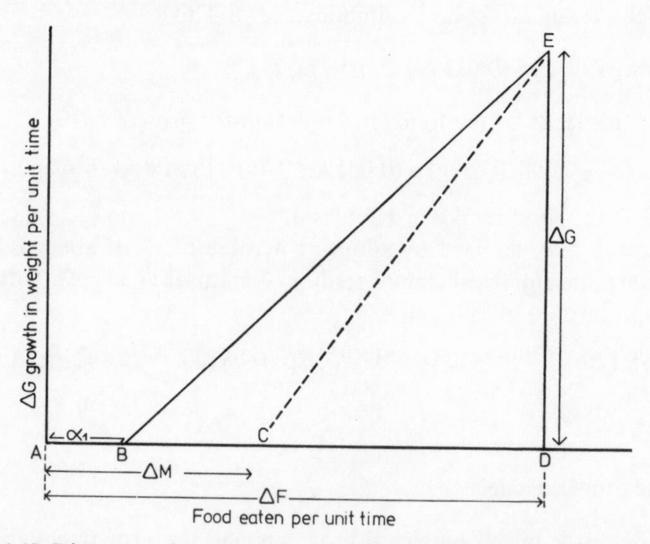

Fig. 11.18. Diagrammatic comparison of experimental and theoretical interpretations of growth.

Equation (11.21) differs from equation (11.9) in two important respects. The first term on the right-hand side of equation (11.21) is $0{\cdot}8\Delta F$ but is $\alpha_2\Delta F$ in equation (11.9). Since values of α_2 are usually in the range 0·2–0·65 these terms are therefore not equivalent. Secondly, Winberg's term for active metabolism (ΔM) is not equivalent to the second term on the right-hand side of equation (11.9) which is equivalent to $\alpha_1 . \alpha_2$.

The reason for this is that Winberg's term ΔM represents the total metabolism of the active growing fish, and this is itself likely to be a function of food intake and growth. Measurements of oxygen consumption, for example, show that after food intake, there is an increase in heat production known as specific dynamic action (Brody 1964, Warren & Davis 1966). Therefore, M in equation (11.21) is not a constant, but can be expected to increase with increasing food consumption and growth rate (Edwards *et al.* 1969, 1972).

The difference between equations (11.9) and (11.21) may be further illustrated with reference to Figure 11.18. Here it is assumed that a fish is consuming food at a rate ΔF ($\equiv AD$) and that it is growing at a rate ΔG ($\equiv DE$). The empirical method (equation 11.9) arrives at E on the graph by direct observation. For a given body weight, α_1 is a constant, and the experimental results for intermediate feeding and growth rates fall on the line BE. The Winberg method on the other hand, regards point E by first calculating $\Delta M = 0{\cdot}8\Delta F - \Delta G$, and then constructing the line CE. The intercept ΔM is therefore not constant, but has to be calculated separately for each level of feeding and growth in such a way that the points for intermediate feeding and growth rates fall on the observed line BE. This view, it should be noted, is not dependent on the line BE necessarily being linear.

It is instructive to make use of the empirical equation (11.9) to obtain an expression for ΔM as a function of growth rate.

Thus, from (11.21)

$$\Delta M = 0{\cdot}8\Delta F - \Delta G \qquad (11.22)$$

also from (11.9)

$$\Delta F = (\Delta G/\alpha_2) + \alpha_1 \qquad (11.23)$$

Substitution for ΔF from (11.23) into (11.22), gives:

$$\Delta M = \Delta G\ [(0{\cdot}8/\alpha_2) - 1] + 0{\cdot}8\alpha_1 \qquad (11.24)$$

This expression shows how the metabolic rate can be expected to increase with the rate of growth (ΔG) and also, since α_1 is a function of body weight, with the weight of the fish.

To allow for this effect, Winberg assumed, after reviewing a good deal of evidence, that on average, the active metabolic rate under natural conditions is about twice the resting or maintenance level of metabolism. Thus, if, as in Figure 11.13, the symbol α_1 is used for the maintenance level of food intake, equation (11.21) could be rewritten in the form:

$$\Delta G = 0{\cdot}8\Delta F - 2\alpha_1 \qquad (11.25)$$

This is still not the same as equation (11.9) however and the assumption that ΔM is twice the maintenance metabolic level is therefore only a useful approximation. To calculate exact values, or an exact function for ΔM, it is first necessary to construct the line BE from experimental data and then to determine each value of ΔM experimentally. For this reason, the empirical equations (11.9), (11.17) and (11.18) seem more appropriate for practical considerations, whereas the Winberg equation may be more suitable for describing the underlying metabolic processes.

Another important metabolic growth model is due to Pütter (1920) who considered the equation:

$$dW/dt = HW^{2/3} - kW \tag{11.26}$$

$HW^{2/3}$ represents the anabolic processes and kW catabolic processes. If $HW^{2/3} > kW$, surplus energy will be available for growth.

In equation (11.26) it is assumed that food is not limiting and that the rate of food intake is regulated by the total area of absorbing surface of the gut. This is assumed to be proportional to the body weight to the power 2/3. The second term (kW) represents the rate of breakdown occurring everywhere in the body and is therefore assumed to be proportional to body weight. For the further development of this equation the reader is referred to work by von Bertalanffy (1934, 1938, 1949), Beverton and Holt (1957), Taylor (1958, 1962) and Ursin (1967). Of particular interest is the solution when the body weight is proportional to the cube of the body length so that one has:

$$W = g\,L^3 \tag{11.27}$$

$$dL/dt = 1/3\;Hg^{-1/3} - 1/3\;kL = E - KL \tag{11.28}$$

This leads to the solution:

$$l_t = L_\infty\,(1 - e^{-K(t-t_0)})$$

where $L_\infty = E/K$

and $W_t = W_\infty\,(1 - e^{-K(t-t_0)})^3$

Equation (11.26) therefore leads to the extremely useful von Bertalanffy equation that has been used for describing growth in the early part of this chapter.

A more general formulation of the Pütter equation can be written in the form:

$$dW/dt = H'W^m - k'W^n \tag{11.29}$$

Comparison of equation (11.29) with the empirical equation (11.17), however, immediately suggests that: $K'W^n \alpha W^{0 \cdot 8}$ so that $n = 0{\cdot}8$.

Values of m can be estimated empirically by finding a value of m such that the term $H'W^m$ in (11.29) becomes equivalent to $\alpha_2 \Delta F$ in (11.17). Ursin (1967) summarizes data for a large number of species and arrives at an estimate of $m = 0{\cdot}56$.

With $n = 0{\cdot}8$ and $m = 0{\cdot}56$, equation (11.29) then becomes equivalent to the empirical equation (11.17). It does not lend itself to a simple algebraic solution in the way that equation (11.26) does. However, computer simulations of growth curves using equation (11.17), or equation (11.29) with $n = 0{\cdot}8$ and $m = 0{\cdot}56$, appear superficially very similar to von Bertalanffy curves. Although not mathematically identical, the differences are frequently too small to be important if one is primarily concerned with fitting curves to length at age data in which there is a normal statistical scatter of observations.

In the case of equations (11.26) and (11.29) the second terms on the right-hand side of equations are usually treated as being independent of food intake.

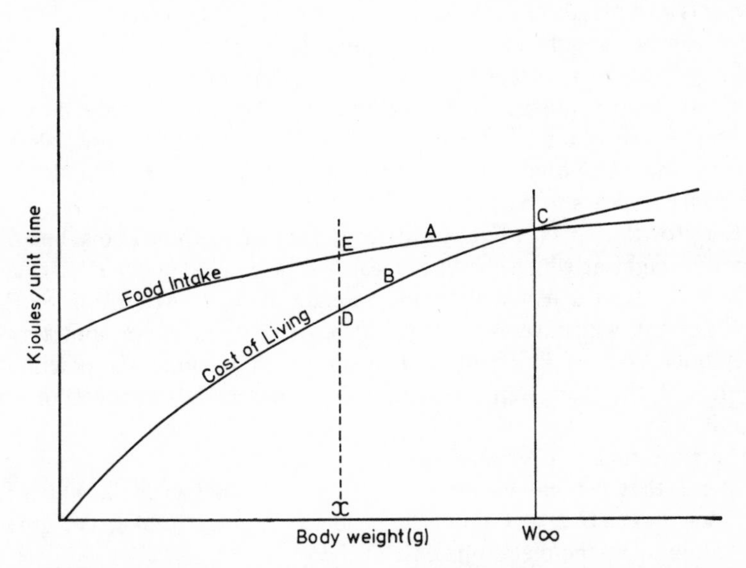

Fig. 11.19. Diagrammatic representation on the relationships between food intake and rate of metabolism.

These are, therefore, really only pseudo-metabolic models and are really more akin to the empirical equations (11.9), (11.17) and (11.18) than to the Winberg model (11.21).

Discussion

From the evidence it is clear that there are many aspects of fish growth that are still unresolved. For example, the fact that some species (e.g. tuna) are invariably larger than others (e.g. gobies) suggests that there is some degree of internal control of growth. However, from the evidence that the growth rates of individual species vary so much either at different times or at different geographical locations it is clear that whatever the mechanisms of growth control, it is not nearly so precise as it is in the higher vertebrates. Eddy and Carlander (1940) suggested that heredity could account for differences in growth rate between species and races but that within species it is the environment that is important in determining the growth rate.

In the higher vertebrates, growth is controlled within relatively narrow limits and this suggests a situation in which food intake is regulated to achieve a predetermined level of growth. Consider for example the situation depicted in Figure 11.19. This shows two lines plotted against the body weight (W). Line A represents food intake as a function of body weight, and line B represents the metabolic cost of living, also as a function of body weight. Line B can be thought of as representing the total metabolic cost of living including the energy needed for active metabolism as well as the energy needed for food capture and for predator avoidance. The evidence suggests that these curves are both functions of body weight to some power less than unity, i.e. the metabolic cost of living appears to be a function of $W^{0 \cdot 8}$ while food intake appears to be a function of body weight to some power less than 0·8. For growth to occur, line A will presumably have to be above line B, and the body weight at which the lines meet will be equivalent to W_{∞}.

Consider then a fish with a body weight $W = x$. According to Figure 11.19, this fish will have an energy intake of E Kjoules per unit time and a metabolic cost of living of D Kjoules per unit time. Its potential for growth will therefore depend on the difference (E–D) Kjoules of energy per unit time.

There are then two possibilities:

(a) One is that fish eat whatever is available so that growth, which is dependent on (E–D), is simply dependent on food intake, i.e. growth $\propto$ food intake – the metabolic cost of living.

(b) The alternative is that the growth rate is genetically predetermined, so that (E–D) at any body weight is predetermined. This means that the food intake ought to be dependent on the growth, i.e. food intake $\propto$ growth + the metabolic cost of living.

Although these equations are mathematically equivalent, their biological implications are very different.

Evidence to distinguish between these alternatives is rather poor. With regard to the first possibility it would appear that some species (e.g. certain species of Coregonids) can indeed exhibit very considerable fluctuations in growth dependent on their food supply. In favour of the alternative hypothesis, however, is the fact that there are other species, such as for example the haddock, that can experience very large fluctuations in year class strength, such as might be expected to influence the food available for each individual, and yet exhibit relatively small variations in growth rate. Beukema (1968) has shown that the food intake of sticklebacks is partly influenced by the hours of deprivation of food. His work suggests that there may be at least partial regulation of food intake in this species due to internal factors as well as due to the quantity of food available.

The evidence is conflicting. It may be that fishes have evolved part-way towards the situation found in higher vertebrates in which growth rates are genetically predetermined within relatively narrow limits. If so, it may be that different species have evolved in this direction to different extents.

PART IV. YIELD FROM THE SEA

Every year nearly 70 million tons of fish are yielded to fishermen by the productive engine of the sea (see Chapter 12). There is a considerable gap, however, between estimates of yield made from global measurements of primary production ($\sim$2000 million tons) and those made somewhat more carefully for particular fisheries in particular regions. One reason for the discrepancy (if it can be so called) is that there are very large reserves of small fish in the deep ocean, beyond the continental shelves, which cannot be exploited at all with present methods of capture. These fish are ten or fifteen cm in length and there is an average of one in every thousand cubic metres of water; to catch the 500 tons needed to make an ocean-going trawler pay, more than 10 km^3 of water would have to be filtered. A considerable technological step would thus be needed to catch such fish.

Another reason for this possible discrepancy, moreover, is our general lack of understanding of how plants are transformed into fish, for remarkably little is known about this grey area of the dynamics of secondary production and food chain efficiency. Most previous phytoplankton studies have ignored the herbivores and most previous fishery studies have failed to consider the sources of energy flux to the fish stocks. Recently integrated, multi-discipline studies of selected marine ecosystems have tentatively begun to address this problem directly and it is possible that future texts may contain a more quantitative exposition of these relationships.

Chapter 12. Production and Catches of Fish in the Sea*

J. A. Gulland

Introduction: World fisheries
Fish population dynamics
Analyses of catch and effort
Analytic models
The world's potential fish production
Discussion

Introduction: World fisheries

Fishing is one of the oldest activities of man, and the harvest from naturally occurring fish stocks is still one of the most important uses that man makes of the oceans. Since 1945 the total catch of marine fish (excluding whales) has increased at around 6 per cent per year, reaching 60 million tons in 1972. Since 1969 there appears to have been some slowing down in the rate of increase in catches, though this has been confused by the pattern of catches of anchovy by Peru. Omitting these, the pattern has been of rather slower, but steadier, expansion (see Table 12.1).

Part of this expansion has been achieved by the increase of industrial-scale fishing by the more developed countries, of which Japan and U.S.S.R. are the prime examples. The long-range fleets of these countries, consisting of powerful trawlers up to 5000 gross tons, supported, in the case of U.S.S.R., by a variety of auxiliary vessels such as tankers, refrigerated transports, etc., are found in all oceans of the world. The pressure on the limited resources of the stocks of fish capable of supporting these fleets, and the desires of coastal states to reserve these resources for their own use, most obviously expressed in the number of countries which have recently extended their territorial limits, or limits of jurisdiction over fishing, of which the increase by Iceland from 19·2 km to 80 km is probably the best known, but not the most extensive, are likely to limit the opportunities for these long-range fleets in the future. The monster super-trawlers, equipped with highly sophisticated equipment for locating, catching and processing fish may well go the way of the monster

* The views expressed are those of the author and not necessarily those of FAO.

Table 12.1. World catches of fish (million tons), including shellfish, but excluding whales. (From FAO, Yearbooks of Fishery Statistics.)

	1960	1961	1962	1963	1964	1965	1966	1967	1968	1969	1970	1972	1973	1974
Total	40·2	43·6	47·1	48·4	52·9	53·7	57·5	61·1	64·3	62·9	69·3	65·5	65·7	69·8
Marine fish	34·6	37·9	41·3	42·5	46·2	46·7	50·4	53·9	57·0	55·5	61·7	55·8	55·9	60·0
Anchoveta	3·5	5·3	7·1	7·2	9·8	7·7	9·6	10·5	11·3	9·7	13·1	4·8	2·0	4·0
Marine fish, less anchoveta	31·1	32·6	34·2	35·3	36·4	39·0	40·8	43·4	45·7	45·8	48·6	51·0	53·9	56·0

dinosaurs, except in some areas, such as the Bering Sea, or the Scotia Sea in the Antarctic, where large-scale local fisheries are unlikely to replace them. For the moment they are impressive examples of the application of modern technology to an ancient craft.

At the same time, the application of technology at a more modest level has contributed to impressive growth of fisheries, especially among the poorer countries. The most spectacular example is the growth of the Peruvian anchovy fishery from a few thousand tons in 1950 to over 10 million tons in 1967, in terms of weight the biggest single-species fishery in the world, and making Peru the number one fishing country. All the anchovy (*Engraulis ringens* (Jenyn)) has been converted into fish meal and oil, mostly for feeding broiler chickens and other animals, and the bulk of the produce exported from Peru, chiefly (but not entirely) to North America and western Europe. A less well-known, but almost equally spectacular growth has been the trawl fishery of Thailand. Here, nearly all the catch is consumed in the country, and since about half of this is used for direct human consumption (the rest has been used for feeding ducks and, particularly recently, for fish-meal production) the value of the Thai catch—which grew from 300,000 tons in 1961 to over 1·6 million tons in 1971—slightly exceeds that of Peru. Falling catch rates in the northern Gulf of Thailand have encouraged the expansion of fishing into more distant waters, and Thai trawlers now work as far afield as the Bay of Bengal and northern Borneo.

This expansion is strikingly similar to the development of English trawling in the late nineteenth and early twentieth centuries—first increased local fishing, followed by extension into more distant waters. The basic causes are much the same—good markets, due to improved distribution and a growing and increasingly better-off population, improved catching methods, and a supply of manpower that is capable of using the new methods, but without such well-paid alternative jobs ashore as make fishermen unwilling to put up with the discomfort of going to sea on fishing boats. In fact, the combination of circumstances needed to trigger off large-scale commercial fishing frequently arises as countries develop from a primitive economy, and the fishing history of modern Thailand and nineteenth-century England can be matched on a large scale by post-war U.S.S.R. and Japan (though high wage rates are leading to a falling off in direct Japanese fishing activity), and on a smaller scale by such countries as Korea and Ghana.

The catches of some of the more important fishing countries since 1960 are set out in Table 12.2. This table shows how some countries have enormously increased their catches in recent years, often from very low levels (though the reported statistics of early years should be treated with some degree of caution), while catches of others, such as the U.K. and U.S.A. have remained at a constant level. The table shows, even if roughly, how important fisheries are to some countries; additional data on trade, or supply of protein,

Table 12.2. Catches of fish by some major fishing countries (thousand tons). (From FAO, Yearbooks of Fishery Statistics.)

	1938	1948	1955	1960	1965	1970	1971	1972	1973
Peru	23	84	236	3,727	7,632	12,612	10,611	4,768	2,299
Japan	3,678	2,519	4,908	6,193	6,908	9,308	9,895	10,273	10,702
U.S.S.R.	1,523	1,485	2,495	3,051	5,100	7,252	7,337	7,757	8,619
Norway	1,128	1,422	1,813	1,543	2,307	2,980	3,075	3,163	2,975
U.S.A.	2,260	2,416	2,790	2,814	2,669	2,755	2,819	2,649	2,670
Spain	408	547	770	970	1,340	1,497	1,505	1,617	1,570
Thailand	161	161	213	221	615	1,448	1,572	1,679	1,692
U.K.	1,198	1,206	1,100	924	1,047	1,099	1,107	1,081	1,144
Chile	32	65	214	340	708	1,161	1,487	792	664
Iceland	327	478	480	593	1,199	734	685	726	906
Poland	32	47	127	183	298	469	518	544	580
Ghana	No data		25	32	72	187	216	281	195

show this more clearly. It is well known that Iceland depends almost entirely on her fisheries for export trade. It is less well known that in several recent years over one-third of Peru's export earnings have been from fish meal. At a more basic level, many countries, especially in south-east Asia, obtain a large proportion of their supply of protein, and particularly of animal protein, from fish.

This dependence, economic or nutritional, is naturally much more marked in communities or regions rather than entire countries. Fishing is by its nature concentrated either close to the fishing grounds—for the less advanced fisheries—or for the more advanced fisheries, at a few ports with the necessary landing, processing and distribution facilities. Many large and small communities, from the small fishing towns of the Moray Firth in north-east Scotland to Chimbote in Peru, the fish-meal capital of the world, are almost entirely dependent on fishing. For them a better understanding of the fish stocks, how they are affected by fishing and other events in the sea, and how changes in the stocks will be reflected in the catches, are matters of vital concern.

This understanding occurs at two levels—at the day-to-day tactical level, e.g. where to fish, with what sort of fishing gear—and at the strategic level, e.g. how much fish can be taken this year, without impairing the ability of the stock to produce fish for next year's harvest. At the tactical level the practical fisherman has, of necessity, gained a knowledge of many aspects of marine ecology—of the distribution and behaviour of fish, the relation between the occurrence of preferred species and certain types of bottom, or between different species—that cannot as yet be matched by the professional marine scientist. It is in the strategic field that the scientist has the greatest opportunity to help the fisherman. Here the scientist's aim is the understanding of the effects on the fish stocks of the removals being made by man, and hence the ability to advise either on the possibility of increasing catches (with some estimate of how much they can be increased) or on the need to apply some regulation or conservation measure to ensure the future well-being of the fishery.

Fish population dynamics

In response to this need, fishery scientists have built up an understanding of the dynamics of fish stocks which is probably as good as that of any other natural stock. Apart from the practical need for this knowledge, the fishery scientists' work has been greatly helped by the information that can be derived from the fishery itself. The catch statistics describe the removals from the stock, by one of the major causes of mortality, often in considerable detail. For example, the records of the Bureau of International Whaling Statistics give information on the date and place (strictly the noon position of the factory ship to which the catcher which actually killed the whale was attached; this will only exceptionally be more than about 80 km from the place of capture) of death of the majority of all blue whales (*Balaenoptora musculus* (Linnaeus)) that have died since 1930, together with the size when killed, the sex, and, if female, whether pregnant, and the size and sex of any fetus present. These records are exceptional in their detail and accuracy. In many other fisheries, including most of those in tropical and sub-tropical waters, the only data easily available are the gross catch, not infrequently without much good information on the species composition. However, this latter information, as well as information on the sizes and condition of fish caught can be obtained comparatively easily by regular observations on the fish markets.

Besides information on what is removed, the fishery also gives some good guidance on what is in the sea. The sizes and ages of fish landed will provide a reasonable estimate of the composition of the fish in the sea, above a certain size, from which growth and mortality rates and other characteristics of the population can be estimated. Even more valuable is information on the catch per unit effort, e.g. catch per haul or catch per day's fishing, which can be used as a measure of the relative abundance of the fish stock. This is mainly used to study year-to-year changes, especially in relation to changes in the amount of fishing, but spatial differences can also be examined. Figure 12.1 illustrates the differences in distribution of some important bottom-living fish (cod, *Gadus morhua* (Linnaeus), haddock, *Melanogrammus aeglefinus* (Linnaeus), plaice, *Pleuronectes platessa* (Linnaeus), and sole, *Solea solea* (Linnaeus)) as shown by the average catch taken in one hour of fishing by English trawlers in the North Sea.

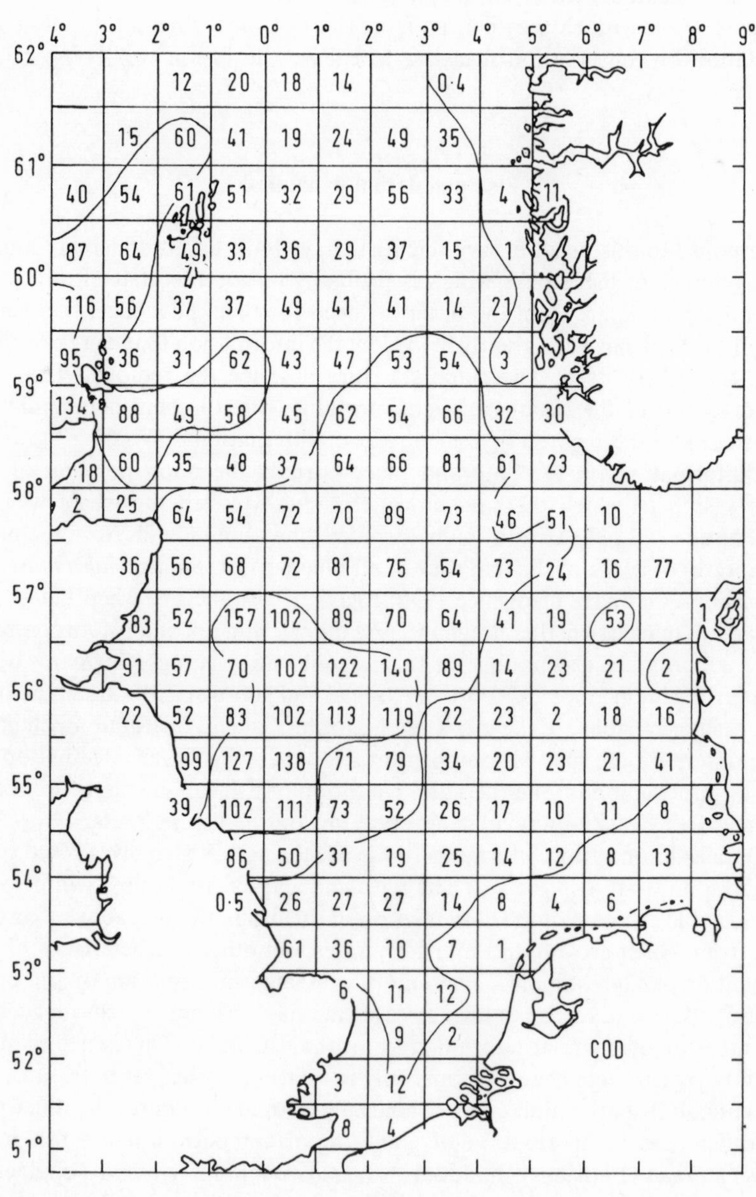

12.1a

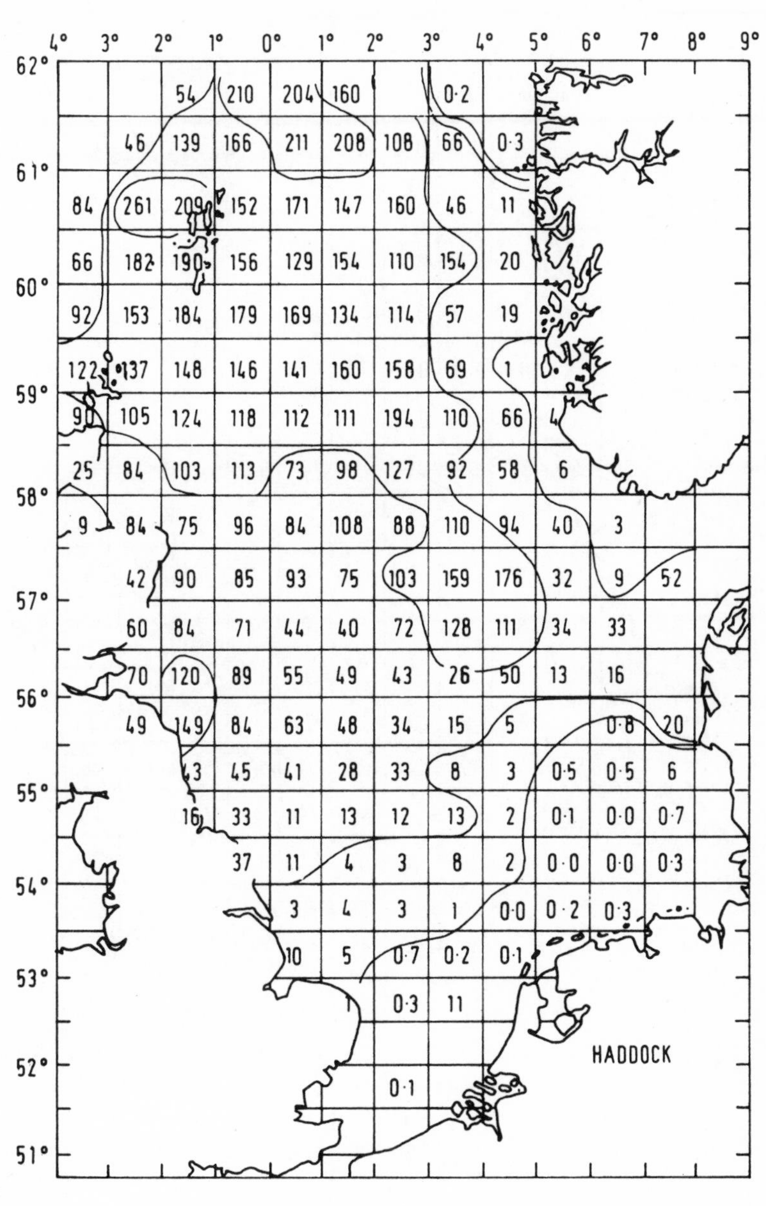

12.1b

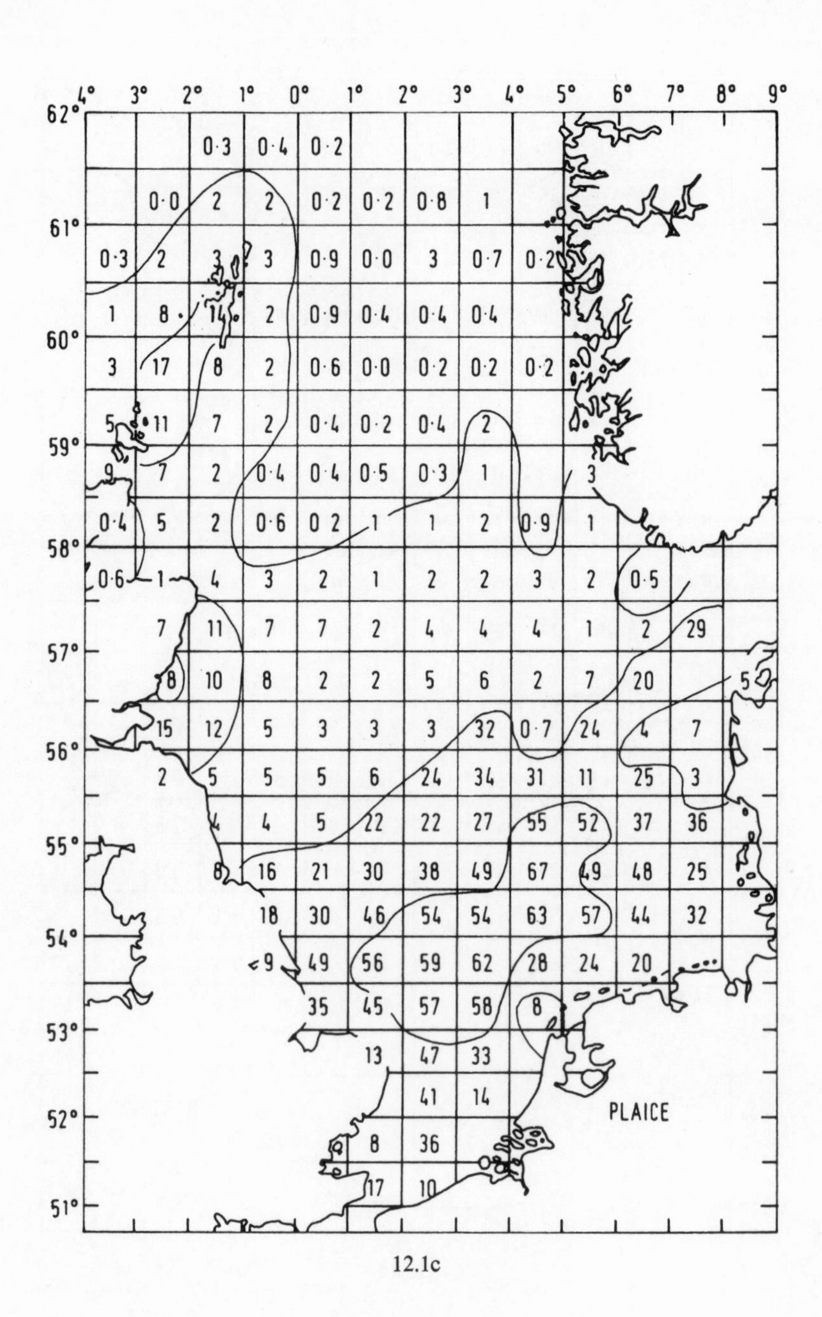

12.1c

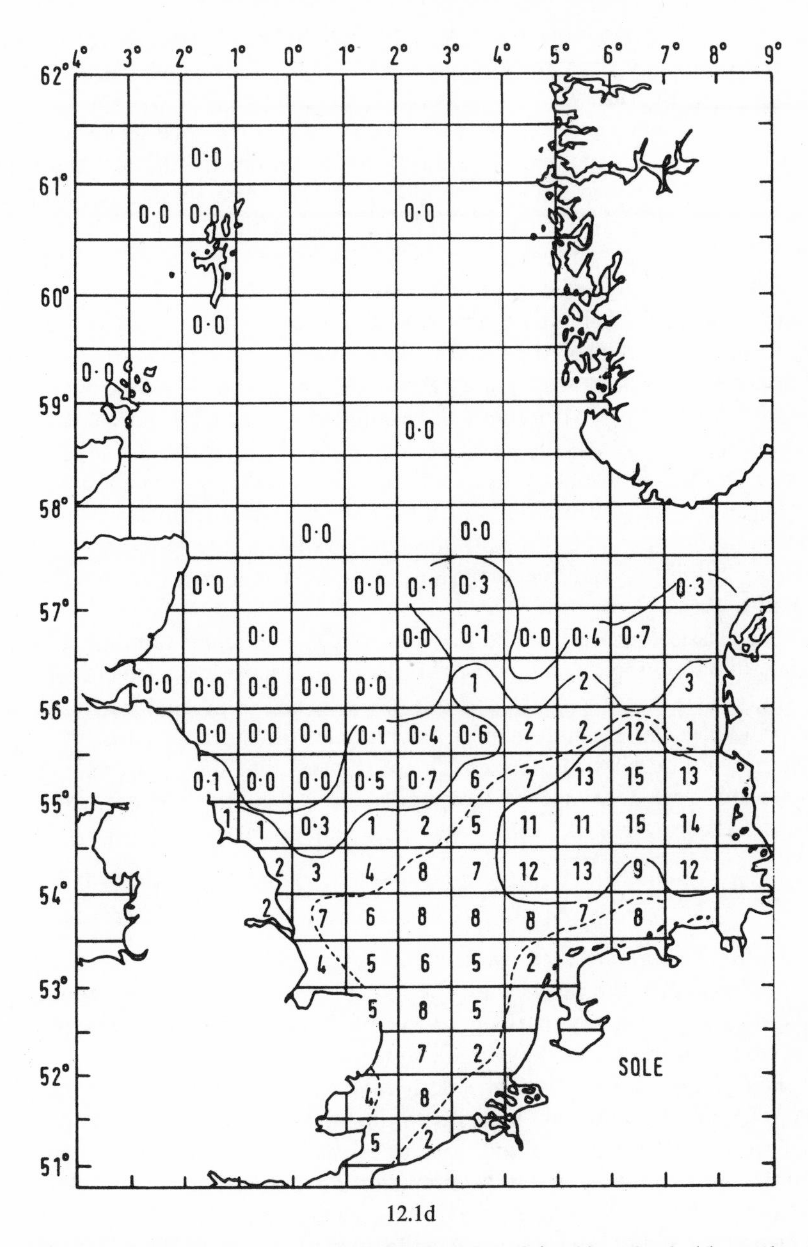

12.1d

Fig. 12.1. Quantities (cwt) of cod (a), haddock (b), plaice (c) and sole (d) caught per 100 h fishing in the North Sea averaged during the decade 1951–60. Note the concentration of haddock and sole in the northern and southern parts respectively, and the more even distribution of cod and plaice. (Data supplied by Fisheries Laboratory, Lowestoft.)

The expense of going to sea, and the difficulties of making quantitative observations once there, have limited the range of data other than from the commercial fisheries which have been used in studying the dynamics of fish stocks. With minor exceptions—such as whales which can be seen when they surface to breathe, or salmon going upstream to spawn—the animals of interest are not easy to count directly or to sample directly in a quantitative manner. Recent developments in the use of acoustic devices such as echo-sounders and sonars, in which the echoes returned from individual fish are measured and counted, can provide absolute counts in favourable circumstances—when the fish are not too near the surface or bottom, are not shoaling too closely and when the species of fish concerned can be clearly identified (Forbes & Nakken 1972; Cushing 1964). Otherwise the only time during the life of a brood of fish when they can be counted is during the egg stage. These cannot dodge the nets being used to sample them, and with appropriate measurements or assumptions of the duration of the egg stages, mortality during this stage, and fecundity of the female fish, data from surveys of eggs can be used to estimate the absolute numbers of mature females. The cost, however, is high, and the precision, even with extensive surveys, is low (English 1964).

The classical indirect method of estimating abundance, familiar to terrestrial animal ecologists, is tagging. Marks or tags have been attached to fish to study their movements, migrations, growths and numbers for a long time. In principle, the estimation of total numbers in the population is easily obtained if a number of marked fish are released into the population and at some later date a sample of the population is examined to determine the numbers bearing marks, being given by the equation:

$$\text{population numbers} = \frac{\text{numbers marked } x \text{ number in later sample}}{\text{number of marked individuals in later sample}}$$

Unlike his terrestrial equivalent, the fishery scientist is seldom in a position to recapture the marked fish he releases, but relies on the commercial fishermen to return any marks that occur in his catches. The later sample in this case corresponds to the whole commercial catch.

This simple formulation is seldom valid in practice. It requires that the marked fish are completely mixed with the unmarked animals in the population, and that there are no additions or losses between the time of marking and the later sampling. These requirements are usually not satisfied. Further, marks may fall or wear off the marked fish, which may be killed by the shock of being caught and tagged, or may suffer excess mortality afterwards, e.g. by increased vulnerability to predators, and tags may not be returned to the scientists even after the fish are caught. Adjustments can be made for these and other sources of error, but even with these corrections reliable estimates of abundance are difficult to obtain from tagging experiments. These have proved more useful in the study of movements, and especially in the degree of mixing or separation between groups or populations of the same species (ICNAF 1963).

Most estimates of the abundance or productivity of fish stocks have come from the study and analysis of data from commercial fisheries. An essential step in this study is the determination of the correct measure of the amount of fishing, or fishing effort. If this is properly measured, the fishing mortality (the fraction of the stock that is caught per unit time) will be proportional to the fishing effort, and the catch per unit effort (c.p.u.e.) will be proportional to the abundance of the stock (or more strictly density, i.e. abundance per unit area).

The fishing effort is generally considered as the product of the fishing power of the vessel concerned (as determined, for example, by the size of the net used), the time spent fishing (or number of discrete operations) and the degree to which fishing is done at the times and places where fish are particularly abundant. Unfortunately for the scientist all this tends to increase from year to year as fishermen become more skilful and call to a greater extent on the products of modern technology. Boats become bigger, with more powerful engines and improved nets, they are less delayed by bad weather and spend a smaller proportion of the time steaming to and from the grounds, and echo sounders and other equipment make it increasingly easy for the fisherman to detect the best concentrations of fish.

Unless corrections are made for these improvements, the simple records of the amount of fishing that can be obtained from commercial data, e.g. number of ships operating, number of landings made, will become progressively misleading. The amount of fishing will be underestimated and the abundance of the stock in later years overestimated. This may often result in a significant (in both the statistical and operational senses) trend of falling abundance being obscured in the simple data. For example, during the first half of this century the standard unit of fishing time for English trawlers was the number of days at sea. At high densities a larger proportion of the time at sea was either spent steaming to and from the grounds, or handling the fish on deck. Less time was spent actually fishing. Thus, the catch per day underestimated the changes in the stock abundance, which was better measured by catch per hour fishing (Fig. 12.2).

Measurement of the various increases in efficiency and the incorporation of appropriate corrections in the statistics of fishing effort is therefore an important basic task in the study of fish populations. Calibration of the fishing power of different vessels is relatively straightforward (Gulland 1956, Beverton & Holt 1957, Robson 1966), provided the changes and improvements in the vessel and gear are sufficiently distinct. Increasing size of vessel is easy to notice, and its effect can be measured, but more subtle improvements in the rigging of the gear can occur without due account being taken—for

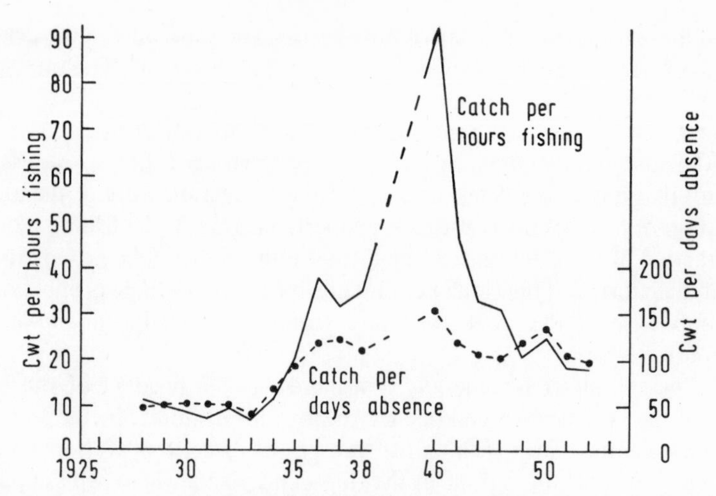

Fig. 12.2. Changes in the abundance of cod off the Norwegian coast, as estimated by the catch per day at sea and the catch per hour fishing by English trawlers, showing the damped fluctuations in the former measure. (From Gulland 1956.)

example, changes in the spacing between hooks on the longlines used for halibut fishing.

Obtaining the correct measure of fishing time is a matter of first deciding, from an understanding of how the fishery operates, what should be the best measure to use (e.g. number of hours during which the net of a trawler is actually on the bottom of the sea), and then obtaining appropriate records from the fishermen, possibly from logbooks or from interviews when they land. In either case the procedure is straightforward once the interest and confidence of the fishermen have been gained. The more intractable questions concern changes in the ability of fishermen to concentrate on the higher densities of fish. This is being steadily increased by a range of electronic fish-finding and navigational devices—however featureless the surface of the sea may appear, variations in bottom topography can cause trawl catches to vary very greatly over distances of one km or less. The improvement in position-fixing achieved by equipment such as Decca has changed North Sea trawling to the extent that all position of wrecks and other obstructions on the bottom are always referred to by the green, red and purple Decca coordinates, rather than by latitude and longitude. The most serious change may occur with shoaling fish, for which a decrease in the total abundance of the stock may be hidden by the increasing ability of the fishermen to find those shoals that remain. In theory, if the size of shoal remains unchanged, and only the number of shoals declines, some measures of catch per unit effort, such as average catch taken by a purse seiner in a set of the net can remain at more or less the same level until the last shoal is caught. These effects are not easy to measure and correct. The most direct measure would be the comparison of the abundance of fish in areas where fishing has been carried out, and in other areas. The best that can be done is to examine carefully how the fishery operates, and to obtain good records of where fishing is done, and how the time of each trip is divided between steaming to and from the chosen grounds, searching for fish on these grounds, and the actual catching operations. For example, in the case of a declining stock, the time spent by the fishermen in searching for the few remaining shoals would increase, and the catch per unit searching time could correctly reflect the changes in the stock. With detailed knowledge of the fishery it is generally possible to know how much reliance can be placed on the effort and c.p.u.e. data, and the likely degree, if any, to which they may be biased by changes in searching efficiency.

Analyses of catch and effort

Given statistics of catch, effort and c.p.u.e., the simplest methods of studying the fish stock and its productivity are to analyse these statistics. In so doing, these two assumptions are usually made; first, that the c.p.u.e. provides an adequate index of the population abundance. The second is more fundamental and is that the population is fully described by its total abundance without taking into account the structure of the population. In these and most other analyses, only that part of the population consisting of the larger individuals (usually corresponding to fish that can be caught by the normal commercial gear) is considered. This is often a practical necessity, since the abundance of the smaller fish—the pre-recruits—may be unknown, but it is also ecologically reasonable. The eggs, larvae and post-larvae, are exposed to an entirely different set of influences—type of food, predators, etc.—than the larger fish, and though there is not always a sharp division, it is usually fairly easy and conceptually convenient to fix some size or age (e.g. the time when young plaice move offshore into the main North Sea fishing grounds from their coastal nursery grounds in which they spend their first few years of life) as that at which the fish recruit to the exploited stock.

In the absence of fishing, and under average environmental conditions the stock (of fish above the age of size or recruitment) will tend to stabilize at some equilibrium abundance. If the abundance is less than this level, the population will tend to increase and the assumption usually made in this type of analysis is that the rate of increase is determined solely by the current abundance, i.e., denoting by B the abundance, or biomass of the stock

$$\frac{dB}{dt} = f(B) \tag{12.1}$$

This approach is particularly associated with the work of Schaefer (1954,

1957), who showed that the simplest form of $f(B)$, to satisfy the conditions of zero growth at zero population, and at the limiting population abundance (B_∞), was $f(B) = aB(B_\infty - B)$. Schaefer also showed that the change in population during a year ΔB, taking into account the catch during the year, C, was given by $\Delta B = f(\bar{B}) - C$, where $\bar{B}$ = average abundance during the year, or, writing $B = {}^1/q\ U$

where U = catch per unit effort;

q = constant, the ratio of c.p.u.e. to abundance

$$\Delta U = f(\bar{U}) - qC \tag{12.2}$$

This approach gives a suitable method of analysing catch and effort data when removals by man are the chief cause of changes in population. This condition was reasonably well satisfied in the case of the post-war blue whale stock in

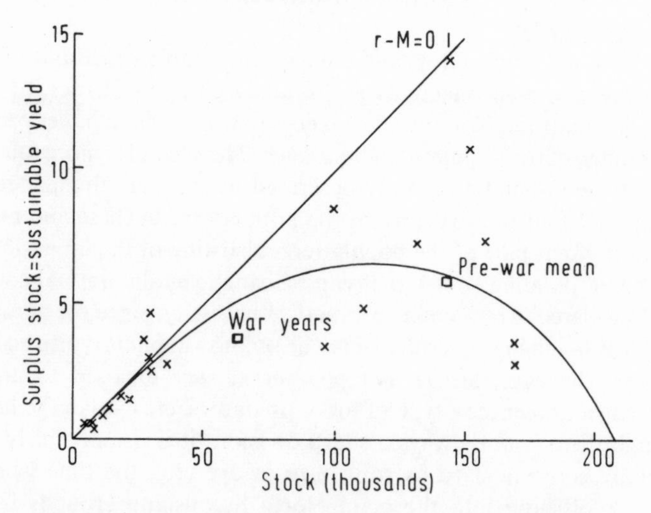

Fig. 12.3. The sustainable yield (net rate of natural increase) of blue whales in the Antarctic, as a function of population size. (From Chapman 1964.)

the Antarctic (see Fig. 12.3, from Chapman 1964). In this case since the curve is quite well determined by the observed points of natural increase in population (catch less observed decrease), and by the original unexploited population of some 220,000 whales, the curve could be drawn by eye. More objective procedures, using computer programs to allow for a range of forms for $f(B)$ have been derived (Pella & Tomlinson 1969). Their procedure does not need independent estimates of the relation between c.p.u.e. and abundance, but determines the value of q that minimizes the differences between expected and observed catches.

Other forms of the relation between the three interdependent variables—stock abundance (or c.p.u.e.), total catch (which in a steady-state situation will be equal to the natural rate of increase of the population) and amount of fishing (fishing effort or fishing mortality)—can be used which are less vulnerable to fishery-independent fluctuations in abundance. For practical purposes of advising administrators it is convenient to consider either c.p.u.e. or catch as the dependent variable, and relate one or other to the amount of fishing. By considering the c.p.u.e. in one year as a function of the effort over some previous period some account can be taken of the delays with which changes in the amount of fishing take effect on the fish population (see Fig. 12.4 from Gulland 1961).

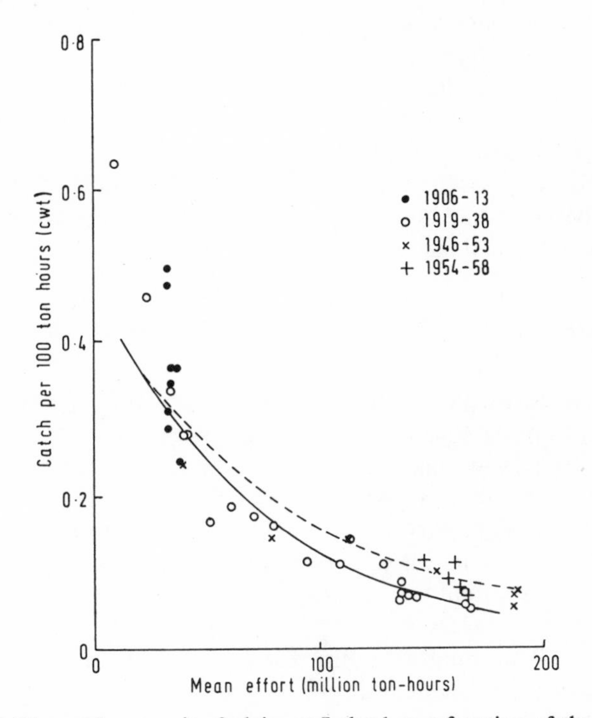

Fig. 12.4. The catch per unit of plaice at Iceland as a function of the average fishing effort in the current year and the previous years. From 1954 to 1958 the catch per unit of effort for a given effort has been higher, probably because fewer small fish are being caught. (From Gulland 1961.)

Though Figures 12.3 and 12.4 were largely derived from the simple theories that ignored the detailed structure of the population, they present completely valid outlines of the facts of life of fish populations which are relevant to rational exploitation of the stocks. A large stock, close to the average unexploited abundance, can support only a small catch without being decreased (see the right-hand part of Fig. 12.3); in other words, any increase in the amount of fishing will always decrease the abundance (c.p.u.e.) (broken line in Fig. 12.4). As the amount of fishing increases, the catch at first increases almost in proportion to the amount of fishing but then less quickly and, as for the Iceland plaice, the catch reaches a maximum; for higher levels of fishing the stock is reduced to a level where the rate of natural increase, and hence the sustainable catch, is lower.

Where these curves can be determined with sufficient precision from the statistics of catch and effort, they are sufficient for most advice to the policy-makers, without recourse to more detailed models. Fortune is seldom so kind, and even when it is, provision of advice is not the only scientific objective. The models above give a description of the reaction of fish populations to fishing, but little insight as to how this reaction occurs. There is also little opportunity for using other information on the stocks or on the environment in which the fish live, or to consider variations in the pattern of fishing (e.g. the sizes of fish caught) rather than the total amount of fishing. For these the analytical models associated particularly with the work of Ricker (1948, 1958) and Beverton and Holt (1957), which deal with the growth and mortality of the individual fish, are necessary.

Analytic models

These models start with the consideration of the history of a brood, or year-class, of fish from the time they recruit until the last survivor dies (or at least until there are so few survivors that they can be ignored for all practical purposes). Clearly in a steady-state situation the total catch in one year, from all year-classes present during that year, will be equal to the yield from any one year-class during its whole life. Other characteristics of the population, e.g. average weight, total biomass, can almost equally easily be derived from the history of a year-class.

During its life the numbers of the year-class will be reduced by fishing and other natural causes. The number of deaths per unit time can be described by the two equations:

$$\left(\frac{dN}{dt}\right) \text{fishing} \qquad = -F_t N_t$$

$$\left(\frac{dN}{dt}\right) \text{natural causes} = -M_t N_t$$

where N_t = numbers of fish alive at time t;
F_t = fishing mortality coefficient applying to fish of age t;
and M_t = natural mortality coefficient applying to fish of age t;

thus the total death rate is $(F_t+M_t)N_t$, and hence writing the equation as $\frac{dN}{N_t} = -(F_t+M_t)dt$ and integrating,

$$\log_e N_t = -\int(F_t+M_t)dt$$

or, writing R = number of recruits (numbers in the year-class at the time they recruit, at age r)

$$N_t = R\exp\left(-\int_r^t (F_t+M_t)\right)dt \tag{12.3}$$

which if F and M are constant, becomes

$$N_t = Re^{-(F+M)(t-r)}$$

The total numbers caught will be given by

$$C = \int_r^\infty F_t N_t dt$$

and the weight caught by

$$Y = \int_r^\infty F_t N_t W_t dt \tag{12.4}$$

where W_t = average weight of a fish of age t.
These are perfectly general expressions. To be of practical value some assumption has to be made concerning the form of F, M and W. The most simple is that F and M are constant, and that W has some form that makes the algebra easy, of which the two most usually employed are the exponential (Ricker 1958) and the von Bertalanffy (von Bertalanffy 1938, Beverton & Holt 1957)

$$W_t = W_0\, e^{g(t-t_0)}$$

and

$$W_t = W_\infty\,(1-e^{-K(t-t_0)})^3$$

respectively.

These give the expressions for the yield as

$$Y = FRW_0\, e^{(g-F-M)(T-r)} \text{ (for the yield of fish up to age } T)$$

$$\text{and } Y = FRW_\infty\left(1-\frac{3e^{-K(r-t_0)}}{F+M+K}+\frac{3e^{-2K(r-t_0)}}{F+M+2K}-\frac{e^{-3K(r-t_0)}}{F+M+3K}\right) \tag{12.5}$$

respectively.

These or other similar equations enable the relations between the amount of fishing and the catch or stock abundance to be determined. By an empirical division of the whole life span into sufficiently short periods such that within each period the weight and numbers do not change much, they can be readily modified to take account of possible variations of fishing mortality with age. An important special case is that when it is zero below some particular age (often referred to as the age at first capture) which may be controlled, e.g. by the use of appropriate sized meshes in the nets. The snag is the practical determination of the values of the different parameters. When, as often occurs, the age of the individual fish can be determined, e.g. from scales or otoliths, growth presents few problems. Given sufficient samples it is merely a matter of determining a growth formula that fits the observations to an acceptable degree, and can be combined with the equations without incurring computational difficulties. Determination of mortalities (especially the division of the total between fishing and natural mortality), and of the relation between fishing mortality and the available units of fishing effort, present greater problems. There is a considerable literature, particularly among the reports of the various international committees, sub-committees and working groups set up by the ICES (the International Council for the Exploration of the Sea) and ICNAF (the International Commission for the Northwest Atlantic Fisheries) and other bodies to study particular fish stocks, on methods of analyzing data on age and size composition of catches, tagging data, etc., to estimate mortalities. (Such methods and the literature are summarized in Gulland 1969.)

Though the simple equations above assume that natural mortality and growth are constant, and are not affected by changes in the stock, it is only a matter of more or less laborious computation to include some appropriate formula relating natural mortality (or some parameter of growth, such as W_∞ of von Bertalanffy) to abundance (Beverton & Holt 1957). The results, with any probable relation between growth or mortality and stock abundance, will be more realistic, but differ little from those based on constant parameters.

More fundamental problems are raised by the determination of recruitment and its possible variation with stock abundance. Recruitment is difficult to measure and is not often estimated explicitly. If an estimate of the average recruitment in recent years is required, it is usually obtained by calculating the yield per recruit (Y/R) from equation (12.5) or its equivalent, using the observed values of growth and mortality, and comparing this yield per recruit with the observed yield. This is one reason why results of assessments of the state of fish stocks are often expressed as yield per recruit. Another is that year-class strength in many fisheries is highly variable for reasons that seem quite independent of fishing or of the abundance of the stock, so the effect of changes in the pattern of fishing on the yield per recruit can be predicted, but not the effect on total yield.

So far as the assumptions of the model are concerned this strictly implies no more than an expression of a state of ignorance of the possible relation between the abundance of the adult stock and the strength of subsequent recruitment. Unfortunately, this ignorance has often turned into an assumption that there is no relation, or rather that the average recruitment is the same for all likely values of adult stock. This can have serious results, since the conclusions as to the effect of high levels of fishing reached under the assumption of constant recruitment can be very different from what actually occurs with heavy fishing if the recruitment varies.

The assumption of recruitment being independent of the adult stock is not as unreasonable as might appear at first glance. Most female fish produce enormous numbers of eggs, ranging from some tens of thousands in the case of anchovies, up to some millions in the case of large cod. Since most fish stocks do not change greatly in number from one generation to the next, of these thousands or millions of eggs only two, on the average, can survive to maturity, i.e. there is a mortality of 99·998 per cent (if 100,000 eggs are produced per female) between spawning and maturity. If the adult stock were halved it would only require the reduction of this mortality to 99·996 per cent for the number of fish reaching maturity in the second generation to remain the same. Clearly, this very minor reduction in mortality rate is quite a feasible response to a halving of the initial number of young fish, with the consequent reduction in competition for food, etc. In fact, until recently there was little clear evidence, except for some stocks of salmon (Ricker 1954), of the average recruitment being affected by variations in adult stock, and for some stocks, e.g. the plaice in the North Sea (Beverton 1962), it was clear that if the average recruitment did change over the observed range of adult stock, the change was very small compared with a sixfold range of adult stocks.

The study of the relation between stock and recruitment has been handicapped by the greater annual variation in recruitment that occurs in many stocks. Relatively small changes in the mortality between birth and recruitment caused by any one, or a combination, of a wide range of environmental factors, can result in very big changes in the numbers of recruits. A typical example is the haddock in the North Sea, where year-classes can vary in strength by a factor of a hundred or more (Fig. 12.5). With this extraneous variation it is difficult to establish significant correlations between stock and recruitment, especially since only one pair of values can be obtained for each year, so the series is usually short. Here it is easy to fall into a statistical trap. The obvious null hypothesis to make when examining the correlation between stock and recruitment is that they are independent. Statistical analysis will often result in there being no reason, at the chosen level of significance, to reject this null hypothesis. However, this does not mean that the null hypothesis should be accepted, or that it is in any way established that recruitment is independent of adult stock—though this is the interpretation sometimes placed on the results. There are a number of other possible null hypotheses, e.g. that recruitment increases with adult stock, though somewhat less than proportionally, that could equally well be put forward.

The whole problem of the relation between stock and recruitment has been recognized as being of major importance to the practical matters of

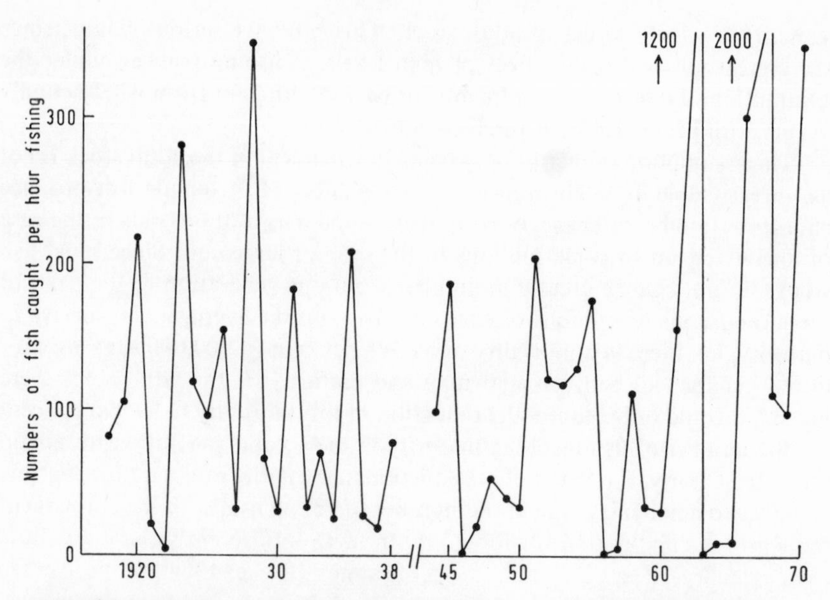

Fig. 12.5. Variations in the strength of haddock year-classes in the North Sea. (Numbers of fish caught when 1-year-old in one hour trawling, by Scottish research vessels—*Explorer*, 1918–58; *Scotia*, 1959–70. Data from Marine Laboratory, Aberdeen.)

managing fisheries, and to the scientific understanding of how the abundance of fish stocks is controlled, and is receiving increasing attention (Ricker 1954, Parrish 1973). At present it seems that the resolution of the problem will require a combination of several approaches, including a better knowledge of the population dynamics of the younger, pre-recruit stages of fish including the effects of their predators, and their food requirements—needing both field observations and laboratory observations—and, hopefully, the establishment of typical forms of the stock-recruit relation to be expected from different stocks, possibly depending on the basic biology of the species concerned, such as the fecundity of the individual female (Cushing & Harris 1973).

The previous sections deal explicitly only with a single species or stock of fish, but no species lives in isolation. Consideration of other species is essential for proper understanding of what is happening, and is becoming of increasing practical importance as fisheries exploit a wider range of species in any one area. The analytic models provide mechanisms for studying interspecific effects, at least provided there is a sufficient supply of data, and of scientific time and talent to study the data. The parameters of growth and mortality, which it has already been suggested can be allowed to vary with the abundance of the species concerned, can also be related to the abundance of other species, particularly potential food species in the case of growth, and of predators in the case of natural mortality. These interactions can include those between the younger stages; for example it appears that the survival of Californian sardine (*Sardinops caerulea* (Girard)), between egg and recruitment is related to the combined abundance of young sardines and anchovies (Gulland 1971a).

This approach is fine in principle, and can even be practicable if the number of species is small. In the Gulf of Thailand, where there is an important trawl fishery (Menasveta 1968) (see Table 12.2) the number of commonly occurring species approaches a hundred. The possible interactions, even between single pairs of species, runs into thousands. Since the interactions can affect growth, mortality or recruitment, and may either become apparent immediately, or only after a delay of one or more years, it is impossible to untangle all the possible inter-relationships.

An alternative approach that at least solves the practical problems is to consider that which is actually observed in the sea is the integration of all the possible interactions, and that when the prime needs are to determine what is happening, and what are the net effects of changes in the amount of fishing (including the indirect effects through interaction between species), rather than why and how these effects take place, it is sufficient to return to the simple, basic statistics of catch and effort.

In the Gulf of Thailand these statistics give a clear answer as to what is

Table 12.3. Commercial catches, catches per unit effort by survey trawlers and estimated total effort in the Gulf of Thailand. (From Shindo 1973.)

Year	Total catches (0·000 tons)		Catch per hour (kg)	Estimated effort (0·000 hours)	Number of trawlers
	All marine fish	Trawl fishery			
1961	No data	123	298	413	
1962	No data	151	277	545	976
1963	323	277	256	1,082	2,026
1964	493	372	226	1,646	2,360
1965	529	393	179	2,196	2,396
1966	635	449	131	3,428	2,695
1967	762	583	115	5,070	3,077
1968	1,004	784	106	7,396	3,182
1969	1,180	908	103	8,816	3,185
1970	1,336		97		3,114

happening (Table 12.3) (from Shindo 1973). In this particular fishery there is very good information on the c.p.u.e., obtained from regular surveys by special vessels, but the statistics of total catch are not so reliable; the best index of total effort is obtained by dividing catch by the survey c.p.u.e. Though there are data on the number of trawlers, they do not take into account increases in the size and efficiency of the average Thai trawler. The statistics, simple though they are, show quite clearly that fishing has greatly reduced the abundance of fish in the Gulf of Thailand, and further increases in fishing will probably decrease the total catch. The practical need is to limit and, if possible, reduce the total fishing effort. The simple statistics, however, do conceal great changes in the species composition of the fish in the Gulf

Table 12.4. Catches per hour taken by survey vessel *Pramong II* in the Gulf of Thailand during the early stages of development of the Thai trawl fishery, showing different changes of various species. (From Tiews *et al.* 1967.)

	Area I			Area II		
	1964	1966	1956 as % of 1964	1964	1966	1966 as % of 1964
Total	340·5	112·9	33	229·7	57·6	25
Fish for human consumption	192·7	72·4	38	165·6	40·1	24
Industrial fish	147·8	40·5	27	64·1	17·5	27
Sciaenids	21·9	1·9	9	18·3	0·2	1
Leiognathus	116·5	27·6	24	63·5	13·5	22
Carangids	31·6	8·5	27	25·6	4·9	19
Squids	5·4	3·6	67	5·3	9·8	185
Nemipterus	36·4	7·4	20	15·5	3·6	23

(Tiews *et al.* 1967, Ritsraga 1971) (Table 12.4). Most of these changes can reasonably easily be ascribed to the direct or indirect effects of fishing, e.g. the large decline of large long-living species, of high market value (e.g. Sciaenids), the lesser decline of species of lower market value, or less vulnerable to the trawl, through being well off the bottom, and increases of some lightly fished species through less predation from larger fish, or less competition (e.g. squids).

The world's potential fish production

The study of fish stocks began in the few areas—the North Sea, the north-east Pacific—where the local stocks (plaice, cod, halibut and salmon) because of their high value or the ready access to large markets, were suffering from heavy fishing. Some fifty years ago these and a few others were the only stocks being fully exploited. Since then, and particularly in the last couple of decades, a much larger number of stocks in all parts of the world, e.g. the Peruvian anchovy and the bottom fish in the Gulf of Thailand, have become heavily fished and there are a steadily decreasing number of stocks that still offer clear opportunities for taking increased catches. Considerable attention is therefore being given to the question of the magnitude of the total catches that could be taken from the world's oceans.

A large number of estimates of the potential world catch have been made, including two recent extensive studies by Moiseev (1969) and Gulland (1971b). A variety of methods have been used, which fall into two basic groups, the study of individual fish stocks, with extrapolation to areas where assessments have not yet been made, and estimation of the total primary productivity, and the production at successive links in the food chain. The latter is best illustrated by the table produced by Schaefer (1965) (Table 12.5).

Table 12.5. Potential annual harvest (million tons) (taken as half annual production from the sea, at different trophic levels and for different assumptions of ecological efficiency). (Adapted from Schaefer 1965.)

Trophic level	Ecological efficiency factor		
	10%	15%	20%
Herbivores	9,500	14,000	19,000
1st stage carnivores (e.g. copepods)	950	2,100	3,800
2nd stage carnivores (e.g. herring)	95	320	760
3rd stage carnivores (e.g. tuna)	9·5	48	152

Though detailed assessments have only been made for a minority of the world stocks, reasonable assessments cover a wide range of areas and types of fish. Extensive studies have been made in the North Atlantic by the two international bodies concerned, and are contained in the reports of the Liaison Committee of ICES (especially of the various working groups set up to study individual stocks) and of the Research and Statistics Committee of ICNAF (published in the series of ICES Cooperative Research Reports, and ICNAF Red Books). In other areas the responsible regional bodies (the General Fisheries Council for the Mediterranean, the Indo-Pacific Fisheries Council and other subsidiary bodies of FAO, as well as independent bodies such as the International Commission for the Conservation of Atlantic Tunas) are undertaking similar studies, but have not, in general, advanced so far. Particular mention should, however, be made of the older bodies on the American west coast, especially the Inter-American Tropical Tuna Commission, which, unlike the other bodies, does not rely on research by scientists supported by member national governments, but maintain its own independent scientific staff. I-ATTC (Inter-American Tropical Tuna Commission) in particular has a long record of distinguished work including the early evaluation of the stock of yellowfin tuna in the eastern tropical Pacific (Schaefer 1957). While many of the most important fish stocks are exploited by a number of countries, and have to be studied on an international basis, there are many other stocks, notably the anchovy off Peru (Anon. 1970, 1972), which are exploited wholly or virtually wholly by a single country, and have been the subject of national studies.

All these studies enable a good impression to be gained of the nature of fish stocks of the types exploited at present, and of their magnitude in several areas. Their magnitude in other areas can be estimated by comparison of general ecological features, primary production, relative catches by comparable fishing gears, etc. There is now general agreement, based on this sort of approach, that the total potential world catch of fish of the familiar kinds is close to 100 million tons per year (Gulland 1971b).

Table 12.6. Potential annual harvest of the familiar types of marine fish, including crustaceans, excluding molluscs and whales (million tons). (From Gulland 1971b.)

Region	Large pelagic (a)	Demersal	Shoaling pelagic	Crustaceans
Atlantic north-west	(a)	3·6	2·8	0·1
north-east	(a)	6·6	6·7	0·2
eastern central	} 1·1	0·9	1·9	0·1
western central		2·0	3·0	0·3
south-west	(a)	3·8	5·0	0·1
south-east	(a)	1·1	3·2	(+)
Mediterranean and Black Sea	(a)	0·4	0·8	(+)
Indian Ocean western	} 0·7	4·4	4·0	0·1
eastern		3·0	2·0	0·1
Pacific north-west	0·3	1·7	3·2	} 0·4
north-east	0·3	1·8	2·5	
eastern central	} 1·9	1·4	3·7	0·1
western central		11·0	4·0	0·7
south-west	(a)	1·5	2·0	(+)
south-east	(a)	0·6	11·9	0·1
Total	4·3	43·8	56·7	2·3

(a) Apart from salmon in the North Pacific, the major species are the highly migratory tunas, whose potentials are grouped in the central, tropical areas.
(+) Value less than 100,000 tons.

Table 12.6 gives details (adapted from Gulland 1971b) of how this total is divided according to types of fish and major ocean areas (using the areas

Table 12.7. Recent catches and potential sustainable yields of some major fish stocks

Species	Area	Catches (thousand tons)				Main countries (% of 1971 catches)
		Potential*	1965	1968	1971	
Anchoveta	Peru current	12,000†	7,681	11,272	11,060	Peru 93, Chile 7
Cod	N.E. Atlantic	2,000	1,307	2,065	1,775	Norway 26, U.K. 20, Iceland 14, U.S.S.R. 10
	N.W. Atlantic	1,400–1,800	1,458	1,802	1,056	Spain 24, Canada 23, Portugal 14, U.S.S.R. 11
Alaska Pollack	North Pacific	No estimate	1,042	2,201	3,637	Japan 74; U.S.S.R. 24
Herring	N.E. Atlantic	3,000	3,735	2,339	1,385	Denmark 24, Norway 17, U.K. 10, U.S.S.R. 10
	N.W. Atlantic	300–1,000	266	871	739	Canada 57, U.S.S.R. 15, Poland 12
Mackerels	N.W. Pacific	800‡	693	1,062	1,400	Japan 89
	N.E. Atlantic	600	302	905	413	Norway 49, France 10
Haddock	N.E. Atlantic	465	499	391	459	U.K. 45, U.S.S.R. 18, Norway 10
Menhaden	U.S. Gulf and East coasts	750	783	624	994	U.S.A. 100
Cape Hakes	S.E. Atlantic	620	332	712	761	U.S.S.R. 44, Spain 21, S. Africa 14
Large tunas§	Warm Oceans	650–850	680	742	741	Japan 40, U.S.A. 14
Pacific salmon¶	N. Pacific	500	431	386	428	U.S.A. 34, Japan 32, U.S.S.R. 20, Canada 15

* Potential yields taken from Gulland 1971b.
† More recent studies suggest a better figure would be about 10 million tons.
‡ Estimated by Fukuda (in Gulland 1971b) that 1964 catches could be increased by 100–200,000 tons.
§ Includes yellowfin, albacore, bigeye and bluefin tunas.
¶ Includes sockeye, pink, chum, chinook and coho salmons.

adopted by FAO in the Yearbook of Fishery Statistics). Table 12.7 gives more details for some of the more important species, including recent catches, and the contribution to these catches of some of the main countries concerned.

These tables show that the potential is dominated by two groups of fish—the shoaling pelagic fish such as herrings and anchovies, and the bottom-living animals such as cod, plaice, snappers and shrimps. In both cases the behaviour of the fish makes it easier for the fisherman to extract the production from a given volume of water—in the first case by pursuing a particular shoal of fish, rather than setting his net at random, and in the second, by having to work in two dimensions only, rather than three.

The same is true of the other less important types of resource, with the extreme example being the large whales, where in the short time required to hunt down and kill a single whale a whale catcher can harvest the accumulated production over a wide area over a period of decades. Large tuna also allow the production over a wide area to be harvested simply; a typical Japanese tuna vessel may set a line 80 km long, which with hooks at intervals of some 60 m may catch most of the fish crossing the line, which, since tuna is a very active fish, might be a fair proportion of the fish within a mile of the line when set.

Geographically, the distribution of fish potential of the familiar types of fish, is also restricted. Demersal fisheries, for bottom-living fish, are naturally confined to the relatively shallow waters of the continental shelves, mostly in depths of less than 100 fathoms. Soviet trawlers and others have fished in much greater depths, down to 500 fathoms, and have had quite promising catches, though these may be of old fish and represent the accumulated production of several years. Beyond this, commercial trawling is not feasible, and in any case the production per unit area is likely to be very low.

Although the shoaling pelagic fish are not so obviously restricted to particular geographical areas, they do need favourable conditions, of which the availability of large quantities of food, particularly zooplankton, is necessary to support large shoals, which are in turn necessary for a successful fishery. Most of these fisheries are therefore in areas of high zooplankton standing crop and high primary production. Since the shelf areas are areas of high primary production there is a close relation between the distribution of primary production and of fisheries as shown, e.g. by the Atlas of Living Resources of the Sea (FAO 1972), or by the very close correlation between areas of upwelling, and the distribution of catches of sperm whales by nineteenth-century New England whalers (Townsend 1935, Cushing 1969).

This clear qualitative and spatial relation is not so easily matched in quantitative terms. Schaefer's analysis (Table 12.5) shows, on a world scale, that even when the primary production is known, estimates of potential fish catches depend critically on the number of stages between primary production and fish, and on the efficiency of transfer between each stage. On a smaller scale, Paulik (1971) has shown that although both Cushing (1969) and Ryther (1969) both obtained estimates of total fish production in the Peru Current system as 20 million tons (in good agreement with the potential harvest of around 10 million tons), the agreement between the two calculations is not too reassuring when comparing the elements in the equations (area of current system × production of carbon by primary producers × number of stages × efficiency at each stage × conversion from carbon to wet weight). There is little agreement for any of these elements and as Paulik points out 'if Cushing had used Ryther's 1½ step ecological efficiency factor of 0·12 (rather than a two-step efficiency of 0·01) he would have obtained about 240 million tons, rather than 20 million tons'.

This uncertainty in an area which is reasonably well known compared with many other parts of the world, suggests that our knowledge of the quantitative links between primary production and fish is not yet good enough to estimate potential fish harvest from data on primary production alone. Although Ryther (1969) in considering the potential of different parts of the ocean is undoubtedly right in pointing out, as have others, that most fish catches have in the past and will in the future come from a relatively small proportion of the whole ocean (principally the continental shelves, and the upwelling areas), his figures for the potential of each zone may depart quite widely from the true value. The figure for the whole ocean is subject to variation through doubts concerning the productivity per unit area within each zone, and on the limits chosen for each zone—the latter being critical since the difference in productivity between the zones as used by Ryther is very high.

One should not be too pessimistic about establishing links between primary production and fish. It is clearly an important element in the study of marine ecology to establish and quantify such links and even though this is not made easier by the difficulty of quantifying either the standing crop or production of the intermediate stages of the pelagic community (quantitative sampling of the larger individuals in the benthic community is easier and has been carried out from the time of Petersen (1918) onwards) one can hope that in the not too distant future the links will be well understood.

Some results are already emerging from the better-studied areas, suggesting that either the links in the food chain between commercial fish and primary production are fewer, or the efficiency at each stage is higher than generally believed. Steele (1965) and Gulland (1971b) have examined the data for the North Sea. In this area fishing is so intense and the possibilities of alternative species for harvesting so carefully examined, that with known exceptions (e.g. dabs, *Limanda limanda* (Linnaeus)), one can be fairly sure that the commercial catches represent all major stocks of fish. The extent of the area is reasonably well defined and almost wholly enclosed, so that migration of fish into and out of the area (which could confuse the relation between local primary production and local fish catches) is relatively small, and for practical purposes (if the substantial catches of spawning herring that used to be caught in the Straits of Dover and nearby part of the eastern Channel are included in the North Sea) confined to the north. Here there is some movement of haddock between the North Sea and the waters to the north and west of Scotland (Jones 1959), and the Atlanto–Scandian stock of herring (*Clupea harengus* (Linnaeus)) which feeds mainly in the open ocean between Iceland and Norway, moves to the Norwegian coast to spawn and is caught in what is technically the North Sea (ICES 1970). These latter catches can be omitted from the North Sea and are now in any case very small since the stock is at a very low level.

Much other relevant information, on primary production, the food habits of the fish, etc., are known or readily available, and the North Sea should be a very favourable area for study. Nevertheless, there are surprises. Steele (1965) noted that the catches of demersal fish (principally cod, haddock, whiting (*Merlangius merlangus* (Linnaeus)) and plaice) had remained remarkably constant (at around 400 thousand tons per year) from 1906 up to 1960 (except for the period of the two wars). He concluded that this quantity represented the maximum potential of these stocks, as determined, in the last instance, by the primary production. This tended to be confirmed by the analysis of the population dynamics of the individual stocks, which showed (e.g. Beverton & Holt 1957) that they were fully exploited.

Almost as though at a signal on the publication of Steele's paper, the North Sea demersal catches began to rise steadily from 1962 onwards reaching over 2 million tons in 1971. The proximate cause of this increase has been a succession of good year-classes. Those of haddock in 1962 and 1967 have been outstanding, causing the haddock catch in 1969 and 1970 to rise to over 600 thousand tons, compared with a low figure of only 53 thousand tons in 1962. Year-classes of cod and plaice fluctuate less widely than haddock and none of the recent ones have been as outstanding as the 1962 and 1967 haddock year-classes, but the average of recent year-classes has been noticeably above the earlier average.

The reasons for the outstanding year-classes in particular years and for the general increase in the average strength are unknown, but it is tempting to relate them to the changes in the pelagic fisheries and in the stocks of pelagic fish, particularly herring. Over the past couple of decades there has been a steady increase in the fishing effort on herring, due to changes in gear, from driftnet to trawl, and then to purse seine, as well as the development of high volume fisheries specifically for meal and oil. Although herring stocks as a whole were still quite lightly exploited up to the time Steele made his analysis, or not much earlier, they have since become heavily exploited and the abundance now is much less than it has been, the Atlanto–Scandian stock and the Downs stock of herring, spawning in the southern North Sea and eastern

English Channel, having been reduced to a very low level. The reduced predation by herring (adults, juveniles or larvae) on zooplankton might be expected to leave more zooplankton available for the larval and post-larval haddock and cod, thus improving their survival.

It cannot be assumed that the high demersal catches of the last few years can be taken at the same time as the high catches of pelagic fish that could have been maintained if the Downs stock had not been so seriously reduced. For comparison with the primary production, it seems better to use the figures for pre-1965 condition, i.e. potential catches of one million tons of demersal fish (the main species being cod: 150 thousand tons; haddock: 150 thousand tons; and plaice: 100 thousand tons). The catch-potential of pelagic fish is less easy to estimate with any precision, but a reasonable estimate is around 2·3 million tons (Gulland 1971b).

These figures are of the potential catch by man. The actual production, allowing for other causes of death, including deaths before reaching a marketable size, must be considerably greater, of the order of 5 million tons per year. The average primary production is around 100 gC/m^2 per year, which over the area of the North Sea (including the Skagerrak) of some 600,000 km^2 gives a carbon fixation of 60 million tons per year. For the commonly used approximation of 10 per cent as the ratio of production at successive stages, this gives a production of herbivores of 6 million tons of carbon and 0·6 million tons of first stage carnivores. The latter is equivalent to about 6 million tons wet weight, depending on the proportion of carbon in the wet weight. The production of fish is therefore very similar to the total production of first-stage carnivores, as predicted on food-chain arguments, although not all North Sea fish are first-stage carnivores (large cod are certainly higher-level predators). The conclusions seem to be, first, that the actual ecological efficiency is better than 10 per cent, and second, that the fish production is not far below the limit set by the primary production.

Discussion

Two important facts emerge from these discussions, namely that world fish catches have been expanding rapidly, and that the potential of the familiar types of fish is limited at around 100 million tons per year. Clearly, the period of expansion cannot last much longer and at the past rate of increase the potential will be reached within a decade. The world's fisheries are faced with two major sets of problems; (1) to manage the traditional fisheries in such a way that at the least no more marine resources go the way of several of the large whales, the Californian sardine (Murphy 1966), or the Atlanto–Scandian herring, which have probably or certainly been reduced by too heavy fishing to very low levels, and preferably the stocks remain at a highly productive level and are harvested in an efficient manner (which is not the same thing, Gulland & Robinson 1973); (2) to promote the harvest of the less familiar stocks, whose potential is very high (Gulland 1971b, Suda 1973).

Most important fish stocks are harvested by more than one country and a large array of international commissions have been set up to promote and institute appropriate management measures (Gulland & Carroz 1968). Those with only a small number of members, particularly when they have very similar interests (e.g. the Pacific halibut fished by Canada and the U.S.A.), have a long history of controlled total catch (or the total amount of fishing). The outstanding example has been the North Pacific Fur Seal Commission, which, under various arrangements, has been successful in rebuilding the fur seal stocks on the Pribilof and other islands, after they had been threatened with extinction. Another example is the International Whaling Commission, which set limits (unfortunately rather too high, and with insufficient allowance for modifications in the light of new information) as long ago as 1946, and which, despite a poor performance in the 1960s, has in earlier and later periods been successful in preventing any stocks in this century being reduced as low as the Right and Gray whales in the nineteenth century, and has maintained the Sei whale stocks at around their most productive level (annual reports of the Scientific Committee of the IWC, Gambell 1972).

Those commissions dealing with a wider range of fish species and a greater variety of countries have, not surprisingly, been slower in introducing controls on the amounts of fishing. Among the most active commissions have been those in the Atlantic—ICNAF in the north-west and NEAFC and its predecessor, the Permanent Commission, in the north-east—which first paid attention to the easier problem of increasing the size at first capture (and hence the catch) by increasing the size of the meshes used in bottom trawls. While this is of some real benefit, it is at best an incomplete answer to the problem of managing the stocks, and in the long run some direct control on the amount of fishing is essential. This has now been achieved by ICNAF through the setting of limits on the total catch that can be taken from several of the more important stocks. Initially, with respect to several depleted haddock stocks, limits were set only on the total catch, but this has undesirable economic side effects as countries or companies increase their efforts at the beginning of the season so as to maximize their share (Gulland & Robinson 1973). To avoid this costly scramble (which occurred, until eliminated by agreements on national shares, in Antarctic whaling, and also in most fisheries managed by a single quota including those for Pacific halibut and yellowfin tuna in the eastern tropical Pacific) shares in later quotas set by ICNAF—including those for most of the important cod and herring stocks—have been allocated to the fifteen member countries on the basis of a formula that takes account of shares taken in recent years, and the special interests of coastal states and of countries with developing fisheries. NEAFC has not progressed so far, partly because the necessary procedural arrangements have not been made, though at the time of writing only one country, Iceland, still had to take the necessary step to give NEAFC power to institute catch quotas. However, some arrangements have already been taken to limit the amount of fishing on the more seriously depleted stocks, including North Sea herring, by a pattern of closed areas and closed seasons. [*Note*: from 1 January 1975 most stocks in the north-east Atlantic have been restrained by quota regulation.—Editors.]

The detailed arrangements for management of marine fish stocks are likely to undergo considerable modification in the next few years, as a result of the United Nations Conference on the Law of the Sea, initiated in 1973. It is quite possible, if not probable, that there will not be an agreement (or that many countries will fail to ratify any agreement reached) on the questions relating to fisheries—especially regarding the width of the territorial sea, or the zone in which the coastal states can have control of the fisheries. However, it seems highly probable that there will at least be a *de facto* extension of these limits—a process that is already proceeding fast, particularly in Latin America and West Africa. Although these extensions may shift responsibility from an international commission to one or more national administrations, the core of the problem—a bigger demand for fish than the resource can supply—remains unchanged. The problems incurred in some of the biggest fisheries that have been largely or wholly under a single national administration—the Californian sardine (Murphy 1966), the South African pilchard, and the Peruvian anchovy (Anon. 1972, 1973)—does not suggest that these administrations have any magic formula for success. The need for much hard work in assessing the status of the stocks and the effect on them of different possible management measures, of choosing between these measures on the basis of their biological, economic and social effects, and of introducing and enforcing the chosen measure, remain.

The development of fisheries for new, less familiar, species offers a similar range of scientific, technical and economic problems. The main problem is probably technical and economic—of developing, on the one hand, methods of catching and processing, and on the other, a market such that they can be caught at a cost that the market will pay. The biologist also has an important role, in determining the kinds of fish and other animals that could provide substantial harvests, and providing some estimate of the quantities that could be caught. One potential that has received much attention is the krill (*Euphausia superba* (Dana)) in the Antarctic. Soviet studies of this resource have been going on for some time (Burukowski 1965, Lyubimova *et al.* 1973), and have reached the stage of pilot scale production of a cheese mixed with krill, which tastes pleasantly of shrimp. Estimates of the potential, based largely on the past consumption by the great whales, range from 50 million tons per year upward. Other groups of fish which appear to offer opportunities (if the technological problems can be solved) of very large catches are the lantern fish and other small oceanic fish, especially in the equatorial upwelling zones, many of which form the food of the large tunas.

Management of existing fisheries and development of new fisheries both require a great volume of scientific, and particularly ecological, studies. Some of this is a matter of progressing along well-established lines. There are well-established techniques for assessing the status of exploited stocks, given a reasonable supply of data on catches (including details of the size and age of fish caught) and on the fishing effort (Gulland 1969). In the first instance, analysis of any newly developed fishery will probably be carried out along these lines. Even if the fishery on one species is considered in isolation, a number of fundamental and as yet unsolved problems are likely to arise, of which the nature of the relation between adult stock and subsequent recruitment is the most obvious. When, as must ultimately be done, each stock is considered in relation to the environment—especially the other fish stocks with which it may interact—in which it lives, the range of unsolved general problems grows—the quantitative relation between different trophic levels, the effect of fishing predators on the stocks of prey species, etc.

Although scientists now know enough about some aspects of fish production in the sea, as discussed in this chapter, to give, on many occasions, useful advice on the management and development of fisheries, much more knowledge is still to be gained. Until it is, the advice will often be incomplete and too late.

PART V. EVOLUTIONARY CONSEQUENCES

Both chapters in this section are hesitant attempts to understand the complexities of the marine system in which a number of subsystems are interlocked. Chapter 13 was originally intended to be a general study of communities, but some, such as mangrove swamps and coral reefs are highly specialized and indeed the biology of the principal animals in such communities is only sketchily known. Such communities have been described, but knowledge of their structure and function is thus limited. Some thought was also given to a discussion of species diversity, equitability, stability, and information theory, but these red herrings were discarded in favour of perhaps more fruitful subjects.

In contrast, the biology of some commercial fishes is known in considerable detail and has thus been used to show how fish might obtain and maintain their reproductive isolation in the establishment of such communities. Because the author's main interest lies in the stock and recruitment problem which is familiar to fisheries biologists, an account of the possible mechanisms involved has only been given in very general terms, although the possibility of extending them to the invertebrate populations in the rest of the ecosystem is considered. Given this possibility the numerical changes in populations due to competition within or between species may then be described.

The study of food chains (Chapter 14) is deliberately restricted to two old and familiar studies and to two more recent ones that reveal some of the lacunae in the oversimplified theory that sets trophic level immutably upon trophic level. Instead, we have tended to use food chain studies to illustrate a few ecological principles. Competition between larval plaice and sand-eels is limited to a very small part of each distribution of sizes for a period as short as a month, demonstrating Gause's principle. Jones' concept of a private food supply for larval fish is tacitly extended to sets of very simple food chains that succeed each other in a given life history.

Although the subject of seral succession and climax communities is not included in the text (because the evidence for it in the sea is a little thin, with the exception of colonization by fouling and littoral communities), there is a sense in which the study of food chains might become a study of succession merely because all animals in the sea grow considerably during their lives. Competition is the mechanism by which evolutionary changes are mediated, yet, like succession, its principles are hard to establish. Hence, there is no explicit treatment of competition in the text; we have come to learn that competition is most likely to occur amongst the larval animals and probably not at all in adult life. As food chains become broken down into a succession of simple episodes of trophic history, the mechanisms of competition may perhaps become revealed.

Chapter 13. Biology of Fishes in the Pelagic Community

D. H. Cushing

Introduction

The biology of fishes in the sea
Life history
The migratory circuit

The biology of fishes in the sea *continued*
The reproductive isolation of fishes
Population processes

The pelagic community

Introduction

The study of communities in the sea has a long history; some of Forbes' work in the Aegean (1843) and off the western coasts of the British Isles (1859) described communities on the sea bed in general terms. During the great descriptive period of the expeditions in the nineteenth century, similar generalizations emerged in the writings of Agassiz (1888), Wyville Thomson (1874) and Murray and Hjort (1912). The first quantitative descriptions of a marine community were Petersen's (1913, 1918), who analysed the fish food community in shallow water near the Danish coast. It is no accident that this was also the first detailed study of a marine food chain because such studies tend to overlap; indeed production biologists examine food chains, whereas communities are analysed by numerical taxonomists and the distinction is as much in discipline as in field of study.

Assemblages of animal species are often called communities, particularly those on the sea bed, where patterns of distribution change with depth or substrate much as the distribution of plants on land changes with soil, height or rainfall. Thorson (1946) has described such benthic communities in detail in Danish waters and has correlated their presence and absence with particular types of sea bed such as mud, sand or gravel and this environmental correlation is determined by the substrates on which the larvae settle. A much more extensive survey of epibenthic communities led by Zenkevitch (1963) charted the distribution of such animals in the seas around the U.S.S.R. and was presented in the form of charts of food available for fish.

In recent years, attempts have been made to assess the distance between different components in a community in a quasi-taxonomic sense, using statistical techniques. Because benthic animals are sedentary or at least restricted in their movements, methods that have been developed for the study of plant distributions can be applied to them. They can also be applied to the distribution of bottom-living fishes, if the movement of the fish is small in relation to the area surveyed. For example, Fager and Longhurst (1968) analysed the results of the Guinean Trawling Survey from this point of view. On this considerable expedition, the West African continental shelf was examined by trawlers between the mouth of the Congo and Cape Roxo in Senegal, which is a distance of 2,500 miles. Sixty-three transects of the shelf were made and trawls were hauled at eight standard depths. Five major assemblages had previously been recognized in this region, estuarine sciaenids, offshore sciaenids, sparids above the thermocline, sparids below it and a deep-water community. The valid distinction between these categories was confirmed by a method based upon the presence or absence of a particular animal at a given position; similarities were established by common scores of presence between different species. The advantage of the statistical method was that existence of the five groupings was established in an objective manner.

Similar assemblages may exist in the pelagic community, but this is difficult to establish with the same methods because the pelagic species cannot be grouped by position. The pelagic community within the depths lit by the sun is the largest in biomass and the most extensive in distribution in the sea. It has often been described as a food chain or food web, but, in Petersen's original view, there was little difference between the study of such webs and the study of communities. There is one environmental factor which dominates the pelagic community, the penetration of light: the production of algae in the photic layer depends upon irradiance and the diurnal vertical migration of animals towards the surface to feed at night is light dependent. In the future, the assemblages of species might be studied in depth at midday when the animals lie in layers by size, the bigger ones deeper in the water.

There are specialized communities in the sea, in coral reefs (Wells 1957), in mangrove swamps (McNae 1968), on rocky shores (Doty 1957) and in estuaries (Emery *et al.* 1957) in addition to those on the sea bed mentioned above. The predominant community, the pelagic, generates all the food and energy needed to sustain not only itself but also in part these dependent communities as well, some of which live in these specialized environments. The pelagic community is the driving force, the one in which living material is created and passed up the Eltonian pyramid of numbers. Other communities subsist on the matter the pelagic community passes over and return material to the sea. Benthic animals depend upon the rejected material from the community above and their function in the whole system is one of recycling. There is some midwater assemblage in estuarine and inshore communities but it is much less predominant and communities in mangrove swamps are independent of the offshore pelagic community to some degree (Mann 1969).

The part played by fishes in the pelagic community is considerable in that they comprise most of the carnivores. There are carnivorous copepods and arrow worms, medusae and ctenophores which also have carnivorous habits, but the larger carnivores are usually fish. The top predators in the system are probably mammals and sharks, although in particular areas, the larger tuna or even the older cod may fulfil the same role, if temporarily. However, much more important than their final position in the chain structure is the series of positions that fishes occupy during their life histories. Some may even start as herbivores, but most grow into the roles of primary, secondary or tertiary carnivores. Hence the biology of fishes has considerable relevance to the study of the pelagic community as a whole.

To understand the processes that sustain any community or food web, the population mechanisms involved must be disentangled, i.e. one must determine how the numbers in a cohort are generated and regulated and how competition is maintained. Studies of this nature have not been carried out in ecosystems, food chains or in communities although some studies have been made on the stability of predator/prey relationships; however, some preliminary work has been done on some fish populations. In the following section, the biology of fishes is examined in some detail leading on to the results of some of the work on population processes. Subsequently, the pelagic community is described in the same terms.

The biology of fishes in the sea

Life history

The marine fish are very fecund ($5 \cdot 10^3$ eggs in salmon to $10 \cdot 10^6$ in a large cod) and the eggs are small, about 1 mm across (Hiemstra 1962) (except those of halibut and salmon which are larger by nearly two orders of magnitude in volume). They are nearly always laid in midwater and drift with the plankton; exceptions to this rule are the demersal eggs of herring, salmon, sand-eel (*Ammodytes marinus* (Raitt)) and lumpsucker (*Cyclopterus lumpus* (Linnaeus)). Some subtropical fish like sardines hatch in about 48 h (Ahlstrom 1943, Cushing 1957), but the eggs of most temperate species develop in about three weeks (Apstein 1909, Ryland 1964) and the rate is an inverse power function of temperature; the salmon, however, lays its eggs in fresh water and these hatch about five months later. All fish larvae, whether from pelagic or demersal eggs, live in the plankton. They are small, 3–10 mm in length, translucent and they swim by wriggling through the water. They feed on copepod nauplii (Lebour 1918, 1919 a, b, Hardy 1924), and mollusc larvae if available, but baby plaice in the southern North Sea feed on *Oikopleura* amost exclusively except in cold winters, when they switch to copepod nauplii (Shelbourne 1957, Wyatt, personal communication). Initially, the fish larvae may use large diatoms for training attacks (J. D. Riley, personal communication) and the smallest often have green algal remains in their guts. In the first stages of their lives (about two weeks) they subsist on yolk as they learn to feed, but if they fail to find food in the first two weeks after exhaustion of the yolk sac they die; Blaxter and Hempel (1963) have established a point of no return beyond which the larvae die even if food becomes available again. Plaice (*Pleuronectes platessa* (Linnaeus)) larvae die at a rate of 5 per cent/d and grow at nearly 6 per cent/d (Harding & Talbot 1973, Ryland 1966) and haddock larvae die at 10 per cent/d and grow at 12 per cent/d (Jones 1973). The larvae feed avidly, grow quickly and their biomass grows increasingly rapidly once they start to feed. At the end of larval life they metamorphose, that is, acquire the nature of adult fishes by growing fins and putting on scales; in flatfish there is also a dramatic change of shape. As they metamorphose, they enter a different environment, the nursery ground.

On the nursery ground, which is nearly always in shallow water, the fish grow from their first summer to the age at which they join the adult stock, two or more years later, or many years in some slow-growing species. Postage-stamp sized plaice come ashore through a band of plaice one year older and live out their first summer in less than three fathoms of water (Riley & Corlett 1966); turbot and herring (*Clupea harengus* (Linnaeus)) at this stage live close to the very tide line itself, but dabs (*Limanda limanda* (Linnaeus)) live in slightly deeper water (Riley, personal communication). Cod (*Gadus morhua* (Linnaeus)) live on the offshore banks in 20–30 fathoms or so, sometimes quite far from the coast. In the deep ocean, young fish live in shallower layers than the older ones. By a similar rule (Heincke's Law 1913), as fish grow bigger they move into deeper water. Here they join the adult stocks that feed in summer and early autumn and spawn in late winter and early spring in temperate waters. There are also autumn and winter spawners, particularly in herring (Parrish & Saville 1966), summer spawners and in tropical and subtropical waters, spawning is either linked to upwelling (Cushing 1971b) or is spread over all seasons (Matsumoto 1966).

Fish continue to grow throughout their adult lives, some by an order of magnitude or more, for example, cod and tuna. Their total growth, like their fecundity, is of three to seven orders of magnitude (for example in salmon and cod) and although most must die by being eaten, a residue must die of old age. Perhaps because they grow a lot, they live a long time; sprats (*Clupea sprattus* (Linnaeus)) live for four to five years (Johnson 1970), herring for 10 or 20 yrs (Hodgson 1934, Lea 1911), plaice for 20 to 40 yrs (Cushing 1975) and cod up to 20 yrs (Rollefsen 1954). Much of the production in the sea, of herbivores for example, comprises animals that live for much shorter periods.

Other animals in the sea may not live as long or have specialized nursery grounds, but they are also relatively fecund. Their eggs are even smaller, about 0·1 mm across (as in *Calanus* or *Oikopleura*) and so such animals also grow through several orders of magnitude (1 μg to 10 mg), a property that they share with fish. An animal like *Calanus* lays several hundred eggs (Marshall & Orr 1952) and the growth rates and death rates of the nauplii may be comparable to those of the fish larvae.

The migratory circuit

Fishermen know that fish appear regularly at certain positions in particular seasons and that this regularity is associated with their migration pattern. Harden Jones (1968) has expressed this regularity in diagrammatic form (Fig. 13.1) The larval drift from spawning ground to nursery ground is the

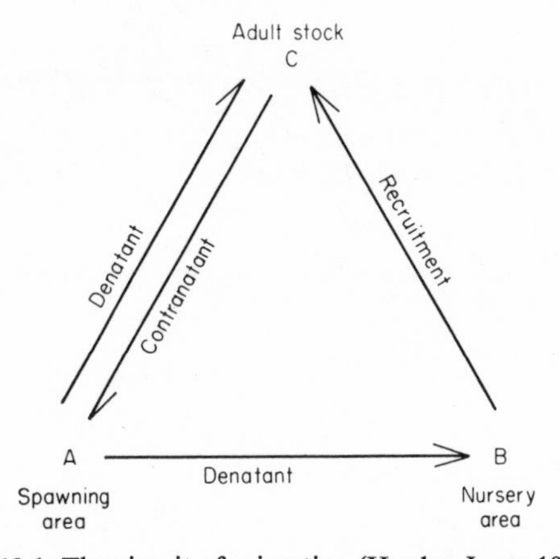

Fig. 13.1. The circuit of migration (Harden Jones 1968).

base of the triangle and the apex is the adult feeding ground. Immature fish are recruited to the adult stock on the feeding ground and mature fish migrate from feeding ground to spawning ground and back again each year. The adult migrations are repeated from year to year in most long-lived fish stocks but the larval drift and immature migration happens only once in the life history and is a property of each year class as it grows. Because the larvae often drift in the same current as the spent fish (if not quite at the same time), this part of the migration is sometimes described as *denatant*, or down current which implies that migration to the spawning ground is upcurrent or *contranatant*. Fish swim up rivers against the current, but to do so in the sea may often require an external referent which is not always immediately obvious; some migrations, like that of the Arcto-norwegian cod, are accomplished in dark water in winter. It has been suggested (Harden Jones 1968) that fish make use of countercurrents to migrate to the spawning ground, an apparently contranatant movement against the surface current that carries larvae and spent fish.

The larval drift of one well-known species, the plaice of the southern

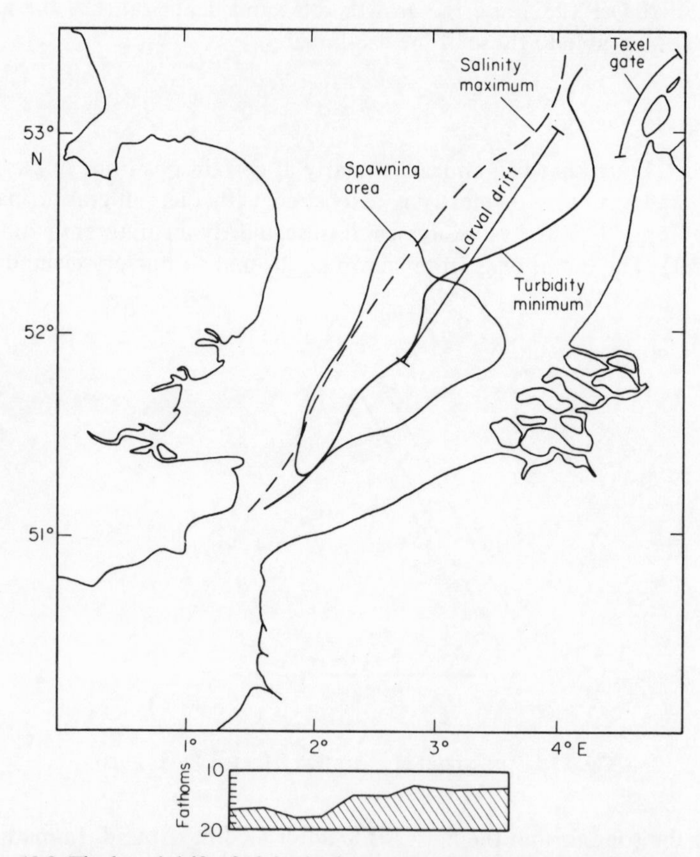

Fig. 13.2. The larval drift of plaice in the southern North Sea from the spawning ground between the Thames and the Rhine to the Texel Gate where they enter the Waddensee, the site of their nursery ground (Cushing 1972); the inset shows the depth of water along the larval drift from south to north.

North Sea, is shown in Figure 13.2. This spawning has been studied since 1911 and the average position of the spawning ground shown in the figure was plotted by Beverton (1962). From the work of Harding and Talbot (1973) and some unpublished work by D. Harding, it can be shown that the larvae drift in the clearer, high saline water (from the English Channel) towards Texel Island (Cushing 1972). There is a well-defined boundary between the offshore water and the vertically stratified Dutch coastal water, and Dietrich (1954) has shown that where the tidal streams flow parallel to such a boundary, there is an inshore flow near the bottom and an offshore flow at the surface, at certain states of the tide. Such conditions occur off Texel Island and the major nursery ground is found on the flat beaches on the Waddensee inside the island (Zijlstra 1972) and off Esbjerg (on the Danish coast) where the larvae of the German Bight stocklet may come ashore. A third nursery ground is found in the estuary of the Rhine and Maas; a recent paper by Ramster *et al.* (1973) has suggested that the inputs to the different nursery grounds might depend upon the amount of flow through the Straits of Dover.

Fish live on their nursery grounds in shallow water for two, three or more years. They are probably secure from predation to some degree and they grow quickly; indeed it is possible that the rate of change of biomass is greatest on the nursery ground. Their movement away into deeper water has been described by Beverton and Holt (1957) quite properly as a diffusion. Certain other migrations have been described as a diffusive process biased slightly in a particular direction (see Harden Jones 1968 for a fuller description).

In adult life, fishes in the sea migrate over considerable distances. The winter flounders (*Pseudopleuronectes americanus* (Walbaum)) off the coast of Rhode Island, U.S.A., move offshore and back again, a distance of about 15 miles (Saila 1961). Sprats move between the Wash, an estuary on the east coast of England, and an offshore area, a distance of about 50 miles (Johnson 1970). Plaice in the southern North Sea migrate about 200 miles from feeding to spawning ground and back again (de Veen 1962, 1970). Herring in the North Sea may cover as much as 450 miles back and forth on their annual migrations (Cushing & Bridger 1966). In the Norwegian Sea, from the Norwegian coast north of Bergen to Jan Mayen, Iceland, and the Faroe Islands, they migrate on a round trip of 1200 miles (Devold 1963). Maslov (1944) showed that the Arcto–norwegian cod travels 1500 miles between the Vestfjord in Northern Norway and Novaya Zemlaya. The Pacific hake (*Merluccius productus* (Ayres)) moves between southern California and the coasts of Oregon and Washington, a distance of 800–1000 miles (Alverson 1967). The Pacific salmon migrate from the rivers of North America half way across the Pacific, south of the Aleutians (Clemens *et al.* 1939, Milne 1957) and back again. Albacore (*Thunnus alalunga* (Bonnaterre)) tagged off southern California have been recovered off Japan a year or so later (Otsu 1960). It is clear that the range of migration is considerable and bigger fish tend to travel further than smaller ones.

The North Sea herring move round the North Sea in the direction of the main system of currents (Cushing 1954, Hoglund 1955), as the Atlanto–Scandian herring migrate round the Norwegian Sea in the main direction of water movement (Devold 1963). They probably drift with the current. The

Arcto-norwegian cod drift away from their spawning ground in the Vestfjord towards Spitzbergen in the west Spitzbergen current and to Novaya Zemlaya in the North Cape current; on their return a year later, they either swim against the same current system or drift in a countercurrent that exists below it and somewhat further offshore (Dickson & Baxter 1972). The fish may be said to board or leave the currents at points that secure the constancy of the migratory circuits although the physiological or behavioural clues are as yet unknown. In this sense a fish stock is contained within a current structure which flows regularly from the position of the spawning ground to that of the nursery ground.

Benthic animals release larvae into the plankton and a primitive form of larval drift and migratory circuit may also exist among such groups. In the plankton itself animals migrate vertically in a diurnal and seasonal manner and within the illuminated layer the older animals usually live at greater depths. Hence there is a migratory circuit in depth which may be analogous to the more complex circuit shown in fishes.

The reproductive isolation of fishes

To maintain itself, any stock needs to secure an isolation in its reproductive mechanisms. In fishes, there is a complex system by which isolation is achieved and, as might be expected, this is probably based on the migratory circuit. In temperate waters, fish tend to spawn on fixed grounds and this characteristic is a most important one in isolating reproduction; for example, herring in the River Blackwater (off the Thames estuary) spawn each year on a little gravel bank called the Eagle Bank. Bolster and Bridger (1957) showed that the herring spawning ground near the Sandettié Bank (in the Straits of Dover) was 2000 m × 500 m, or 1 km^2, in less than 40 m of water; Parrish *et al.* (1959) and Stubbs and Lawrie (1962) showed that the spawning ground on Ballantrae Bank in the Firth of Clyde was also restricted. Runnstrøm (1936) from grab samples, characterized herring spawning grounds off the island of Utsira (off the coast of Norway) as 'early February', 'late February', 'early March', etc. The three observed groups cited have one fact in common, namely that they spawn at the same position each year; for example, there are about six such grounds in the Southern Bight of the North Sea and in the eastern English Channel, where the Downs herring used to spawn exclusively (Cushing & Bridger 1966). Similarly, the position of the plaice spawning ground in the southern North Sea has been monitored since 1911 as noted above. The Arcto-norwegian cod has spawned in the Vestfjord for centuries (Rollefsen 1955); in 1955, the area was 20 miles long on the northern edge of the fjord (Bostrøm 1955) and at the present time only 3 per cent of the stock spawn to the south (from tagging experiments in the Barents Sea, Trout 1957). Elsewhere, in the North Sea, off Iceland and off Greenland, the cod spawning grounds are somewhat more extensive; those of the North Pacific tuna lie in the north Equatorial current between mid ocean and the Philippines (Matsumoto 1966).

Linked to the idea of a fixed spawning ground is that of return to the parent or natal stream, a behaviour pattern which has been established in the Pacific salmon (Gilbert 1916, Rich 1937). The work of Foerster (1937) and Pritchard (1948) showed that the stray to streams other than the parent might be as low as 0·3 per cent (from tagging experiments on pink salmon, *Oncorhynchus gorbuscha* (Walbaum)) although it is higher for steelhead trout (*Salmo gairdneri* (Richardson)) (Taft & Shapovalov 1938). A large tagging experiment has been carried out by firing tags into the peritoneal cavities of the British Columbian herring. The tags were recovered in fish meal factories. Recaptures were listed by statistical areas, and in the first year up to 20 per cent were recovered from the area next to that in which tagging took place; this proportion decreased with age of the fish (Harden Jones 1968). Thus, salmon that spawn on restricted local sites return to their parent streams and herring probably return to their ground of first spawning and do so more with increasing age. De Veen (1962) showed that plaice tagged on two of the three spawning grounds in the southern North Sea were recovered on the same grounds one year later, but not on either of the other two. Herring and salmon spawn on the bottom on restricted sites, whereas cod and plaice spawn in midwater over somewhat more extensive areas. Thus, the spawning ground is in the same position from year to year and fish tend to return to the ground of first spawning and in the case of salmon to their parent grounds.

In temperate waters, the peak date of spawning is also fixed within rather close bounds, the standard deviation for plaice spawning being a week or less. Of fifty trial streams studied in which Pacific salmon spawn, only one showed a trend with time. The standard deviation of the peak spawning date for the rest was less than a week. For the Atlanto–Scandian herring, the standard deviation of the peak date over a period of a few years was again less than a week. For a period of nearly seventy years, a trend in the date of peak catch in the Vestfjord was detected in the Arcto-norwegian cod stock, but the standard deviation about the trend was low and the total delay during the period was of the order of a week. Summarizing, the peak date of spawning for the species observed is about the same each year and has a low standard deviation. In contrast, the standard deviation of the peak date of spawning of the Californian sardine is expressed in months (from about ten years' observations) and no peak date can be easily extracted from the observations on the occurrence of tuna larvae in the North Pacific Ocean. This account is taken from Cushing (1971). The conclusion is that where there is a discontinuous cycle of production, as in temperate waters, fish spawn at a fixed season and their offspring grow with their food (as Jones 1973, has suggested). Where food is available at all times as in upwelling areas or in subtropical

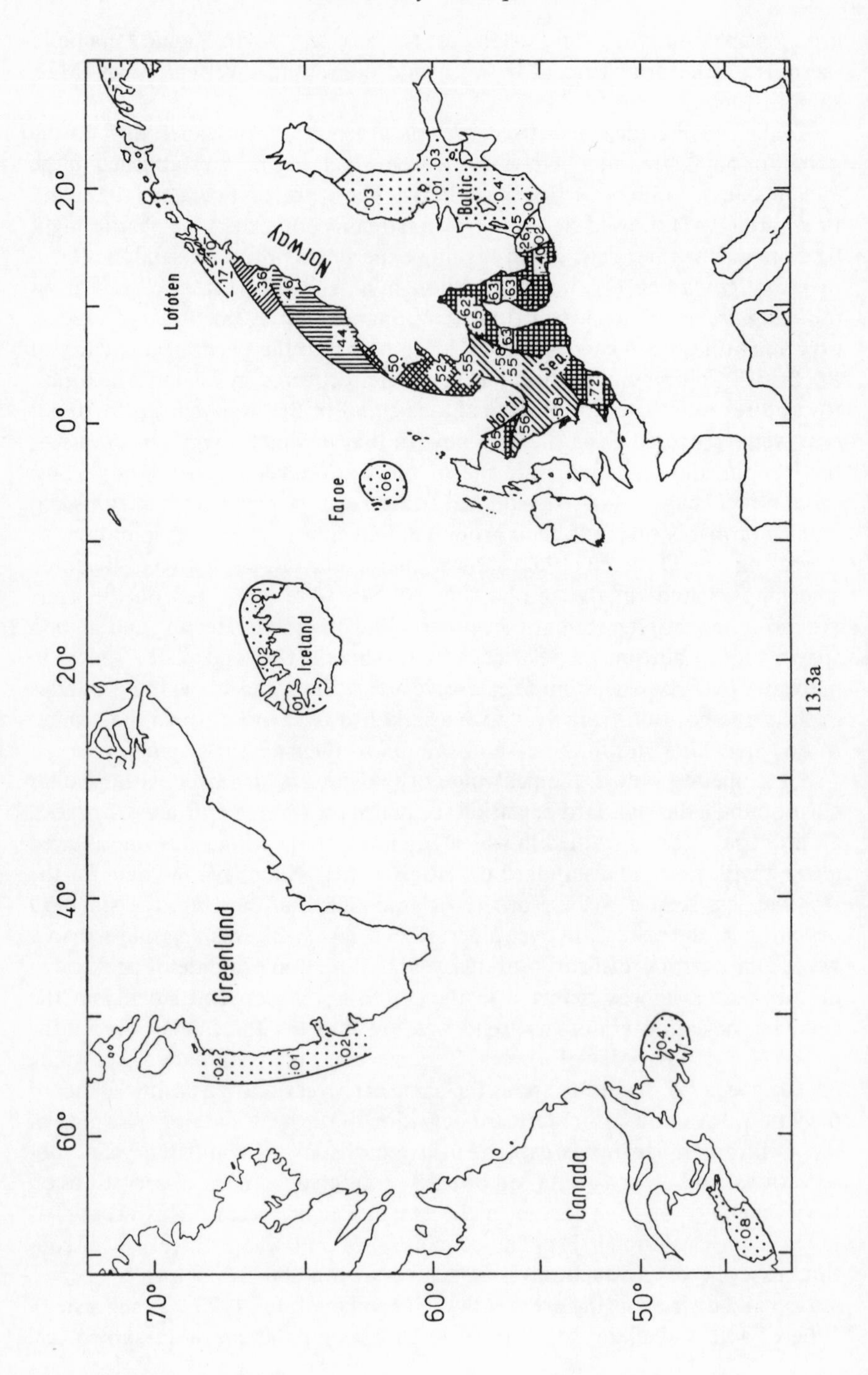
Lofoten
Norway
Baltic
North Sea
Faroe
Iceland
Greenland
Canada
20°
0°
20°
40°
60°
70°
60°
50°

13.3a

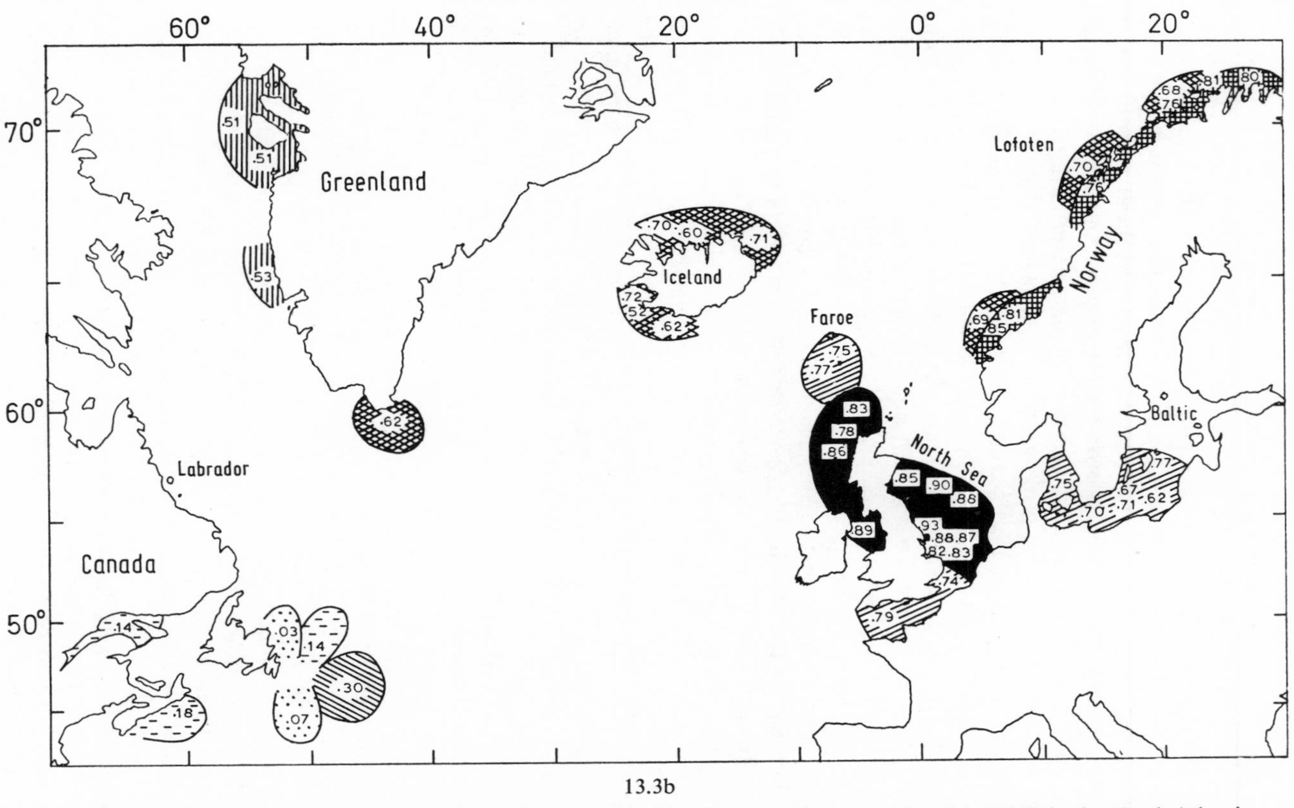

13.3b

Fig. 13.3. The genetic structure of the North Atlantic cod: (a) Distribution of a haemoglobin allele (HbI') in the North Atlantic cod stocks (de Ligny 1969); (b) Distribution of a transferrin allele (Tf^C) in the North Atlantic cod stocks (de Ligny 1969).

and tropical seas, they spawn with less precise timing and perhaps with little seasonal difference.

However, in many temperate species, the spawning season is long. In herring, it probably lasts for ten days or so on each restricted ground (Ancellin & Nédèlec 1959) and for three months in the whole stock area (Runnstrøm 1936). In plaice, the season lasts for three months and in cod it may even be slightly longer. Whether the fish spawns at local sites for short periods, as in herring or salmon or in a more dispersed manner as do plaice (spawning female plaice are 80 m apart on average, Beverton 1962), the total spawning season is long and peaks at the same time each year.

If the spawning ground is fixed from year to year, then so is the starting position of the larval drift and given a regular current system, so is the position of the nursery ground. In this sense, the larval drift may be said to form the geographical base of the stock. As noted above, larval plaice drift from a point between the Thames and Rhine to a position off Texel Island. Herring larvae drift from the Straits of Dover and from the Dogger Bank to the German and Danish coasts (Bückmann 1942). Eel larvae drift from the Sargasso Sea to the mouths of the major European rivers and to the Nile (Schmidt 1922). The range of the larval drift can, but need not, be considerable, and the greater the range, the greater the spread by diffusion (except in a special case like that of the southern North Sea plaice larvae that migrate past Texel Island to the Waddensee). The spawning area is usually large so the patch of larvae retains its identity against diffusive processes during the larval drift, which ensures that a nursery ground does in fact exist. Because of the long spawning season the geographical area reserved as a nursery ground is continuously occupied by larvae from the same stock, which deny access to any competing stock of the same species. Because spawning grounds of different stocks are a considerable distance from each other, the chance of two stocks mixing before their larvae develop is remote; for example, any plaice larvae from the Moray Firth spawning in the northern North Sea would metamorphose before they reached the southern North Sea. Thus, reproductive isolation is obtained by the fixed larval drift and by the long and fixed spawning season.

If such isolation really works, the stocks should be genetically distinct. In recent years, geneticists have made considerable studies on fish stocks; for example, three groups have been established in the Californian sardine (Sprague & Vrooman 1962), two groups have been differentiated in the herring stocks off the Gulf of Maine (Sindermann & Mairs 1959) and a separation has been established in Pacific tuna populations (Fujino & Kang 1968). However, the most spectacular work has been on the North Atlantic cod. Jamieson has separated three haemoglobin and seven transferrin alleles. Figure 13.3 shows the distribution of one of the haemoglobins and one of the transferrins across members of this species (de Ligny 1969). There is obviously a trend in distribution across the North Atlantic and the 'taxonomic' distance between populations obviously increases with the physical distance between stocks. For one character, the probability of significant difference between adjacent stocks may be calculated in a 2×2 matrix, and in many cases the difference is not great enough to preclude mixture with only one allele. However, with such an array of characters at one's disposal, a $(2 \times n)$ table shows the probability of mixture between most of the North Atlantic cod stocks to be as low as 1 in 10^4.

The cod stocks of the North Atlantic are isolated from each other (except off Iceland and west Greenland), but Li (1955) has suggested that a large population needs an array of spawning groups (or 'stocklets') in order to maximize variation. Some exchange is permitted between generations in stocklets as shown for example in British Columbian herring, but none between stocks. It is possible that the groups in the southern North Sea plaice or the restricted spawning groups characteristic of herring populations represent stocklets between which mixture is expected to some degree.

Population processes

There are three processes by which populations sustain themselves, by generating first, an adequate magnitude of recruitment, second, through the density-dependent processes that regulate numbers and third, in maintaining competition. Ottestad (1942) correlated catches for a period of nearly seventy years in the Lofoten cod fishery with periodicities elicited from observations on the width of pine tree rings in the same region; the correlation was arranged between the periodicities and catches set back in time by a number of years so that differences from year to year were in effect differences in year-class strength. If climatic factors are to affect recruitment directly, then this is achieved through influence on the production cycle, particularly its timing. If fish spawn at fixed seasons in temperate waters, where the production cycle may vary considerably in onset from year to year and therefore the amount of available food, the production of fish larvae may be matched or mismatched to that of their food. If the magnitude of year-class strength does depend upon the availability of food in such a way, then there is a reasonable basis for Ottestad's correlation. Templeman (1965, 1971) has shown that for thirty years the strongest year-classes of gadoids were common to stocks across the whole North Atlantic, which suggests that recruitment was related to factors as pervasive as climatic ones.

The stabilization mechanisms operating in a population of fish depend upon the fecundity of the parent stock and on density-dependent growth and mortality during the early part of the life history. Density-dependent growth declines with age and differences in fecundity are not great enough on their own to establish control. It is now generally conceded that density-dependent mortality occurs early in the life of fishes, on the salmon redds (McNeil 1963), amongst first feeding larvae that may starve, amongst older larvae that may suffer greater predation if they do not feed well, and in the first summer on the beaches, when metamorphosed larvae settle from the volume of the sea to the surface of the beach. If the magnitude of recruitment is determined by the availability of food, then it would be parsimonious to suggest that the stabilization mechanism was also rooted in the quantity of food available. Larvae that eat well, and thus grow well and swim well, avoid predation. If they eat poorly the reverse is true (Cushing & Harris 1973). Then a greater percentage of larvae must be more vulnerable to predation, that is, the mortality is density dependent.

However, the predatory process is itself density dependent. Predation-time may be defined (perhaps too simply) as comprising the time to search for prey, and then to handle it. In the case of the plankton, at least, the handling time is a very small proportion of the searching time (Cushing 1968), so predatory mortality varies inversely with searching time and is, therefore, directly proportional to numbers of fish larvae. Holling (1965) has also pointed out that the time to search varies inversely as the ratio of predator speed to prey speed. Thus, there are not two distinct density-dependent processes but one which can be looked at from either the point of view of the predator or its prey.

Fish grow considerably during their lives (from 1 mg to 1 kg, for example). As they grow, they pass through a series of predatory fields, within which each predator is larger but much less numerous than its predecessor. The position in the sequence depends upon size, which itself is a function of age. As growth depends on food, it is inversely proportional to the density-dependent component of mortality. Thus, density-dependent mortality itself may be expressed as a function of age. It has been shown that mortality in the larval stages decreases with age (Pearcy 1962), as might be expected if such a mechanism played an important part.

Any cohort (which is the same as a year-class), in a multi-aged stock must replace itself if the annual recruitment to that stock merely equals the annual losses by death. The biomass of the cohort is accumulated by an excess growth over death and by a *critical age* the specific growth rate equals the specific death rate. The size of any cohort represents the maximum attainable biomass at one time in the face of loss of numbers during its life history and thus when growth rate and death rate are equal the maximization is complete. Then mortality due merely to old age, might be expected to start. Let us suppose that density-dependent mortality is a function of age as suggested, and that any reduction in numbers due to density-independent causes subsequently influences biomass in a density-dependent manner as the cohort ages. It can be shown that $N_t = N_0/(1+M_0 t)$, where N_0 is the initial number of larvae, M_0 the initial mortality and N_t the number at time t (Cushing

1975). Figure 13.4 shows the curve fitted to observations on the decline in numbers with age of the southern North Sea plaice (Cushing 1975), the Boothbay harbour herring and the Baltic cod (Cushing 1974b). Figure 13.4(a) is based on the observation that the plaice larvae die at 5 per cent/d (from all causes, density dependent or otherwise); the open circles show the 40 per cent/month mortality observed on reaching the beaches, the 10 per cent/month mortality observed after the first winter of life and the 10 per cent/yr mortality observed between the ages of five and fifteen. Figures 13.4(b) and 13.4(c) for the herring and the cod show the curves fitted to the numbers observed. The three figures suggest that mortality from larval stage to adult life can perhaps be considered as a function of age in the rather special way described above.

If these mechanisms operate as described above, mortality must be a function of food lack. Figure 13.5 is a diagram showing the downward trend of mortality with age. At a given age, mortality is reduced as the larvae feed well on rich food; during the period of rich food, the rate of change of mortality is lower than it would have been had there been no patch of rich food. The food patch disappears and mortality increases to a relatively higher level because more larvae survived than had been expected before the patch of food appeared. Obviously, food is continuously variable in time and such a mechanism will absorb density-independent effects as well as density-dependent ones like that described above. The diagram represents the life of a cohort in the simplest way. The cohort may be said to have an exploratory function, by which the environment both physical and biological is continuously sampled and assessed. The cohort's biomass expresses the success of that assessment. Then the stock, as the sum of the cohorts, may be said to have a conservative function in that successive explorations made by the cohorts are averaged.

Jones (1973) has suggested that there is a single process by which the larval fish grow with a private food supply in the production cycle of temperate waters. The environmental influence on year-class strength may well be exerted through the match or mismatch of larval production to that of food. The density-dependent control is probably started during the larval drift and continues with less strength through the later life history. Competition can be expressed at any point in the life history as the ratio of specific growth rate to specific mortality rate; in this sense, the most avid competition will occur during the larval drift, but if it is expressed in terms of stock biomass from generation to generation, then competitive processes might persist in a steadily diminishing manner until the critical age.

The question arises as to how well these speculative principles can be applied to the rest of the pelagic ecosystem. In animals other than fishes, fecundity is relatively high, and growth rates and death rates, high. Perhaps more important, growth is also high; four orders of magnitude in *Calanus*.

Thus, it is possible that as nauplii, for example, grow into copepodites they pass through a series of predatory fields, as was suggested for fish. Any death at one trophic level provides growth in the one above. One might imagine that the ratio of predator speed to prey speed depends on food at each trophic level above the first in the pelagic system; as the ratio is inversely proportional to searching time, the same mortality structure in age put forward for fishes might occur elsewhere in the pelagic ecosystem.

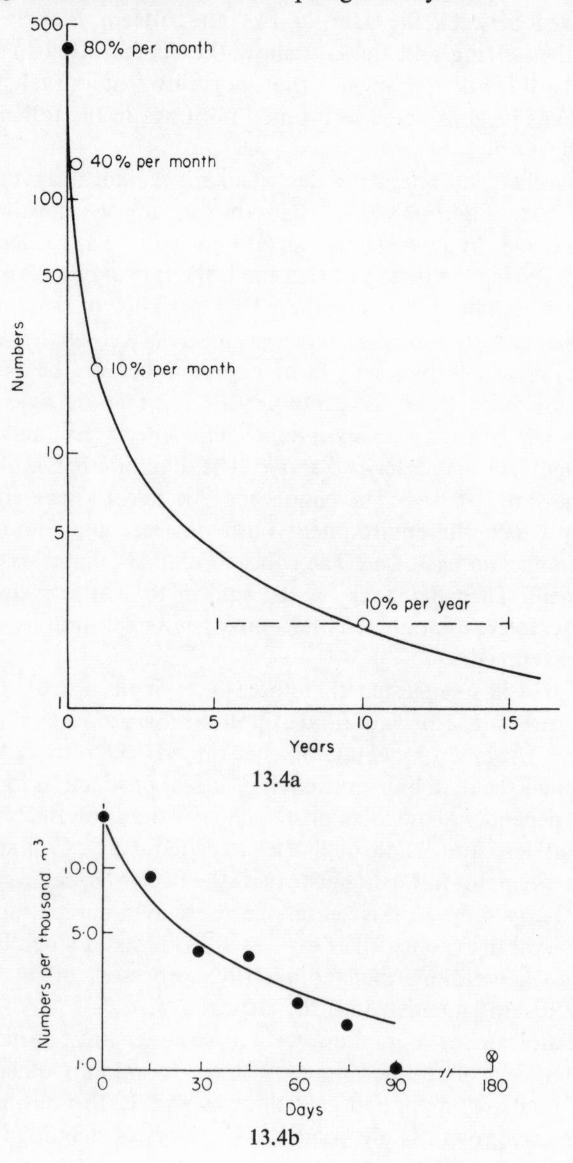

13.4a

13.4b

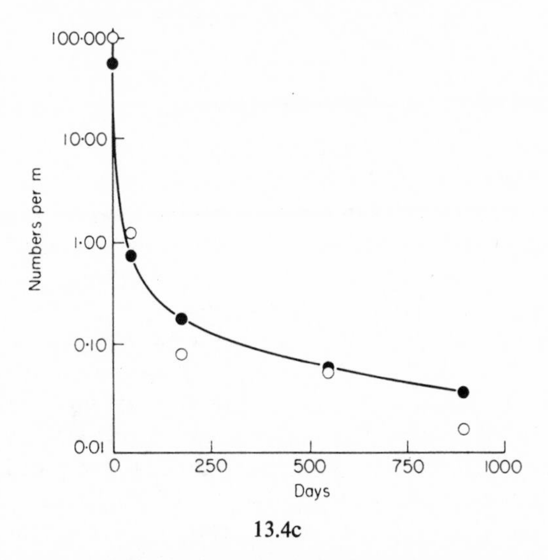

13.4c

Fig. 13.4. The trend of natural mortality with age; (a) southern North Sea plaice (Cushing 1975); (b) Boothbay harbour herring (Cushing 1974b from Graham *et al.* 1972); (c) Baltic cod (Cushing 1974b, from Poulsen 1931).

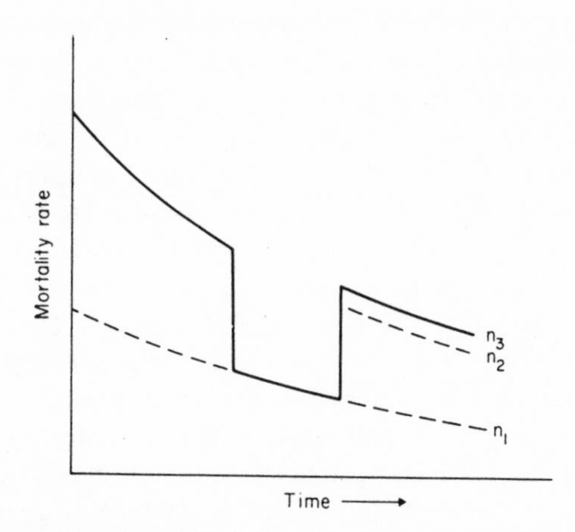

Fig. 13.5. The possible effect of food upon the trend of mortality with age. A patch of rich food appears for a period during which the rate of change of mortality is less than it would have been had the patch not appeared; at the end of the period, when the patch disappears, numbers are greater than expected and so is mortality. At any point in time a density-independent mortality might occur, which is subsequently modulated in a density-dependent manner. Thus, the cohort exploits the environment continuously.

The pelagic community

The biology of fishes, particularly where population control mechanisms are concerned, has been described in some detail as a possible example of a more general phenomenon in the pelagic community. In this sense fishes may represent the pelagic animals even if some parts of their lives are specialized and highly developed. For example, animals other than fish may not migrate thousands of miles in a year in their journeys to and from their spawning grounds as some of the larger fishes do, yet *Euphausia superba* (Dana) maintains itself in the same region by migrating seasonally between two currents that flow in opposite directions, one above the other (Mackintosh 1937). Most fishes live for one or two decades, whereas copepods may produce a number of generations within a single year. Common to both, however, is the maximization of biomass in the recruiting year-class or the new generation. The difference between them is that the stock in fishes is composed of many cohorts that must average the annual variability of year classes; the copepods might achieve the same effect in the overwintering generation which is the residual from three or more generations throughout the year. Thus, some of the principles obtained from the study of fishes can be applied in a general way to other populations in the pelagic community.

The community is limited to the depth of the sun's penetration, at most about 1000 m, and for food, depends on the production of algae in the photic layer; within this depth-range, the bigger animals live in the deeper layers in daytime and most tend to rise towards the upper layers at night (see Chapter 6). The community extends across the ocean from the antarctic glaciers to the arctic floes and (perhaps even beneath them), from the shelves of one continent to those of another. Estuarine and inshore environments have structures in common with that of the open sea but they tend to be specialized, and dominated by local physical processes. Within this enormous field, there are many subdivisions bounded by the different current systems, the North Sea, the Alaska gyral, and the Arabian Sea, for example. There are probably oceanographic structures within such regions, but their biological relevance is not yet understood, as their influence is on a small scale.

Visibility in the sea is not great because light is scattered (see Chapter 2). Detection depends on contrast against the scattered background. Both prey and predator seek to reduce their own contrast and the range over which they are visible. The distance from which a predator can attack depends on its acceleration, for which the white rather than the red muscle (which is employed in cruising) is used; Bainbridge (1960) has shown that a fish reaches its maximum speed of about 10 lengths/sec within half a second and so the attack distance, the effective successful range, is of the order of ten lengths. The predator must take its prey before it disappears into the murk, and equally, the same prey must try to vanish as soon as the attack is detected.

The attack distance is a function of visual range and so the larger predators are found in the clearer water, tuna in the transparent ocean but whiting (*Merlangius merlangus* (Linnaeus)) in the silty coastal water. Hence, there is a relationship between size of predator and the degree of scattering of light in the water. Both predator and prey live in three-dimensional space, as specified volumes of water; that of the predator is the product of its course and attack range, whereas that of the prey is a sphere about its position with the predator's attack range as radius. Such volumes for the predators might be referred to as pelagic territories.

On land, a territory is well defined as the area defended by an animal against competitors of the same species in order to secure a food supply, particularly for its offspring. Such territories may be established by benthic animals or by reef fishes. The phrase, pelagic territory, as used here, is a little different in that it is defined in terms of attack distance and the limited visual range in the sea. It is a volume exploited by a predator to secure its food supply rather than an area defended for the same purpose. The latter is fixed in position, but the former shifts with the predator on its inconsequential course.

There are large predators in the sea, the marine mammals (excluding the baleen whales), the sharks and giant squids. Except for the sperm whale (*Physeter macrocephalus* (Linnaeus)), which dives to a thousand fathoms or more, the marine mammals are limited to the surface layers. From the evidence of the Japanese pelagic long line catches in the deep ocean, the large sharks also live in the surface layers. Here the visual range is maximal and some of the top predators may increase their attack distances with acoustic signalling. These predators also take care of their young or they lay only few eggs, whereas the teleost fishes on which they feed are very fecund. The latter have to stabilize their populations within the growth and death structure of the food chain, whereas the top predators can achieve stable populations independently of the food-chain structure.

Ursin (1973) has shown that the predator–prey ratio in weight for the demersal commercial fishes is about 100:1. The length ratio is 4·63:1 and, by extension, the following simple chain of predators may be conceived:

100·00 cm
21·40 cm
4·62 cm
1·00 cm
0·21 cm

The largest predator in this chain might be a cod or a tuna and the smallest animal is the size of *Calanus*, which is, of course, a herbivore, although some herbivores, such as euphausiids are larger. From herbivore upwards, there are thus four or five links in the chain, and these correspond roughly to the number of links estimated by other methods (see Chapter 14). Whatever the cruising speeds or attack speeds of the animals, the differences in length between trophic levels might be of the order given above. However, any such differences must be distributed. For a short period a predator might be linked to a particular prey of the appropriate size, but because the latter grows relatively faster the period must be short; in general, the smaller animal grows out of one particular predatory field into another.

Not all animals in the sea are predators in the sense described above. Many such as herring, the baleen whales, manta rays (*Dicerobatis eregoodoo* (Cant)) and basking sharks (*Cetorhinus maximus* (Gunnerus)) are filter feeders. A herring may select its food by size (Battle *et al.* 1936) but it filters the selected prey. During a day it may make a thousand or more successful encounters (Cushing 1964b). On the other hand, a large cod may only eat a few whiting a day in order to satisfy the daily ration. The difference in weight between a large cod and a whiting is perhaps two orders of magnitude, but that between a herring and a *Calanus* is perhaps five orders. The latter difference in length is about two orders of magnitude, which also happens to be the difference in speed. Hence, the herring needs no attack speed for feeding, its cruising speed being so much greater than that of its prey. The high speed of about ten lengths/sec is perhaps used only for escape. Whereas the cod may make a dozen attacks a day at 10 lengths/sec for very short periods of two or three seconds each, the herring must feed for much longer periods at cruising speed in order to satisfy its daily ration. On the other hand, cod and hake sometimes feed on euphausiids and then presumably behave as filterers, and when herring feed on sand eels they may use attack speeds. Such occasional exceptions do not invalidate the general rule that the clupeids are filterers and the gadoids predators.

It has long been observed that the clupeids, the herring-like fishes, form pelagic shoals and such shoals have been described with sonar and with the recording echo sounder (Cushing 1973). A series of clupeid shoals was analysed with the ARL scanner (a very high resolution sonar) and it was found that the higher the packing density, the bigger the shoal (Cushing 1974). Such a result can be explained if the fish swim more slowly at high packing density to a limiting density; as speed increases so does the variance in relative position between fishes and they have to be further apart to retain a relatively constant station-keeping angle. Spawning shoals of herring are larger than feeding ones (Cushing 1955) and in dense concentrations of food, speed may be reduced and vice versa. If some fishes were to accelerate from 3 to 10 lengths/sec in half a second, the shoal of which they were a part might not retain its identity. The clupeids are the most important filterers in the economy of the sea and it is a reasonable generalization that the filterers are shoaling fish that feed at cruising speeds or less.

Some of the largest animals in the sea, such as baleen whales and the basking sharks filter their food, and their daily rations may be reckoned in tons. They are not predators at all in the sense defined above, yet exceed the top predators of the food webs in the pelagic community by more than an order of magnitude in weight. Secure from predation they do not shoal at all, although they may live in small groups of individuals. It has long been thought that the clupeids shoal to minimize predation and this principle was stated formally by Brock and Riffenburgh (1960): let the number of fish in a shoal be n_f. Then

$$n_f = (4\pi/3)(r^1/c)^3$$

where r^1 is the radius of the shoal;
c is the interfish distance.

Then the number eaten, $n_e \ll r^3 n_f / [r + c(3n_f/4\pi)^{1/3}]^3$
where r is the sighting range of the predator.

The numbers of prey (i.e. in a shoal) were calculated for different prey densities and sighting ranges. When the sighting range is greater than the distance between shoals, or when the prey density is high, the number eaten is high, a disadvantage to the prey. Hence, the advantage of shoaling is nullified at short ranges, i.e. at night or in turbid water. From acoustic evidence, fish shoals are less frequently observed at night than during the day time (Cushing 1973).

Saila and Flowers (1969) applied the search theory of submarine warfare to the problem of shoaling. They concluded that the greater the speed of the prey, the more come into range than escape, so a wise predator works during periods of prey activity, i.e. at dawn and at dusk, the periods of active vertical migration. It is in general true that animals are active at dawn and dusk, when they do most of their feeding. Another conclusion was that if the predator's search field was restricted to a zone ahead, the predator need only cruise at about twice the speed of the prey for detection, which makes sense of Weihs' (1973) calculation on hydrodynamic grounds, that fish cruise at 1 length/sec.

If we contrast the predators and the filterers, we observe that speed difference between predator and prey may decrease with the relatively faster growth rate of the prey during a period of time, hence the death rate of the prey might be expected to decrease during that period. On the other hand, the speed difference between filterer and prey is large and would not decrease much over the same period and so the death rate of the prey would be expected to remain constant. However, because the clupeids select their prey as they filter it, they may slow down to feed and so they may congregate (for example, herring on *Calanus*, Cushing 1955). For this reason the death rate of the prey would increase during the period and subsequently decrease as the filterers disengaged from the thinned food patches. Under predation, the death rate of the prey should only decrease within any predatory field, but congregation of filterers will lead to an increase in death rate, followed by a decrease. Herring eat adult *Pseudocalanus* and *Calanus* (Savage 1937) and leave the larval forms for the planktonic predators. Perhaps predators eat growing juveniles, whereas filterers eat fully-grown adults.

A most important characteristic of fishes that might be shared by other animals in the pelagic community is that of considerable growth and decreasing mortality rate during the life history, resulting in the maximization of biomass within the cohort. Each juvenile might, as little fish were said to do, grow through a series of predatory fields, each larger and much less numerous than its predecessor. By the same token, each predator grows through a succession of prey densities, each less than its predecessor but comprised at each stage of larger animals.

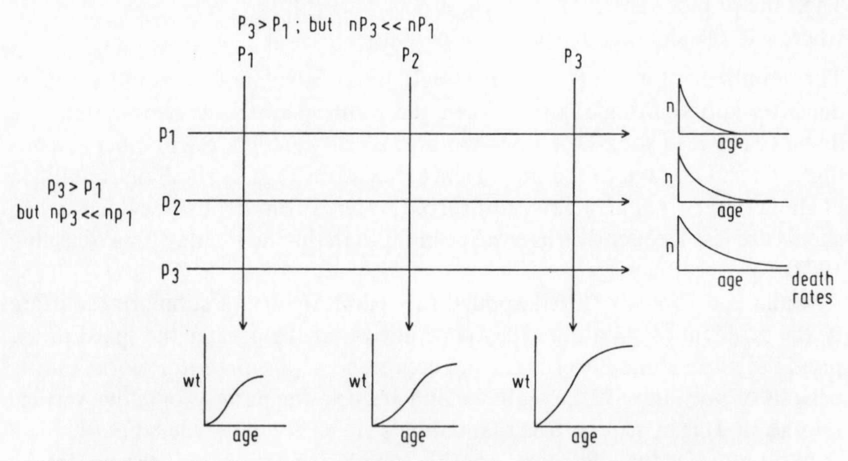

Fig. 13.6. The net of growth and death; P_1, P_2 and P_3 are predators, each larger and much less numerous than its predecessor in the sequence; p_1, p_2 and p_3 are prey animals, also each larger and much less numerous than its predecessors in the sequence. The death rates and growth rates are shown in diagrammatic form.

Figure 13.6 shows the network of growth and death for such a series of predators and prey. There are three predators, P_1, P_2 and P_3 and three prey p_1, p_2 and p_3. Because $P_3 > P_2 > P_1$, as p_1 grows, its predators succeed each other in order of size, P_1, P_2, and P_3. Similarly, $p_3 > p_2 > p_1$ and as the predator grows, it feeds on prey of increasing size, p_1, p_2 and p_3. The horizontal and vertical lines represent these relationships in age. The horizontal ones indicate the mortality of the prey in the three predatory fields and the actual death rate as a function of age is given in the inset on the right of the figure. The vertical lines indicate the growth rates of the predators as they feed on the prey organisms in increasing order of size and the actual growth rates are given in insets on the bottom of the figure. Although each larger predator kills more effectively than a smaller one, there are fewer of them and so the death rate of the prey declines with age. For the three predators shown, the death rates of prey at any given age of prey decline with size of prey. The larger prey are much less numerous than the smaller, but fewer are needed to satisfy the daily ration of the appropriate predator. As all the prey within the field of the largest predator are bigger than those within that of the smallest predator, the growth rates of the predators increase with predator size. If death rates and growth rates are density dependent in the manner suggested for some fish populations, then the stabilization mechanisms of each population are linked within the network of growth and death.

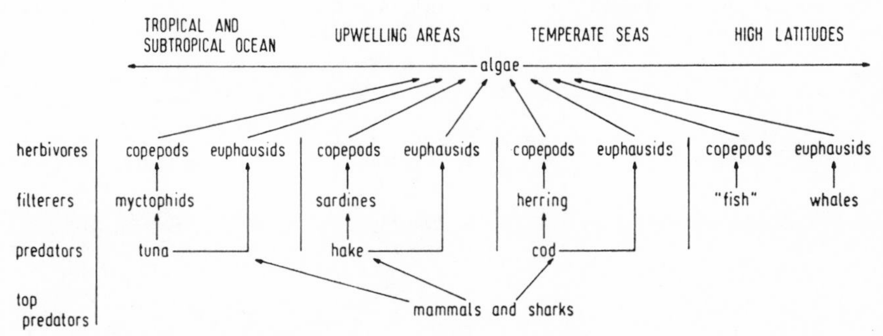

Fig. 13.7. The pelagic community as a structure of herbivores, filterers and predators in four major subdivisions.

Figure 13.7 is a diagrammatic representation of the pelagic community. The community is divided into four groups by area, the deep tropical and subtropical ocean outside the upwelling areas, the upwelling areas themselves, the temperate seas and the high latitudes, exemplified by the southern ocean. In each area, a food chain is fitted into a system of herbivores, filterers, predators and top predators. Any real food chain is more complex, with different links and changes of roles within the life history of any animal. The simplification in Figure 13.7 has the advantage that the analogous roles played by filterers and predators are accented, herring, sardines and myctophids, or cod, hake and tuna. The herbivores are classed as copepods and euphausiids; the latter are exploited by the predators when they are not eating the filterers. In the southern ocean, euphausiids are exploited by the baleen whales and copepods are taken by various fishes. The top predators are mammals and sharks and it is interesting to notice that both groups are less fecund than the fishes by many orders of magnitude; indeed, mammals rely upon parental care of a few offspring and sharks upon a few tens of eggs each endowed with very large quantities of yolk for the survival of their species.

The biology of fishes has been studied in this chapter in some detail with the purpose of extending some characteristics to parts of the pelagic community as a whole. Because animals other than fish are also very fecund and because growth is high, the extension seems at first sight reasonable. In teleological terms, a cohort seeks to maximize its biomass in the face of competitors, food availability and other environmental effects. The structure put forward in speculative terms in Figure 13.6 relies upon two important factors; first, a sequence of private food supplies (as Jones 1973, suggested) and secondly, upon the dependence of growth and death on the availability of food in the particular way described. The mechanism suggested may be at the root of the processes of stabilization in a fish population. It is possible that on a more extensive scale the same sort of process is responsible for the observed stability of an ecosystem, as broad and as economically important as the pelagic community.

Chapter 14. Food Chains in the Sea

T. Wyatt

Introduction

Some examples
The North Sea Herring
Petersen's Limfjord
Southern Bight fish larvae

Some examples *continued*
Black Sea plankton

The lengths of food chains

The problem of top predators

The role of bacteria

Introduction

The idea of a food chain in which one species is fed upon by another, and this in turn by a third, appears simple. This apparent simplicity may account for the fact that although the food chain concept and related ideas have a long history, they have until recently not been examined in a precise way. Writers on the subject of food chains have often been content to point out that simple linear food chains are infrequently observed in nature, and that we should think instead in terms of food meshes or food webs, in which populations have trophic (i.e. feeding) connections with a variety of food species and a variety of predators, rather than with one of each. A second reason for this long neglect of a very basic ecological idea may stem from the introduction of the 'trophic–dynamic' approach to communities of organisms, due especially to Lindeman (1941, 1942). Subsequent studies, in which productivity at different trophic 'levels' rather than at specific trophic links has been emphasized, have largely ignored the fact that few if any animals spend their whole lives within the confines of a single trophic level, even if this level is broadly defined.

The simplest imaginable food chain consists of two species, one of which feeds upon the other (Fig. 14.1a). This is the classical predator–prey system, which has stimulated so many important theoretical and experimental studies following the classical work of Volterra (1928) and Lotka (1925). Volterra's equations are as follows, where N_1 is the number of prey and N_2 the number of predators. The rate of change of prey with respect to time is given by

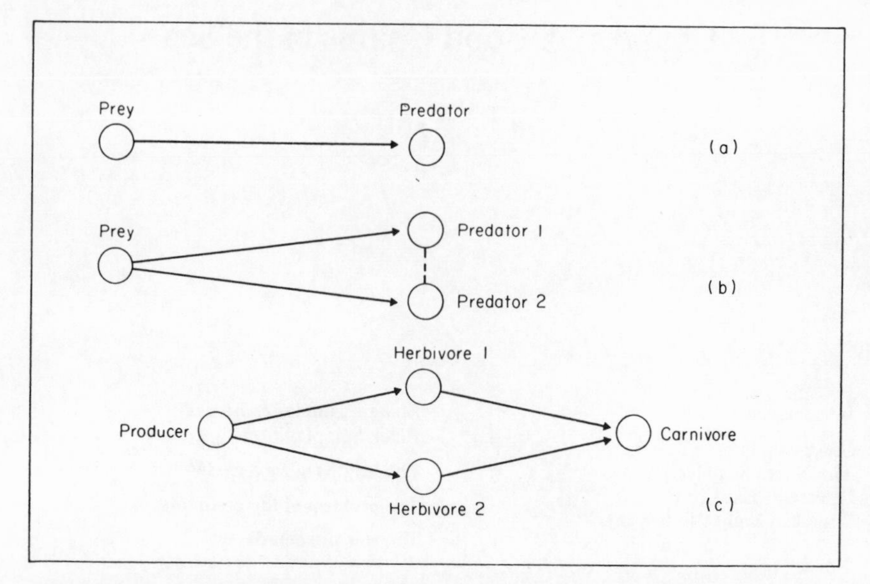

Fig. 14.1. Diagrams illustrating trophic links. (a) Simple predator–prey system. (b) Prey with two predators. The broken line represents competition. (c) Two herbivores feed on the same producer, and in turn are preyed on by a single carnivore. Competition may be eliminated (see text).

$$\frac{dN_1}{dt} = N_1 (b_1 - k_1 N_2) \tag{14.1}$$

where b_1 is the growth coefficient of the prey, and k_1 is a predation coefficient equivalent to a constant mortality per predator. The predator equation is

$$\frac{dN_2}{dt} = N_2 (k_2 N_1 - d_2) \tag{14.2}$$

where k_2 is a coefficient of food assimilation, and d_2 is a constant death rate. From the point of view of the predator, the arrow which joins the two species represents food, while to the prey it usually represents death. Increase in prey numbers is regulated by the abundance of predators, and increase in predator numbers is controlled by the abundance of prey. Such a model in theory has an equilibrium solution where predator and prey numbers remain unchanged, but in general gives rise to oscillating numbers of the two species, with first the prey and then the predator in the ascendant. Some strategies available to predator and prey within the framework of this model are discussed by McArthur and Connell (1966). Hiding places for the prey, and selection of the largest prey by the predators are considered important. Most of the better known models which have been used in attempts to improve our understanding of food chain relationships in the sea are technically derivable from Volterra's formulation of the problem (see Section 6).

Two predators share the same food in Figure 14.1b. If the number of prey restricts growth in numbers of the two predator populations, then it is possible to say that competition exists between them. A system of this kind can be shown to be inherently unstable (Kerner 1961). Laboratory experiments by Gause (1932, 1934) demonstrated that the end result is likely to be extinction of one or other predator, at least in a homogeneous environment. The system in Figure 14.1b can be stabilized by adding a third link as in Figure 14.1c. Here a carnivore controls herbivore numbers, and effectively reduces the competition between them (Parrish & Saila 1970). If we imagine the carnivore eating whichever herbivore is more abundant, then neither of them need become extinct. From the behaviour of this system, it would appear that an increased number of species at one trophic level in a food web might be sustained by the addition of links to another trophic level. Coexistence of two predators can also be achieved experimentally by reducing the food available to the prey (Luckinbill 1973). There is also here, in simple terms, the possibility that increase of the herbivore populations may be so heavily regulated that the producer population is relatively unrestricted by them, and must therefore be limited in some other way. The main point is that even this simple food chain model can exhibit quite diverse behaviour patterns, and at the same time is sufficiently complex to be experimentally intractable. There is the further point that food chains and food webs are central to several other important ecological ideas; competition, diversity and stability have already been mentioned. Trophic links, therefore, form conceptual connections between populations and the communities to which they belong.

While there was then nothing new about the food chain concept, it was Elton (1927) who first exploited it to describe the relations between the populations within a community. He discussed several important consequences for community structure, beginning with the observation that, in his own characteristically guileless words, 'each stage in an ordinary food-chain has the effect of making a smaller food into a larger one . . .' it follows that large animals must be fewer in number than small ones, so here is the origin of the pyramid of numbers. Elton also emphasized that the number of links possible in a linear food chain is quite small, usually not more than five, since there is an inevitable loss of biomass when one species is made into another. This point has been discussed in a well-known essay by Hutchinson (1959). A third generalization made by Elton was the limited size range of organisms which an animal might feed on. This is another observation which, like the food chain itself, has not attracted as much attention as it merits, but Ivlev (1961) has given a short discussion of predator/prey size relations in freshwater fishes.

Some examples

The North Sea herring

Hardy's famous study of the North Sea herring (*Clupea harengus* (Linnaeus)), published fifty years ago (Hardy 1924), is still one of the most detailed accounts available of the feeding relationships of a marine animal. His diagram of such relationships, reproduced here as Figure 14.2, is widely quoted to demonstrate the complexity of a typical food web. However, the complexity shown is misleading. The figure summarizes data for ten months in each of two consecutive years, and for fish collected from a wide range of localities in the North Sea. Although feeding at any particular time and place is rarely confined to a single item, the whole range of links shown do not occur simultaneously. The figure shows that young herring in the 12–42 mm size-group feed mainly on *Pseudocalanus* (86 per cent by numbers), but what it fails to show, for example, is the fact that during March and April, the adult herring's diet consists almost exclusively of juvenile *Ammodytes*. From an energetic point of view, the remaining links in the food web of the adult herring during these two months are of small consequence. The number of links in a particular locality may be further restricted by the range of food organisms available there. Conversely, the number of links in Figure 14.2 can be increased (and their relative importance altered) by sampling herring from a more extended geographical range. This probably explains the rather different emphasis placed on items in the herring's diet by Savage (1937). The choice of food organisms to any one stage in the life history is probably much less than the diagram indicates, and may never be more than two or three. Figure 14.2 thus summarizes a great deal of information very conveniently, but obscures the dynamic nature of the trophic relationships between the herring and its prey.

This example suggests that the complexity of food webs has been overemphasized, and in fact implies that rather simple food chains can be found in nature. This is not to deny the importance of more extended feeding relations for long-term community integration and stability, nor the existence of species whose associated food links are always several. It is, however, often possible to view the feeding trajectory of an animal as passing through a series of linear food chains during its life history. A rather similar idea has been expressed by Jones and Hall (1973) who use the phrase 'private food supply' when speaking of the way in which haddock and cod larvae eat progressively larger organisms as they grow up. Two processes are at work here. As an animal grows, it is able to capture larger food organisms, and this will tend to make it a member of successively higher trophic levels in the food web. The smallest post-larvae of the herring feed on small herbivores, the juveniles on larger herbivores, and the adults on both herbivores and carnivores. There is an important corollary to this process. As growth proceeds, the smaller prey organisms are released from predation by a particular predator as larger ones

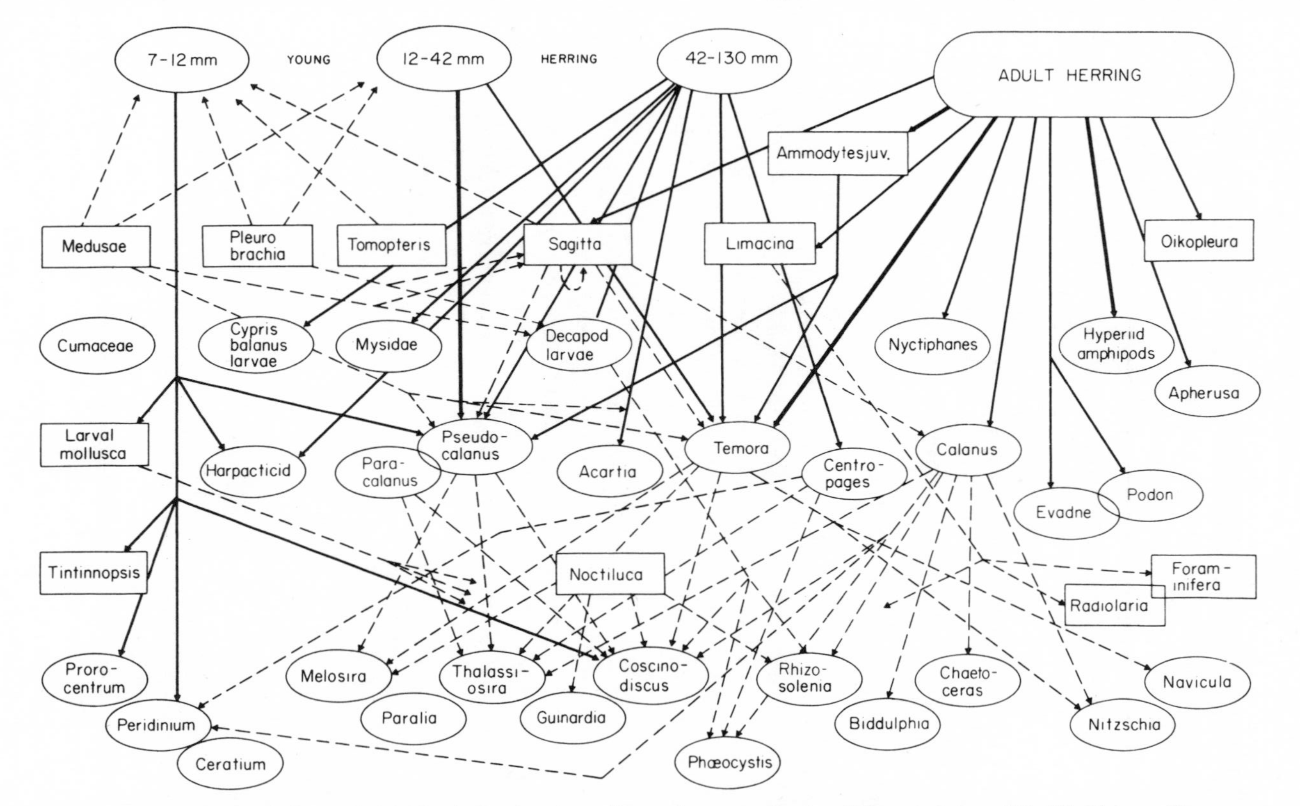

Fig. 14.2. Feeding relations of the North Sea herring, *Clupea harengus*, during different stages of its life history. (From Hardy 1924.)

become subject to it. It further suggests that the sort of situation shown in Figure 14.1b where interspecific competition exists may be very transient or even infrequent in natural communities. This is explicit in Gause's axiom which states that if two sympatric noninterbreeding populations occupy the same niche, one will eventually eliminate the other (Hardin 1960). When a link in a food chain is vacated in this way, a potential source of unexploited food is created, which can then be colonized by another species. This suggests one reason why different species might spawn at different times of the year rather than simultaneously, and is a further aspect of succession. Plaice and sand-eels, which hatch a few weeks apart in the Southern Bight of the North Sea, provide a possible illustration of this process, and are discussed below. Secondly, the food species available are constantly changing in abundance and kind as succession proceeds. Indeed, succession itself must in part result from this process.

The increase in the average size of food organisms found in Hardy's herrings is a general phenomenon supported by data on a wide variety of fish, though it is rarely possible to calculate numerical values of the change from published information. Examinations of the food lists given by many authors for different sizes of a particular fish species almost always indicate a decline with increasing predator size of the numbers of small organisms eaten, and a corresponding increase in the numbers of larger organisms. Nor is this phenomenon confined to fish. The feeding habits of sagittae have been studied by Rakusa-Suszczewski (1969), and data for *Sagitta elegans* (Verrill) is summarized in Figure 14.3. The food organisms are arranged across the top of the figure in order of increasing length. There is a remarkably clear-cut increase in the size of food organisms with increase in the length of the sagittae. A similar but much less detailed story has been given for *Calanus finmarchicus* (Gunner) by Marshall and Orr (1955). The same principle emerges from a bird's-eye view of whole groups of animals. The pelagic copepods are a good example. If highly specialized families like the Sapphirinidae are excluded, they can be arranged according to their feeding habits in a series ranging from filter feeders to raptorial carnivores (Wickstead 1962). Along the same series there is an increase in body size.

Before leaving the herring, one more aspect of food chains can be illustrated by reference to Figure 14.2. Adult herring exploit the two preceding trophic levels (primary carnivores and herbivores) in the food web, while the juvenile fish feed mainly on herbivores. In general, animals high in a food chain are less likely to confine their feeding activity solely to the trophic level immediately beneath them (Lindeman 1942). The smallest herring larvae appear to contradict this trend, but the numbers of phytoplankton cells eaten are in fact very small (about 1 per cent) and energetically insignificant. It is this trend which dissolves the distinctions between herbivores, carnivores, and other categorizations based on feeding habits. An excellent illustration of

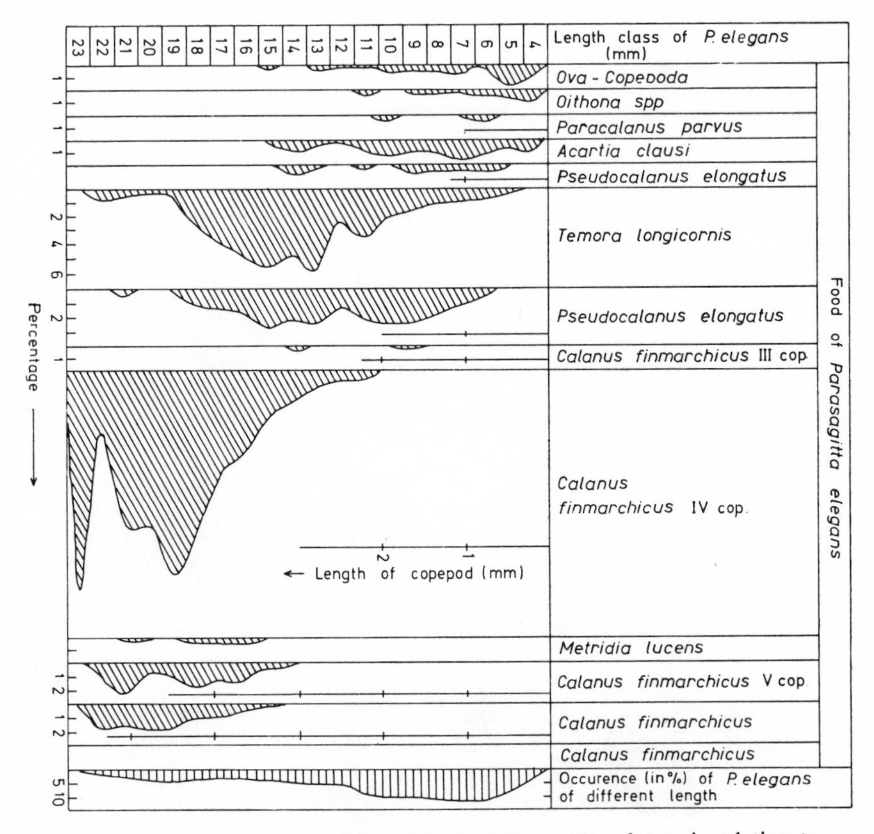

Fig. 14.3. Percentage composition of food of *Parasagitta elegans* in relation to predator size. The food organisms listed at the top of the diagram are in order of increasing size. (From Rakusa-Suszczewski 1969.)

the limited value of these categories has been provided by Roger (1973). His figure showing the relative quantities of phytoplankton and zooplankton eaten by sixteen species of tropical euphausiids is reproduced here as Figure 14.4. The first four species in this figure are readily classified as carnivores, and the last two as herbivores. But the remaining ten species are mixed feeders. In studies of production based on the trophic level concept, the importance of this concept will depend on the relative abundance of animals which cannot be easily classified.

There is amongst the plankton, though it is not distinguished clearly in Hardy's figure, a microfauna consisting of the youngest copepod nauplii, veligers and other larval forms, tintinnids and so on, which on account of their small size must be largely herbivorous. This fauna is not readily separated by routine sampling procedures from the larger-celled fraction of the phytoplankton, and was on the whole neglected by ecologists, until recently (see

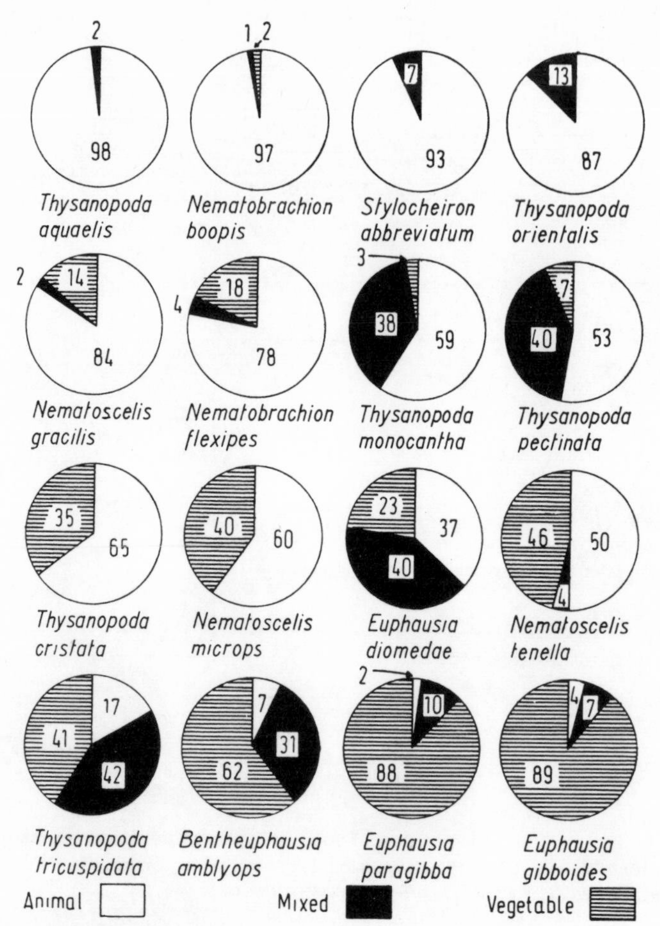

Fig. 14.4. Relative quantities of phytoplankton and zooplankton eaten by sixteen species of tropical euphausiids. (From Roger 1973.)

Beers & Stewart 1969, 1971), so that the part it plays in food-chain dynamics is not yet established. The guts and food vacuoles of these animals generally contain greenish or brownish material which may be presumed to consist of nanoplankton and perhaps detritus and bacteria as well. The ecological role might then be to make the smallest organic particles (living and non-living) in the sea available to the larger herbivores. Here again we see the distinctions between trophic levels becoming blurred.

Petersen's Limfjord

The base of the food web to which the herring belongs is occupied by phytoplankton, which is fed on directly by the herbivores. To illustrate a different food web, in which the links between the producers and herbivores may be less direct, a classical study has again been chosen. Petersen spent 35 years studying the bottom fauna in the shallow seas surrounding Denmark. In one of these, the Limfjord, the main plant formation consisted of meadows of eel-grass, *Zostera* (Petersen 1918). These meadows, however, did not directly provide food for the bottom communities in which Petersen was interested. The leaves and flowering parts of *Zostera* become detached in the autumn and through mechanical and bacterial action are converted to a finely divided detritus. It was this material, suspended in the water or deposited on the bottom, which together with its bacterial population was thought by Petersen to form the base of the major benthic food chains in the Limfjord. The next trophic level in this food web therefore consists of filter feeders and mud eaters such as bivalve molluscs and polychaetes. Phytoplankton was thought to play a very minor part here. There is, as in the pelagic environment, a fauna of much smaller animals, such as protozoa, harpacticoids and nematodes, which live in the bottom deposits, and on its surface. These animals feed on the bacteria and detrital material which settles from the water column, and on the epipelic flora of diatoms and other algae at the mud-water interface which is also partly derived from pelagic populations. Some of the larger members of this fauna, nematodes and turbellarians especially, are predatory.

The relative importance assigned to eel-grass and phytoplankton by Petersen and his colleagues must be revised in the light of subsequent information. No remarkable changes occurred in the Danish inshore fisheries following massive destruction of the eel-grass communities in the early 1930s, though much of the fauna at intermediate trophic levels was replaced by other species (Rasmussen 1973). It has also been shown that phytoplankton production in Danish coastal waters is very much higher than Petersen perhaps imagined (Steeman Nielsen 1944), and probably greatly exceeded eel-grass production even when the latter was abundant. On the other hand, bay scallop production on the Virginia coast was severely affected by the disappearance of eel-grass.

The lamellibranchs and polychaetes, then, probably feed on this micro- and meiobenthos as well as on detritus, although such a relationship has not yet been effectively demonstrated (McIntyre 1971). They are fed on by predatory crustaceans, gastropods, echinoderms and fish. The latter in turn are eaten by larger predators such as cod (and birds in intertidal areas). From this brief account, we see that the broad structure of the food web in shallow-water benthic communities is much like that of the pelagic zone, but that detritus and bacteria play a more important role. We note also that there is an important flow of material into the community from the plankton, and this flow must become more important as one moves offshore, and away from the zone of attached plants.

The upper trophic levels of the benthic community are also linked to the pelagic food web. Cod feed on herring in addition to benthic invertebrates, suggesting that the benthic community is not fully able to support predators so many links removed from the base of the food web. It is also noteworthy that the larvae of the vast majority of benthic animals are planktonic. At the time when their food requirements are highest, these larvae leave the community which will ultimately support them, and migrate to the more productive pelagic regions. (This of course is not the only pressure favouring the production of pelagic larvae.) Thus, both upper and lower levels of the benthic food web, and a suspension-feeding intermediate level, are linked to the pelagic community.

The general structure of the food chains studied by Petersen is found in shallow water and estuarine environments throughout the world. The most striking differences between these communities and pelagic communities are the relatively greater importance in the former of detritus derived from macroscopic plants (together with its bacteria and a benthic microflora), and the replacement of the herbivorous zooplankton by a group of much larger filter-feeders and mud eaters. A similar picture is found in many tropical estuarine environments where mangrove or other vegetation provides large amounts of detritus, and mullets, milkfish, and killifish are prominent members of the herbivorous or detritophagous trophic level. Detritus is assigned an important role in a tropical estuary studied by Wyatt and Qasim (1973).

Mann (1972) has reviewed the information available on productivity in these shallow water communities, and on the production and utilization of detritus. He concludes that over 90 per cent of the primary production of marine macrophytes enters the food web as detritus. This generalization applies both to the mangrove coasts of the tropics as well as the kelp beds of temperate and boreal regions, and to eel- and turtle-grass communities. In all these instances the detritus is first colonized by fungi and bacteria which in turn support protozoa, nematodes and microcrustaceans. This mixture of macrophyte detritus and microscopic plants and animals is eaten by the detritophagous macrofauna. The larger plant fragments appear in the faeces undigested, and the sequence repeats itself. Mann thus sees the ecological role of detritus feeders as the reduction of particle size. The smaller particles ($<50\mu$) are collected by filter feeders, both benthic and planktonic or settle to the bottom and form the food of mud eaters.

Southern Bight fish larvae

In discussing the North Sea herring, it was pointed out that simple linear food chains can be identified if the space and time scales are sufficiently restricted. A good example is found in the Southern Bight of the North Sea where larvae of plaice (*Pleuronectes platessa* (Linnaeus)) and sand-eels

(*Ammodytes marinus* (Raitt) and *A. tobianus* (Linnaeus)) feed on *Oikopleura dioica* (Fol) for a period of nearly two months (Wyatt 1971, 1974). Between January and April, these species account for most of the fish larvae in the Southern Bight, but at all times they are outnumbered within their own broadly defined trophic level by chaetognaths.

Plaice and sand-eel larvae both feed initially on phytoplankton, but they soon start to capture *Oikopleura*, and it is the food pellets of the latter which are recognized in the larval fish stomachs (Shelbourne 1957, Ryland 1964). In sand-eels, the change from phytoplankton to *Oikopleura* is gradual, but by the time they are 10–11 mm long, very little of the former is captured. Between 10 mm and 15 mm, copepods and other organisms are added to the diet, but *Oikopleura* constitutes the main food during this time (Wyatt 1971). In plaice larvae, the switch from phytoplankton to *Oikopleura* is more marked, and plant cells cease being eaten almost as soon as the larvae have graduated from the yolk sac stage. Normally, plaice larvae continue to feed on *Oikopleura* until metamorphosis two months later. These species of course may feed differently outside the Southern Bight as the herring do in different regions of the North Sea. These two species of larvae accounted for most of the mortality of the *Oikopleura* populations in the Southern Bight during a period of several weeks in 1968 (Wyatt 1971), when they were abundant. But the mortality of the prey did not change much before and after this period, so clearly other predators relinquished this food source as these larvae began to exploit it.

That these two species of fish larvae feed on the same prey species suggests competition exists between them. There is evidence though that the ranges of prey densities at which the two species feed are different. The prey densities at which the highest feeding rates are seen in plaice larvae are low, and when the prey are sufficiently abundant for the sand-eels to feed, the plaice capture fewer (Wyatt 1974). This unexpected finding needs to be checked, but the depressed feeding of plaice larvae at high prey densities cannot be attributed to competition between plaice and sand-eel larvae, since it is found even when no sand-eels are present. There may of course be other competitors. Nor is there any obvious reason why sand-eel larvae seem to stop feeding at low-prey densities which are optimal for plaice larvae. It almost seems as if the two species are cooperating.

Three other species of fish larvae are common in the Southern Bight during the same weeks when plaice and sand-eel dominate the larval ichthyofauna. These are flounder (*Platichthys flesus* (Linnaeus)), dab (*Limanda limanda* (Linnaeus)) and cod (*Gadus morhua* (Linnaeus)). Together, these five species form 85 per cent of all fish larvae found between January and April. Flounder larvae feed mainly on the larger diatoms like *Biddulphia* and *Coscinodiscus*, and also on *Nitzschia*, while other abundant diatoms are not eaten. They also eat *Phaeocystis*. The diet of dab larvae consists almost entirely of copepodids and nauplii, largely of *Temora*, and lamellibranch larvae. Cod larvae are mixed feeders and eat mainly copepod nauplii especially *Pseudocalanus* but also considerable numbers of *Coscinodiscus*. In all these larvae, there is a clear increase in the size range of food which is acceptable with increase in larval size, as expected. If their distribution in time and space is taken into account, there is little or no overlap in their feeding habits, and hence no competition between them. Again, there may be other competitors not accounted for, and if so, chaetognaths might be chief amongst these. These are very numerous, and also ubiquitous.

Plaice and flounder larvae have rather restricted feeding habits in normal conditions. Cod larvae, by contrast, eat a variety of different food items when at a given stage of development. This is a distinction separate from that already made between the more mixed feeding habits of animals at higher trophic levels. This recalls the distinction made by entomologists between specialists and generalists as regards food requirements. It is the latter which are likely to show the greatest fluctuations in numbers since their population growth is not linked to a single food source. But the specialized plaice larvae, with their strong preference for *Oikopleura*, are by no means unadaptable. On one recorded occasion when *Oikopleura* were virtually absent from the Southern Bight during the time of occurrence of the larval plaice and sand-eel populations, the diet of the plaice consisted largely of *Coscinodiscus*, together with other phytoplankton, and small numbers of fish eggs and copepod nauplii (Wyatt 1973, 1974). The sand-eels ate nauplii and copepodids, eggs, and phytoplankton. Both species were therefore able to adapt to the failure of their normal prey, and became much less specialized in their feeding habits when forced to by circumstance. Changes of this kind can also occur in the short term. Lasker (1966) has shown that the crustacean *Euphausia pacifica* (Hansen) can change from filter feeding to raptorial feeding, and Parsons and le Brasseur (1970) suggest this change is necessary offshore where the phytoplankton cells are smaller than in coastal regions. The same thing has been noted in two species of *Centropages* by Anraku and Omori (1963). In deep-water upwelling areas in the Antarctic where the average size of plant cells is large as in neritic environments, *Euphausia superba* (Dana) (krill) feeds mainly on colonies of the diatom *Fragilariopsis antarctica* (Swartz) (Marr 1962). It appears then that the absolute dimensions of the primary producers determine the trophic level of the larger zooplankton, and hence the relative importance of the microzooplankton which feeds on the smallest phytoplankton. Further discussion of this point is found in Chapter 3.

Behavioural changes of the kind outlined above are clearly important from at least two points of view. As already indicated, they can influence the number of trophic levels between the producers and the top carnivores (see below). In addition, they must play a role in the stability of the food web. If this stability is defined in some way related to the survival of and magnitude of fluctuations in the component populations of the food web, then the modeller would need to introduce switching mechanisms to prevent his predators becoming extinct when their preferred prey was insufficiently abundant. In the Southern Bight example, the absence of *Oikopleura* has an obvious effect on the feeding of plaice and sand-eel larvae. Volterra's model would have predicted the death of these populations, but a behavioural change allows them to survive.

The absence of *Oikopleura* was probably caused by unusual meteorological conditions (Wyatt 1973), which apparently had no direct effect on the other zooplankton populations of the Southern Bight. This absence had an obvious effect on the larval plaice and sand-eels, but may also have had effects on other members of the food web here. There is as yet no information on this point. In a quite different environment, an intertidal rocky shore community lost seven out of fifteen species as a result of the experimental removal of one of them (Paine 1966). Such experiments have not been devised in the pelagic environment, though they may be taking place inadvertently along heavily industrialized coasts as a result of eutrophication processes.

Black Sea plankton

So far, in discussing the herring and the Southern Bight larvae, little detail has emerged concerning the links between these fish and the phytoplankton. The general nature of these intermediate links in the pelagic food web can be found in Russian studies in the Black Sea. Much of this work has been summarized by Petipa, Pavlova and Mironov (1970). Although the Black Sea is geographically very different from the North Sea the epiplanktonic fauna is almost identical in species composition. Petipa *et al.* distinguished six different ecological groups of herbivores, and significantly, these are largely based on the size of plant cells consumed. Three of these categories consist of different nauplii and copepodid stages of *Oithona minuta*, *Paracalanus parvus* (Claus), *Acartia clausi* (Giesbrecht), and pontellid copepods, and each of these three feeds on a characteristic size range of algal cells. *Oikopleura dioica* forms a category by itself and eats only the smallest algae and detritus. The last two groups are polychaete and molluscan larvae which feed on all sizes of algae and detritus, and the adult and pre-adult stages of *Calanus* and *Pseudocalanus* which eat medium and large algae.

An intermediate trophic level called 'mixed food consumers' has been introduced to classify those animals which eat both plant and animal food. Adult and pre-adult *Acartia*, *Centropages*, *Oithona*, and pontellids fall in this group. The food consists of small and medium-sized algae and nauplii and juveniles of other copepods. Primary carnivores in the Black Sea plankton are adult *Oithona minuta* and pontellids which feed on small copepods and cladocerans; sagittae, which here feed on adult copepods, *Oikopleura* and their own young, are secondary carnivores, and *Pleurobrachia pileus* (O.F. Müller) which eats sagittae and copepods is a tertiary carnivore. Two points emerge from this brief summary. Five distinct trophic levels are recognized in the Black Sea zooplankton by Petipa and her co-workers. This is rather more than most ecologists would have allowed, but the importance of these distinctions was underlined by their subsequent detailed studies of energy flow in each of these categories separately. The second point is that several species are found in more than one category. Adult *Oithona minuta*, for example, are classified both as mixed feeders and primary carnivores, and its nauplii and juvenile stages are found in several categories also. This reflects the observational data, and is a realistic approach to the drawbacks of the trophic-level concept.

The lengths of food chains

It has been noted in the introduction to this chapter that food chains are restricted in length to about five or six trophic levels on physiological grounds. We have seen also that an intermediate trophic level may be required to allow large zooplankton to take advantage of small-celled algae, since the food collecting apparatus of the latter cannot capture particles below a certain size. *Euphausia pacifica* was given as an example. This suggests that there are restrictions on the relative size of predator and prey, and that these restrictions in addition to the physiological constraints play a part in determining the number of links in a food chain. Ursin (1973) has determined the size of prey organisms preferred by cod and dab of given dimensions, and if such predator/prey size relationships are assumed to be general at all trophic links, then the approximate number of links in a whole food chain can be predicted (see Chapter 13). Adaptive mechanisms to the absence of a preferred food can also shorten a food chain as in the case of plaice larvae which choose to eat algae rather than other zooplankton when *Oikopleura* is absent.

Ryther (1969) has argued, mainly on the basis of predator/prey size relationships, that oceanic, coastal and upwelling areas are in general characterized by food chains of different lengths. His argument leads to a minimum estimate of five trophic levels for the oceanic environment where the algae are characteristically relatively small, and are made available to the primary carnivores of the plankton by protozoans and larval crustaceans. In coastal regions with large-celled algae, there are three trophic levels, as described for the North Sea herring, and in upwelling areas where the fish are partly herbivores, there are one or two. These estimates are up to the link useful to man, and the top predators in each case have been neglected. This general view summarizes the main pathways of energy flow in the three major environments of the oceans. Longer food chains can of course be identified in each area, but these are not likely to be quantitatively important. Shortening of food chains on the other hand is important since it allows a species to gain the advantages of large size and consequent relative freedom from predation with an abundant food supply. The best known examples of this strategy are the baleen whales which feed on krill in the Antarctic. The basking sharks and sun-fishes (*Mola*) of temperate regions, and the whale sharks and manta rays of the tropics are other examples, all very large animals which feed mainly on planktonic crustaceans. The many filter-feeding members of the benthic and intertidal communities are less dramatic examples of the same strategy, as are the herbivorous salps and doliolids of tropical oceans. The opposite development has also taken place. Ceratioid and stomiatoid fishes of the deep sea are able to swallow other fish bigger than themselves, clearly an adaptation to an environment where food is thinly spread. This bizarre habit presumably lengthens the food chains in the deep sea.

For the sake of completeness, three other ways in which the constraints on food-chain dynamics are in a sense modified may be mentioned here. These are parasitism, symbiosis and commensalism. If any free-living members of a food chain support parasite populations, then the latter effectively add a new dimension to that chain without lengthening it. At the same time they may influence the free-living members in a number of important ways through their control of host growth rate and fecundity. Commensal relations, in which two species co-operate to their mutual benefit, usually result in one member exploiting a food resource which might otherwise remain unused. Cleaning symbiosis is an example. Symbiosis is a widely occurring but poorly understood phenomenon where unicellular algae live intracellularly in various animal tissues. These zooxanthellae are a notable component of many planktonic protozoans, and of the hermatypic corals, but are found also in a wide variety of invertebrates including the giant clam *Tridacna*. The algae are thought to obtain their nutrient requirements from the animal tissue, but it is not yet clear whether the latter simply digests the algae or achieves some more subtle benefit. The subject is reviewed by Yonge (1957).

The problem of top predators

The uppermost trophic levels in the sea are occupied by fish, sharks, mammals (including man), birds, and squids. Some of the larger jellyfish perhaps also belong here. In benthic communities, the largest crustaceans, echinoderms and molluscs might be added to this list. It is generally agreed that at all lower trophic levels, the numbers of individuals in populations must be controlled locally by availability of food and by predation, though it is not clear how these are interrelated. Since the top predators do not suffer significantly from predation, it can be argued that their numbers are governed by their food supply. There is, however, no evidence for this argument one way or the other. Two other processes between them can regulate these populations. The first regulates them before they ever reach the ultimate links of the food web. Their larval and juvenile stages must grow through the food web like those of the herring, so that their numbers are regulated by the processes which operate at all trophic levels, before they cease to be preyed upon. It is this necessity which dictates the very high fecundities found in fish, since mortality is extremely high, particularly during the larval stages. Larval mortality has in fact long been regarded as the most important single factor in the control of fish numbers (Hjört 1914, Cushing 1972). Species which seek to escape this high mortality can only do so at the expense of fecundity, by producing offspring large enough (hence fewer in number for a given reproductive effort) to omit the early stages of their climb through the food web. The marine environment does not seem to have encouraged this approach in teleost fishes, though it is found in freshwater fishes, and of course in the birds and mammals which feed in the sea. Sharks more closely resemble tetrapods than teleosts in this respect.

The second way in which top predators are regulated is through a gradual reduction in their metabolic efficiency. A large part of the energy young fish gain from their food is used for growth. As they become bigger, this proportion declines, until eventually it approaches zero. All the intake is then used in maintenance and for reproduction, and bodily growth practically ceases. While fish may be very long lived, ageing processes and disease take their toll, fecundity declines, and death eventually occurs.

The role of bacteria

Food chains make small organisms into larger ones, as Elton said. Hand in hand with this process, there is a vast production of excretory products and faecal materials. The total quantities of nonliving organic matter in the sea greatly outweigh the living. Strickland (1965) has roughly estimated that for every gram of living carbon in the water column, there are 10 grams of dead particulate organic carbon, and 100 grams of dissolved organic carbon. In coastal waters where production is higher, and where run-off from the land also contributes an organic load, these values are higher, and show seasonal variations (Duursma 1960). In the deep sea, these quantities decrease with depth to reach steady values at 1000–1500 metres. Pütter (1909) claimed that dissolved organic matter was used directly by the zooplankton in such amounts as to fulfil their energetic needs. This view has found little favour, and it is now generally thought that this material must be made available through the intermediary of bacteria. The particulate matter is of course directly available to filter feeders, as we have seen in discussing Petersen's Limfjord. The dissolved organic matter is thought to be first adsorbed on to mud particles, or on to bubbles (Sutcliffe *et al.* 1963, Riley 1963b). Such organic aggregates attract bacteria and other small particles, and are then again available to filter feeders. These aggregates may be an important source of food to the herbivorous zooplankton when phytoplankton is scarce.

Sorokin (1971, 1973) has measured bacterial production in the Western tropical Pacific, and found it to be greater than phytoplankton production. He speculated that the source of this production was organic material transported in deep water from high latitudes. This bold hypothesis adds fuel to the long-standing discussion concerning the relative importance of bacterial and autotrophic production in the oceanic environment. More detailed studies on the flux rates of organic matter at all depths in the deep sea may help to resolve this problem.

Conclusions

An attempt has been made in the preceding pages to isolate the main features of marine food chains. These may be summarized as follows:

1 Food chains are simple and short-lived. As animals grow, the nature of their diet changes in accordance with the size and availability of prey organisms, so that food chains are continually forming and disappearing. It is the summation of these simple units, over longer periods of time, which gives rise to the apparent complexity of food webs. The changing diets of animals as they grow through the food web limits the usefulness of the trophic concept as applied to individual species.
2 Food chains are limited in length to five or six links by physiological constraints, but some animals have overcome this limitation by the evolution of specialized feeding mechanisms, and specialized ways of life.
3 Detritus and bacteria are important food sources in shallow marine environments, and probably also play some role in deeper water. They may also have the effect of damping out fluctuations in food supply caused by seasonal changes in plant growth.
4 A particular prey is likely to be exploited successively rather than simultaneously by different predators, so that competition between two species for food may be limited to the brief period of change-over, and to rather small fractions of each population.
5 Top predators are regulated by low metabolic efficiency and senescent processes, and by reduced fecundity.
6 Succession is in part an expression of the dynamic nature of food-chain processes.

Attention has been drawn to the fact that most food-chain models can be derived from Volterra's equations. Early models by Riley (1946), Riley *et al.* (1949), and Steele (1958) form the main starting points of more recent work, and have been reviewed by Riley (1963a). These models consist essentially of four groups of equations, which are:

1 An equation describing the exponential decay of radiant energy with increasing depth in the water column. This generally takes the form

$$I_z = I_0 \exp(-kz)$$

where I_0 is the radiation reaching the surface, Iz that reaching depth z, and k is the extinction coefficient.

2 Equations describing the growth of the phytoplankton in relation to light and a limiting nutrient such as phosphate, and the loss of phytoplankton due to grazing by the herbivorous zooplankton and other processes such as respiration and sinking. Together these correspond to Volterra's prey equation. Riley (1946) used the equation

$$\frac{dP}{dt} = P\,(P_h - R - G)$$

where P_h, R, and G are photosynthetic, respiratory, and grazing coefficients respectively, and P is the total phytoplankton population. P_h was integrated over the whole water column in relation to the radiant energy available, and the concentration of phosphate was made to depend on temperature. $(P_h - R)$ corresponds to b_1 in Volterra's prey equation. The grazing coefficient was made a simple linear function of the herbivore population.

3 Equations describing the nutrient concentration in terms of uptake, supply by regeneration, and by mixing of surface and deep water. The Michaelis–Menten expression borrowed from enzyme kinetics is now widely used to control nutrient concentration in these models (see Chapter 7).

4 Equations describing the growth rate of zooplankton correspond to Volterra's predator equation. Equations for higher trophic levels may also be added.

The essential structure of such a model is thus a set of first-order differential equations, and an associated matrix of coefficients which can be varied by subsidiary equations. Operationally, such a model 'can be visualized as an iterative process, running continuously and leading to a steady state' (Margalef 1968). One of the great values of such models is that they allow us to assess the relative importance of different parameters on population numbers, although it is not always easy to distinguish those properties of the models which result from simplification or misunderstanding of ecological processes, from those implicit in the mathematical language. Yet simplification is an essential feature of models. If they are made too complex, their behaviour may soon become as daunting as that of the ecosystems whose workings they seek to enlighten. It is clear from the account of food chains in this chapter, that the general family of models outlined in this section are a compromise between reality and simplicity. The detailed construction of one such model is described in Chapter 16.

PART VI. THEORY

Marine ecology is a science lacking in theory; indeed it might be said that little advance has been made on the masters of the twenties and thirties, Volterra, Elton and Gause. When we look at the data masses and the technical advances through which the theory of physics has been elaborated, biologists feel envious. The last two chapters of this text describe two developments in technique that might make theoretical advancement possible. These are the statistics of patch sampling at sea (Chapter 15) and the establishment of biological modelling within a hydrodynamic framework (Chapter 16). They are obviously very closely linked and the real point is that successive samples in time can be taken from a population in such a way that the biological model can be evaluated.

The model used has two virtues which may illustrate why we have been a little chary of the higher analysis of marine systems or why we like biological models, especially if they are little ones. The first virtue is that the quantities are expressed in nutrients throughout the ecosystem and can be used in a running analysis to check that the budget is the same, i.e. a conservation of nutrients. The second virtue is that observations and model appear to march together in three separate ecosystems. The point is not that the data fit in any statistical sense, but that the first steps in generality are being taken, the steps towards a theory.

Chapter 15. Sampling the Sea

J. C. Kelley

Introduction

The sampling problem
Difficulty 1
Difficulty 2
Difficulty 3

Some solutions to the sampling problem
Condition 1
Condition 2
Distribution of discrete samples
Random sampling techniques
Adaptive and heuristic sampling

Some solutions to the sampling problem *continued*
Areal sampling
Synoptic and quasi-synoptic sampling techniques
High-altitude remote sensing
Low-altitude remote sensing
Shipboard discrete station sampling
Moored arrays
Continuous shipboard sampling

Summary

Introduction

The ocean is spatially and temporally inhomogeneous and displays its inhomogeneity through a broad spectrum of space and time scales. Although many of the variables used to describe the ocean are continuous in space and time, some are not, particularly those used to describe its biological properties.

In empirical science, conclusions regarding the nature of a phenomenon are drawn from the analysis of samples taken from that phenomenon. Since these samples are only representative of the phenomenon, the conclusions drawn may be of varying validity (they also depend upon the skill and sagacity of the investigator who develops them), but it is syllogistically demonstrable that these conclusions cannot reflect the nature of the phenomenon more faithfully than do the samples themselves. A rather large part of oceanographic research, including most of that in biological oceanography, can properly be called empirical, and the foregoing argument applies.

If the ocean were spatially and temporally homogeneous, a single sample, taken anywhere in the time-space domain over which it is defined, would represent it completely. A second sample would be redundant; it would provide no additional information on the phenomenon in return for the work expended (N.B. there is no assurance that two identical samples will yield identical measurements of a given variable, and certainly none that duplicate conclusions based on the two samples will agree, but these are technological and personality problems, respectively, and do not vitiate the sufficiency of the samples.) To the degree that the ocean is inhomogeneous, an increasingly large number of samples is required to be representative. A large sample size alone, however, does not ensure that the samples are representative. For this to be the case the samples must be distributed spatially and temporally so that all significant scales of variability which characterize the phenomenon are reflected. In practice the sample size is likely to be hundreds of orders of magnitude smaller than the number of points at which the ocean is defined. The investigator must therefore exercise his discretion in choosing from the following alternatives. He may (1) make 'simplifying assumptions' and consciously choose to ignore phenomenological complexities which he cannot hope to resolve (but which are still present in his samples); (2) greatly limit the scope of his investigation by, for example, studying the phenomenon in only one dimension, say time, or by choosing a very small part of the domain, in which the phenomenon *is* homogeneous; (3) suppress previously acquired information to the contrary and proceed as though the ocean is 'really very simple'. If the investigator chooses any combination of these three alternatives, he may be ready to address the sampling problem, for having redefined his subject until it is tractable with at least a finite number of samples, he may now wish to ask which set of samples will give the greatest return for the effort expended in their acquisition.

The sampling problem

As outlined above, the empirical oceanographer finds himself facing a complex entity, the ocean, which is defined over a large time-space domain. The phenomenon can be conceptualized at one point in this domain, P_i: (x_i, y_i, z_i, t_i) (where x, y, z, and t are the usual four dimensions: longitude, latitude, depth, and time) as a set of measurable variables, the vector, $\mathbf{A}$:

$$\mathbf{A}(A_1, A_2, \ldots, A_k) \qquad (15.1)$$

Where A_1 might be temperature, A_2 salinity, and so on, and where k might be quite large. Although the sampling problem is simpler when scalar quantities alone are considered, most oceanographic observations are multivariate (e.g. STD measurements). It is, therefore, preferable to develop these concepts in a multivariate framework from the outset (Kelley 1971a). A *realization* of the phenomenon can be defined as a set of specific measurements of these variables at some point, P_i, at which the phenomenon is defined. (Points and vectors are defined in the same notation. Since a point can be thought of as the end of a vector emerging from the origin, they will be considered equivalent.) Let us refer to the set of specific measurements of the k variable as another vector, $\mathbf{a}$:

$$\mathbf{a}_i(a_{1i}, a_{2i}, \ldots, a_{ki}) \quad (15.2)$$

where a_{1i} is the temperature at point P_i in the time-space domain. (One can conjure up conceptual difficulties with this definition in the oceanographic context. For example, if P_i is vanishingly small and a_{ji} is the number of blue whales at P_i and a_{ji} happens to be 1, what are the values of the other $\mathbf{a}_i$'s? We will find later in this chapter that in practice this causes no difficulty.)

Let us refer to this vector as a state vector because it describes the state of the phenomenon at P_i. The set of all points, P_i ($i = 1, 2, \ldots, N$) at which the phenomenon is defined is the time-space dimension of the ocean, and N is a *very* large number. All possible *realizations* of the phenomenon $\mathbf{a}_1$ at points P_i ($i = 1, 2, \ldots, N$) comprise the *theoretical population.* A *sample* can be defined as a proper subset of the population: $\mathbf{a}_i$ ($i = 1, 2, \ldots, n; n < N$) which will actually be measured. From combinatorial theory we know that N things can be combined n at a time in exactly $C(N,n) = N!/n!(N-n)!$ ways. That is, $C(N,n)$ samples of size n can be chosen from a population of size N. Since usually $N \gg n$, the *sampling problem* is simply: which sample will optimize the information returned on the population for the amount of effort expended in obtaining the sample.

Let us refer to the time-space domain over which the phenomenon, the ocean, is defined as the population space. In fact, it is a continuous volume in the x, y, z, t hyperspace. Following the nomenclature of Krumbein and Graybill (1965), we can refer to that part of the population space which is of interest in a particular investigation as the *target population.*

The purpose of sampling is to examine some aspects of one or more of the variables in the state vectors which exist for the target population. Typically, an investigator must settle for a very small part of the target population from which to draw his samples. This subset is called the *sampling population* and it is important that it is clearly identified conceptually by the investigator because it is only this small subset of the target population to which conclusions drawn from the sample apply. Extrapolation of the results to the target population depend upon the validity of two assumptions: (1) that the sampling population contains all relevant characteristics of the target population, i.e. that it is sufficiently large and contains all types of variability that appear in the target population and (2) that it is an *unbiased* subset of the target population (i.e. that there is nothing in the criteria which define the sampling population which is related prejudicially to the values which appear in the state variables of the sampling population). It is acceptable to have some members of the target population carry a greater probability of appearing in the sample than others. It is unacceptable to have members of the target population, which display high, low, or average values of the measured variables, appear with greater probability. The validity of these assumptions assures that the sampling population is *representative* of the target population—but their validity is seldom demonstrable and most often is no more than a scientific article of faith. It may be helpful here to give some specific examples of sampling difficulties which mitigate the representativeness of the sampling population.

Difficulty 1

Limited resources (e.g. only one ship available for sampling) require that samples over different parts of the population be taken sequentially rather than synchronously (in oceanography this is called the synopticity myth). The variables in the oceanic state vector seldom exhibit distributions which are constant in the time domain. In fact, much of current oceanographic research is directed at the time-dependent behaviour of a number of variables (Stewart 1967). In biology, diel variability has long been of interest, but recently the added (and numerically comparable) effects of internal waves in the ocean are being addressed by biological oceanographers (Kamykowski 1974). A good deal of work in biological oceanography has been focused on the seasonal time scale at which biological variability is quite pronounced. Short-term fluctuations, which can be of the same size as seasonal ones, have yet to be investigated fully (Kelley 1975).

It has been traditional for oceanographers to go to sea on a single ship for a period of perhaps a month, occupy a number of stations distributed over some geographic area and separated in time by hours or days, and then to view the data measured on these stations 'synoptically'. In synoptic presentation the data are analysed as a set, ignoring the fact that they are separated in time, and the propriety of such presentation depends upon the validity of the assumption that the variable is constant in time during the entire sampling period (i.e. one month). This assumption is surely reasonable for variables such as mineralogical properties of deep sea sediments or for the chemistry of deep water, but it is not at all clear that it is ever valid in the euphotic zone or in the mixed layer of the ocean. If the 'synopticity assumption' cannot be made, then the sampling operation must be modified. We shall discuss alternative sampling schemes in a later section.

Difficulty 2

All of the target population is unavailable for sampling or refuses to be sampled. This difficulty, called 'non-response', is a classical statistical problem and takes its name from the well-known situation in sociological research such as market analysis or political poll-taking in which the population of interest is sampled by mailed questionnaire or by telephone. Many individuals refuse to return questionnaires or to respond to the questions of telephone inquisitors and it is always possible that the behaviour is related in some way to the individual's social, economic, religious, or political position, and consequently to the variables under investigation. In oceanography examples of non-response are sediments which are too coarse or well indurated to be collected by the sampling gear, net avoidance by fish or zooplankton, and those parts of the total population occupied by storms which preclude sampling activity. In such examples, that part of the population which is non-responsive is also peculiar in some way that is likely to be of interest in the investigation. Thus, the sampling population is not representative of the target population and the sample is *biased.* The amount of bias is a function of the size and peculiarity of the non-responding part of the target population, but, in general, non-response implies bias in some degree.

Difficulty 3

Samples do not represent vanishingly small points in the population space, but volumes. This problem is brought on by the inability to resolve the position of the sample in the space domain; usually it can be resolved in the time dimension with ease.

The technology used to determine a ship's position in the open ocean is designed to meet navigational, rather than scientific standards, and thus the positions given for oceanographic station locations are not exact, but typically carry an associated error ellipse within which the true position is expected to lie with some level of confidence. If one cannot resolve points in the x, y plane which are closer than some distance, d, the samples taken at a station in fact represent an area in the plane which is at least $\pi d^2/4$ in size. Typical values for d in the open ocean, with satellite navigation, are at least 0·5 km. (In very nearshore or inshore waters, positioning is often done by radar and the precision may be much better than the open ocean figure, if good radar targets are available. It is unusual to find values for d, even inshore, which are demonstrably less than 0·1 km.) Therefore a station can only be said to lie somewhere within an area of about 0·2 km^2. If the variable of interest, as in Difficulty 1, can be assumed constant over a horizontal distance of 0·5 km, Difficulty 3 need not disturb us. Similar problems occur in the vertical plane. In many oceanographic sampling activities, sampling depth is measured by the length of wire out, although more accurate estimates can be obtained with pingers, recording pressure transducers, or unprotected thermometers. Ship roll, elasticity of the wire, and the effects of current shear on the wire all degrade the depth estimate. Since vertical gradients in the ocean are typically much stronger than horizontal gradients, small errors in depth determination may be as serious as much larger errors in positioning. In any case the problem exists, and the level of resolution in any variable of interest is ultimately limited by the spatial resolving capability.

Some solutions to the sampling problem

The sampling problem can be addressed and, in specific contexts, solutions can often be obtained. It is useful first to review the conditions on the solutions which obtain.

Condition 1

If one wishes to sample a variable a, which is a continuous function of some other variable, t, and if it is possible to make certain assumptions about the

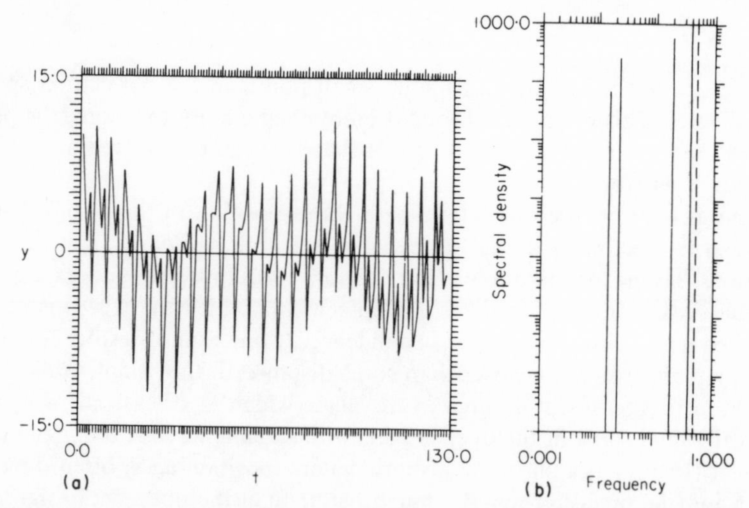

Fig. 15.1. (a) Time series generated by a Fourier series with four (sin plus cosine) terms. (b) Line spectrum generated by the Fourier transform of the series in Fig. 15.1a. Dashed line indicates the Nyquist frequency. Ordinate is $64a_j^2$.

functional relationship, the Nyquist sampling theorem provides the maximum allowable sample spacing in the t domain, or the minimum in the frequency domain, to ensure adequate representation of the population (i.e. it provides the cheapest sufficient sampling structure). The theorem was developed in the specific context of time-series analysis (Blackman & Tukey 1957), but can provide useful guidelines even in problems to which it does not exactly apply.

Let us first consider an example of its application. Consider the time series in Figure 15.1a. The function, y, is the sum of four terms in a Fourier series.

$$y_k = \sum_{j=1}^{4} a_j \left[cos\,(2\pi kf_j)+sin\,(2\pi kf_j)\right] \; k = 0, 1 \ldots, N-1$$

where $N = 128$, $a_j = j$ *and* $f_1 = 0{\cdot}015, f_2 = 0{\cdot}023, f_3 = 0{\cdot}2, f_4 = 0{\cdot}4$.

The terms differ in the frequency of the arguments and in the coefficient with which the term enters the general equation. The series is 'band-limited'; i.e.

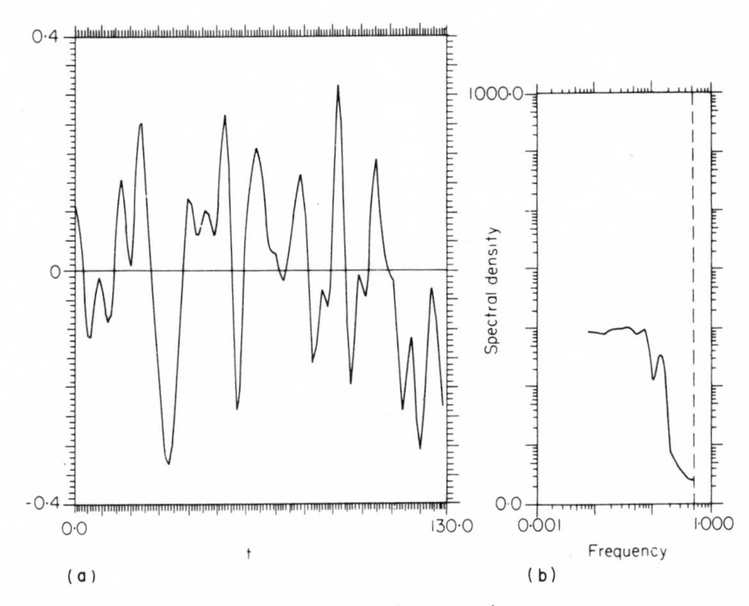

Fig. 15.2. (a) Time series generated from the expression

$$y_k = \frac{1}{21} \sum_{j=9}^{9} w_j\, u_{k+j}$$

where u_k is independent uniform noise on the interval $(-0{\cdot}5, 0{\cdot}5)$ and the w_j are cosine bell weights. (b) The spectrum calculated from the time series in Fig. 15.2a. The dashed line indicates the Nyquist frequency. Ordinate is $64a^2$.

the lowest frequency is $l = 0{\cdot}015$ and the highest frequency is $h = 0{\cdot}4$, and thus is contains no frequencies outside the 'band' $\pm B$ where $B = (h-1)/2$. This can be clearly seen by switching to the frequency domain by computing the Fourier transform of the time-series which yields the power spectrum of the series (Fig. 15.1b). Since only four discrete frequency terms comprise the series, the spectrum is only defined at those four frequencies. The power, or spectral density, associated with each frequency is related to the size of the coefficient with which the term is included in the series. This type of spectrum is called a line spectrum (Jenkins & Watts 1968). It can be shown that any continuous time series can be expressed as a Fourier series. In general, it is

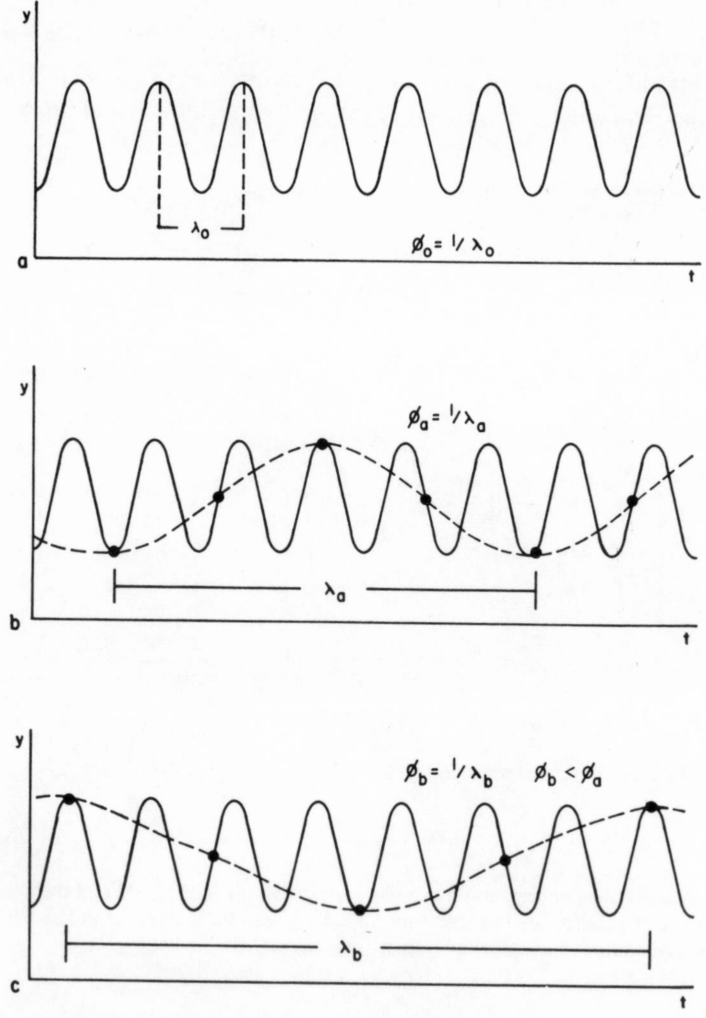

Fig. 15.3. (a) A sinusoidal function, y, with characteristic frequency ϕ_0. (b) The alias (dashed line) created by sampling at frequency ϕ_a. (c) The alias (dashed line) created by sampling at frequency ϕ_b.

unusual for a function to be represented as a sum of a few discrete terms, as above, and the spectra calculated from Fourier series approximations to most functions are continuous functions such as that shown in Figure 15.2a. If, however, the spectral curve drops to the abscissa at some frequency, ϕ_1, and remains there at all higher frequencies $\phi \geqslant \phi_1$, the function is still band-limited (Fig. 15.2b). (The low frequency band limit is often determined by

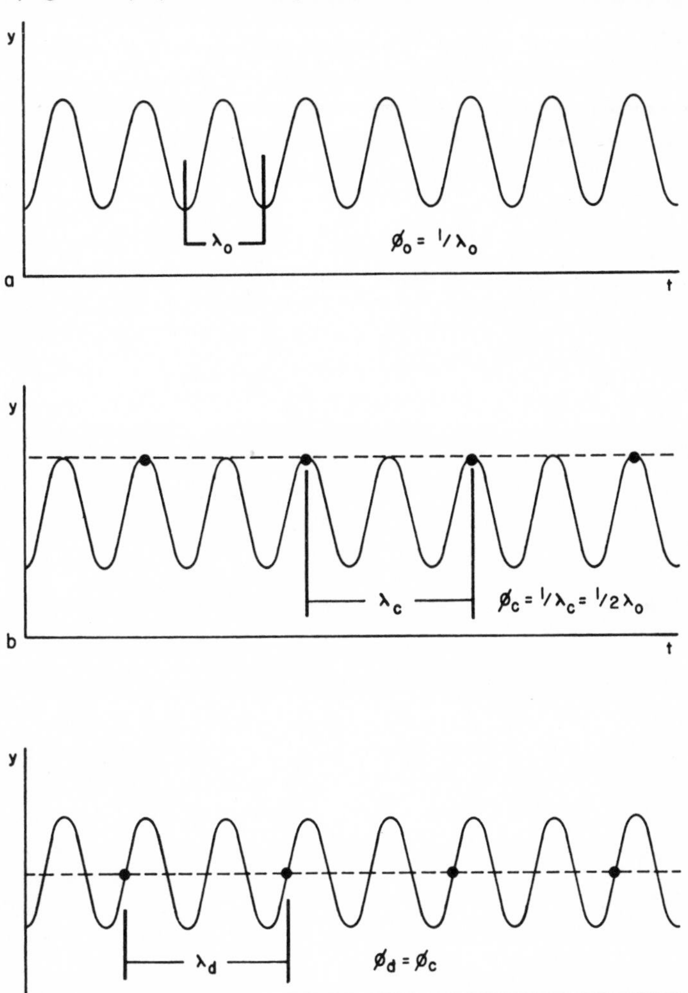

Fig. 15.4. (a) A sinusoidal function, y, with characteristic frequency ϕ_0. (b) The alias (dashed line) created by sampling at frequency ϕ_c. (c) Another alias (dashed line) created by sampling at frequency $\phi_d = \phi_c$.

the length of the time series.) In this case the Nyquist theorem states that the function, f, can be reconstructed from samples taken not more than $1/2\ \phi_1$ apart, that is, two samples per wavelength at the highest frequency component of the Fourier series. The results of the theorem can be interpreted in another way. If one wishes to recognize the presence of a component of a function of frequency ϕ_0, one must take samples at a frequency of at least $2\phi_0$. This condition is necessary, but not sufficient, however, because if the function is not band-limited at ϕ_0 and above, spurious results are likely to occur from an effect known as *aliasing*. Consider the situation shown in Figure 15.3 in which a function which varies at frequency ϕ_0 is sampled at lower frequencies ϕ_a and ϕ_b. The three samples indicate the presence of curves at different frequencies (dashed lines), although none exists. In the two special cases (Fig. 15.4) of the harmonic frequency $\phi_c = \phi_d = \phi_0/2$ particularly curious results can occur.

It should also be clear that although we have been discussing functions in the time domain, in deference to convention in spectral analysis, the only requirement on the independent variable is that it be monotonically increasing and that the function be continuous and single-valued. These criteria apply in the four dimensions of the population space (x, y, z, and t). The lesson is then that one must identify the highest frequency of variability, in the variables of interest, which *occurs* in each of the four dimensions and sample at twice that frequency. If that is impossible, one must accept the risk of aliasing which, as shown in Figures 15.3 and 15.4, usually results in an overly simplistic representation of the phenomenon. While simplicity may be appealing in science, accuracy is a high price to pay for it.

Condition 2

Cost-benefit considerations. In many oceanographic sampling problems, the investigator assumes that he is sampling from a population whose elements are distributed in the variable space, according to some frequency function, $f(a_i)$. This implies that in the population space the variable, a_i, can be described as a stochastic process relative to one or more coordinates. (A stochastic process is a set of random variables $\{x(t), t \epsilon T\}$ where T is the part of the time (or space) domain of interest.) For example, if the concentration of dissolved nitrate at some point in space (x_i, y_i, z_i) varies randomly, according to some probability distribution, in time (at least over the sampling period), or if the variable is distributed stochastically in the x, y plane over the distance of the positioning uncertainty in either dimension, the assumption would be applicable. If the assumption is made, some method of random sampling (in the dimension in which the stochasticity is assumed) is appropriate. The mechanics of random sampling are well presented elsewhere (Cochran 1963, Krumbein & Graybill 1965) and are reviewed below, and a number of considerations which dictate the minimum sample size required have been covered in the literature (Cochran 1963, McIntyre 1963, Kelley & McManus 1970, Kelley 1971a). Only the general principle of sample size determination will be reviewed here. In addition to the stochasticity which occurs in nature, the collection of samples and the measurement of variables introduces variability in the appearance of the variable in the sample. This variability is usually called 'error' and is usually assumed to occur randomly. (The stochastic variability in nature is often also called 'error', although it is a real characteristic of the phenomenon. As Prof. D. B. McIntyre (personal communication) points out, the use of the term 'error' for this variability raises theological questions which will not be addressed herein.) Variability is measured by the variance σ^2. When variances from several sources occur simultaneously in the determination of a particular variable they are said to be confounded and the variances are added together. That is,

$$\sigma_T^{\,2} = \alpha_1^{\,2}\,\sigma_1^{\,2} + \alpha_2^{\,2}\,\sigma_2^{\,2} + \ldots + \alpha_n^{\,2}\,\sigma_n^{\,2} \qquad (15.3)$$

where $\sigma_T^{\,2}$ is the total variance, $\sigma_j^{\,2}$ $(j = 1, 2, \ldots n)$ are the variances introduced by n sources and the a_j's are proportionality constants, where

$$y = \alpha_1\,x_1 + \alpha_2 x_2 + \ldots + \alpha_n x_n \qquad (15.4)$$

and y is the measured variable whose variance is $\sigma_T^{\,2}$ and the x_j's are the effects of the n sources of variability. One result of equation (15.3) which is of particular interest is the situation in which replicate samples, x_j, are used to estimate the mean, μ, of a population distribution. Recalling that μ is estimated by $\bar{x}$ and that

$$\bar{x} = \sum_{j=1}^{n} \frac{x_j}{n} \qquad (15.5)$$

for a typical sample mean, we can see that $a_j = \frac{1}{n}$ $(j = 1, 2, \ldots, n)$. Then, the variance of the sample mean, $s_{\bar{x}}^2$, is estimated by

$$s_{\bar{x}}^2 = \sum_{j=1}^{n} \frac{1}{n^2}\,s_j^2 = \frac{1}{n^2} \sum_{j=1}^{n} s^2 \qquad (15.6)$$

where the $s_j^2 = s^2$ $(j = 1, 2, \ldots, n)$ the sample variance. Or,

$$s_{\bar{x}}^2 = \frac{s^2}{n} \qquad (15.7)$$

which we recognize as the equation for the standard error of the sample mean. It is also of interest to note as a result of the Central Limit Theorem (Freund 1962) the distribution of the variable, y, in equation (15.4) will be the Normal Distribution whatever the distributions of the x_j's provided that $n > 1$.

The result in equation (15.7) is interesting because it shows that the population mean μ can be estimated arbitrarily closely ($\pm 2s_{\bar{x}}$ with 95 per cent confidence) by expending a particular amount of effort (i.e. taking n samples). Consider a population, normally distributed, with mean μ and variance σ^2. The cost of sampling this population is a function of n, the sample size, and the benefit received is a function of $\frac{1}{n}$, in terms of reduction in the standard error which results from increasing n. Figure 15.5 shows the relationship between $\frac{1}{n}$ and n. The ordinate $\{\frac{1}{n}\}$ is the benefit axis and the abscissa is the cost axis. It is obvious that for low values of n, great benefits are realized by small increases in the sample size, but that there is a point of diminishing returns in the region $n = 30$ beyond which very large costs must be incurred to achieve small benefits. If, in a given context, the level of precision required to solve the problem is well beyond the point of diminishing returns, it may be useful to question the wisdom of embarking toward the solution in the first place. At a minimum, the cost-benefit analysis should be considered. For problems of hierarchical sources of variability see, for

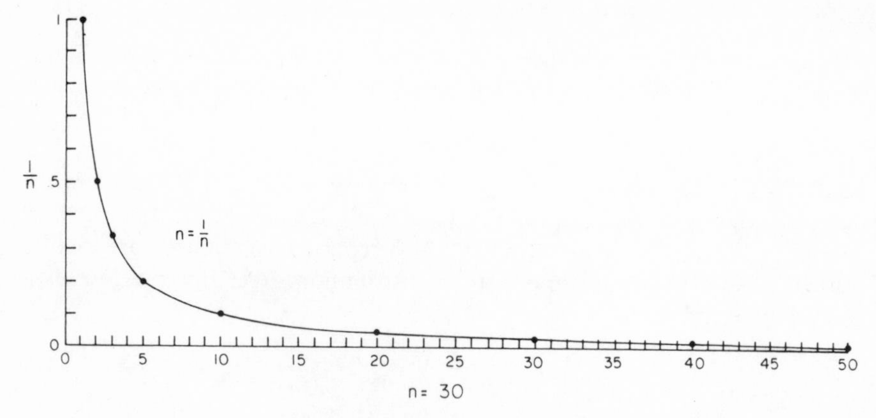

Fig. 15.5. The relationship $n = \frac{1}{n}$.

example, Kelley and McManus (1970) and for multivariate optimization, Kelley (1971a).

Distribution of discrete samples

In the previous section we discussed the problem of random sampling for a variable in the variable space. Let us then consider the distribution of samples in the population space. This problem is of critical importance because it is here that the technological aspects of sampling very often dictate the properties of the sample. In general, the central criterion for a good sample is that it be representative of the population. In random sampling, this is often stated in a slightly different way: every element in the population must have a known chance of appearing in the sample (Cochran 1963). A number of writers have addressed the random sampling problems (Cochran 1963, Krumbein & Graybill 1965, Sokal & Rohlf 1969), but it may be helpful to cover them quickly here.

Random sampling techniques

Let us briefly review three commonly used random sampling techniques. These techniques are called probability sampling or statistical sampling, as compared with purposeful sampling methods used to find a particular result of interest, and are distinguished by the fact that they include a randomization procedure.

In simple random sampling a sample of size n is drawn from a population of size N. Typically n random numbers are chosen from the uniform distribution on the interval between 1 and N. All samples of size n have an equal chance of being chosen. Simple random sampling has the advantage that it is conceptually straightforward, but in practice it is sometimes difficult to

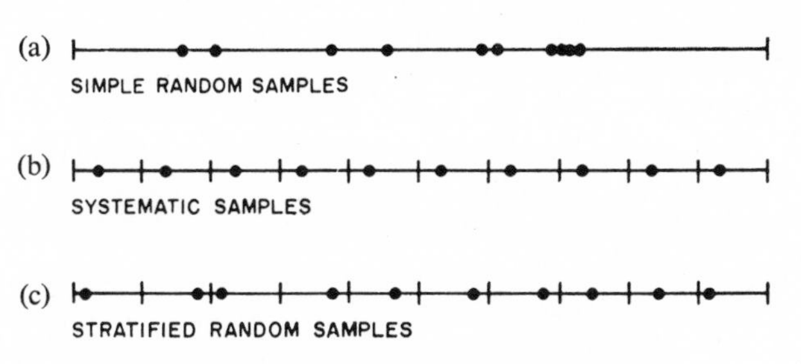

Fig. 15.6. Random sampling techniques.

execute properly and often the vagaries of random numbers cause some samples to be clumped together while large parts of the population are not sampled at all (Fig. 15.6a) (Krumbein & Grayhill 1965).

There are two common ways to improve on the distribution problems which typically occur in simple random sampling. Both begin with the investigator deciding that the population can be divided into sections, from each of which at least one sample is to be drawn. This decision may be not more than saying that the n samples should be evenly distributed over the population. This implies that every kth sample will be drawn, where $k = n/N$. The population is enumerated, divided into n segments of k samples each and one random number between 1 and k is chosen. Thereafter, samples are chosen at every kth interval. This technique is called systematic sampling. It has the advantage that samples are evenly distributed over the population and that it is easy to carry out correctly in practice (Fig. 15.6b). Potential problems can arise when systematic sampling is applied to a periodically varying function as a result of aliasing.

The second technique for ensuring that samples are spread more equitably over the population is to divide the population into strata or subpopulations, from each of which at least one sample is to be drawn. The strata are typically unequal in size and are designated on the basis of information known *a priori*, before the experiment is initiated. Subpopulations may be separated because they are of different size, because they exhibit differing internal variability, or because sampling problems differ from one to the other. Within-strata samples are chosen by simple random sampling. This technique is called stratified random sampling and it has the advantage that it covers the population very efficiently because it can adjust the sample size to the sampling problems in each subpopulation (Fig. 15.6c).

Adaptive and heuristic sampling

In some sampling problems, efficiency can be improved by taking advantage of recently acquired information on the phenomenon. Random statistical sampling is not intended to be blind, but only to be fair and unbiased. By observing the recent behaviour of the phenomenon it may be possible to decide that sampling frequency or sample size should be increased or decreased. The application of these considerations to modify the sampling design is called adaptive sampling. Sometimes it is possible to develop a simple rule or algorithm to control sampling frequency, for example, in an unattended sensor-recorder. When this rule is based on the real-time behaviour of the phenomenon and each point is accepted or rejected on the basis of the algorithm, the technique is called heuristic sampling. For example, an algorithm might test whether the next data point contributes new information to the sample already in hand and reject or accept it upon that basis, where 'new information' is operationally defined. Some examples of heuristic criteria are discussed by Kelley (1971b).

Areal sampling

Let us now address a particular aspect of random sampling, the areal distribution of samples. When sampling in the x, y plane, all the principles of one-dimensional random sampling apply, but are complicated both by operational constraints and by time dependency in the phenomenon itself. In general, the goal in areal sampling is to cover the x, y plane with a sufficient number of stations to provide a representative sample, in as short a time as possible. If the distribution of the variables of interest is essentially smooth and displays no preferred 'grain' in any direction, samples can simply be distributed uniformly, for example, on a regular grid over the area of interest until a sufficient number of samples has been acquired. Randomization can be achieved by choosing the origin of the grid at random and grid spacing can be chosen according to the Nyquist frequency. This scheme is an example of two-dimensional systematic sampling. It is also possible to use a two-dimensional simple random scheme, but operationally it is more complicated, especially if ship-handling is involved (certainly it is less psychologically appealing to ship operators), and there is little statistical or theoretical advantage to recommend it, except that it will avoid aliasing should the grid size turn out to be poorly chosen. Systematic sampling has the added advantage that it is simple to explain and execute correctly, whereas in simple random sampling, the pure caprice of the sample locations demands a good deal of dedication on the part of the sampler to faithfully proceed from one obviously totally 'arbitrary' location to the next. It is also important to remember that 'random', except in the limiting case, does not mean 'uniform'. A few points chosen randomly from the uniform distribution on the unit interval seldom cover the interval uniformly. As the sample size increases, $n \rightarrow \infty$, the coverage, by definition, approaches uniformity.

If some 'grain' is to be expected in the phenomenon (for example, on the continental shelf, we often expect longshore gradients to be lower than offshore gradients) the systematic sampling scheme can be modified by using a rectangular grid (see also Kelley & McManus 1970, for another variation). If the area of interest can be divided into sub-areas on the basis of criteria other than the measured variables, it may be appropriate to employ some stratified sampling scheme. The general rules for stratified sampling are that those sub-areas are sampled more heavily which are: (1) larger; (2) more internally variable; or (3) less expensive to sample (Cochran 1963). Within strata (sub-areas) either simple or systematic random sampling may be employed.

The simple fact that it is possible to specify an areal distribution of samples which satisfies the Nyquist constraint does not mean that operationally everything will fall into place. The sampling must still be completed in a period during which the phenomenon remains constant with respect to the measured variables. If the variables are sedimentary or deep ocean or, perhaps even benthic parameters, there may be little problem; but in the upper levels of the ocean and in estuaries and coastal regions where major changes may occur in phenomenon on a scale of hours or days, we must seek sampling techniques which more nearly satisfy the synopticity assumption.

In order to examine the implications of these sampling considerations in a marine ecological study, we can refer to a study carried out in a coastal upwelling cruise off the coast of Oregon (Kelley 1975). Considerable biological interest has been directed to experiments on phytoplankton sampled near the bottom of the euphotic zone (1 per cent light level) (see Chapters 4, 7, and 8). The depth of the one per cent light level is determined by Secchi disc or light meter, but the water found at a particular time at that depth may, itself, have experienced quite large vertical displacements due to internal waves and tides. Figure 15.7 from Kelley (1975) shows the vertical variability in σ_t during a 25-hour time series station on the Oregon continental shelf. Figure 15.8a presents a long-term temperature time series from a depth of 21·6 m at a moored buoy in the same location as the vertical time-series (Halpern *et al.* 1974). Figure

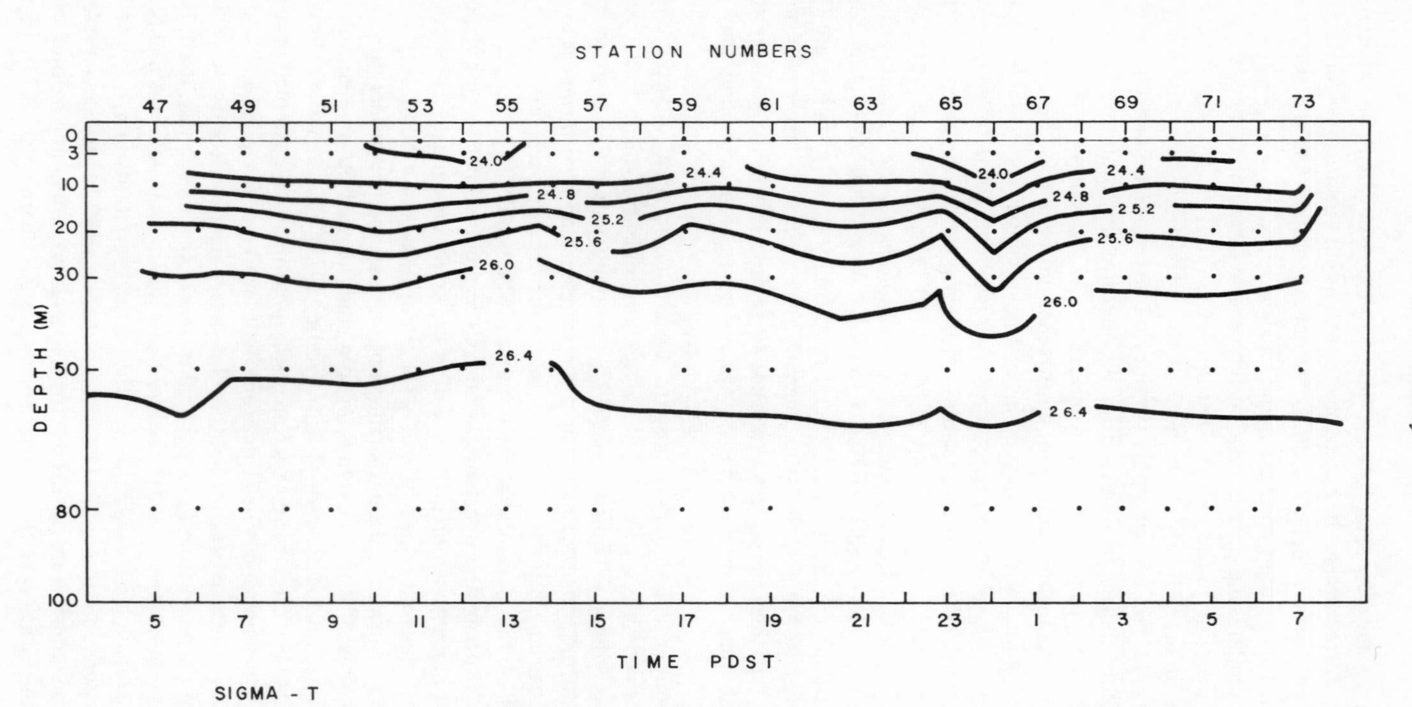

Fig. 15.7. Vertical time series of σ_t during a 25-h time-series station on 21–2 July 1973 at buoy *B* off the Oregon coast.

15.8b shows the spectrum calculated from the temperature time series. The time series data were taken every 3·75 min. The spectrum appears to be band-limited at approximately 1 cycle/h. Significant energy in the spectrum occurs at frequencies less than about 0·12 cycles/h. These frequencies correspond to periods of one and eight hours. The Nyquist criterion would, therefore, require

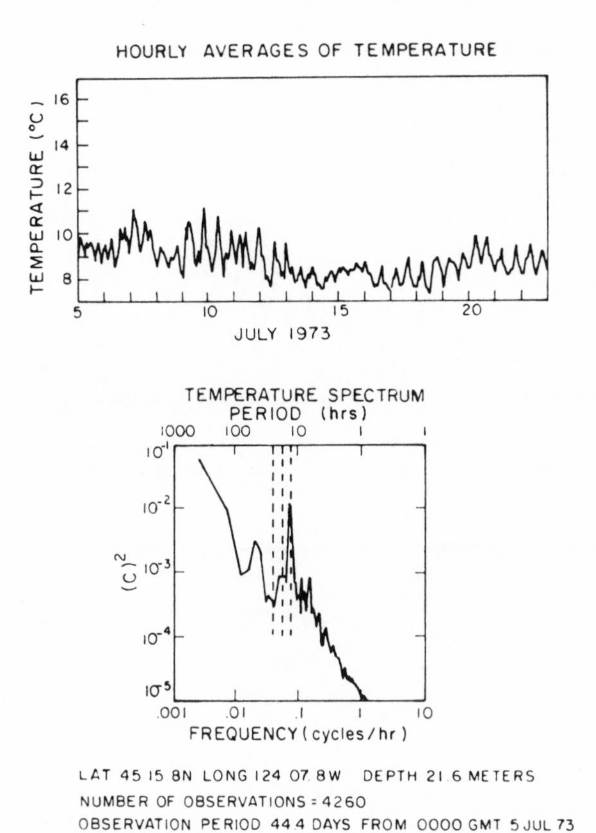

Fig. 15.8. 44-day time series and spectrum of temperature at 21·6 m at buoy *B* off the Oregon coast.

samples to be taken at least every four hours to adequately sample the observed variability below 0·12 cycles/h.

In space, the density of samples required to capture the variability is more complicated to estimate. In the example shown above, a current meter at 20 m (Pillsbury *et al.* 1974) suggested an average current speed of 27·8 cm/sec in a direction of approximately 265°T. Extrapolating the four-hour time figure to the horizontal (based on the displacement of a parcel of water in four hours by this current) suggests a sampling distance of 4 km. This would provide spatial density commensurate with the temporal density (remember that this is a fairly large current speed, which decreases the sampling density). Thus, an 'adequate' sampling scheme would require stations taken every four hours on all points on a 4 km grid. For an area 20 km on a side (certainly not a large area by oceanographic standards) this is equivalent to 150 stations a

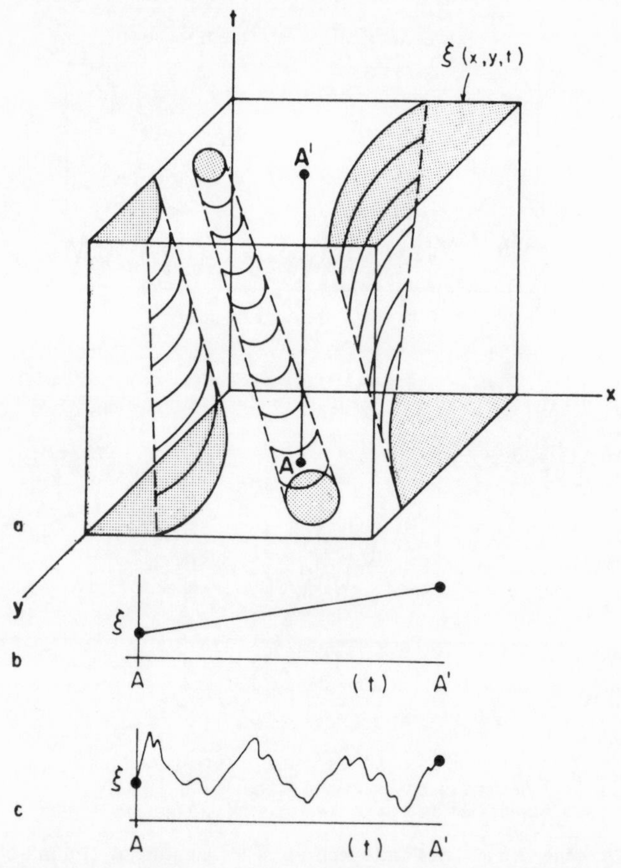

Fig. 15.9. (a) The sea surface distribution of some variable, ξ, contoured (stippling) on the x–y plane at two points in time. Dashed lines indicate a linear interpolation between the two planes. (b) A simple linear trend in ξ between points A and A'. (c) A more complicated path which ξ might follow between A and A'.

day! Even if ships could take an average of one station every hour, the experiment would require at least five ships. In order to insure against bias, the stations would have to be allocated randomly among the ships. The result would be five ships steaming around in a 20 km square between random locations, an arrangement hardly designed to win the admiration of the ships' operating personnel.

Of course, the experiment outlined above is fanciful, but the exercise emphasizes the disparity between oceanographic practice and the level of sampling which is desirable. It is always useful to carry out such an exercise to clarify the simplifying assumptions implicit in the sampling scheme actually used.

Synoptic and quasi-synoptic sampling techniques

An areal sampling technique which satisfies the synopticity assumption must be rapid enough to sample from the entire area (or volume) of interest before any significant change occurs in the phenomenon. Consider the situation

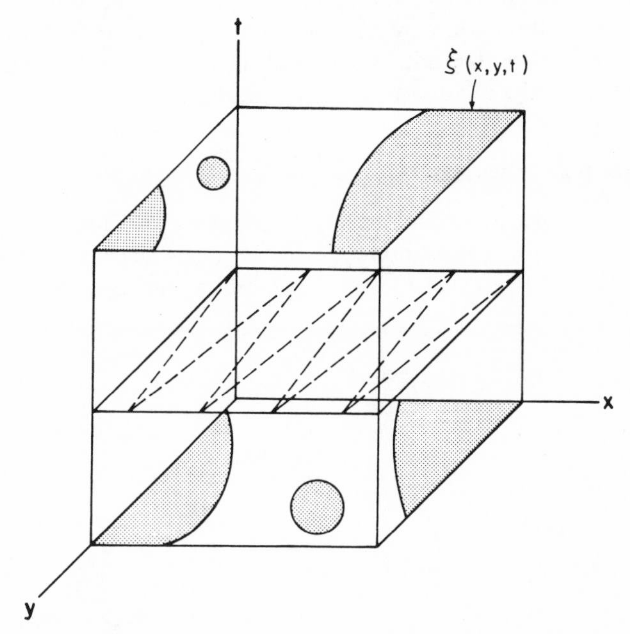

Fig. 15.10. A synoptic sample (horizontal plane) in the sampling space for ξ.

shown in Figure 15.9a in which the distribution of some variable, ξ, measured on a single plane (e.g. sea surface temperature) is shown to be slowly varying in time. Such a change could be simple, for example, linear drift due to solar heating or horizontal advection of surface water across the plane, or it might be complex, resulting from high frequency internal waves or the development of biological patchiness such as phytoplankton blooms or fish schools (Fig. 15.9b, c). The volume represents the target population and samples will be taken at specific points within it. To be perfectly synoptic, the samples should be synchronous; all the sampled points should lie in a horizontal plane

(Fig. 15.10). Some types of remote sensing techniques meet this criterion quite well.

High-altitude remote sensing

A single infra-red image from a satellite or high-altitude aircraft represents a fairly dense matrix of sample points, the loci of which lie on a single time plane. While a photograph represents a truly horizontal plane (normal to the time axis), an image obtained by a satellite scanner is on a slightly inclined plane because it is a single sensor moving through time, albeit for a very short interval. Although these techniques satisfy the synopticity assumption, their coarse horizontal resolution and problems of precision and accuracy of measurement by the sensors makes their application of limited usefulness at present. In the future they will undoubtedly come into their own as essential and routine oceanographic measurements (Szekielda 1972, 1973a, b).

Low-altitude remote sensing

Another remote-sensing technique which samples on a more highly inclined plane (Fig. 15.11), but with improved resolution is the use of low-altitude aircraft (<1000) ft elevation). Instrumentation has been developed and

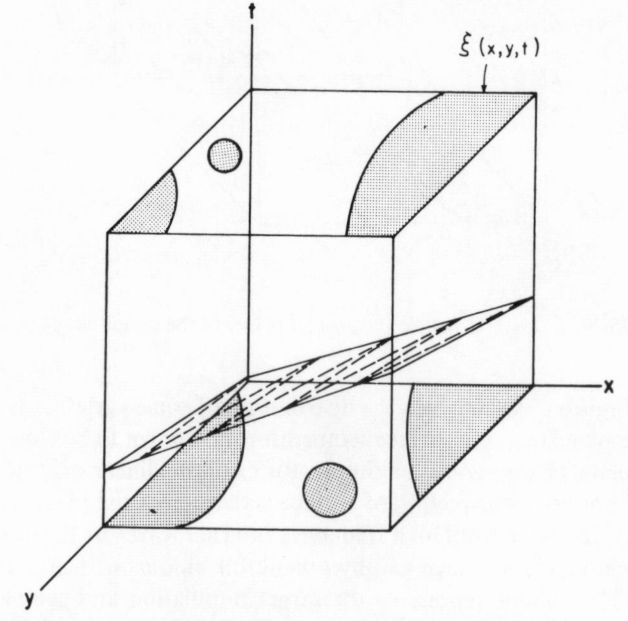

Fig. 15.11. A nearly synoptic sample (slightly inclined plane) in the sampling space for ξ.

successfully used on a number of field experiments for the measurement of sea surface temperature and chlorophyll content of sea water (Arvesen *et al.* 1971, 1973). In addition, photographs taken obliquely or vertically from low altitude aircraft are useful synoptic 'measurements' of visible surface features (slicks, fronts, fish schools, colour boundaries, etc.) and infra-red radiation. An example of a sea-surface temperature map produced from aircraft data is shown in Figure 15.12. (O'Brien 1972).

Shipboard discrete station sampling

Although both high- and low-altitude remote-sensing techniques provide very nearly synoptic representations of sea-surface data, they are limited in the number of variables that can be measured by the technological state-of-the-art. Many of the variables of interest in oceanographic research can at present only be measured by *in situ* instrumentation or by analysis of sampled sea water. Discrete sea water samples, for example, from hydrographic casts, provide a minimum of synopticity because of the ship handling involved, and the time required to complete the cast. Simultaneous casts made by different ships do, of course, yield synoptic data, but the cost is quite high. Shipborne *in situ* sensors, such as STD or CTD profilers and bathythermographs, shorten the station time (or eliminate it altogether in the case of XBT's). Nearly synoptic three-dimensional representations are possible with multi-ship surveys as was seen for example in the early Operation Cabot (Fuglister & Worthington 1949) and recent CUE II experiments (Smith 1974, Halpern 1974). Vertical profiling systems are still largely limited by technological development to the measurement of temperature, pressure and conductivity of sea water. The development of specific ion electrodes to determine a number of the dissolved constituents of sea water, as well as the development of new *in situ* instrumentation for measuring particulate matter will eventually greatly broaden the applications of profiling systems, but at the current rate of sensor development it will be a few years before routinely useable and reliable profiling systems for sea-water chemistry and particulates become commercially available. For the present, *in situ* pumping systems are in use (Curl & Becker 1972) but they do not represent a qualitative improvement over hydrographic cast technology. For analyses which require volumes of water, pumping systems provide a solution, but because of mixing problems in the hose, vertical resolution will usually be poorer than with hydrographic samples.

Moored arrays

One area of marine research which provides truly synoptic data is that of moored instrument arrays. An example is the moored current meter arrays

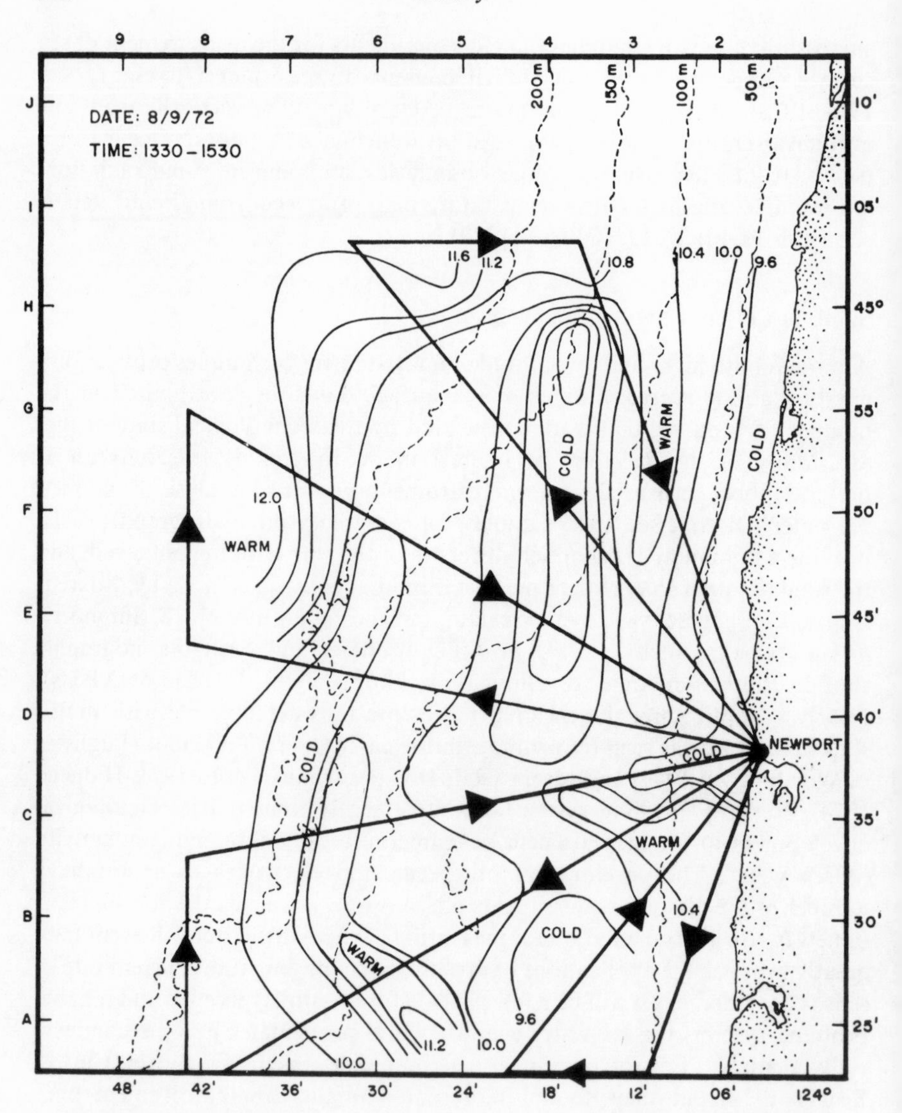

Fig. 15.12. A sea-surface temperature map taken off the Oregon coast from an aircraft.

used in typical circulation studies to resolve the advective field. A good example of such an array was that used in the CUE I experiment off the Oregon coast in 1972. The arrays were equipped with current meters, recording thermographs and meterological sensors and provided a truly synoptic picture of the shelf circulation (Pillsbury *et al.* 1974). The difficulty with moored arrays is that they provide very low horizontal resolution and they must be carefully designed to catch horizontal complexities in the field. Some thoughts on current meter array design have been presented by Mooers and Allen (1973). Most biological problems do not lend themselves to the moored array sampling methods, although devices such as sediment traps and even lobster traps could be considered moored arrays which give a synoptic determination of an integrated sample.

Continuous shipboard sampling

The area of greatest improvement in synopticity has been in the development of sampling systems which operate with ships underway. This field includes

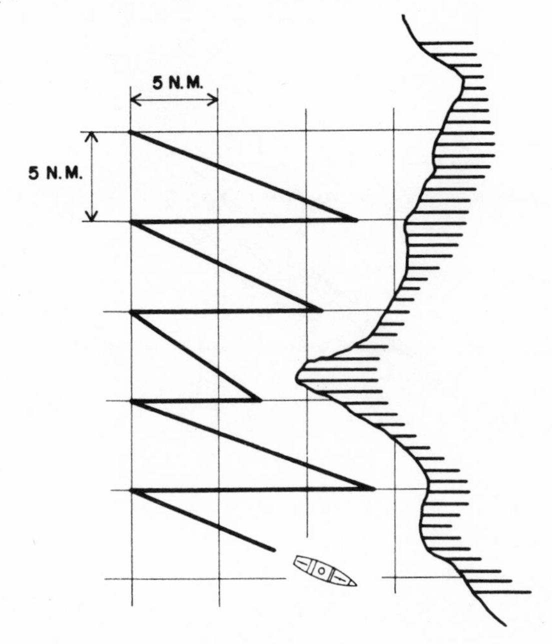

Fig. 15.13. Sawtooth mapping path for sea-surface measurements.

hull-mounted sensors and sampling systems as well as towed systems. Hull-mounted systems include acoustic transducers and hydrophones for depth measurement, fish location and sub-bottom profiling, many of which are taken for granted at present, but which represent a tremendous improvement over earlier measuring systems which required dedicated use of the ship. These measurements are now compatible with other ship uses and can be used on ships-of-opportunity. A similar development is in the area of hull-mounted pumping systems which provide a continuous stream of near-surface water to the ship's laboratories from a through-hull sea-water inlet. The sample stream can be sampled discretely or analysed continuously. For example, continuous records of nutrient chemistry or fluorescence can be obtained by AutoAnalyzer[R] and continuous flow fluorometry. Temperature and salinity can be measured continuously by sensors placed in the flow stream near the inlet. Any number of continuous flow cells can be developed for continuous measurement of pH, enzyme activity (T.T. Packard personal communication), and some attempts have been made at carrying out taxonomic identification in continuous flow systems (Margalef *et al.* 1966).

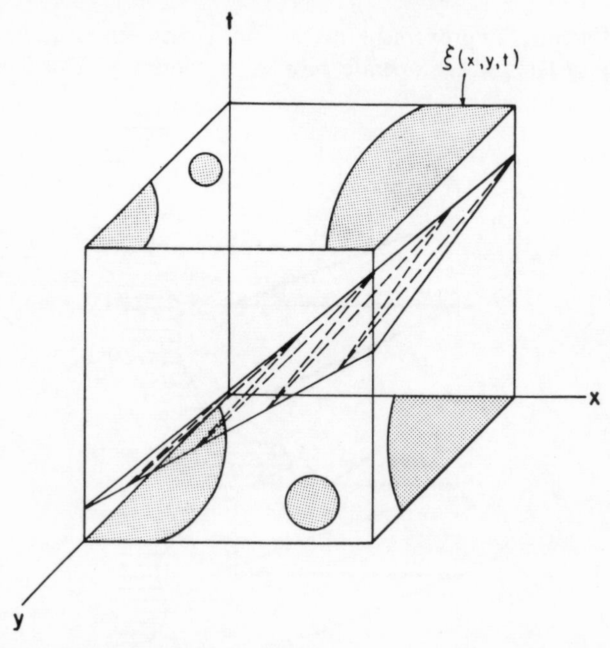

Fig. 15.14. A sample which, though continuous, requires considerable time (highly inclined plane) in the sampling space for ξ.

If the ship is deployed along a cruise track which provides fairly dense coverage of an area of the sea surface, data from continuous surface measurements can be combined to produce a surface map of the value of a variable of interest (Fig. 15.13). These maps are similar to those produced by remote sensing, but represent a much more highly inclined plane in the sampling space (Fig. 15.14). Although they are, therefore, less synoptic, they can be produced for a large number of variables, and the analyses can be carried out with shipboard laboratory instruments and intercalibrated directly with discrete sample analyses. Considerable experience has been accumulated in this mapping activity for coastal studies and some results are shown in Figure 15.15a, b, c. The data can be recorded graphically, but if a shipboard

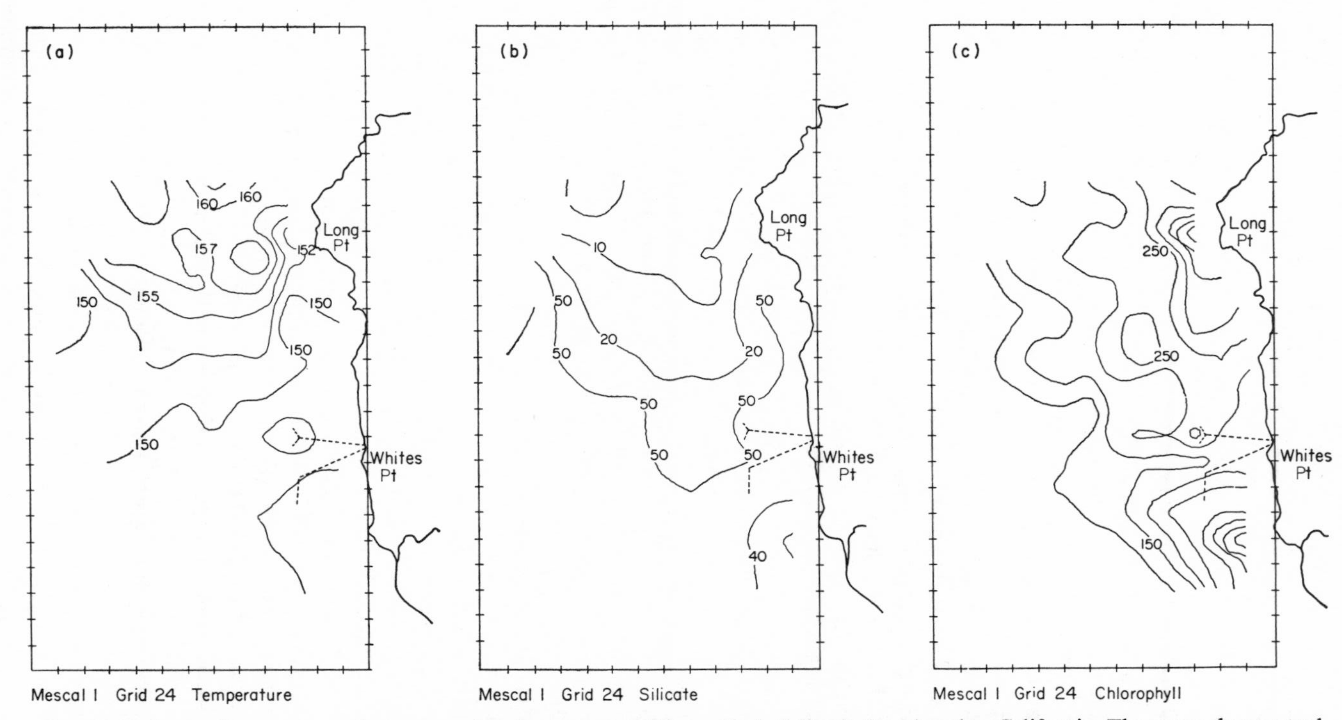

Fig. 15.15. Sea-surface maps made near a subsurface sewage diffuser (dashed lines), Los Angeles, California. The maps show actual computer output with the coast-line superimposed later. Tic marks on map border are 1 km apart. (a) Temperature ($\times 10°C$). (b) Silicate (μg-at/l). (c) Chlorophyll (arbitrary fluorescence units).

data acquisition system is available, they can be multiplexed or sequentially scanned, and recorded in analogue or digital form. The examples shown here were digitized and edited on-line with a shipboard computer system and the maps were produced by the computer in a short time after the data were accumulated. This technique allows the use of the maps to plan further sampling activities. This use of very recent information to adjust the sampling to the phenomenon is an example of adaptive sampling. It is important to recognize that if the sampling scheme is to retain random qualities, the information used adaptively must be used in the *a priori* sense, for example, to stratify the population. If the sampling plan is modified to search for a particular feature, which had been previously observed, the sampling activity becomes purposeful. Results from repeated purposeful samples cannot be compared or combined to further analyse the phenomenon because purposeful samples are by their nature biased.

Continuous analyses of surface water provide useful representation of the x, y fields at the depth of the sampling inlet, but information on the vertical distribution of the variables is still largely dependent upon hydrographic station samples which are typically separated in time by several hours from the surface map data. Research is underway in a number of laboratories to develop vertically profiling systems which will operate with the ship underway. Many studies involving towed sensor arrays taken to depth by passive depressors (weights, fins, etc.) have been carried out (O'Connell & Leong, 1963; Beers *et al.* 1967). For the measurement of variables for which *in situ* sensors have not been developed, research is underway in towed pumping systems using both active and passive depressors (Becker *et al.* 1973, Ballester *et al.* 1972).

In another area, an underway zooplankton counter has been developed by Boyd and Johnson (1969) in a towed fish which can also be equipped with other sensors whose signals are multiplexed and transmitted over a single conductor. This counter represents a very significant improvement over the earlier Longhurst-Hardy plankton recorder which collects zooplankton samples underway to provide a time history of the tow (Haury 1973).

Some interesting work is also underway in dye diffusion research which has led to the development of important new technological capabilities such as *in situ* fluorometry and which will certainly affect research in marine ecology (Ewart & Bendiner 1972, 1973; Ewart *et al.* 1972).

All of these techniques are developing rapidly and underway profiling in the vertical will become much more routine in the future, but the problem involves many difficulties which must be overcome. When the methods do become routinely used, especially as new specific ion and particle sizing technology is developed, they will introduce a change in oceanographic field operations as major as that which has resulted from the development of towed acoustic arrays and other sensors in marine geophysics. The developments in this area will make a nearly synoptic three-dimensional representation of the variable fields possible for the first time. The representation will provide new intuitive understanding of the spatial–temporal complexities and scales on which they dominate and will thus produce a dimensional framework in which to interpret results from discretely sampled data.

Summary

The ecological purpose of striving for synoptic representation of the marine environment is to understand the framework in which marine ecosystems operate. It is impossible to understand the dynamics of the ecosystem unless the variability in space can be separated from the temporal variability of the system. This separation can only be accomplished with synoptic sampling. Without synoptic samples, changes in the space and time domains are always confounded by problems such as aliasing. Separation of spatial and temporal variability and the development of synoptic three-dimensional field representations will represent a major improvement in marine ecological research and understanding.

Chapter 16. Models of the Sea

J. J. Walsh

Introduction

Modelling philosophy

An ecosystem model
Selection of state variables
Definition of important processes
Specification of the state equations

Model interpretation
Peru
Baja California
North-west Africa

Model implications

Conclusions

Appendix

Introduction

The *Challenger* era of oceanography (1872–6) initiated numerous observations on the winds, currents, sediments, chemistry, and organisms of the marine habitat. Attempts to synthesize these and subsequent data over the last hundred years have led to word models or conceptual hypotheses about the way measurable quantities interact in the sea to produce their observed distributions in space and time. As sophisticated instrumentation and analysis methods have replaced the crude techniques of this early expedition (Thompson 1878), more and more information has been generated about the oceans. Over the past century, these primitive word models have been slowly translated into mathematical models of varying complexity as further definition of our increased understanding of how natural systems behave.

These mathematical models were usually restricted to those individual components of marine systems with large data bases, and little effort was made to build comprehensive ecosystem models (Walsh 1972). The initial development of models of oceanographic phenomena thus tended to be along disciplinary lines, with the amount of theoretical work an inverse function of the number of variables measured. No attempt is made to review the literature here, but rather more recent studies are cited as examples of ongoing work, additional references being provided in previous chapters of this text. The meteorological (Manabe *et al.* 1970) and physical subsystems (Bryan

& Cox 1972) of the ocean are most strongly coupled to each other (Manabe & Bryan 1969) and such models do not depend on chemical, geological, or biological processes; the development of simulation has thus been most rapid in these independent studies of the physical variables of the marine habitat. Chemical and geological processes tend to be less complex than biological phenomena, and models of these subsystems are also fairly well developed (Keeling & Bolin 1967, Grill 1970, Harbaugh & Bonham-Carter 1970). Models of individual biological subsystems of varying conceptual isolation from the marine habitat have also been built for phytoplankton (Grenney *et al.* 1973), zooplankton (Sameoto 1971) and fisheries (Riffenburg 1969).

Patten (1968) has reviewed the early attempts to cross disciplinary lines with construction of plankton models involving various levels of physical and chemical sophistication. More recent mathematical analyses of estuarine (Takahashi *et al.* 1973), coastal (Walsh & Dugdale 1971), enclosed (Steele 1974a), and open ocean ecosystems (Vinogradov *et al.* 1972) suggest that inclusion of physical, chemical, and biological variables in an ecosystem model leads to reasonably realistic results in comparison with field observations. The development of these complex ecosystem models reflects the conviction of some scientists (Cushing 1971b, Walsh 1972, Ketchum 1972) that mathematical models are necessary as complex hypotheses to deal with the enormous amount of data generated by recent international and interdisciplinary studies of varying ecological emphasis such as the IIOE (International Indian Ocean Expedition), CalCOFI (California Cooperative Fisheries Investigations), IBP (International Biological Program), CINECA (Cooperative Investigations of the North East Central Atlantic), and CUEA (Coastal Upwelling Ecosystem Analysis) programs.

Modelling philosophy

Marine ecosystems are physically dominated phenomena in a chemical medium and attempts to understand the biological dynamics of these systems should reflect such properties of the marine habitat. Ultimately, the autecology of an organism or the synecology of a community can only be properly related to the environment with systems ecology as the eventual focal point of marine studies (Walsh 1972). The central role of an ecosystem model as a heuristic tool to focus data synthesis and acquisition has been discussed by Van Dyne (1966), Watt (1968), Patten (1971), Odum (1971), and Steele (1974a). At present models are more efficient in demonstrating what data are lacking and what processes are poorly understood than in predicting the future state of a system. The unambiguous, rigorous nature of a mathematical hypothesis is of immediate value, however, in the cyclic generation and testing of ecological hypotheses.

The level of resolution of a mathematical model depends on the types of questions to be solved and the data bank that is available, for no model is a perfect representation of the 'real world', assuming one can define the real world. The only one-to-one correspondence of the real world is the real world itself, and every theoretical construct is a homomorph, or a many-to-one correspondence with the real world. An implied ecological Heisenberg uncertainty principle in our measurements further suggests that not all states and transitions of biological variables in a marine ecosystem will ever be known (see Chapter 15). The detail of abstraction in the physical, chemical, and biological sub-models that constitute any model will thus depend on the choice of state variables, which are the subject of the analysis, and the processes necessary to describe the state variables in response to specific questions or hypotheses about the spatial and temporal properties of marine ecosystems.

It is relatively easy to qualitatively catch, catalogue, and preserve marine organisms. Studies of their phenotypic, and, more importantly, genetic variability (which leads to greater inherent variance of biological systems in contrast to physical or chemical ones) have placed much emphasis on the species concept. Species, however, may not always be appropriate theoretical units for an ecosystem context of biological organization. Most models involve, instead, the rather broad trophic level concept (Lindeman 1942) of ecosystem structure with the state variables chosen to be simple-minded producers, herbivores, carnivores, or detritivores, for example (see Chapters 4 & 14).

An intermediate abstraction of ecosystem structure can perhaps be obtained with selection of functional units of the dominant species as state variables at each trophic level, such as silica-requiring groups, vitamin B_{12} producers, spiny inedible organisms, fast or slow growers, and diurnal migrators. Recent studies of size distribution (Parsons & Takahashi 1973a) and prey size selection (Wilson 1973) suggest that size fractionation (see Chapter 4) may be an additional means of subdividing trophic level or grouping species level state variables in future models. The units of the biological variables are usually in terms of elemental composition or caloric value (Mann 1969) which smear over species variation to yield presumable properties of a system which have less variance than its component species populations (Mann 1975).

If the objectives of the model, its state variables, and the relationships of the physical, chemical, and biological processes linking each state variable can be defined, the mathematical formulation and solution of the problem is relatively simple. The methodology of modelling is only relatively simple, however, for there are many options. Depending upon the problem, one must decide whether steady state or time dependent studies (McNider & O'Brien 1973) with simulation and/or statistical models (Walsh 1971), involving the analytical or numerical solution (Acton 1970) of differential or difference equations (Innis 1972), made of linear or non-linear terms (Bledsoe 1975), should be solved on analog or digital computers (Ma *et al.* 1971) in relation to spatial or single-point considerations (Walsh 1975) and exogenous or endogenous driving functions (Kowal 1971) within a stochastic or deterministic parameter space (May 1973b). It is not the purpose of this chapter to define or discuss these approaches in any detail, and the reader is referred to the cited work; obviously selection is a function of each hypothesis as demonstrated in the numerous theories and equations of the preceding chapters. Instead, the approach utilized in a comparative study of upwelling ecosystems off Peru, Baja California, and north-west Africa will be used as an example of one of many alternative approaches to modelling processes in the sea.

An ecosystem model

Selection of state variables

Quantitative sampling of the higher trophic levels of upwelling or other marine ecosystems with traditional nets or trawls is a difficult task because of the mobility and avoidance behaviour of large organisms (see discussions in Chapter 12). More accurate determinations of nekton distribution with acoustic techniques (Blackburn & Thorne 1974) and of zooplankton distribution with towed particle counters (Boyd 1973) appear to be exciting developments which may soon provide a partial solution to this problem. Continuous methods of sampling have already been developed, however, for routine determination of the distributions of the chemical variables (Armstrong *et al.* 1967) and of the lower trophic levels (Lorenzen 1966) of marine ecosystems. A series of simulation models was thus developed, with respect to this large data base, to predict the spatial distributions of nutrients and phytoplankton as indices of our understanding of their interaction with the circulation and higher trophic levels of upwelling ecosystems. The model involves the state variables, light, nitrate, ammonia, phosphate, silicate, phytoplankton, zooplankton, anchoveta or red crabs, and detritus during those periods of the year when field data were available from the Peruvian, Baja California, and north-west African systems.

Definition of important processes

Differential heating and north–south alignment of the continents leads to eastern boundary currents of the oceans that flow equatorward along North and South America and Africa. These areas are the major upwelling regions of the oceans, constitute 0·1 per cent of the world's surface, and yield 50 per cent of the world's fish catches. The alongshore stress (τ_y) of the equatorward winds induces an acceleration of the surface currents, which drift offshore under the influence of the Coriolis force (f = the Coriolis parameter). This offshore, wind-induced or Ekman transport ($M_x = \tau_y/f$) requires that water is replaced in the surface flow along the coast (see Chapter 2 for more detail of the physical properties of upwelling). Subsurface water, rich in nutrients, thus rises or upwells to replace the water advected seaward; the spectacular biological productivity of upwelling systems results from this accelerated return of nutrients to the euphotic zone.

The three geographical cases of the upwelling model are steady state (time invariant) with respect to season but time varying over a diel period; Walsh (1975) describes the detailed assumptions of the Peru model, which are applied to the other two systems of this discussion as well. Although the field data were taken during March and April in each system, these months represent different temporal stages of upwelling in each area. The Peru ecosystem at 15° S was in a stage of weak offshore Ekman transport during the austral fall, with plume-like structure or localized upwelling observed along the coast (Walsh *et al.* 1971). The Baja California ecosystem at 27° N was in the initial or spin-up phase, from oceanic to upwelling conditions, with plume structure at the beginning of the study. As the winds continued to blow, however, a transition occurred from plume to parallel alongshore structure, indicating upwelling along most of the coast at the end of the study (Walsh *et al.* 1974). The north-west Africa ecosystem at 19° N was fully developed during these months with only parallel alongshore structure observed (Herbland *et al.* 1973).

Within the climatological regimes of each upwelling system, additional factors such as the local bottom topography and wind stress may determine the alongshore and offshore components of Ekman transport as approximately experienced by drogue studies in each of the *Anton Bruun* (Peru), *Thomas Thompson* (Baja California), and *Capricorne* (NW Africa) cruises. The offshore Ekman flow, in turn, effects the vertical velocity field, the duration of a phytoplankton population in the nearshore eutrophic area, the depth from which nutrients are upwelled, and consequently the nutrient concentrations available for phytoplankton growth. If nutrients, other essential elements, and light are not limiting at low latitudes, the daily phytoplankton productivity and their biomass are presumably a function of the grazing pressure as set by the number of food steps leading to the top carnivore.

The interaction of the system variables of Table 16.1 is formulated in a simulation model that attempts to predict the distributions of nutrients, phytoplankton, and detritus as dynamic variables which are a function of the velocity field, light, herbivores, and carnivores. The water parcels are assumed to rise from a depth of 50 m within 10 to 20 km of the coast, and to drift 100 km downstream along the axis of the main flow. Details of the boundary conditions, phytoplankton species composition, herbivore migratory patterns, and grazing stress formulations will be presented in relation to each geographical case. The discussion will be restricted to the distributions of

Table 16.1. System features of the Peru, Baja California, and north-west Africa upwelling ecosystems during March and April.

System variable	Peru (15° S)	Ecosystem Baja California (27° N)	North-west Africa (19° N)
Mean longshore wind (m/sec)	4·6	4·9	5·5
Mean shelf (200 m) width (km)	15	25	45
Mean drogue velocity (cm/sec) at 5–10 m depth	20	25	20
Mean light input (ly/h)	29(33)*	30(37)*	(46)*
Estimated vertical velocity (cm/sec)	1×10^{-2}	1×10^{-2}	1×10^{-2}
Mean nutrient concentration at 40–50 m depth:			
Nitrate (μgat/l)	23	20	21
Silicate (μgat/l)	18	20	14
Phosphate (μgat/l)	1·8	1·8	1·8
Mean chlorophyll a biomass (μg/l) over 0–20 m depth	6	7	12(2·3)†
Mean phytoplankton productivity (gC m^{-2} d^{-1})	5·6	7·1	2·3(3·1)†
Mean zooplankton biomass (mg dw/m^3)	41‡	60–115§	45–105¶
Estimated fish yield of total upwelling ecosystem (10^6 tons/yr) from Walsh (1972)	12·1	5·3	3·7

* Computed radiation from Kimball's (1928) tables.
† Four stations were occupied on a cruise of the Navire Océanographique *Jean Charcot* (MEDIPROD, 1973) during March–April 1971 along the same isobath off Cap Timris as the 1972 *Capricorne* study.
‡ Mean zooplankton biomass off Peru with a dry- to wet-weight conversion factor of 10 per cent.
§ A range in mean zooplankton biomass off Baja California in March–April 1972 and March–April 1973. Both the 1972 and 1973 estimates of zooplankton off Baja California are also computed with a dry- to wet-weight conversion factor of 10 per cent.
¶ A range in mean zooplankton biomass during the November–May 1971 upwelling season off Cape Blanc (Maigret 1972) and during February–May 1974 off Cape Corviero (Clutter personal communication); these values are estimated from zooplankton displacement volume as well, while *Capricorne* results are direct observations.

nitrate and phytoplankton nitrogen over 0 to 10 m of the two-dimensional *x–z* model; the reader is referred to Walsh (1975) for the subsurface and seasonal results of the Peru case.

Specification of the state equations

With the objective, state variables, and their spatial and temporal extent determined (i.e. specification of boundary conditions and interval of calculation) the model consists of the numerical solution on a digital computer of non-linear, second order, partial, differential equations of the form,

$$\frac{\partial NO_3}{\partial t} = -\frac{\partial uNO_3}{\partial x} - \frac{\partial wNO_3}{\partial z} + K_y \frac{\partial^2 NO_3}{\partial y^2} - \frac{V_m(NO_3)(P)}{(K_t + NO_3)} \quad (16.1)$$

for nitrate, where $V_m = (0{\cdot}11 - 0{\cdot}2\ NH_3)\ \sin\ (0{\cdot}2618\ t)$ and the other terms are defined in Table 16.2. This is an example of one of the nine state equations

Table 16.2. Definition of the state equation for nitrate distribution off Peru, Baja California, and north-west Africa

NO_3	is the nitrate concentration (μgat/l)
u	is the water velocity (15 to 20 cm/sec) in the x (downstream) direction
w	is the water velocity (1×10^{-3} to 1×10^{-2} cm/sec) in the z (vertical) direction
Ky	is the eddy coefficient (1×10^6 cm^2/sec) or parameterization of turbulent motion in the y (cross-stream) direction
V_m	is the specific uptake rate (h^{-1}) of nitrate by phytoplankton as a function of nitrate and ammonia
P	is the phytoplankton concentration in units of the limiting nutrient (μgat/l)
K_t	is the half-saturation constant (μgat/l) of the nutrient at which the uptake rate is half the maximal rate
NH_3	is the ammonia concentration (μgat/l)
sin	is the periodic function that allows phytoplankton growth to be time varying on a daily period
t	is the cumulative time in the simulation

of the model, and details of the full set can be found in the appendix. Other examples of these types of equations can be found in either Walsh and Dugdale (1971) or Walsh and Howe (1975).

This model is a simulation model because it involves the numerical solution of differential equations expressing a causal relation (a network of flows that can be described with input–output fluxes) between state variables as a function of these and the abiotic variables (i.e. light, currents, salinity, and temperature). A statistical submodel is inevitably involved, however, because of the above algebraic, empirical expression of V_m for the ammonia inhibition of nitrate uptake. The above equation is non-linear because it involves cross products, such as $V_m \times NO_3 \times P$ to express the dependence of nitrate flux on both the available nutrients and phytoplankton (see Chapter 7). It is time-varying because a periodic forcing function (sine) induces a diel variation in phytoplankton growth. It is of second order because the second derivative of nitrate is used to express the concept of turbulent diffusion of this quantity in the sea. It is solved numerically because most non-linear equations cannot be solved analytically. The above equation is actually written and solved in finite difference form of the partial differential equation because digital computers solve such equations with algorithms (computation instructions to the computer) that approximate the differential in finite increments. The digital computer is used because a limited number of multipliers are available on analog computers to express cross-products, and the storage capacity of analog computers to handle large sets of equations at a number of grid points in space is very limited.

Partial difference equations are involved because the model considers the change of each state variable with respect to the three spatial dimensions as well as time. Space must be considered in the analysis because coastal regimes are not homogeneous, i.e. every m^3 is not representative of the next one. Spatial heterogeneity or patchiness is thought to be one of the stabilizing control functions in marine ecosystems (Walsh 1972) and may explain shifts in ecological efficiencies of these systems (Walsh 1975)—recall the discussions of Chapter 5. Light, for example, is considered to be a diel forcing function of the model (i.e. an exterior input to the system) that is modified with depth and distance off-shore by phytoplankton biomass. The changing spatial boundary conditions of the nutrients at depth and distance offshore are also considered to be exterior forcing functions. Moreover, multiple nutrient regulation by nitrogen, phosphorous, or silica at each hour's time-step (interval of computation) in the interior of the Peru model allows the phytoplankton standing crops to be under nitrogen limitation inshore and silica limitation offshore. Similarly, a herbivore grazing stress, with a positive threshold, is used to simulate sub-grid scale patchiness of the phytoplankton with relatively large grazing losses offshore and small losses inshore. Finally, the herbivores are assumed to have a diurnal vertical migration (see Chapter 6) with a day distribution over 30 to 60 m and a night distribution over 0 to 30 m.

Model interpretation

The general model was constructed with estimates of the circulation field, the nutrient uptake by phytoplankton, and the diel migration, grazing stress, and excretion of the herbivores that are independent of the data sets used to test the cases for Peru, Baja California, and north-west Africa. The applicability or 'validity' of the basic hypotheses can be qualitatively examined by comparing the observed and predicted state variables of each system. The present deterministic analysis is considered mainly to be of heuristic value; however, no goodness of fit criteria have been applied, and no attempt has been made to distribute random error within components of the model.

Peru

The Peru case is assumed to be directed along the 0 to 10 m axis of a eutrophic plume surrounded by an oligotrophic water column down to 50 m depth (recall the temporal stages of upwelling described for each system). A return onshore flow is assumed below the 20 m Ekman layer (depth of the wind drift) as a result of current meter observations and continuity calculations. The predicted nitrate and phytoplankton nitrogen (line segments connected by crosses in Fig. 16.1) match the observed nitrate and particulate nitrogen

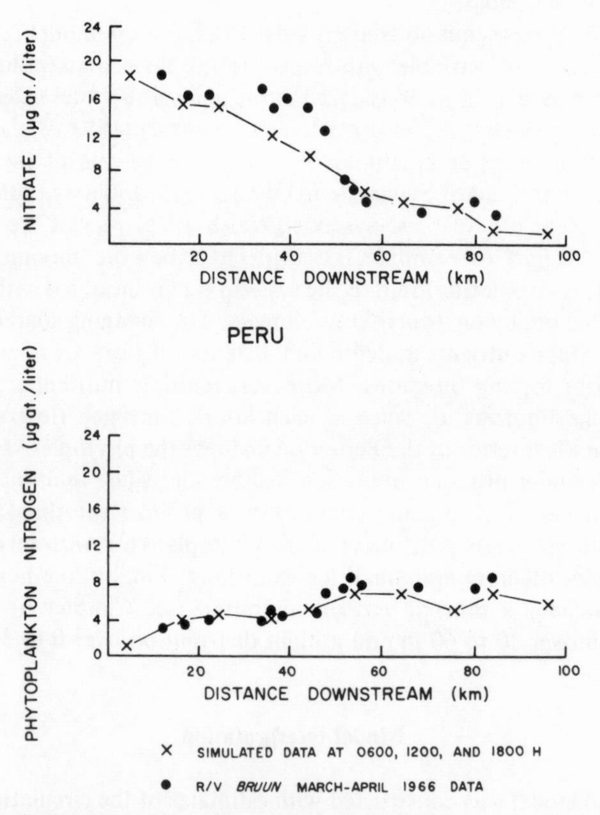

Fig. 16.1. Comparison of the observed and simulated nitrate and phytoplankton nitrogen over 0–10 m in the Peru upwelling ecosystem.

levels of the Peru ecosystem fairly well, where particulate nitrogen has been transformed from particulate phosphorous (PN = 16 PP). The simulated data are taken from the model at the same time of day as the drogue observations (solid circles) made during the 1966 R/V *Anton Bruun* cruise. Anchoveta and zooplankton are both assumed to be herbivorous at this time of year off

Peru, and the phytoplankton are assumed to be diatomaceous. Walsh (1975) discusses in detail the Peru model's suggestions of diel periodicity, of silica limitation at the offshore ecotone, of excretory recycling of nutrients at point sources by migratory herbivores, and of a seasonal shift in the functional role of anchoveta from that of herbivore to carnivore as a function of the intensity of upwelling and of the patchiness of food availability.

Baja California

The Baja California case assumed the same circulation pattern, upwelling velocity, nutrient input, phytoplankton growth rate, nutrient regeneration rate, and grazing stress as the Peru case. A transition from plume to more alongshore structure was observed, however, and it is assumed that, in contrast to Peru, the oligotrophic surrounding water only extends to the 20 m

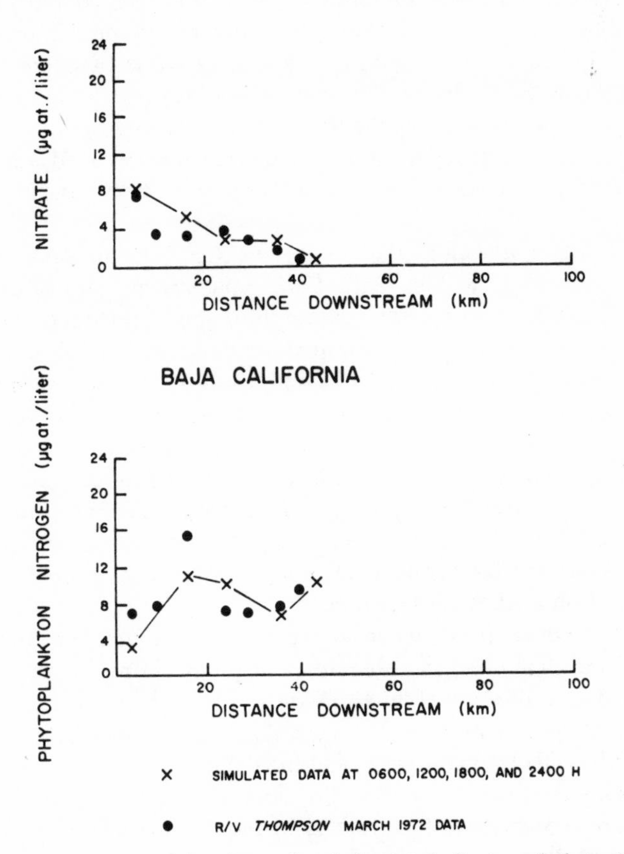

Fig. 16.2. Comparison of the observed and simulated nitrate and phytoplankton nitrogen over 0–10 m in the Baja California upwelling ecosystem.

depth of the Ekman layer. Below this depth, no variation is assumed in the upwelled input along the coast. An additional assumption is made about the functional composition of the phytoplankton. A bloom of *Gonyaulax polyedra* (Stein), a red tide dinoflagellate, was observed to dominate the phytoplankton during the 1972 R/V *Thomas Thompson* cruise. These organisms exhibited a diel migration to possibly obtain nutrients or avoid grazing at depth during the night (Walsh *et al.* 1974); the migratory pattern was simulated by deleting light regulation at depth, allowing dinoflagellates to accumulate in the aphotic layer of the model.

The predicted data are taken from the Baja California simulation at the same time of day as the *Thompson* drogue observations. Diel migration of dinoflagellates is not an explicit part of the model as yet, however, and thus the phytoplankton nitrogen over 0 to 10 m of the model (Fig. 16.2) is obtained by vertically integrating the algal standing crop (measured as particulate nitrogen) over 0 to 10 m at 2400 h, over 0 to 20 m at 0600 and 1800 h, and over 0 to 30 m at 1200 h. The varying depth intervals of integration during the day reflect the presumed migratory pattern of these dinoflagellates, with swarming populations observed in the surface layers at noon and a diffuse distribution found throughout the water column at midnight. The predicted values of nitrate at 0 to 10 m of the model are not summed, for nitrate is assumed to have a passive distribution (Fig. 16.2) in comparison with the dinoflagellates.

There is some question about applying the Peru model's subsurface, poleward circulation to the Baja California system and what nutrient anomalies such an undercurrent assumption might introduce in the Baja California model. Although no current meter data are available from the 1972 *Thompson* cruise, one might expect to find a subsurface, poleward baroclinic flow as a general phenomenon in upwelling systems off the west coast of North America (Hopkins 1974). Current meter data from a 1973 *Thompson* cruise to the same area off Baja California at the same time of year under much stronger winds, do in fact suggest a net subsurface baroclinic northerly flow, when the upwelling winds died and the dominant southerly barotropic flow dissipated; the California Undercurrent (Wooster & Jones 1970) may have come close to shore at this time.

Irrespective of return flow consideration for *prediction* of nutrients in the interior of the Baja case, the observed *and* predicted nitrate of the surface waters off Baja California (Fig. 16.2) are less than half those of Peru (Fig. 16.1). Yet the aphotic nutrient concentrations, the upwelled rate of nutrient input, the phytoplankton biomass, and the daily primary productivity are all approximately the same in the two ecosystems (Table 16.1). The match of the predicted and observed state variables in the surface waters off Baja California suggest that the additional assumptions of phytoplankton vertical migration, separate from the Peru case, may express essential functional

Table 16.3. Consequences of nutrient enrichment (upwelling) in hypothetical food chains of the Peru, Baja California, and north-west Africa ecosystems

System	Food chain (Number of Steps)	Plant biomass and productivity	
		Predicted	Observed
Peru, weak upwelling	Plants→Anchoveta, Zooplankton→ Top Carnivores (2)	high	high
Peru, strong upwelling	Plants→Zooplankton→Anchoveta→ Top Carnivores (3)	low	low
Baja California, early phase	Dinoflagellates→Zooplankton→Red Crabs (2)	high	high
Baja California, intermediate phase	Diatoms→Zooplankton, Red Crabs (1)	low	low
Baja California, mature phase	Diatoms→Zooplankton, Red Crabs→Top Carnivore (Tuna) (2)	high	(?)
North-west Africa	Diatoms→Zooplankton→Sardine→ Top Carnivores (3)	low	usually low

differences between a diatom and a dinoflagellate-dominated ecosystem. The presumed ability of dinoflagellates to intercept nitrate below the euphotic zone, in contrast to diatoms, is one hypothesis to be tested as a result of this model. Phytoplankton growth and nutrient utilization are also a function of the length of food chain; this aspect of the Baja California ecosystem will be discussed in relation to the implications of the model.

The small size of *Gonyaulax polyedra* off Baja California (30–40 microns) implies that the dominant facultative herbivore, *Pleuroncodes planipes* (Stimpson), may not have been able to graze effectively on these blooms, for the minimum prey size for this red crab is supposedly about 25 microns (Longhurst *et al.* 1967). Limited shipboard experiments also suggested that dinoflagellates were not being eaten by the red crabs (Walsh *et al.* 1974). Zooplankton grazing may have had some impact on the dinoflagellate blooms off Baja California, however, for the zooplankton biomass was observed to increase five-fold during the *Thompson* 1972 study and the computed nightly loss of dinoflagellates could be attributed to hypothetical ingestion rates of the dominant copepod, *Calanus helgolandicus* (Claus) (Walsh *et al.* 1974).

The use of the Peru grazing stress in the Baja California model and the resultant fit of the observed and predicted phytoplankton nitrogen in this second case further reinforces the concept of limiting grazing by nekton during this bloom off Baja California. The assumed dinoflagellate strategy of dispersal during the night leads to a phytoplankton biomass per unit volume in the model which, at most grid points, is lower than the assumed grazing threshold of the larger nocturnal herbivores; i.e. the model crabs ate no model algae in this situation. As in the Peru case (Walsh 1975), the assumption of a threshold in the nekton grazing of the Baja California model can be interpreted as an inability to directly simulate the patchiness or small particle size of the prey below a grid scale of 10 km. Copepods were observed to be readily eaten by the red crabs, during the 1972 *Thompson* cruise, and the model suggests that if the red crab population did obtain its daily ration in the water column (rather than on the bottom) the ration must consist of something other than dinoflagellates. These ideas are admittedly speculative, but the value of simulation analysis can again be seen in isolating testable hypotheses for future field studies.

North-west Africa

Application of the Peru model to the north-west Africa system leads to further questions about the comparative trophic structure of the three ecosystems. The only assumption of the north-west Africa case, as distinct from those made of Peru, concerns the structure of the alongshore velocity field. The upwelling zone is assumed to extend along the whole coast with no north–south gradient in the nutrient input. Without a lateral diffusive loss in the north-west Africa case, the predicted nitrate distribution (Fig. 16.3) is higher than that of Peru or Baja California and agrees with the observed values. The same criticism of questionable application of the Peru subsurface circulation to the north-west Africa case is valid, but, again, one might expect to find countercurrents in all the major upwelling areas (Smith 1968).

The phytoplankton in the north-west Africa study are assumed to be diatoms, and the predicted phytoplankton nitrogen agrees with the transformed ($PN = 0{\cdot}9 \times Chl\ a$) particulate nitrogen until a distance of 80 km downstream is reached in the model (Fig. 16.3). Anomalously high values of 40–50 mg chlorophyll a/m^3 were found downstream of Cap Timris, however, in comparison with the low productivity observed on the same stations of the 1972 *Capricorne* drogue study (Table 16.1). The low phytoplankton biomass and productivity found on the 1971 *Charcot* study of the same area at the same time of year (Table 16.1) are more similar to what one might expect from the low ^{14}C measurements made during the *Capricorne* study.

If the *Capricorne* data are correct, however, the Peru grazing stress is probably an over estimate of the north-west Africa system, for it does not allow enough downstream phytoplankton growth in the north-west Africa model. If the *Capricorne* phytoplankton biomass is spurious, and the actual algal standing crop is close to the *Charcot* data, then no grazing would occur in the north-west Africa model as the crop level would be below threshold. In either situation, the third case of the model suggests that the nekton grazing flux may not be as large off north-west Africa as it sometimes is in the Peru ecosystem. Our observations and the results of the Baja California model also

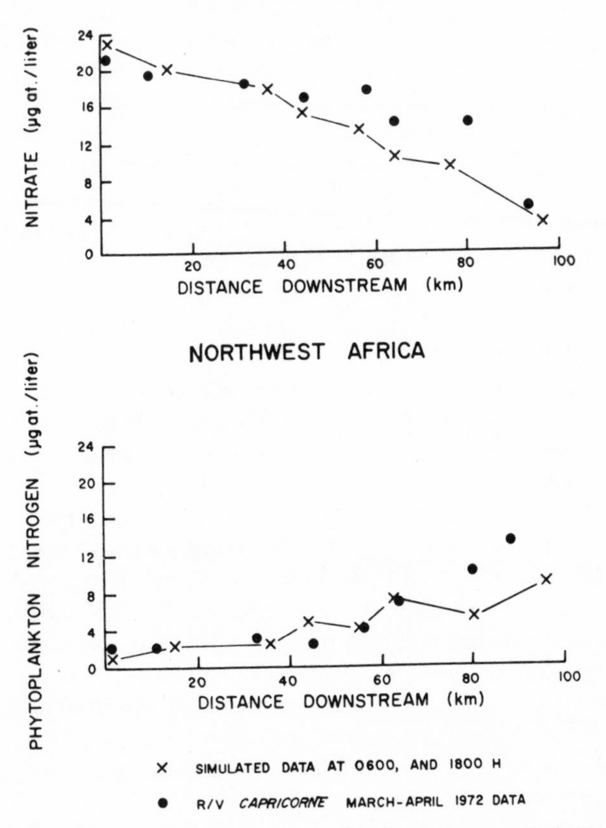

Fig. 16.3. Comparison of the observed and simulated nitrate and phytoplankton nitrogen over 0–10 m in the north-west Africa upwelling ecosystem.

suggest that the nekton grazing flux off Baja California may sometimes be lower than the Peru system.

Model implications

If the three cases of model are correct in suggesting that the nekton were not mainly phytophagous during the Baja California and north-west Africa drogue studies, one might further speculate as to why the trophic relations might be different among the three ecosystems. Upwelling occurs year round off Peru (Maeda & Kishimoto 1970, Wooster & Sievers 1970) in contrast to Baja California (Bakun *et al.* 1974) and north-west Africa (Wooster *et al.* in preparation); thus the upwelling environment is more persistent and more predictable off Peru than California or north-west Africa. Walsh (1972) has suggested that the evolution and yield of clupeids within an upwelling food chain might be related to the duration of intense nutrient input; the number of trophic levels in an upwelling ecosystem may also be related to the frequency of eutrophication.

The Peruvian anchoveta (*Engraulis ringens* (Jenyn)), the south-west African pilchard (*Sardinops ocellata* (Pappé)), and the Panamanian anchoveta (*Cetengraulis mysticetus* (Günther)) may be the only clupeids of upwelling ecosystems that are partly phytophagous (Longhurst 1971) during at least some period of the year (Walsh 1975). The south-west Africa ecosystem is similar to Peru and also has persistent upwelling year round (Wooster *et al.* in preparation). The seasonal stability of these two major systems, notwithstanding El Niño phenomena, may have allowed a phytophagous fish to evolve as a repeatable component of the pelagic community; the Panamanian anchoveta, for example, apparently switches to a detritivore niche on the bottom after upwelling ceases and the rainy season continues for six months in the Gulf of Panama (Smayda 1966).

O'Brien (personal communication) has observed that the farther north one examines climatological wind data along the west coast of North America, the more intermittent the Ekman transport becomes. Accordingly, if one proceeds north of the Gulf of Panama to Baja California, the phytophagous nekton niche is again only partially filled, and in this situation by the red crab (*Pleuroncodes planipes*), which evidently also feeds on the bottom when upwelling does not occur or is very weak (Boyd 1962). In addition, the red crab may only be phytophagous when large diatoms are present, while the northern California anchovy (*Engraulis mordax* (Girard)) appears to feed mainly on zooplankton (Longhurst 1971).

The north-west Africa sardines (*Sardinella eba* (Regan), *S. aurita* (Valenciennes)) are also predators (Longhurst *op. cit.*) and the estimated yield of all fish off north-west Africa is approximately one-third that of Peru (Table 16.1). The large demersal component of the north-west Africa fishery (Gulland 1970) and the food preferences of its clupeids imply a longer food chain in this system than Peru or south-west Africa. It is possible that an evolutionary strategy of forming a short food chain with phytophagous fish off north-west Africa is untenable; the more inefficient, but environmentally damped, three-step food chain, involving zooplanktivorous fish, may be of selective advantage in this more unpredictable habitat.

The length of the food chain can have profound effects on the composition and behaviour of an ecosystem. Field (Hurlbert *et al.* 1972) and model (Smith 1969, Caperon 1975) studies of eutrophication suggest that food chains of an odd number of steps generate low phytoplankton biomass and productivity, while food chains of an even number of steps lead to high phytoplankton biomass and productivity. If one assumes contiguous populations of plants, herbivores, and carnivores, the plants are held in check when only herbivores are present (odd number of steps), but the plants accumulate where carnivores reduce the number of herbivores (even number of steps). An estimate of the number of steps that might exist within food chains at different times of the year in each upwelling ecosystem is presented in Table 16.3.

During periods of strong upwelling off Peru, intermediate upwelling off Baja California, and well-developed upwelling off north-west Africa, food chains of an odd number of steps may occur (phytoplankton→zooplankton→nekton→terminal predator) and low phytoplankton productivity and biomass have been observed. With respect to the high chlorophyll reported from the *Capricorne* study, Lloyd (1971) observed off Cap Blanc, north-west Africa, a similar productivity of 2·08 gC m^{-2} d^{-1} *and* a low chlorophyll biomass of 4.14 mg/m^3. He also thought that these chlorophyll values might be too large as a result of extraction error. During periods of weak upwelling off Peru and Baja California, high productivity and biomass are observed, and there are suggestions of an even number of steps (phytoplankton→nekton→terminal predator) in these food chains.

Spatial properties of aquatic ecosystems complicate this simple-minded parity concept of trophic level interaction, however, for temporal or physical constraints may not allow predators access to their prey. Fertilization of a two-step food chain (phytoplankton→zooplankton→salmon) in Great Central Lake, Vancouver Island, did not immediately generate the expected phytoplankton bloom and eventual increase of salmon yield (Parsons *et al.* 1972, LeBrasseur & Kennedy 1972, Barraclough & Robinson 1972). Instead, low phytoplankton biomass and an increase of zooplankton biomass and productivity occurred after eutrophication. A combination of diel migration and low temperature preferences of the salmon was attributed as the mechanism for spatial isolation of this predator and its prey. Similarly, the increased length of the Peru food chain during strong upwelling (Walsh, 1975) and of the Baja California food chain at the onset of upwelling were both attributed to prey inaccessibility, expressed in the assumption of a grazing threshold in the models. When large diatoms are present and the tuna are unavailable to prey on the red crabs, because of warm temperature preferences of the top carnivore (Blackburn 1969), a one-step food chain with low phytoplankton productivity and biomass might be expected off Baja California (Walsh *et al.* 1974).

Conclusions

The above speculations are, of course, only hypotheses subject to revision after additional field and laboratory experiments, but they serve to illustrate the premise that sufficient data are now becoming available, on at least selected aquatic ecosystems, to consider modelling and eventually understanding them and their component parts. Some marine ecosystems such as upwelling regions or the North Sea are sufficiently restricted in space and time to almost be discrete, well-defined ecosystems. The gradients in weakly developed upwelling situations are sufficiently sharp, for example, that it is possible for fast, well-equipped ships to collect almost all of the data necessary for making and testing ecosystem models.

The 1970s might be considered as the beginning of the *challenged* era of oceanography with the realization that manipulation of non-renewable resources in territorial waters raises questions about the relevancy and the justification of marine science (Cadwalader 1973). Future prediction of hurricanes and associated damage to urban societies (Simpson 1973), of

Table 16.4. Definition of terms in the additional state equations of a model for the Peru, Baja California, and north-west Africa upwelling ecosystems

	The other variables
PO_4	= phosphate (μgat/l)
SiO_4	= silicate (μgat/l)
I_z	= ambient light (ly/h)
r	= extinction coefficient (m^{-1})
Z	= zooplankton nitrogen (μgat/l)
F	= fish nitrogen (μgat/l)
D	= detrital nitrogen (μgat/l)
	The other parameters
K_t	= NO_3 and NH_3 half saturation constant (1·5 μgat/l)
V_s	= maximum nutrient uptake rate (0·08 h^{-1}) on a 12-h-day basis
G_Z	= zooplankton maximum grazing rate (0·02 h^{-1}) on 12-h-night basis
G_F	= fish maximum grazing rate (0·005 h^{-1}) on a 12-h-night basis
K_f	= phytoplankton half saturation constant (1·5 μgat/l)
P_A	= zooplankton grazing threshold (0·5 μgat/l)
P_O	= fish or red crab grazing threshold (3·0 μgat/l)
K_p	= phosphate half saturation constant (0·25 μgat/l)
K_s	= silicate half-saturation constant (0·75 μgat/l)
K_I	= light half-saturation constant (10 per cent of I_O)
I_O	= incident surface radiation (29 ly/h)
w_s	= the sinking velocity of detritus ($1·5 \times 10^{-2}$ cm/sec for zooplankton fecal pellets and $1·5 \times 10^{-3}$ cm/sec for anchoveta pseudo-faeces)
	The boundary conditions (μgat/l)
Lower (50 m) $NO_3 = 33$, $NH_3 = 1$, $PO_4 = 2·5$, $SiO_4 = 25$, $P = 0·5$	
Upper (atmosphere) $NO_3 = NH_3 = PO_4 = SiO_4 = P = 0$	
Onshore (coast) $NO_3 = NH_3 = PO_4 = SiO_4 = P = 0$	
North, South, and Offshore (step function at 5, 15, 25, 35, and 45 m depths)	
NO_3	= 5, 10, 15, 20, 25
NH_3	= 0·5, 0·5, 0·5, 0·5, 0·5
PO_4	= 0·4, 0·8, 1·2, 1·6, 2·0
SiO_4	= 3, 4, 7, 10, 13
P	= 2·5, 2·0, 1·5, 1·0, 0·5

coupled changes in the atmosphere (Namias 1973) and in the currents (Wyrtki 1973) leading to failure of commercial fisheries (Idyll 1973), of available protein (Walsh & Howe 1975), and of the effects of eutrophication for mariculture, involving accelerated natural (Othmer & Roels 1973) or man-transformed (Goldman *et al.* 1973) nutrient input may all be possible goals.

It is rather dubious that mathematical models are a panacea to replace data collection. As suggested in Chapter 1, however, a science of the ecology of the sea has hopefully emerged and benefited from synthetic efforts over the last century to the extent that marine ecologists will more fully utilize mathematical analysis in the future as have their physical colleagues in the past. The diverse store of information presented in previous chapters of this text is an indication of the large data base that continues to be built about marine ecosystems. The *challenge* to future marine ecologists, however, is to stretch their ingenuity and devise new methods to exploit present data banks rather than to simply participate in additions to these banks.

Appendix

The state equations and definition of each variable, of the parameters, and of the boundary conditions of the basic model are presented below. Discussion of the state equation for nitrate has been presented previously in relation to Table 16.2 and will not be repeated here. The state equation for ammonia is,

$$\frac{\partial NH_3}{\partial t} = -\frac{\partial u NH_3}{\partial x} - \frac{\partial w NH_3}{\partial z} + Ky\frac{\partial^2 NH_3}{\partial y^2} - \frac{(V_s)(NH_3)(P)}{(K_t + NH_3)} + (0{\cdot}67)\frac{(G_Z(P-P_A)(Z))}{(K_f + (P-P_A))} + (0{\cdot}67)\frac{((G_F)(P-P_0)(F))}{(K_f + (P-P_0))}$$

where $V_s = (1{\cdot}11)\sin(0{\cdot}2168\,t)$

$$G_Z = \begin{cases} (0{\cdot}03)\cos(0{\cdot}2618\,t + 1{\cdot}571) & \text{if } z < 30 \text{ m} \\ (0{\cdot}03)\sin(0{\cdot}2618\,t) & \text{if } z > 30 \text{ m} \end{cases}$$

$$G_F = \begin{cases} (0{\cdot}008\cos(0{\cdot}2618\,t + 1{\cdot}571) & \text{if } z < 30 \text{ m} \\ (0{\cdot}008)\sin(0{\cdot}2618\,t) & \text{if } z > 30 \text{ m} \end{cases}$$

in which the terms are either defined in Table 16.2 or above in Table 16.4. The fifth and sixth terms of this equation are the excretory return of ammonia to the dissolved nitrogen pool. The periodic functions of the maximum zooplankton (G_Z) and anchoveta (G_F) grazing rates, above and below 30 m, are used to impose diurnal migration of herbivores in the model, leading to mainly a nocturnal grazing stress.

The state equation for silicate is

$$\frac{\partial \mathrm{SiO_4}}{\partial t} = -\frac{\partial u\mathrm{SiO_4}}{\partial x} - \frac{\partial w\mathrm{SiO_4}}{\partial z} + Ky\frac{\partial^2 \mathrm{SiO_4}}{\partial y^2} - \frac{(V_s)(\mathrm{SiO_4})(0{\cdot}67\,\mathrm{P})}{(K_s + \mathrm{SiO_4})}$$

and note that there is no excretory return of silica in this equation. This assumption leads to possible silica limitation in the Peru case (Walsh, (1975), but the implications of silica control are not considered in the Baja California and north-west Africa cases.

The state equation for phosphate is essentially the same as that for ammonia,

$$\frac{\partial \mathrm{PO_4}}{\partial t} = -\frac{\partial u\mathrm{PO_4}}{\partial x} - \frac{\partial w\mathrm{PO_4}}{\partial z} + $$
$$+ Ky\frac{\partial^2 \mathrm{PO_4}}{\partial y^2} - \frac{(V_s)(\mathrm{PO_4})(0{\cdot}067\,P)}{(Kp + \mathrm{PO_4})}$$
$$+ (0{\cdot}13)\frac{((G_Z)(P - P_A)(\mathrm{Z}))}{(K_f + (P - P_A))}$$
$$+ (0{\cdot}13)\frac{((G_F)(P - Po)(F))}{(K_f + (P - Po))}$$

with excretory recycling of phosphate as well. The state equation for phytoplankton nitrogen,

$$\frac{\partial P}{\partial t} = -\frac{\partial uP}{\partial x} - \frac{\partial wP}{\partial z}$$
$$+ Ky\frac{\partial^2 P}{\partial y^2} - \frac{(G_z)(P - P_A)(\mathrm{Z})}{K_f + (P - P_A)}$$
$$- \frac{(G_F)(P - Po)(F)}{K_f + (P - Po)} + VP$$

$$\text{where } V \text{ is the minimum of} \begin{cases} \dfrac{(V_m)(\mathrm{NO_3})}{K_t + \mathrm{NO_3}} + \dfrac{(V_s)(\mathrm{NH_3})}{K_t + \mathrm{NH_3}} \\ \text{or } \dfrac{(V_s)(\mathrm{PO_4})}{K_p + \mathrm{PO_4}} \\ \text{or } \dfrac{(V_s)(\mathrm{SiO_4})}{K_s + \mathrm{SiO_4}} \\ \text{or } \dfrac{(V_s)(I_z)}{K_I + I_z} \end{cases}$$

and where $I_z = I_O e^{-rz}$

with $r = 0{\cdot}16 + 0{\cdot}0053\,P + 0{\cdot}039\,P^{2/3}$

contains the grazing thresholds of zooplankton (P_A) and of anchoveta (P_O); phytoplankton growth as a function of multiple nutrient and light regulation in which the smallest rate is chosen in accordance with an hourly Liebig's law of the minimum; and light regulation with depth (I_z) as a function of a variable extinction coefficient (r) in relation to self-shading of the phytoplankton populations—recall Chapter 8.

The state equation for zooplankton nitrogen,

$$\frac{\partial Z}{\partial t} = (1 - 0{\cdot}67 - 0{\cdot}13 - 0{\cdot}20)\frac{((G_Z)(P - P_A)(Z))}{K_f + (P - P_A)}$$

and the state equation for anchoveta nitrogen,

$$\frac{\partial F}{\partial t} = (1 - 0{\cdot}67 - 0{\cdot}13 - 0{\cdot}20)\frac{((G_F)(P - P_O)(F))}{K_f + (P - P_O)}$$

do not allow temporal changes of these herbivores in the model because of the assumption that change in herbivore biomass is slow compared to phytoplankton variations over the 10-day time scale of the three cases—recall Chapters 9 and 11.

The state equation for detrital nitrogen,

$$\frac{\partial D}{\partial t} = -\frac{\partial uD}{\partial x} - \frac{\partial wD}{\partial z} - w_s\frac{\partial D}{\partial z} + \frac{(0{\cdot}2)(G_Z)(P - P_A)(Z)}{K_f + (P - P_A)} + \frac{(0{\cdot}2)(G_F)(P - P_O)(F)}{K_f + (P - P_O)}$$

assumes that detrital production is only a consequence of an 80 per cent herbivore assimilation efficiency during grazing. The input of detritus to the benthos (Chapter 10) of an upwelling system is not considered in this discussion.

References

ACTON F.S. (1970) *Numerical Methods that Work*. Harper & Row, New York, 541 pp.

ADACHI R. (1972) A taxonomical study of the red tide organisms. *J. Fac. Fish. Pref. Univ. Mie* **9,** 9–145.

ADAMS J.A. & STEELE J.H. (1966) Shipboard experiments on the feeding of *Calanus finmarchicus* (Gunnerus), pp. 19–35. In *Some Contemporary Studies on Marine Science*, ed. Barnes H. Allen & Unwin. London, 716 pp. Vol. I London, Low, Marston, Searl and Rivington Ltd, 314 pp.

AGASSIZ A. (1888) *The Three Cruises of the Blake*. New York and London.

AHLSTROM E.H. (1943) Studies on the Pacific pilchard or sardine (*Sardinops caerulea*) 4. Influence of temperature on the rate of development of pilchard eggs in nature. *Spec. Sci. Rep. U.S. Fish Wildlife Serv. Fisheries, 23* (Old Series) 26 pp.

ALBERS V.M. (1967) *Underwater Acoustics*, Vol. 2. Plenum Press, New York, 416 pp.

ALEXANDER R.MCN. (1967) *Functional Design in Fishes*. Hutchinson Univ. Lib., London, 160 pp.

ALLEN K.R. (1951) The Horokiwi Stream. *Fish. Bull. New Zealand Mar. Dept. No.* 10, 231 pp.

ALLEN, K.R. (1971) Relation between production and biomass. *J. Fish. Res. Bd Can.* **28,** 1573–81.

ALLEN J.S. (1973) Upwelling and coastal jets in a continuously stratified ocean. *J. Phys. Oceanogr.* **3,** 245–57.

ALM G. (1939) Investigations on growth etc.: by different forms of trout. English summary. *Kgl. Lantbruksstyrelsen Medd.* **15,** 1–93.

ALM G. (1949) Influence of heredity and environment on various forms of trout. *Rep. Inst. Freshwatr. Res. Drottningholm* **29,** 29–34.

ALVARINO A. (1964) Bathymetric distribution of chaetognaths. *Pacif. Sci.* **18** (1), 64–82.

ALVERSON D.L. (1967) Distribution and behaviour of Pacific hake as related to design of fishing strategy and harvest rationale. *FAO Fish Rep.* **62,** 2, 361–76.

ANCELLIN J. & NÉDELÈC C. (1959) Marquage de harengs en Mer du Nord et en Manche orientale (Campagne du 'Président Théodore Tissier', Novembre 1957). *Revue Trav. Inst. (Scient. tech.) Peches. Marit.* **23,** 177–201.

ANDERSON G.C. (1969) Subsurface chlorophyll maximum in the Northeast Pacific Ocean. *Limnol. Oceanogr.* **14,** 386–91.

ANDERSON G.C., FROST, B.W. & PETERSEN, W.K. (1972) On the vertical distribution of zooplankton in relation to chlorophyll concentration. In *M. Biological Oceanography of the northern North Pacific Ocean*, ed. Takenouti A.Y. *et al.* Idtmitsu Shoten, Tokyo, 341–345.

Andrewartha H.G. & Birch L.C. (1954) *The Distribution and Abundance of Animals.* University of Chicago Press.

Andrews K.J. (1966) The distribution and life history of *Calanoides acutus. Discovery Rep.* **34,** 117–62.

Angel M.V. (1968) The thermocline as an ecological boundary. *Sarsia* **34,** 299–312.

Angel M.V. (1969) Planktonic ostracods from the Canary Islands region, their depth distributions, diurnal migrations and community organisation. *J. mar. biol. Ass. U.K.* **49,** 515–53.

Ann Reps Sci. Committee Intern. Whaling comm. (1964) **14.**

Anon. (1969) Report of the north-western working group, 1968. *ICES Co-operative Res. Rep. Ser. A,* **10,** 32 pp.

Anon. (1970) Report of the panel of experts on the population dynamics of Peruvian anchoveta. *Bol. Inst. Mar Perú* **2** (6), 325–71.

Anon. (1972) Report of the second session of the panel of experts on the population dynamics of Peruvian anchoveta. *Bol. Inst. Mar Perú* **2** (7), 377–457.

Anon. (1973) Report of the third session of the panel of experts on the population dynamics of Peruvian anchoveta. *Bol. Inst. Mar Perú* **2** (9), 525–99.

Anraku M. (1964) Some technical problems encountered in quantitative studies of grazing and predation by marine planktonic copepods. *J. oceanogr. Soc. Japan* **20,** No. 5, 19–29.

Anraku M. & Omori M. (1963) Preliminary survey of the relationship between the feeding habit and the structure of the mouth parts of marine copepods. *Limnol. oceanogr.* **8,** 116–26.

Ansell A.D. (1960) Observations on predation of *Venus striatula* (da Costa) by *Natica alderi* Forbes. *Proc. Malac. Soc.* (*Lond.*) **34,** 157–64.

Antia N.J., McAllister C.D., Parsons T.R., Stephen K. & Strickland J.D.H. (1963) Further measurements of primary production using a large-volume plastic sphere. *Limnol. Oceanogr.* **8,** 166–83.

Apstein C. (1909) Die Bestimmung des Alters pelagisch lebender Fischeier. *Mitt. dt. SeefischVer.* **25** (12), 364 pp.

Armstrong F.A.J. (1965a) Phosphorus, pp. 323–64. In *Chemical Oceanography,* Vol. I eds Riley J.P. & Skirrow G. Academic Press, New York. 712 pp

Armstrong F.A.J. (1965b) Silicon, pp. 409–32. In *Chemical Oceanography,* Vol. I eds Riley J.P. & Skirrow G. Academic Press, New York. 712 pp.

Armstrong F.A.J. & LaFond E.C. (1966) Chemical nutrient concentrations and their relationship to internal waves and turbidity off Southern California. *Limnol. Oceanogr.* **11,** 538–47.

Armstrong F.A.J., Stearns C.R. & Strickland J.D.H. (1967) The measurement of upwelling and subsequent biological processes by means of the Technicon Auto-analyzer and associated equipment. *Deep-Sea Res.* **14** (3), 381–9.

Arvesen J.C., Millard J.P. & Weaver E.C. (1973) Remote sensing of chlorophyll and temperature in marine and fresh waters. *Astronautica Acta* **18,** 229–39.

Arvesen J.C., Weaver E.C. & Millard J.P. (1971) Rapid assessment of water pollution by airborne measurement of chlorophyll content. *Joint Conf. on Sensing of Environ. Pollutants, AIAA paper* 71-1097, pp. 1–7.

Atkins W.R.G. (1925) Seasonal changes in the phosphate content of sea water in relation to the growth of algal plankton during 1923 and 1924. *J. Mar. Biol. Assn. UK.* N.S. **13,** 700–20.

Backiel T. (1971) Production and food consumption of predatory fish in the Vistula River. *J. Fish. Biol.* **3,** 369–405.

Backus R.H., Clark R.C. & Wing A.S. (1965) Behaviour of certain marine organisms during the solar eclipse of July 20, 1963. *Nature (Lond.)* **205,** 989–91.

Bagenal T.B. (1969) The relationship between food supply and fecundity in brown trout *Salmo trutta. J. Fish. Biol.* **1,** 167–82.

Bainbridge R. (1957) The size and shape and density of marine phytoplankton concentrations. *Biol. Rev.* **32,** 91–115.

Bainbridge R. (1960) Speed and stamina in three fish. *J. Exp. Biol.* **37,** 129–53.

Bainbridge R. (1961) Migrations. In *Physiology of Crustacea* **2,** 431–63. New York.

Bainbridge V. (1972) The zooplankton of the Gulf of Guinea. *Bull. Mar. Ecol.* **8,** 61–97.

Baird R.C., Wilson D.F. & Milliten D.M. (1973) Observations on *Bregmaceros nectabanus* Whitley in the anoxic, sulfurous water of the Carioca Trench. *Deep-Sea Res.* **20** (5), 503–4.

Bakun A. (1973) Coastal upwelling indices, west coast of North America, 1946–71. U.S. Dept. of Commerce, *NOAA Tech. Rpt.* NMFS SSRF-671, 103 pp.

Bakun A., McLain D.R. & Mayo F.V. (1974) The mean annual cycle of coastal upwelling off western North America as observed from surface measurements. *Fish. Bull. U.S.*

Ballester A., Cruzado A., Juliá A., Manríquez M. & Salat J. (1972) Análisis automático y continuo de las caracteristicas físicas, químicas, y biológicas del mar. *Publs. Tecnicas* **1,** 1–72 *Inves. Pesq.*

Banse K. (1964) On the vertical distribution of zooplankton in the sea. *Prog. Oceanog.* **2,** 53–125.

Barham E.G. (1963) Siphonophores and the deep scattering layer. *Science* **140,** 826–8.

Barnes H. & Marshall S.M. (1951) On the variability of replicate plankton samples and some applications of 'contagious' series to the statistical distribution of catches over restricted periods. *J. mar. biol. Ass. U.K.* **30,** 233–63.

Barraclough W.E., LeBrasseur R.J. & Kennedy O.D. (1969) Shallow scattering layer in the subarctic Pacific ocean. Detection by high frequency echo sounder. *Science* **166,** 611–13.

Barraclough W.E. & Robinson D. (1972) The fertilization of Great Central Lake. III. Effect on juvenile Sockeye salmon. *Fish. Bull, U.S.* **70** (1), 37–48.

Bartlett M.S. (1960) *Stochastic population models in ecology and epidemiology.* Methuen, London, 362 pp.

Bary McK.B. (1967) Diel vertical migrations of underwater scattering mostly in Saanich Inlet, B.C. *Deep-Sea Res.* **14,** 35–50.

Battle H.I., Huntsman A.G., Jeffers A.M., Jeffers G.W., Johnson W.H. & McNairn N.A. (1936) Fatness, digestion and food of Passamaquoddy young herring. *J. Fish. Res. Bd. Can.* **2,** 401–29.

Baylor E.R. & Smith F.E. (1957) Diurnal migration of plankton crustaceans. *Recent Adv. in Invert. Physiol.* (Univ. of Oregon Publ.), 21–35.

Bé A.W.H. (1962) Quantitative multiple opening–closing plankton samplers. *Deep Sea Res.* **9,** 144–51.

Beamish F.W.H. (1964) Respiration of fishes, with special emphasis on standard oxygen consumption. II. Influence of weight and temperature on respiration of several species. *Can. J. Zool.* **42,** 177–88.

Beamish F.W.H. & Mookherjii P.S. (1964) Respiration of fishes with special emphasis on standard oxygen consumption. I. Influence of weight and temperature on respiration of goldfish, *Carassius auratus* L. *Can. J. Zool.* **42,** 161–75.

Beckman W.C. (1943) Further studies on the increased growth rate of the rock bass *Ambloplites rupestris* (Rafinesque), following the reduction in density of the population. *Trans. Amer. Fish. Soc.* 1942 **72,** 72–8.

BEERS J.R., STEVENSON M.R. EPPLEY R.W. & BROOKS E.R. (1971) Plankton populations and upwelling off the coast of Peru, June 1969. *Fish. Bull.* **69,** 859–76.

BEERS J.R. & STEWART G.L. (1969) The vertical distribution of micro-zooplankton and some ecological observations. *J. Cons. int. Explor. Mer.* **33** (1), 30–44.

BEERS J.R. & STEWART G.L. (1970) Numerical abundance and estimated biomass of micro-zooplankton. *Bull. Scripps Inst. Oceanogr.* **17,** 67–87.

BIERS J.R. & STEWART G.L. (1971) Micro-zooplankters in the plankton communities of the upper waters of the eastern tropical Pacific. *Deep-Sea Res.* **18,** 861–83.

BEERS J.R., STEWART G.L. & STRICKLAND J.D.H. (1967) A pumping system for sampling small plankton. *J. Fish. Res. Bd. Canada* **24,** 1811–18.

BEKLEMISHEV C.W. (1957) The spatial interrelationship of marine zooplankton and phytoplankton. *Trudy Inst. Okeanol., Akad. Nauk., SSSR.* **20,** 253–78.

BELAEV G.M. & USCHAKOV P.V. (1957) Some regularities in the quantitative distribution of the benthic fauna in Antarctic waters. *Dokl. Akad. Nauk, SSSR Biol. Sci. Sect.* **112,** 116–19.

BERNER A. (1962) Feeding and respiration in the copepod *Temora longicornis* (Muller). *J. mar. biol. Ass. U.K.* **42,** 625–40.

BERTALANFFY L.VON (1934) Untersuchungen über die Gesetzlichkeit des Wachstums. 1. Teil. Allgemeine Grundlagen der Theorie; mathematische und physiologische Gesetzlichkeiten des Wachstums bei Wassertieren. *Arch. Entro Mech. Org.* **131,** 613–52.

BERTALANFFY L.VON (1938) A quantitative theory of organic growth. *Hum. Biol.* **10** (2), 181–213.

BERTALANFFY L.VON (1949) Problems of organic growth. *Nature* (*Lond.*) **163,** 156–8.

BERTELSEN E. (1957) The ceratioid fishes, ontogeny, taxonomy, distribution and biology. *Dana Rep.* **39,** 184 pp.

BEUKEMA J.J. (1968) Predation by the three-spined stickleback (*Gasterosteus aculeatus* L.), the influence of hunger and experience. *Behaviour* **31** (Pt 1–2), 1–126.

BEVERTON R.J.H. (1962) Long-term dynamics of certain North Sea fish populations, pp. 242–64. In *The Exploitation of Natural Animal Populations*, ed. Le Cren E.D. & Holdgate M.W. Blackwell Scientific Publications, Oxford, 399 pp.

BEVERTON R.J.H. & HOLT S.J. (1957) On the dynamics of exploited fish populations. *Fishery Invest.* (*Lond.*) *Ser.* 2 **19,** 533 pp.

BEVERTON R.J.H. & HOLT S.J. (1959) A review of the lifespans and mortality rates of fish in nature, and their relation to growth and other physiological characteristics, pp. 142–80. In *Ciba Foundation Colloquia on Ageing*. 5. *The Lifespan of animals*, eds Wolstenholme G.E.W. & O'Connor M. J. & A. Churchill Ltd, London, 324 pp.

BIEN G.S., RAKESTRAW N.W. & SUESS H.E. (1965) Radiocarbon in the Pacific and Indian oceans and its relation to deep water movements. *Limnol. Oceanogr.* 10 (Suppl.), R25–R37.

BIERSTEIN J.A. (1957) Certain peculiarities of the ultra-abyssal fauna exemplified by the genus *Storthyngura* (Isopoda Asdlota). *Zool. Zhurnal* **36,** 961–85.

BIRKETT L. (1959) Production in benthic populations. *Cons. L. Int. Explor. Mer.* (Unpubl. Rep. C. M., 1959, No. 42).

BLACKBURN M. (1969) Conditions related to upwelling which determine distribution of tropical tunas off western Baja California. *Fish. Bull. U.S.* **68** (1), 147–76.

BLACKBURN M. & THORNE R.E. (1974) Composition, biomass, and distribution of pelagic nekton in a coastal upwelling area off Baja California, Mexico. *Tethys.* **6,** 181–190.

BLACKMAN R.B. & TUKEY J.W. (1959) *The Measurement of Power Spectra.* Dover, 190 pp.

BLAXTER J.H.S. (1973) Monitoring the vertical movements and light responses of herring and plaice larvae. *J. mar. biol. Ass. U.K.* **53,** 635–47.

Blaxter J.H.S. & Currie R.I. (1967) The effect of artificial lights on acoustic scattering layers in the ocean. *Symp. zool. Soc.* (*Lond.*) **19,** 1–14.

Blaxter J.H.S. & Hempel G. (1963) The influence of egg size on herring larvae (*Clupea harengus* L.). *J. Cons. int. Explor. Mer.* **28,** 211–40.

Blaxter J.H.S. & Parrish B.B. (1965) The importance of light in shoaling, avoidance of nets and vertical migration by herring. *J. Cons. int. Explor. Mer.* **30,** 40–57.

Bledsoe L.J. (in press) Linear and nonlinear approaches for ecosystem dynamic modeling. In *Systems Analysis and Simulation in Ecology, Vol. IV*, ed. Patten, B.C. Academic Press, New York.

Blegvad H. (1933) Plaice transplantations. *J. Cons. int. Explor. Mer* **8** (2), 161–80.

Boden B.P. & Kampa E.M. (1967) The influence of natural light on the vertical migrations of an animal community in the sea. *Symp. zool. Soc.* (*Lond.*) **19,** 15–26.

Bogdanov Yu.A. (1965) Suspended organic matter in the Pacific. *Oceanology* **5,** 77–85.

Bogdanov Yu.A. Lisitsyn A.P. & Romankevitch Ye.A. (1971) Organic matter in suspension and bottom sediments of seas and oceans. (In Russian), pp. 35–104. In *Orgonicheskoye veshchestvo sovremennykh i iskopayemykh osadkov*. Moskow, Akad. Nauk SSSR. Otd. Nauk Zemle. Kom. Osad. Porod.

Bogorov V.G. (1941) Diurnal vertical distribution of zooplankton under Polar conditions (The White Sea). *Trudȳ Polyarnauchno-issled. mersk. Inst. Rȳb. Khoz. Okeanog.* **7,** 287–311.

Bogorov V.G. (1946) Diurnal vertical migration of zooplankton in Polar seas. *Trudȳ Inst. Okeanol. An SSSR.* **I,** 151–8.

Bogorov V.G. & Vinogradov M.E. (1960) Distribution of the zooplankton biomass in the central part of the Pacific Ocean. *Trudȳ Vses. Gidrobiolog. Obshch.* **10,** 208–23.

Bolin B. & Stommel H. (1961) On the abyssal circulation of the world ocean—IV. Origin and rate of circulation of deep waters as determined with aid of tracers. *Deep-Sea Res.* **8,** 95–110.

Bolster G.C. & Bridger J.P. (1957) Nature of the spawning area of herrings. *Nature* (*Lond.*) **179,** 638.

Bostrøm O. (1955). 'Peder Ronnestad' Ekkolodding-og-meldetjeneste ar Skreiforekomstene i Lofoten i tiden 1st March–2nd Apr. 1955. Praktiske fiskeforsok 1954 og 1955. *Årsberet. vedkomm. Norges Fisk.* **9,** 66–70.

Bourne N. (1964) Scallops and the offshore fishery of the Maritimes. *Bull. Fish Res. Bd. Can.* No. 145. 60 pp.

Bowden K.F. (1964) Turbulence. *Oceanogr. Mar. Biol. Ann. Rev.* **2,** 11–30.

Bowden K.F. (1965) Horizontal mixing in the sea due to a shearing current. *J. Fluid Mech.* **21,** 83–95.

Boyd C.M. (1962) The biology of a marine decapod crustacean, *Pleuroncodes planipes* Stimson, 1860. Ph.D. dissertation, University of California, San Diego, 123 pp.

Boyd C.M. (1973) Small scale spatial patterns of marine zooplankton examined by an electronic *in situ* zooplankton detecting device. *Neth. J. Sea Res.* **7** (4), 555–666.

Boyd C.M. & Johnson G.W. (1969) Studying zooplankton populations with an electronic zooplankton counting device and the Ling-8 computer, pp. 83–90. *App. of Seagoing Computers, Mar. Tech. Sec.*, eds Mudie J.D. & Jackson C.B.

Brandt K. (1899) Über den Stoffwechsel im Meer. *Wiss. Meeresunters. NF. Abt. Kiel*, NF. IV–V, 215–30.

Brett J.R. (1964) The respiratory metabolism and swimming performance of young sockeye salmon. *J. Fish. Res. Bd Can.* **21,** 1183–226.

Brett, J.R. (1965) The relation of size to rate of oxygen consumption and sustained swimming speed of sockeye salmon (*Oncorhynchus nerka*). *J. Fish. Res. Bd Can.* **22** (5), 1491–501.

Brett J.R. & Sutherland D.B. (1965) Respiratory metabolism of pumpkinseed (*Lepomis gibbosus*) in relation to swimming speed. *J. Fish. Res. Bd Can.* **22,** 405–9.

Brinton E. (1962) The distribution of Pacific euphausiids. *Bull. Scripps Instn. Oceanogr.* **8** (2), 51–270.

Brock V. & Riffenburgh R. (1960) Fish schooling: a possible factor in reducing predation. *J. Cons. int. Explor. Mer.* **25,** 307–17.

Brocksen R.W (1970) Analysis of trophic processes on the basis of density-dependent functions, pp. 468–98. In *Marine Food Chains*, ed. Steele J.H. Oliver & Boyd, Edinburgh.

Brody S. (1964) *Bioenergetics and growth.* Reinhold, New York, 1023 pp.

Brooks J.L. & Dodson S.I. (1965) Predation, body size, and composition of plankton. *Science* **150,** 28–35.

Brown M.E. (1946) The growth of brown trout (*Salmo trutta* Linn.). II. The growth of two-year-old trout at a constant temperature of 11·5°C. *J. exp. Biol.* **22,** 130–44.

Brown M.E. (1971) Experimental studies on growth, pp. 361–400. *The Physiology of Fishes*, Vol. I, ed. Brown M.E. Academic Press, New York, 447 pp.

Bryan K. & Cox M.D. (1967) A numerical investigation of the oceanic general circulation. *Tellus* **19,** 54–80.

Bryan K. & Cox M.D. (1968) A nonlinear model of an ocean driven by wind and differential heating, parts I and II. *J. Atmos. Sci.* **25,** 945–78.

Bryan K. & Cox M.D. (1972) The circulation of the world ocean: a numerical study. Part I, a homogeneous model. *J. Phys. Oceanogr.* **2** (4), 319–35.

Buchanan J.B. (1967) Dispersion and demography of some infaunal echinoderm populations. *Symp. zool. Soc.* (*Lond.*) **20,** 1–11.

Buchanan-Wollaston H.J. (1923) The spawning of plaice in the southern part of the North Sea in 1913–14. Fishery Invest. (*Lond.*) **II,** 5, 2.

Bückmann A. (1942) Die Untersuchungen der Biologischen Anstalt über die Ökologie der Heringsbrut in der südlichen Nordsee. *Helgoländer, wiss Meeresunters.* **3,** 1–57.

Burd A.C. & Cushing D.H. (1962) I Growth and recruitment in the herring of the southern North Sea II Recruitment of the North Sea herring stocks. *Fishery Invest.* **2,** 23(5) 71 pp.

Burke M.C. & Mann K.H. (1974) Productivity and production: biomass ratio of bivalve and gastropod populations in an eastern Canadian estuary. *J. Fish. Res. Board. Can.* **31,** 161–77.

Burnet A.M.R. (1970) Seasonal growth in brown trout in two New Zealand streams. *N.Z. Jl. Mar. Freshwat. Res.* **4,** 55–62.

Burns C.W. (1968) The relationship between body size of filter feeding Cladocera and the maximum size of particle ingested. *Limnol. Oceanogr.* **13,** 675–8.

Burukowski R.N. (ed.) (1965) Antarkticheskii krill. Biologiiai promysel. (Antarctic krill. Biology and industry) *Antarkticheskii Nauchno-issledovatel'skii Institut Rybnogo Khoziaistva i Okeanografii*, Kaliningrad, 92 pp.

Cadwalader G. (1973) Freedom for science in the oceans. *Science* **182,** 15–20.

Caperon J. (1975) A trophic level ecosystem model analysis of the plankton community in a shallow-water subtropical estuarine emboyment. *Proc. Second Int. Estuarine Res. Conf.* Academic Press.

Caperon J. & Meyer J. (1972a) Nitrogen-limited growth of marine phytoplankton. II. Uptake kinetics and their role in nutrient limited growth of phytoplankton. *Deep-Sea Res.* **19,** 619–32.

Caperon J. & Meyer J. (1972b) Nitrogen-limited growth of marine phytoplankton. I. Changes in population characteristics with steady-state growth rate. *Deep-Sea Res.* **19,** 601–18.

CARPENTER E.J. (1972) Nitrogen fixation by *Oscillatoria* (*Trichodesmium*) *thiebautii* in the southwestern Sargasso Sea. *Deep-Sea Res.* **20,** 285–8.

CARPENTER E.J., REMSEN C.C. & WATSON S.W. (1972) Utilization of urea by some marine phytoplankters. *Limnol. Oceanogr.* **17,** 265–9.

CASPERS H. (1957) The Black Sea and the Sea of Azov. In *Treatise on Marine Ecology and Palaeoecology*, ed. Hedgpeth J. *Mem. Geol. Soc. Am.* **67,** 803–90.

CASSIE R.M. (1960) Factors influencing the distribution pattern of plankton in the mixing zone between Oceanic and Harbour waters. *N.Z.J. Sci.* **3,** 26–50.

CASSIE R.M. (1963) Microdistribution of plankton. *Oceanogr. Mar. Biol. Ann. Rev.*, ed. Barnes H. **1,** 223–52.

CASWELL H. (1972) On instantaneous and finite birth rates. *Limnol. Oceanogr.* **17** (5), 787–91.

CHAPMAN D.G. (1964) Special Committee of Three Scientists. Final report. *Rep. Int. Whaling Comm.* **14,** 39–92.

CHINDONOVA YU.G. (1959) Nutrition of certain groups of deepwater macroplankton in the north-west Pacific Ocean. *Trudy. Inst. Okeanol. USSR* **30,** 166–89.

CLARKE G.L. (1934) Factors affecting the vertical distribution of copepods. *Ecol. Monog.* **4,** 530–40.

CLARKE G.L. (1936) On the depth at which fish can see. *Ecology* **17,** 452–6.

CLARKE G.L. (1971) Light measurements. *Martek Mariner* **3** (3), 1.

CLARKE G.L. & DENTON E. (1962) Light and animal life, pp. 456–68. In *The Seas*, ed. Hill M.N. I. Interscience Publs, New York and London, 864 pp.

CLARKE M.R. (1970) Function of the spermaceti organ of the sperm whale. *Nature* (*Lond.*) **228,** 873–4.

CLAYDEN A.D. (1972) Simulation of the changes in abundance of the cod (*Gadus morhua* L.) and the distribution of fishing in the North Atlantic. *Fishery Invest.* (*Lond.*) *Ser.* 2 **27** (1), 58 pp.

CLEMENS W.A., FOERSTER R.E. & PRITCHARD A.L. (1939) The migration of Pacific salmon in British Columbian waters. *Publs. Amer. Adv. Sci.* **8,** 51–9.

CLEVE P.T. (1895–7) Microscopic marine organisms in the service of hydrography. *J. Mar. Biol. Ass. UK.* NS. **4,** 381–5.

COCHRAN W.G. (1963) *Sampling Techniques*. John Wiley & Sons, New York, 413 pp.

COMMITTEE ON POLAR RESEARCH (1974) *Southern Ocean Dynamics*. Nat'l. Acad. Sci. U.S.A., Washington, D.C., 52 pp.

CONOVER R.J. (1959) Regional and seasonal variation in the respiratory rate of marine copepods. *Limnol. Oceanogr.* **4,** 259–68.

CONOVER R.J. (1960) The feeding behaviour and respiration of some marine planktonic crustacea. *Biol. Bull. mar. biol. Lab., Woods Hole*, **119** (3), 399–415.

CONOVER R.J. (1962) Metabolism and growth in *Calanus hyperboreus* in relation to its life cycle. *J. Cons. int. Expl. Mer.* **153,** 190–7.

CONOVER R.J. (1964) Food relations and nutrition of zooplankton. *Proceedings of the Symposium on Experimental Marine Ecology. Occasional Publication No.* 2, Graduate School of Oceanography, Univ. Rhode Is.

CONOVER R.J. (1966a) Feeding on large particles by *Calanus hyperboreus* (Kroyer), pp. 187–94. *Some Contemporary Studies in Marine Science*. ed. Barnes H. Allen & Unwin, London. 716 pp.

CONOVER R.J. (1966b) Assimilation of organic matter by zooplankton. *Limnol. Oceanogr.* **11,** 339–45.

CONOVER R.J. (1966c) Factors affecting the assimilation of organic matter by zooplankton and the question of superfluous feeding. *Limnol. Oceanogr.* **11,** 346–54.

CONOVER R.J. (1968) Zooplankton—Life in a nutritionally dilute envitonment. *Am. Zool.* **8,** 107–18.

CONOVER R.J. & CORNER E.D.S. (1968) Respiration and nitrogen excretion by some marine zooplankton in relation to their life cycles. *J. mar. biol. Ass. U.K.* **48,** 49–75.

CONOVER R.J. & FRANCIS V. (1973) The use of radioactive isotopes to measure the transfer of materials in aquatic food chains. *Marine Biology* **8** (4), 272–83.

CONWAY H.L. (1974) The uptake and assimilation of inorganic nitrogen by *Skeletonema costatum* (Grev.) Cleve. Ph.D. dissertation, University of Washington, 126 pp.

COOPER L.H.N. (1967) Stratification in the deep ocean. *Sci. Prog.* **55,** 73–90.

CORKETT C.J. (1970) Techniques for breeding and rearing marine calanoid copepods. *Helgoländer Wiss. Meeresunters* **20,** 318–24.

CORNER E.D.S. (1961) On the nutrition and metabolism of zooplankton. I. Preliminary observations on the feeding of the marine copepod *Calanus helgolandicus* (Claus). *J. mar. biol. Ass. U.K.* **41,** 5–16.

CORNER E.D.S., COWEY C.B. & MARSHALL S.M. (1967) On the nutrition and metabolism of zooplankton. V. Feeding efficiency of *Calanus finmarchicus*. *J. mar. biol. Ass. U.K.* **47,** 259–70.

CORNER E.D.S. & DAVIES A.G. (1971) Plankton as a factor in the nitrogen and phosphorus cycles in the sea. *Adv. Mar. Biol.* **9,** 101–204.

CORNER E.D.S., HEAD R.N. & KILVINGTON C.C. (1972) On the nutrition and metabolism of zooplankton. VIII. The grazing of *Biddulphia* cells by *Calanus helgolandicus*. *J. Mar. biol. Ass. UK.* NS. **52,** 847–61.

COULL B.C. (1973) Estuarine meiofauna: a review: trophic relationships and microbial interactions, pp. 499–512. In *Estuarine microbial ecology*, eds Stevenson L.H. & Colwell R.R. University of South Carolina Press.

COX R.A. (1963) The salinity problem. *Progress in Oceanogr.* **1,** 243–61.

COX R.A., CULKIN F. & RILEY J.P. (1967) The electrical conductivity/chlorinity relationship in natural sea water. *Deep-Sea Res.* **14,** 203–20.

CRAIG H. (1969) Abyssal carbon and radiocarbon in the Pacific. *J. Geophys. Res.* **74,** 5491–506.

CRISP D.J. & BARNES H. (1954) The orientation and distribution of barnacles at settlement with particular reference to surface culture. *J. Anim. Ecol.* **23** (1) 141–62.

CURL H. & BECKER P. (1972) Mini computer system for shipboard sampling. *Ocean. Int.* **7,** 1. pp. 34–6.

CUSHING D.H. (1951) The vertical migration of planktonic crustacea. *Biol. Rev.* **26** (2), 158–92.

CUSHING D.H (1955) On the autumn spawned herring races of the North Sea. *J. Cons. int. Explor. Mer.* **21,** 44–60.

CUSHING D.H. (1955) Production and a pelagic fishery. *Fishery Invest.* (*Lond.*) *Ser. II* **18** (7), 104 pp.

CUSHING D.H. (1957) The number of pilchards in the Channel. *Fishery Invest.* (*Lond.*) *Ser. II* **21** (5), 27 pp.

CUSHING D.H. (1959) On the nature of production in the sea. *Fishery Invest.* (*Lond.*) *Ser. II* **22,** 40 pp.

CUSHING D.H. (1961) On the failure of the Plymouth herring fishery. *J. mar. biol. Ass. U.K.* **41,** 799–816.

CUSHING D.H. (1962) An estimate of grazing from field observations. *Rapp. P-v Réun. Cons. int. Explor. Mer.* **153,** 198–9.

CUSHING D.H. (1964a) The counting of fish with an echo sounder. *Rapp. P-v Réun. Cons. int. Explor. Mer.* **155,** 190–5.

CUSHING D.H. (1964b) The work of grazing in the sea. In *Grazing in Terrestrial and Marine*

Environments, ed. Crisp D.J., pp. 207–25. Blackwell Scientific Publications, Oxford, 322 pp.

Cushing D.H. (1968) Grazing by herbivorous copepods in the sea. *J. Cons. int. Explor. Mer.* **32,** 70–82.

Cushing D.H. (1969) Upwelling and fish production. *FAO Fish. Tech. Pap.* **84,** 40 pp.

Cushing D.H. (1971a) The dependence of recruitment on parent stock in different groups of fishes. *J. Cons. int. Explor. Mer.* **33,** 340–62.

Cushing D.H. (1971b) Upwelling and the production of fish. *Adv. Mar. Biol.* **9,** 255–334.

Cushing D.H. (1971c) The regularity of the spawning season in some fishes. *J. Cons. int. Explor. Mer.* **33** (3), 340–62.

Cushing D.H. (1972) The production cycle and the numbers of marine fish. *Symp. Zool. Soc. (Lond.)* (1972) **29,** 213–32.

Cushing D.H. (1973) Recruitment and parent stock in fishes. *Washington Sea Grant Publ.* 73–1. 197 pp.

Cushing D.H. (1974a) Observations on fish shoals with the ARL scanner. *Symposium on Acoustic methods in Fisheries Research* **39.** 8 pp.

Cushing D.H. (1974b) The possible density dependence of larval mortality and adult mortality in fishes. pp. 103–12. In *The Early Life history of fish*, ed. Blaxter J.H.S. Springer-Verlag, Heidelberg. 765 pp.

Cushing D.H. (1975) The natural mortality of the plaice. *J. Cons. int. Explor. Mer.* **36** (2), 150–157.

Cushing D.H. & Bridger J.P. (1966) The stock of herring in the North Sea and changes due to fishing. *Fishery Invest. (Lond.) Ser. II* **25** (1), 1–123.

Cushing D.H. & Harris J.G.K. (1973) Stock, and recruitment, and the problem of density dependence. *Rapp. P.-V. Réun. Cons. int. Explor. Mer.* **164,** 142–155.

Cushing D.H. & Tungate D.S. (1963) Studies on a *Calanus* patch. I. The identification of a *Calanus* patch. *J. mar. biol. Ass. U.K.* **43,** 327–37.

Cushing D.H. & Vucetic T. (1963) Studies on a *Calanus* patch. III. The quantity of food eaten by *Calanus finmarchicus*. *J. Mar. Biol. Ass. UK.* NS. **43,** 349–71.

Cuvier Le Baron (1817) *La règne animale* 2 (*Poissons*). Paris, 532 pp.

Davis C.O. (1973) Effects of changes in light intensity and photoperiod on the silicate-limited continuous culture of the marine diatom *Skeletonema costatum* (Grev.) Cleve. Ph.D. dissertation, University of Washington, 123 pp.

Dawes B. (1931) Growth and maintenance in the plaice (*Pleuronectes platessa* L.), Part I. *J. mar. biol. Ass. U.K.*, NS. **17,** 103–74.

Dayton P.K. (1971) Competition, disturbance and community organization: the provision and subsequent utilization of space in a rocky intertidal community. *Ecol. Monogr.* **41,** 351–89.

Dayton P.K. (1973) Dispersion, dispersal, and persistence of the annual intertidal alga, *Postelsia palmaeformis* Ruprecht. *Ecology* **54,** 433–8.

Devold F. (1963) The life history of the Atlanto-Scandian herring. *Rapp. P.-V. Réun. Cons. int. Explor. Mer.* **154,** 98–108.

Dickie L.M. (1955) Fluctuations in abundance of the giant scallop *Placopecten magellanicus* (Gmelin), in the Digby area of the Bay of Fundy. *J. Fish. Res. Bd. Can.* **12,** 797–857.

Dickson R.R. (1971) A recurrent and persistent pressure pattern as the principle cause of intermediate-scale hydrographic variation in the European Shelf Seas. *Deutsche Hydrographische Zeitschrift* **24**(2), 97–119.

Dickson R.R. & Baxter G.C. (1972) Monitoring deep water movements in the Norwegian Sea by satellite. *Int. Cons. Explor. Sea. C.M.* 1972. C9, 9 pp. (mimeo).

Dietrich, G. (1954) Verteilung, Ausbreitung und Vermischung der Wasserkörper in der südwestlichen Nordsee auf Grund der Ergebnisse der 'Gauss-Fahrt' im Februar–März 1952. *Ber. deutsch. wiss. Komme Meeresforsch* NF B. **13** (2), 104–29.

DIETRICH G. (ed.) (1972) Upwelling in the ocean and its consequences. *Geoforum* **11,** 3–71.

DIGBY P.S.B. (1961) The vertical distribution and movements of plankton under midnight sun conditions in Spitzbergen. *J. anim. Ecol.* **30** (1), 9–25.

DITTMAR W. (1884) Report on research into the composition of ocean water, collected by H.M.S. *Challenger. Challenger Reps., Physics and Chem.* **1,** 1–251.

DOTY M.S. (1957) Rocky intertidal surfaces. *Mem. Geol. Soc. Am.* **67** (1), 535–85.

DOW R.L. (1969) Cyclic and geographic trends in seawater temperature and abundance of American lobster. *Science* **164,** 1060–3.

DROOP M.R. (1968) Vitamin B_{12} and marine ecology. IV. The kinetics of uptake, growth and inhibition in *Monochrysis lutheri. J. Mar. Biol. Ass. U.K.* **48,** 689–733.

DROOP M.R. (1973) Some thoughts on nutrient limitation in algae. *J. Phycol.* **9,** 264–72.

DUGDALE R.C. (1972) Chemical oceanography and primary production in upwelling regions. *Geoforum* **11,** 47–61.

DUGDALE R.C., GOERING J.J. & RYTHER J.H. (1964) High nitrogen fixation rates in the Sargasso Sea and the Arabian Sea. *Limnol. Oceanogr.* **9,** 507–10.

DUGDALE R.C. & GOERING J.J. (1967) Uptake of new and regenerated forms of nitrogen in primary productivity. *Limnol. Oceanogr.* **12,** 196–206.

DUGDALE R.C. (1967) Nutrient limitation in the sea: Dynamics, identification and significance. *Limnol. Oceanogr.* **12,** 685–95.

DUGDALE R.C. & GOERING J.J. (1970) Nutrient limitation and the path of nitrogen in Peru. Current production. Scientific Results of the Southeast Pacific Expedition, *Anton Bruun Report No.* 5, 5.3–5.8.

DUGDALE R.C. & MACISAAC J.J. (1971) A computation model for the uptake of nitrate in the Peru upwelling region. *Inv. Pesq.* **35,** 299–308.

DUNTLEY S.Q. (1963) Light in the sea. *J. Opt. Soc. Am.* **53,** 214–33.

DUSSART B.H. (1965) Les différentes catégories de plancton. *Hydrobiologia,* **26,** 72–74.

DUURSMA E.K. (1960) Dissolved organic carbon, nitrogen, and phosphorus in the sea. *Netherl. J. Mar. Res.* **1,** 1–148.

DYRSSEN D. & JAGNER D. (eds) (1972) *The changing chemistry of the oceans.* Wiley-Interscience, New York, 365 pp.

EDDY S. & CARLANDER K.D. (1940) The effect of environmental factors upon the growth rates of Minnesota fishes. *Proc. Minn. Acad. Sci.,* **8,** 14–19.

EDMONDSON W.T. (1960) Reproductive rates of rotifers in natural populations. *Mem. Ist. Ital. Idrobiol.* **12,** 21–77.

EDMONDSON W.T. (ed.) (1966) Ecology of invertebrates. In *Marine Biology, Vol.* 3, N.Y. Acad Sci., New York, 313 pp.

EDMONDSON W.T. & WINBERG G.G. (1971) A manual on methods for the assessment of secondary productivity in fresh waters. *IBP Handbook Bo.* 17. Blackwell Scientific Publications, Oxford. 358 pp.

EDWARDS R.R.C., FINLAYSON D.M. & STEELE J.H. (1969) The ecology of O-group plaice and common dabs in Loch Ewe. II. Experimental studies of metabolism. *J. exp. mar. Biol. Ecol.* **3,** 1–17.

EDWARDS R.R.C., FINLAYSON D.M. & STEELE J.H. (1972) An experimental study of the oxygen consumption, growth and metabolism of the cod (*Gadus morhua* L.). *J. exp. mar. Biol. Ecol.* **8,** 299–309.

EDWARDS R.R.C., STEELE J.H. & TREVALLION A. (1970) The ecology of O-group plaice and common dabs in Loch Ewe. III. Prey–predator experiments with plaice. *J. exp. mar. Biol. Ecol.* **4,** 156–73.

Eibl-Eibesfeldt, I. (1965) *Land of a thousand atolls: a study of marine life in the Maldive and Nicobar Islands.* MacGibbon & Kee, London. 195 pp.

Ekman V.W. (1905) On the influence of the Earth's rotation on ocean currents. *Arkiv f. matematik, Astron. och Fysik* **2** (11), 1–53.

Elton C. (1927) *Animal ecology.* Sidgwick & Jackson, London, 207 pp.

Emery K.O., Orr W.K. & Rittenberg S.C. (1955) Nutrient budgets in the ocean, pp. 299–310. In *Essays in the Natural Sciences in Honor of Captain Allan Hancock.* Univ. Southern Calif. Press, Los Angeles.

Emery K.O. & Stevenson R.E. (1957) Estuaries and Lagoons. I. Physical and Chemical Characteristics. *Mem. Geol. Soc. Amer.* **67** (1), 673–750.

English T.S. (1964) A theoretical model for estimating the abundance of planktonic fish eggs. *Rapp. P.-V. Réun. Cons. int. Explor. Mer.* **155,** 174–81.

Enright J.T. & Hammer W.M. (1967) Vertical diurnal migration and endogenous rhythmicity. *Science, N.Y.* **157,** 937–41.

Eppley R.W. (1972) Temperature and phytoplankton growth in the sea. *Fish. Bull.* **70,** 1063–85.

Eppley R.W. & Renger E.M. (1974) Nitrogen assimilation of an oceanic diatom in nitrogen-limited continuous culture. *J. Phycol.* **10,** 15–23.

Eppley R.W., Renger E.M., Venrick E.L. & Mullin M.M. (1973) A study of plankton dynamics and nutrient cycling in the central gyre of the North Pacific Ocean. *Limnol. Oceanogr.* **18,** 534–51.

Eppley R.W., Rogers J.N. & McCarthy J.J. (1969) Half saturation constants for uptake of nitrate and ammonium by marine phytoplankton. *Limnol. Oceanogr.* **14,** 912–20.

Esterley C.O. (1911) Diurnal migration of *Calanus finmarchicus* in the San Diego region during 1909. *Int. Rev. hydrobiol.* **4,** 140–51.

Esterly C.O. (1912) The occurrence and vertical distribution of the copepoda of the San Diego region with particular reference to nineteen species. *Univ. Calif. Publ. Zool.* **9** (6), 253–340.

Ewart T. & Bendiner W.P. (1972) Techniques for estuarine and open ocean dye dispersal measurement. *Symp. on Phys. Proc. Responsible for Dispersal of Pollutants in the Sea with Special Reference to the Nearshore Zones AARHUS, Den., July,* 1972.

Ewart T. & Bendiner W.P. (1973) 3-D measurement of estuarine circulation using a tracer dye. *Proc. 3rd Tech. Conf. on Estuaries of Pac. N.W., Oregon State Univ., Corvallis, Oregon, March,* 1973.

Ewart T., Bendiner W.P. & Linger E.H. (1972) Prediction of excess heat distribution using tracer dye techniques. *APL Int. Rept. APL/UW* 7206.

Fager E.W. (1973) Estimation of mortality coefficients from field samples of zooplankton. *Limnol. Oceanogr.* **18** (2), 297–301.

Fager E.W. & Longhurst A.R. (1968). Recurrent group analysis of species assemblages of demersal fish in the Gulf of Guinea. *J. Fish. Res. Bd Can.* **25** (7), 1405–21.

Farmer G.J. & Beamish F.W.H. (1969) Oxygen consumption of *Tilapia nilotica* in relation to swimming speed and salinity. *J. Fish. Res. Bd Can.* **26,** 2807–21.

Feldmann J. (1937) Recherches sur la végétation marine de la Méditerranée. *Rev. Algologique* **10,** 1–339.

Fenchel T. (1969) The ecology of marine microbenthos. IV. Structure and function of the benthic ecosystem, its chemical and physical factors and the microfauna communities with special reference to the ciliated protozoa. *Ophelia* **6,** 1–182.

Fenchel T. & Straarup B.J. (1971) Vertical distribution of photosynthetic pigments and the penetration of light in marine sediments. *Oikos* **22,** 172–82.

Filatova Z.A. (1970) Quantitative distribution of deep-sea benthic fauna. In *The Pacific*

Ocean, Vol. 7. *Biology of the Pacific Ocean*, Pt. 2. *The deep-sea bottom fauna. Pleuston*, ed. Zenkevich L.A. Akademiia Nauk SSSR, Instituut Okeanologie. Translation 487 U.S. Naval Oceanographic Office, Washington, D.C.

Filatova Z.A. & Levenstein R.J. (1961) The quantitative distribution of deep-sea bottom fauna in the north-eastern Pacific *Trudy Inst. Okeanol. Akad. Nauk SSSR* **45,** 190–213. (In Russian).

Fleming R.H. (1939) The control of diatom populations by grazing. *J. Cons. int. Explor. Mer.* **14,** 210–227.

Foerster R.E. (1937) The return from the sea of sockeye salmon (*Oncorhynchus nerka*) with special reference to percentage survival, sex, proportions and progress of migration. *J. Biol. Bd Can.* **3,** 26–42.

Fofonoff N.P. (1962) Physical properties of sea-water. pp. 3–30. In *The Sea, Vol.* 1, ed. Hill M.N. Interscience. Wiley & Sons. New York. London. 864 pp.

Food and Agriculture Organization of the UN Department of Fisheries (1972) *Atlas of the Living Resources of the Seas.* FAO, Rome, 19 pp. and 64 charts.

Forbes E. (1843). Report on the Mollusca and Radiata of the Aegean Sea. *Rep. Br. Ass. Adv Sci.* 130.

Forbes E. (1859) *Natural History of the European Seas.* Van Voorst. 306 pp.

Forbes S.T. & Nakken O. (1972) Manual of methods for fisheries resource survey and appraisal. Part 2. The use of acoustic instruments for fish detection and abundance estimations. *FAO Man. Fish. Sci.* **5,** 138 pp.

Ford E. (1933) An account of the herring investigations conducted at Plymouth during the years from 1924 to 1933. *J. mar. biol. Ass. U.K.*, N.S. **19,** 305–84.

Fournier R.O. (1971) Studies on pigmented microorganisms from aphotic marine environments II. North Atlantic distribution. *Limnol. Oceanog.* **16** (6), 952–61.

Frassetto R., Backus R.H. & Hays E. (1962) Sound scattering layers and their relation to thermal structure in the Strait of Gibraltar. *Deep-Sea Res.* **9** (1), 69–72.

Frazer F.C. (1937) On the development and distribution of the young stages of krill (*Euphausia superba*). *Discovery Rep.* **14,** 1–192.

Freund J.E. (1962) *Mathematical Statistics.* Prentice-Hall, Englewood Cliffs, N.J. 390 pp.

Frosch R.A. (1964) Underwater sound: deep-ocean propagation. *Science* **146,** 889–94.

Frost B.W. (1972) Effects of size and concentration of food particles on the feeding behaviour of the marine planktonic copepod *Calanus pacificus. Limnol. Oceanogr.* **17** (6), 805–15.

Fryer G. & Iles T.D. (1972) *The Cichlid Fishes of the Great Lakes of Africa; Their Biology and evolution.* Oliver & Boyd 641 pp.

Fuchs T. (1882) Beitrage zur Lehre über den Einfluss des Lichtes auf die bathymetrische Verbreitung der Meeresorganismen. *Verh. geol. Reichsanst Aust. Wien.*

Fuglister F.C. & Worthington L.V. (1949) Some results of a multiple ship survey of the Gulf Stream. *Tellus* **3** (1), 1–14.

Fuhs G.W., Demmerle S.D., Canelli E. & Chen M. (1972) Characterization of phosphorus-limited plankton algae (with reflections on the limiting-nutrient concept). *Limnol. Oceanogr. Special Symposia, Vol. I,* 113–33.

Fujino K. & Kang T. (1968) Serum esterase groups of Pacific and Atlantic tunas. *Copeia* 1968 **1,** 56–63.

Gaardner T. & Gran H.H. (1927) Investigation of the production of plankton in the Oslo Fjord. *Repp. P.-V. Cons. int. Explor. Mer.* **42,** 1–48.

Gambell R. (1972) Why all the fuss about whales? *New Scientist* **54** (801) 674–6.

Gardner A.C. (1973) Phosphate production by plankton animals. *J. Cons. int. explor. Mer.* **14,** 220–227.

Garstang W. (1900–3) The impoverishment of the sea. *J. Mar. Biol. Assn. U.K.* **6,** 1–69.

GATTEN R.R. & SARGENT J.R. (1973) Wax ester biosynthesis in calanoid copepods in relation to vertical migration. *Neth. J. Sea Res.* **7**, 150–8.

GAULD D.T. (1964) Feeding in planktonic copepods. pp 239–46. In *Grazing in Terrestrial and Marine Environments. Symposium of the British Ecological Society*, ed. Crisp D.J. Blackwell Scientific Publications, Oxford. 336 pp.

GAUSE G.F. (1932) Experimental studies on the struggle for existence. I. Mixed population of two species of yeast. *J. Exp. Biol.* **9**, 389–402.

GAUSE G.F. (1934) *The struggle for existence*. Baltimore. 163 pp.

GERKING S.D. (1966) Annual growth cycle, growth potential, and growth compensation in the Bluegill Sunfish in northern Indiana lakes. *J. Fish. Res. Bd Can.* **23** (12), 1923–56.

GERKING S.D. (1971) Influence of rate of feeding and body weight on protein metabolism of Bluegill Sunfish. *Physiol. Zool.*, **44** (1), 9–19.

GIBSON M.B. (1955) The effect of salinity and temperature on the pre-adult growth of Guppies. *Copeia* **3**, 241–3.

GILBERT C.H. (1916) Contributions to the life history of the sockeye salmon 3. *Rep. Commnt Fish. Br. Columb.* 1915. 27–64.

GOERING J.J., DUGDALE R.C. & MENZEL D.W. (1966) Estimates of *in situ* rates of nitrogen uptake by *Trichodesmium sp.* in the tropical Atlantic Ocean. *Limnol. Oceanogr.* **11**, 614–20.

GOERING J.J., NELSON D.M. & CARTER J.A. (1973) Silicic acid uptake by natural populations of marine phytoplankton. *Deep-Sea Res.* **20**, 777–89.

GOLD A. (1973) Energy expenditure in animal locomotion. *Science*, **181**, 275–6.

GOLDMAN C.R. (1966) *Primary Productivity in Aquatic Environments.* Univ. of California Press, Berkeley. 464 pp.

GOLDMAN J.C., TENORE K.R. & STANLEY H.I. (1973) Inorganic nitrogen removal from wastewater: effect on phytoplankton growth in coastal marine waters. *Science* **180**, 955–6.

GOODWIN T.W. (1966) *Biochemistry of chloroplasts I* Academic Press, New York and London. 476 pp.

GOODWIN T.W. (1967) *Biochemistry of chloroplasts II* Academic Press, New York and London. 776 pp.

GOSSE P.H. (1853) *A naturalist's rambles on the Devonshire coast.* van Voorst, London, 451 pp.

GRAHAM J.J., CHENOWETH S.R. & DAVIS C.W. (1972) Abundance, distribution, movements and lengths of larval herring along the western coast of the Gulf of Maine. *Fishery Bull. U.S. Dept. Commerce* **70** (2), 307–21.

GRAHAM M. (1924) The annual cycle in the life of the mature cod in the North Sea. *Fishery Invest. (Lond.) Ser.* 2, **6** (6), 77 pp.

GRAHAM M. (1935) Modern theory of exploiting a fishery and application to North Sea trawling. *J. Cons. int. Explor. Mer.* **10**, 264–74.

GRAN H.H. & BRAARAD T. (1935) A quantitative study of the phytoplankton in the Bay of Fundy and the Gulf of Maine (including observations on hydrography, chemistry and turbidity). *J. Biol. Bd Canada* **1**, 279–467.

GRAY J. (1926) The kinetics of growth. *J. Exp. Biol.* **6** (3), 248–74.

GREGG M. (1973) The microstructure of the ocean. *Scientific American* **228** (2), 64–77.

GRENNEY W.J., BELLA D.A. & CURL H.C., JR. (1973) A theoretical approach to interspecific competition in phytoplankton communities. *Amer. Nat.* **107**, 405–21.

GREZE B.S. & BALDINA E.P. (1964) Population dynamics and annual production of *Acartia clausi* Giesbr. and *Centropages kroyeri* Giesbr. in the neritic zone of the Black Sea, *Fish. Res. Bd Can. Translation Series* 893.

GREZE V.N. (1970) The biomass and production of different trophic levels in the pelagic communities of south seas. pp 458–67. In *Marine Food Chains*, ed. Steele J.H. Oliver & Boyd, Edinburgh. 552 pp.

GRILL E.V. (1970) A mathematical model for the marine dissolved silicate cycle. *Deep-Sea Res.* **17,** 245–66.

GRINDLEY J.R. (1964) On the effect of low salinity water on the vertical migration of estuarine zooplankton. *Nature* (*Lond.*) **203,** 4946.

GUILLARD R.L., KILHAM P. & JACKSON T.A. (1973) Kinetics of silicon-limited growth in the marine diatom *Thalassiosira pseudonana* Hasle and Heimdal (= *Cyclotella nana* Hustedt). *J. of Phycol.* **9,** 233–7.

GULLAND J.A. (1956) On the fishing effort in English demersal fisheries. *Fishery Invest.* (*Lond.*) *Ser.* 2, **20** (5), 41 pp.

GULLAND J.A. (1961) Fishing and the stocks of fish at Iceland. *Fishery Invest.* (*Lond.*) *Ser.* 2, **23** (4) 52 pp.

GULLAND J.A. (1969) Manual of methods for fish stock assessment. Pt. 1 Fish population analysis. *FAO Man. Fish. Sci.* **4,** 154 pp.

GULLAND J.A. (1970) The fish resources of the ocean. *FAO Tech Pap.* **97,** 1–425.

GULLAND J.A. (1971a) Ecological aspects of fishery research. *Adv. Ecol. Res.* **7,** 115–76.

GULLAND J.A. (ed.) (1971b) *The Fish Resources of the Ocean.* Fishing News (Books) Ltd., West Byfleet, Surrey, 255 pp.

GULLAND J.A. & CARROZ J.E. (1968) Management of fishery resources. *Adv. Mar. Biol.* **6,** 1–71.

GULLAND J.A. & ROBINSON M.A. (1973) Economics of fishery management. *J. Fish. Res. Bd Can.* **30** (12), Pt. 2, 2042–50.

HAECKEL E. (1866) *General morphology.* 2 vols. G. Reimer, Berlin. Vol. 1, 574 pp.; Vol. 2, 462 pp.

HALPERN D. (1974) Summertime surface diurnal period winds measured over an upwelling region near the Oregon coast. *J. Geophys. Res.* **79,** 2223–30.

HALPERN D., HOLBROOK J.R. & REYNOLDS R.M. (1974) A compilation of wind, current, and temperature measurements: Oregon, July and August, 1973. *Coastal Upwelling Ecosystems Analysis Tech. Rept.* 6, 190 pp.

HANSEN W.J. & DUNBAR M.J. (1970) Biological causes of scattering layers in the Arctic Ocean. *Proc. Int. Symp. biol. sound scattering ocean*, ed. Farquahar G.B. Dept. of the Navy, Washington, 508–26.

HARBAUGH J.W. & BONHAM-CARTER G. (1970) *Computer Simulation in Geology.* John Wiley & Sons, New York, 575 pp.

HARDEN JONES F.R. (1968) *Fish migration.* Arnold, London, 325 pp.

HARDER W. (1968) Reactions of plankton organisms to water stratification. *Limnol. Oceanogr.* **13** (1), 156–68.

HARDIN G. (1960) The competitive exclusion principle. *Science* **131,** 1292–7.

HARDING D. & TALBOT J.W. (1973) Recent studies on the eggs and larvae of the plaice (*Pleuronectes platessa* L.) in the Southern Bight. *Rapp. P.-V. Réun. Cons. int. Explor. Mer.* **164,** 261–9.

HARDY A.C. (1924) The herring in relation to its animate environment. Part I. The food and feeding habits of the herring with special reference to the east coast of England. *Fishery Invest., ser. II*, Vol. 7, no. 3, 53 pp.

HARDY A.C. (1953) Some problems of pelagic life. pp. 101–121. *Essays in Marine Biology* (*Richard Elmhirst Memorial Lectures*), Oliver & Boyd, Edinburgh. 144 pp.

HARDY A.C. (1956) *The Open Sea, its Natural History: The World of Plankton.* Collins, London, 335 pp.

HARDY A.C. & GUNTHER E.R. (1935) The plankton of the South Georgia whaling ground and adjacent waters, 1926–7. *Discovery Rep. 11*, 511–38.

HARGRAVE B.T. (1973) Coupling carbon flow through some pelagic and benthic communities. *J. Fish. Res. Bd Can.* **30,** 1317–26.

HARRIS J.E. (1963) The role of endogenous rhythms in vertical migration. *J. Mar. Biol. Ass. U.K.* **43** (1), 153–66.

HARRISON P.J. (1973) *Continuous culture of the marine diatom* Skeletonema costatum (*Grev.*) *Cleve under silicate limitation.* Ph.D. dissertation, University of Washington, 141 pp.

HART T.J. & CURRIE R.I. (1960) The Benguela Current. *Discovery Rep. 31*, 123–298.

HARVEY H.W. (1937) Note on selective feeding by *Calanus. J. Mar. Biol. Assn. UK. NS.* **22** (1) 97–100.

HARVEY H.W. (1948) Investigations of Measuring Nets of Plymouth. Paper read at Intern. Council. Explor Sea 1948. Mimeo.

HARVEY H.W. (1963) *The Chemistry and Fertility of Sea Waters.* Cambridge University Press, 240 pp.

HARVEY H.W., COOPER L.H.N., LEBOUR M.V. & RUSSELL F.S. (1935) Plankton production and its control. *J. Mar. Biol. Assn. UK.* NS. 20. 407–41.

HAURY L.B. (1973) Sampling bias of a Longhurst-Hardy plankton recorder. *Limnol. Oceanogr.* **18,** 500–6.

HAXO F.T. (1960) The wavelength dependence of photosynthesis and the role of accessory pigments. In *Comparative Biochemistry of Photoreactive Systems*, ed. Allen M.B. Academic Press, New York, pp. 339–60.

HEDGPETH J.W. (1957) Classification of marine environments. In *Treatise on marine ecology and paleoecology, Vol. I, Geol. Soc. Amer., Mem.* **67,** 17–27.

HEINCKE F. (1898) Naturgeschichte des Herings. *Abt. Dtsch. Seefischerei vereins.* **2.** No. 1.

HEINCKE F. (1913) Untersuchungen über die Scholle, Generalbericht I Schollenfischerei und Schonmassregeln. Vorläufige kurze Übersicht über die wichtigsten Ergebnisse des Berichts. *Rapp. P.-V. Réun. Cons. int. Explor. Mer.* **16,** 1–70.

HEINLE D.R. (1966) Production of a calanoid copepod *Acartia tonsa*, in the Patuxent River estuary. *Chesapeake Sci.* **7** (2), 59–74.

HEINRICH A.K. (1962) The life histories of plankton animals and seasonal cycles of plankton communities in the oceans. *J. Cons. int. Explor. Mer.* **27,** 15–24.

HELLERMAN S. (1967) An updated estimate of the wind stress on the world ocean. *Mon. Wea. Rev.* **95,** 607–26.

HENSEN V. (1887) Ueber die Bestimmung des Planktons oder des im Meere treibenden Materials an Pflanzen und Thieren. In Vol. 5, Wiss. Meeresunters 1–106.

HENSEN, V. (1911) Das Leben im Ozean nach zählungen seiner Bewohner. Ubersicht und Resultaten der quantitativen Untersuchungen. *Ergebn. Plankton Expdn der Humboldt Stiftung V.* 406 pp.

HENSEN V. & APSTEIN C. (1897) Die Nordsee-Expedition 1895 des Deutschen Seefischerei-Vereins: Uber die Eimenge der in Winter laichenden Fische. *Wiss. Meeresunters.* **2,** No. 2.

HENTSCHEL E. (1936) Allgemeine Biologie des Süd-Atlantischen Ozeans. *Wiss. Ergebn dt. atlant. Exped. 'Meteor'* **11,** 1–343.

HERBERT D. (1959) Some principles of continuous culture. In *Recent Progress in Microbiology*, ed. Tunevall G. *Int. Congr. Microbiol.* **8,** 331–96.

HERBLAND A., LE BORGNE R. & VOITURIEZ B. (1973) Production primaire, secondaire et regeneration des sels nutritifs dans l'upwelling de Mauretanie. *Doc. Scient. Centre Rech. Oceanogr. Abidjan* **IV** (1), 1–75.

HERMAN S.S. (1963) Vertical migration of the opossum shrimp, *Neomysis americana. Limnol. Oceanogr.* **8,** 228–38.

HERON A.C. (1972) Population ecology of a colonizing species: the pelagic tunicate *Thalia democratica.* II. Population growth rate. *Oecologia* **10,** 294–312.

HERSEY J.B. & BACKUS R.H. (1962) Sound scattering by marine organisms. In *The Sea, Vol. 1*, ed. Hill M.N., Interscience, New York, pp. 498–539.

HERSEY J.B. & MOORE H.B. (1948) Progress report on scattering layer observations in the Atlantic Ocean. *Trans. Am. Geophys. Un.* **29** (3), 341–54.

HESSE H. (translation by H. Rosner, 1951) *Siddhartha.* New Directions, New York, 122 pp.

HESSLER R.R. & JUMARS P.A. (1974) Abyssal community analysis from replicate box cores in the central North Pacific. *Deep-Sea Res.* **21,** 185–209.

HEWITT E.J. (1957) Some aspects of micronutrient element metabolism in plants. *Nature (Lond.)* **180,** 1020–2.

HEYERDAHL T. (1950) *The Kon-Tiki Expedition.* Allen & Unwin, London, 235 pp.

HICKEL W., HAGMEIER E. & DREBES G. (1971) *Gymnodinium* blooms in the Helgoland Bight (North Sea) during August, 1968. *Helgoländer wiss. Meeresunters.* **22,** 401–16.

HIEMSTRA W.H. (1962) A correlation table as an aid for identifying pelagic fish eggs in plankton samples. *J. Cons. int. Explor. Mer.* **27** (1), 101–8.

HILE R. (1936) Age and growth of the Cisco, *Leucichthys artedi* (Le Sueur) in the lakes of the Northeastern Highlands, Wisconsin. *Bull. Bur. Fish., Wash.* **48** (19), 211–317.

HIMMELMAN J.H. & STEELE D.H. (1971) Foods and predators of the green sea urchin *Strongylocentrotus droebachiensis* in Newfoundland waters. *Mar. Biol.* **9,** 315–22.

HJORT J. (1914) Fluctuations in the great fisheries of Northern Europe viewed in the light of biological research. *Rapp. P.-V. Réun. Cons. int. Explor. Mer.* **20,** 1–228.

HOBSON L.A. & LORENZEN C.J. (1972) Relationship of chlorophyll maxima to density structure in the Atlantic Ocean and Gulf of Mexico. *Deep-Sea Res.* **19,** 297–306.

HODGSON W.C. (1934) *The natural history of the herring of the southern North Sea.* Arnold, London, 120 pp.

HÖGLUND H. (1955) Swedish herring tagging experiments, 1949–1953. *Rapp. P.-V. Réun. Cons. int. Explor. Mer.* **140** (2), 19–29.

HOLLAND W.R. (1971) Ocean tracer distributions. *Tellus* **23,** 371–92.

HOLLING C. (1965) The functional response of invertebrate predators to prey density. *Mem. ent. Soc. Can.* **48,** 1–86.

HOLMES R.W. (1957) Solar radiation, submarine daylight and photosynthesis. In *Treatise on Marine Ecology and Paloeocology*, ed. Hedgpeth J.W. *Geological Soc. Amer.* **1,** 109–28. Memoir 67, Waverly Press, Maryland, U.S.A.

HOLT S.J. (1959) Water temperature and cod growth-rate. *J. Cons. int. Explor. Mer.* **24** (2), 374–6.

HOOD D.W. (ed.) (1971) *Impingement of man on the oceans.* Wiley-Interscience, New York, 738 pp.

HOPKINS T.S. (1974) The circulation in an upwelling region, the Washington coast. *Tethys.* **6** (1–2), 375–394.

HORNE R.A. (1969) *Marine chemistry: the structure of water and the hydrosphere.* Wiley-Interscience, New York, 568 pp.

HULBERT E.M., RYTHER J.H. & GUILLARD R.R.L. (1960) The phytoplankton of the Sargasso Sea off Bermuda. *J. Cons. Int. Explor. Mer.* **25,** 115–28.

HURE, J. & SCOTTO DI CARLO B. (1971) Importance quantitative et distribution verticale des Copepodes pélagiques de profondeur de la Mer Tyrrhénienne et la l'Adriatique Méridionale. *Rapp. P.-V. Réun. Comm. int. Explor. Scient. Mer Méditerr.* **20,** 401–4.

HURLBERT S.H., ZEDLER J. & FAIRBANK D. (1972) Ecosystem alteration by mosquito fish (*Gambusia affinis*) predation. *Science* **175,** 639–41.

HUTCHINSON G.E. (1959) Homage to Santa Rosalia. *Amer. Nat.* **93,** 145–59.

HUTCHINSON G.E. (1967) *A treatise on limnology*, 2. Wiley, New York, 1115 pp.

HUYER A., SMITH R.L. & PILLSBURY R.D. (1974) Observations in a coastal upwelling region during a period of variable winds. *Tethys* **6** (1–2), 391–404.

IDYLL C.P. (1973) The anchovy crisis. *Scientific American* **228** (6), 22–9.

IKEDA T. (1970) Relationships between respiration rate and body size in marine plankton animals as a function of the temperature of the habitat. *Bull. Fac. Fish. Hokk. Univ.* **21,** 91–112.

ILES T.D. (1973) Dwarfing or stunting in the genus *Tilapia* (Cichlidae), a possibly unique recruitment mechanism. *Rapp. P.-V. Réun. Cons. int. Explor. Mer.* **164,** 247–54.

INNIS G. (1972) The second derivative and population modeling: another view. *Ecology* **53** (4), 720–3.

INTERNATIONAL COMMISSION FOR THE NORTHWEST ATLANTIC FISHERIES (1970) North Atlantic fish marking symposium. *Spec. Publ. Int. Comm. N.W. Atlant. Fish.* 4, 370 pp.

INTERNATIONAL COUNCIL FOR THE EXPLORATION OF THE SEA (1970) Report of the working group on Atlanto-Scandian herring. *Coop. Res. Rep. ICES* (A), 17, 43 pp.

ISAACS J.D. (1969) The nature of oceanic life. *Sci. Amer.* **221,** 146–62.

ISAACS J.E. & SCHWARTZLOSE R. (1965) Migrant sound scatterers: interaction with the sea floor. *Science* **150,** 1810–13.

IVANENKOV V.N. & ROZANOV A.G. (1961) Hydrogen sulphide contamination of the intermediate waters of the Arabian Sea and the Bay of Bengal. *Okeanologia* **1** (3), 443–9.

IVANOV A.V. (1955) The main features of the organization of Pogonophora. On external digestion in Pogonophora. On the assignment of Class Pogonophora to a separate phylum of Deuterostomia–Brachiata A. Ivanov, phyl. nov. *Syst. Zool.* **4,** 171–77.

IVLEV V.S. (1939) Utilisation of fat and carbohydrate oxidation energy by poikilothermic animals. *Byull. mosk. Obshch. Ispyt. Prir.* **48,** 70–8.

IVLEV V.S. (1961) *Experimental Ecology of the Feeding of Fishes.* Translated by D. Scott. Yale Univ. Press, New Haven, pp. 302.

JENKINS G.M. & WATTS D.G. (1968) *Spectral Analysis and its Applications.* Holden-Day, 525 pp.

JERLOV N.G. (1957) Optical studies of ocean waters. *Rep. Sweden Deep Sea Exped.* **3,** 1–59.

JERLOV N.G. (1963) Optical oceanography. *Oceanogr. mar. Biol. An. Rev.* **1,** 89–114.

JERLOV N.G. (1968) *Optical oceanography.* Elsevier, Amsterdam, 194 pp.

JERLOV N.G. & STEEMANN-NIELSEN E. (1974) *Optical Aspects of Oceanography.* Academic Press, New York, 494 pp.

JOHANNES R.E. (1964) Phosphorus excretion and body size in marine animals: microzooplankton and nutrient regeneration. *Science* **146,** 923–4.

JOHANNES R.E. (1965) Influence of marine protozoa on nutrient regeneration. *Limnol. Oceanogr.* **10,** 434–42.

JOHNSON L. (1966) Experimental determination of food consumption of pike, *Esox lucius*, for growth and maintenance. *J. Fish. Res. Bd Can.* **23,** 1495–505.

JOHNSON P.O. (1970) The Wash sprat fishery. *Fishery Invest.* (*Lond.*) *Ser* 2 **26** (4), 77 pp.

JOHNSTON R. (1969) On salinity and its estimation. *Oceanogr. Mar. Biol. Ann. Rev.* **7,** 31–48.

JOHNSTONE J. (1924) Report on: 1. The Irish Sea cod fishery. 2. The cod as a food fish. 3. Parasites and diseases of the cod. *Fishery Invest.* (*Lond.*) *Ser* 2 **6** (7), 27 pp.

JONES R. (1959) A method of analysis of some tagged haddock returns. *J. Cons. int. Explor. Mer.* **25** (1), 58–72.

JONES R. (1973) Density dependent regulation of the numbers of cod and haddock. *Rapp. P.-V. Réun. Cons. int. Explor. Mer.* **164,** 156–73.

JONES R. (1974) The rate of elimination of food from the stomachs of haddock, cod and whiting. *J. Cons. int. Explor. Mer.* **35.**

JONES R. (in press (b)). Some theoretical observations on the energy requirements of haddock under natural conditions.

JONES R. & HALL W.B. (1973) A simulation model for studying the population dynamics of some fish species. pp. 35–39. In *The mathematical theory of the dynamics of biological populations*, eds Bartlett M.S. & Hiorns R.W. Academic Press, London and New York. 347 pp.

JØRGENSEN C.B. (1966) *Biology of suspension feeding*. Pergamon, Oxford, 357 pp.

JOSEPH J. & SENDNER H. (1958) Über die horizontale diffusion in Meere. *Dt. hydrogr. Z.*, **11,** 49–77.

JUDAY C. (1940) The annual energy budget of an inland lake. *Ecology* **21** (4), 438–50.

KALMIJN A.J. (1971) The electric sense of sharks and rays. *J. Exp. Biol.* **55** (2), 371–83.

KAMPA E. (1955) Euphausiopsin: a new photosensitive pigment from the eyes of euphausiid crustaceans. *Nature* (*Lond.*) **175,** 996–8.

KAMPA E.M. (1970) Photoenvironment and sonic scattering. *Proc. Int. Symp. biol. Sound Scattering in the Ocean*, ed. Farquhar G.B. Dept. of Navy, Washington, D.C., pp. 51–59.

KAMYKOWSKI D. (1974) Possible interactions between planktonic organisms and semi-diurnal internal tides. *Abs. 37th Ann. Mtg. of ASLO, Seattle, June* 1974.

KASHKIN N.I. (1962) On the adaptive significance of seasonal migrations of *Calanus finmarchicus*. *Zool. Zhurnal.* **41** (3), 342–57.

KAWAJIRI M. (1928) On the studies of the population-density of cultured fishes. I. On the influence of population-density of fishes upon the survival-rate and the rate of growth. *J. Imp. Fish. Inst., Tokyo* **24** (1), 8–11.

KEELING D.C. & BOLIN B. (1967) The simultaneous use of chemical tracers in oceanic studies. I. General theory of reservoir models. *Tellus* **19** (4), 566–81.

KELLEY J.C. (1971a) Multivariate oceanographic sampling. *Math. Geol.* **3,** 43–50.

KELLEY J.C. (1971b) Sampling considerations in upwelling studies. *Inves. Pesq.* **35,** 251–60.

KELLEY J.C. (1975) Time-varying distributions of biologically significant variables in the ocean. *Deep-Sea Res.* **22** (10), 679–688.

KELLEY J.C. & MCMANUS D.A. (1970) Hierarchical analysis of variance of shelf sediment texture. *J. Sed. Pet.*, pp. 1335–9.

KENT W.S. (1893) *The Great Barrier Reef of Australia; Its Products and Potentialities.* W.H. Allen London, 387 pp.

KERNER E.H. (1961) On the Volterra-Lotka principle. *Bull. Math. Biophys.* **23,** 141–57.

KERR S.R. (1971) Analysis of laboratory experiments on growth efficiency of fishes. *J. Fish. Res. Bd Can.* **28** (6), 801–8.

KETCHUM B.H. (1939) The absorption of phosphate and nitrate by illuminated cultures of *Nitzschia closterium*. *Am. J. of Bot.* **26,** 399–407.

KETCHUM B.H. (1972) *The Water's Edge; Critical Problems of the Coastal Zone.* MIT Press, Cambridge, 393 pp.

KHMELEVA N.N. (1972) Intensity of generative growth in crustaceans. *Dok. Akad. Nauk SSSR*, **207,** 707–10.

KIERSTEAD H. & SLOBODKIN L.B. (1953) The size of water masses containing plankton blooms. *J. mar. Res.* **12,** 141–7.

KIMBALL H.H. (1928) Amount of solar radiation that reaches the surface of the earth on the land and on the sea, and the methods by which it is measured. *Mon. Wea. Rev.* **56,** 393–9.

KING J.E. & HIDA T.S. (1954) Variations in zooplankton abundance in Hawaiian waters. *Spec. scient. Rep. U.S. Fish Wild. Serv. Fisheries* **118,** 1–66.

KINNE O. (1960) Growth, food intake, and food conversion in a euryplastic fish exposed to different temperatures and salinities. *Physiol. Zool.* **33** (4), 288–317.

KINZER J. (1969) On the quantitative distribution of zooplankton in deep scattering layers. *Deep-Sea Res.* **16,** 117–25.

KINZER L. (1970) On the contribution of ephausiids and other planktonic organisms to deep scattering layers in the Eastern North Atlantic. *Proc. Int. Symp. biol. Sound scattering Ocean.* Dept. of Navy, Wash. D.C. (5), 476–89.

Kirpichnikov V.S. (1970) Selective breeding of carp and interspeciation of fish breeding in ponds. *Izv. Gos. Nauch. Iss. Inst. Oz. i Rech. Ryb. Kohz.* **61,** 249 pp.

Kitching J.A. & Ebling F.J. (1961) The ecology of Lough Ine. XI. The control of algae by *Paracentrotus lividus* (Echinoidea). *J. Anim. Ecol.* **30,** 373–83.

Kleiber M. (1961) *The Fire of Life, An Introduction to Animal Energetics.* John Wiley & Sons, New York, 454 pp.

Koblentz-Mishke O.J., Volkovinsky V.V. & Kabanova J.G. (1970) Plankton primary production of the world ocean, pp. 183–93. In *Scientific Exploration of the South Pacific*, ed. Wooster W.S. National Academy of Sciences, Washington.

Koczy F.F. (1954) A survey on deep-sea features taken during the Swedish deep-sea expedition. *Deep-Sea Res.* **1** (3), 176–84.

Kohler A.C. (1964) Variations in the growth of Atlantic cod (*Gadus morhua* L.). *J. Fish. Res. Bd Can.* **21** (1), 57–100.

Kowal N.E. (1971) A rationale for modeling dynamic ecological systems. In *Systems Analysis and Simulation in Ecology, Vol. I*, ed. Patten B.C. Academic Press, New York, pp. 123–94.

Kraus E.B. (1972) *Atmosphere–ocean interaction.* Oxford University Press, 275 pp.

Krefft G. (1968) Ergebnisse der Forschungsreisen des FFS 'Walther Herwig' nach Südamerika IV *Luciosudis* Fraser Brunner, 1931, ein valides Genus der Familie Scopelosauridae (Osteichthyes, Alepisauroidei). *Arch. Fisch. Wiss.* **19** (213), 95–102.

Krogh A. (1931) Dissolved substances as food of aquatic organisms. *Biol. Rev.* **6,** 412–42.

Krumbein W.C. & Graybill F.A. (1965) *An Introduction to Statistical Models in Geology.* McGraw-Hill, New York, 475 pp.

Kullenberg G. (1972) Apparent horizontal diffusion in stratified vertical shear flow. *Tellus* **24,** 17–28.

Kumlov S.K. (1961) Plankton and the feeding of baleen whales. *Tr. Inst. Okeanol. Akad. Nauk. SSSR* **51,** 142–56.

Kuo H.H. & Veronis G. (1970) Distribution of tracers in the deep oceans of the world. *Deep-Sea Res.* **17,** 29–46.

Kuo H.H. & Veronis G. (1973) The use of oxygen as a test for an abyssal circulation model. *Deep-Sea Res.* **20,** 871–88.

Lack D. (1954) *The Natural Regulation of Animal Numbers.* Clarendon Press, Oxford, 343 pp.

LaFond E.C. (1963) Detailed temperature structures of the sea off Baja California. *Limnol. Oceanogr.* **8,** 417–25.

Lamb H.H. (1972) *Climate: present, past and future. Vol. I. Fundamentals and Climate now.* Methuen, London, 613 pp.

Langmuir I. (1938) Surface motion of water induced by wind. *Science* **87,** 118–23.

Lasker R. (1960) Utilization of carbon by a marine crustacean. Analysis with Carbon-14. *Science* **131,** 1098–100.

Lasker R. (1966) Feeding, growth, respiration and carbon utilization of a euphausiid crustacean. *J. Fish. Res. Bd Canada* **23,** 1291–1317.

Lauff G.H. (ed.) (1967) *Estuaries.* Amer. Ass. Advan. Sci., Washington, D.C., Publ. No. 83, 757 pp.

Lawry J.V. (1974) Lantern fish compare downwelling light and bioluminescence. *Nature (Lond.)* **247,** 155–7.

Lea E. (1911) A study of the growth of herrings. *Publs Circonst. Cons. int. Explor. Mer.* **61,** 35–57.

Lea E. (1913) Further studies concerning the methods of calculating the growth of herrings. *Publs Circonst. Cons. int. Explor. Mer.* **66,** 3–36.

Leach J.H. (1970) Epibenthic algal production in an intertidal mudflat. *Limnol. Oceanogr.* **15,** 514–21.

Lebour M.V. (1918) The food of post larval fish I. *J. Mar. Biol. Ass. NS UK* **11,** 433–69.

Lebour M.V. (1919a) The food of post larval fish II. *J. Mar. Biol. Ass. NS UK* **12,** 22–47.

Lebour M.V. (1919b) The food of post larval III. *J. Mar. Biol. Ass. NS UK* **12,** 261–324.

Le Brasseur R.J. & Kennedy O.D. (1972) The fertilization of Great Central Lake. II. Zooplankton standing stock. *Fish. Bull., U.S.* **70** (1), 25–36.

Le Cren E.D. (1951) The length-weight relationship and seasonal cycle in gonad weight and condition in the perch (*Perca fluviatilis*). *J. Anim. Ecol.* **20,** 201–19.

Lee A.J. (1952) The influence of hydrography on the Bear Island cod fishery. *Rapp. P-v Réun. Cons. int. Explor. Mer.* **131,** 74–102.

Lee R.F., Hirota J. & Barnett A.M. (1971) Distribution and importance of wax esters in marine copepods and other zooplankton. *Deep-Sea Res.* **18** (12), 1147–66.

Lee R.F., Nevenzel J.C. & Paffenhöfer G. (1971) Importance of wax esters and other lipids in the marine food chain: phytoplankton and copepods. *Mar. Biol.* **9** (2), 99–108.

Legand M. (1958) Étude sommaire des variations quantitatives diurnes du zooplankton autour de la Nouvelle Calédonie. *Inst. Franc. d'Océanie Nouméa, Océanogr. Rapp. Sci.* **6,** 1–42.

Lendenvald R.V. (1901) Planktonuntersuchungen in Grossteiche bei Hirschberge. *Biol. Cblt.* **21,** 182–8.

Leslie P.H. (1948) Some further notes on the use of matrices in population mathematics. *Biometrika* **35,** 213–45.

Lewis J.R. (1964) *The ecology of rocky shores.* The English Universities Press Ltd. London. 323 pp.

Li C.C. (1955) *Population genetics.* Univ. Chicago Press, Chicago and London, 366 pp.

Liebermann L.N. (1962) Other electromagnetic radiation. In *The Sea, Vol.* 1, ed. Hill M.N., Interscience, New York, pp. 469–75.

de Ligny W. (1969) Serological and biochemical studies on fish populations. *Oceanogr. Mar. Biol. Ann. Rev.* **7,** 411–513.

Lindeman R.L. (1941) Seasonal food-cycle dynamics in a senescent lake. *Amer. Mid. Nat.* **26,** 636–73.

Lindeman R.L. (1942) The trophic-dynamic aspect of ecology. *Ecology* **23,** 399–418.

Lissmann H.W. (1958) On the function and evolution of electric organs in fish. *J. Exp. Biol.* **35** (1), 156–91.

Lloyd I.J. (1971) Primary production off the coast of North-west Africa. *J. Cons. int. Explor. Mer.* **33** (3), 312–23.

Lohmann H. (1909) Untersuchungen zur Feststellung des Vollständigen Gehaltes des Meeres an Plankton. *Wiss. Meeresunters. Abt. Kiel. NF* **10,** 129–370.

Lohmann H. (1933) Appendiculariae. *Kükenthal u Krumbach's Handbuch der Zool.* **5** (2), 3–192.

Longhurst A.R. (1967a) Vertical distribution of zooplankton in relation to the eastern Pacific oxygen minimum. *Deep-Sea Res.* **14,** 51–63.

Longhurst A.R. (1967b) The pelagic phase of *Pleuroncodes planipes* in the California current. *CalCOFI Rpts.* **11,** 142–54.

Longhurst A.R. (1968) Distribution of the larvae of *Pleuroncodes planipes* in the California Current. *Limnol. Oceanog.* **13** (1), 143–55.

Longhurst A.R. (1971) The clupeid resources of tropical seas. *Oceanogr. Mar. Biol. Ann. Rev.* **9,** 349–85.

Longhurst A.R., Lorenzen C.J. & Thomas W.H. (1967) The role of pelagic crabs in the grazing of phytoplankton off Baja California. *Ecology* **48** (2), 192–200.

Longhurst A.R., Reith A.D., Bower R.E. & Seibert D.L.R. (1966) A new system for the collection of multiple serial plankton samples. *Deep-Sea Res.* **13**, 213–22.

Longuet-Higgins M.S. (1965) Some dynamical aspects of ocean currents. *Quant. J. Roy. Meteorol. Soc.* **91**, 425–51.

Lorenzen C.J. (1966) A method for the continuous measurement of *in vivo* chlorophyll concentration. *Deep-Sea Res.* **13**, 223–7.

Lorenzen C.J. (1967) Vertical distribution of chlorophyll and phaeopigments: Baja, California. *Deep-Sea Res.* **14**, 735–46.

Lorenzen C.J. (1971) Continuity in the distribution of surface chlorophyll. *J. Cons. int. Explor. Mer.* **34**, 18–23.

Lorenzen C.J. (1972) Extinction of light in the ocean by phytoplankton. *J. Cons. int. Explor. Mer.* **34**, 262–7.

Lotka A.J. (1925) *Elements of physical biology*. Williams & Wilkins Coy, Baltimore. 460 pp.

Love C.M. (ed.) (1970) EASTROPAC Atlas, *Circ. Fish Widl. Serv. Wash.* **330** (1), 1–12+ plates.

Luckinbill L.S. (1973) Coexistence in laboratory populations of *Paramecium aurelia* and its predator *Didinium nasutum*. *Ecology* **54**, 1320–7.

Lund J.W.G. (1950) Studies on *Asterionella formosa* Hass II. Nutrient depletion and the spring maximum. Part I. Observations on Windermere, Esthwaite and Blelham Tarn. Part. 2. Discussion. *J. Ecol.* **38**, 1–35.

Lyubimova T.G., Naumov A.G. & Lagunov L.L. (1973) Prospects of the utilization of krill and other non-conventional resources of the world ocean. *J. Fish. Res. Bd Can.* **30** (12, Pt. 2), 2196–201.

Ma Y.H., Peura R.A. & Brown M.E. (1971) An analog and digital simulation of the germination of spores. *Proc.* 1971 *Summer Computer Simulation Conference, Board of Simulation Conferences, Denver, Colorado*, pp. 763–8.

McAllister C.D. (1969) Aspects of estimating zooplankton production from phytoplankton production. *J. Fish Res. Bd Can.* **26**, 199–220.

McAllister C.D. (1971) Some aspects of nocturnal and continuous grazing by planktonic herbivores in relation to production studies. *Tech. Rep. Fish. Res. Bd Can.* **248**, 1–281.

McAllister C.D., Parsons T.R. & Strickland J.D.H. (1960) Primary productivity at Station 'p' in the northeast Pacific Ocean. *J. Cons. int. Explor. Mer.* **25**, 240–59.

McArthur R.H. (1962) Some generalized theories of natural selection. *Proc. Nat. Acad. Sci.* **48**, 1893–7.

MacArthur R.H. (1965) Fluctuations of animal populations and a measure of community stability. *Ecology* **36**, 533–6.

MacArthur R.H. & Connell J.H. (1966) *The biology of populations*. John Wiley & Sons, New York. 200 pp.

McCarthy J.J. (1972) The uptake of urea by natural populations of marine phytoplankton. *Limnol. Oceanogr.* **17**, 738–48.

McCarthy J.J. & Whitledge T.E. (1972) Nitrogen excretion by anchovy (*Engraulis mordax* and *E. ringens*) and jack mackerel (*Trachurus symmetricus*). *Fish. Bull.* **70**, 395–401.

Macfadyen A. (1964) Energy flow in ecosystems and its exploitation by grazing. pp. 3–20. In *Grazing in Terrestrial and Marine Environments. Symposium No.* 4, *British Ecological Society*, ed. Crisp D.J. Blackwell Scientific Publications Oxford. 336 pp.

MacGinitie G.E. & MacGinitie N. (1949) *Natural History of Marine Animals*. McGraw-Hill, New York.

McIntyre A.D. (1969) Ecology of marine meiobenthos. *Biol. Rev.* **44**, 245–90.

McIntyre A.D. (1971) Control factors on meiofauna populations. *Thalassia Jugosl.* **7,** 209–15.

McIntyre A.D., Munro A.L.S. & Steele J.H. (1970) Energy flow in a sand ecosystem. In *Marine Food Chains*, ed. Steele J.H. pp. 19–31, Oliver & Boyd, Edinburgh. 552 pp.

McIntyre D.B. (1963) Precision and resolution in geochronometry. In *Fabric of Geology*. ed. Albritton C.C. Freeman, Stanford, California, pp. 112–34.

MacIsaac J.J. & Dugdale R.C. (1969) The kinetics of nitrate and ammonia uptake by natural populations of marine phytoplankton. *Deep-Sea Res.* **16,** 45–57.

MacIsaac J.J. & Dugdale R.C. (1972) Interactions of light and inorganic nitrogen in controlling nitrogen uptake in the sea. *Deep-Sea Res.* **19,** 209–32.

MacKereth F.H.J. (1953) Phosphorus utilization by *Asterionella formosa* Hass. *J. Exp. Bot.* **4,** 296–313.

Mackintosh N.A. (1937) The seasonal circulation of the Antarctic macroplankton. '*Discovery*' *Rep.* **16,** 385–412.

McLaren I.A. (1963) Effects of temperature on growth of zooplankton and the adaptive value of vertical migration. *J. Fish. Res. Bd Can.* **20,** 685–727.

McNae W. (1968) A general account of the fauna and flora of Mangrove swamps and forests in the Indo-West Pacific region. *Adv. Mar. Biol.* **6,** 74–270.

McNeil W. (1964) Redd superimposition and egg capacity of pink salmon spawning beds. *J. Fish. Res. Bd Can.* **21** (6), 1385–96.

McNider R.T. & O'Brien J.J. (1973) A multi-layer transient model of coastal upwelling. *J. Phys. Oceanogr.* **3** (3), 258–73.

Maeda S. & Kishimoto R. (1970) Upwelling off the coast of Peru. *J. oceanogr. Soc. Japan* **26** (5), 300–9.

Mague T.H., Weare N.M. & Holm-Hansen O. (1974) Nitrogen fixation in the north Pacific Ocean. *Mar. Biol.* **24,** 109–19.

Maigret J. (1972) Rapport de la campagne experimental de pêche de la sardinelle et espèces voisines. Observations concernant l'oceanographie et la biologie espèces. SCET Intern. No. 083, Puteaux, France.

Makarova N.P. & Zaika V.Ye. (1971) Relationship between animal growth and quantity of assimilated food. *Hydrobiol. J.* **7,** 1–8.

Malone T.C. (1971) The relative importance of nanoplankton and net plankton as primary producers in the California current system. *Mar. Biol.* **10,** 285–9.

Manabe S. & Bryan K. (1969) Climate and the ocean circulation. *Mon. Wea. Rev.* **97,** 739–827.

Manabe S., Holloway J.L. jr. & Stone H.M. (1970) Tropical circulation in a time-integration of a global model of the atmosphere. *J. Atmos. Sci.* **27** (4), 580–613.

Mandelstam J. & McQuillen K. (1968) *Biochemistry of Bacterial Growth.* Blackwell Scientific Publications, Oxford, 540 pp.

Mann K.H. (1965) Energy transformations by a population of fish in the River Thames. *J. Anim. Ecol.* **34,** 253–75.

Mann K.H. (1969) The dynamics of aquatic ecosystems. *Adv. Ecol. Res.* **6,** 1–81.

Mann K.H. (1972) Macrophyte production and detritus food chains in coastal waters. *Mem. Ist. Ital. Idrobiol.* **29** (Suppl.), 353–83.

Mann K.H. (1973) Seaweeds: their productivity and strategy for growth. *Science* **182,** 975–81.

Mann K.H. (1975) Relationship between morphometry and biological functioning in three coastal inlets of Nova Scotia. *Proc. Second Int. Estuarine Res. Conf.* Academic Press.

Mann K.H. & Breen P.A. (1972) The relation between lobster abundance, sea urchins, and kelp beds. *J. Fish. Res. Bd Can.* **29,** 603–9.

MARGALEF R. (1955) Los organismos indicadores en la limnologia. *Biologia de las aguas continentales*, **12,** 300 pp.

MARGALEF R. (1963) Succession in marine populations. In *Advancing Frontiers of Plant Sciences*, *Vol.* 2, ed. Vira R., pp. 137–88.

MARGALEF R. (1967) Some concepts relative to the organization of plankton. *Oceanogr. Mar. Biol. Ann. Rev.* **5,** 257–89.

MARGALEF R. (1968) *Perspectives in ecological theory*. Chicago University Press. 111 pp.

MARGALEF R., HERRERA J., STEYAERT M. & STEYAERT J. (1966) Distribution et caracteristiques des comunautés phytoplanctoniques dans le bassin tyrrhénien de la meditérranée et la fin de stratification éstivale de l'année 1963. *Bull. Inst. Royal Sci. Natur. Bel.* **42,** 2–41.

MARR J.W.S. (1962) The natural history and geography of the Antarctic krill (*Euphausia superba Dana*) *Discovery Rep.* **32,** 33–464.

MARSHALL S.M. (1973) Respiration and feeding in copepods. *Adv. mar. Biol.* **11,** 57–120.

MARSHALL S.M. & ORR A.P. (1952) On the Biology of *Calanus finmarchicus*. VII. Factors affecting egg production. *J. mar. biol. Ass. U.K.* **30,** 527–48.

MARSHALL S.M. & ORR A.P. (1955) *The biology of a marine copepod.* Oliver & Boyd, Edinburgh, 195 pp.

MARSHALL S.M. & ORR A.P. (1955b) On the Biology of *Calanus finmarchicus*. VIII. Food uptake, assimilation, and excretion in adult and stage V Calanus. *J. mar. biol. Ass. U.K.* **34,** 495–529.

MARSHALL S.M. & ORR A.P. (1958) On the biology of *Calanus finmarchicus*. X. Seasonal changes in oxygen consumption. *J. mar. biol. Ass. U.K.* **37,** 459–72.

MARTIN N.V. (1970) Long-term effects of diet on the biology of the lake trout and the fishery in Lake Opeongo, Ontario. *J. Fish. Res. Bd Can.* **27,** 125–46.

MASLOV N.A. (1944) Bottom fishes of the Barents Sea and their fisheries. V. Migrations of the cod. *Trans. Knip. Polar Inst. Sea Fish and Oceanogr. Murmansk* **8,** 3–186.

MATSUMOTO W. (1966) Distribution and abundance of tuna larvae in the Pacific Ocean. *Proc. Governors' Conf. on Central Pacific Fishery Resources Honolulu* 1966, ed. Manar T.A. pp. 221–30. U.S. Dept. Interior.

MAUCHLINE J. (1972) The biology of bathypelagic organisms especially crustacea. *Deep-Sea Res.* **19,** 753–80.

MAUCHLINE J. & FISHER L.R. (1969) The biology of euphausiids. *Adv. mar. Biol.* **7,** 1–454.

MAY R.M. (1973a) *Stability and complexity in model ecosystems*. Princeton University Press, Princeton, New Jersey, 235 pp.

MAY R.M. (1973b) Stability in randomly fluctuating versus deterministic environments. *Amer. Nat.* **107,** 621–50.

Mediprod (1973) Géneralités sur la Campagne CINECA–CHARCOT II (15 Mars–29 Avril 1971) *Tethys* **6** (1–2) 33–42.

MEEK A. (1917) *The migrations of fish*. Arnold, London, 427 pp.

MENASVETA D. (1968) Potential demersal fish resources of the Sunda Shelf. In *The Kuroshio: a symposium on the Japan current*, ed. Marr J.C. Honolulu, East–West Center Press, pp. 525–55.

MENZEL D.W. (1960) Utilization of food by a Bermuda reef fish *Ephinephelus guttatus*. *J. Cons. int. Explor. Mer.* **25** (2), 216–22.

MERRIMAN D. (1941) Studies on the Striped Bass (*Roccus saxatilus*) of the Atlantic coast. *Fishery Bull. Fish. Wildl. Serv. U.S.* **35,** 1–77.

MICHAEL E.L. (1911) Classification and vertical distribution of the Chaetognaths of the San Diego region. *Univ. Calif. Publs Zool.* **8,** 21.

MILEIKOVSKY S.A. (1968) Some common features in the drift of pelagic larvae and juvenile stages of bottom invertebrates with marine currents in temperate regions. *Sarsia* **34,** 209–16.

Millenbach C. (1950) Rainbow broodstock selection and observations on its application to fishery management. *Progve Fish. Cult.* **12,** 151–2.

Miller R.J. & Mann K.H. (1973) Ecological energetics of the seaweed zone in a marine bay on the Atlantic coast of Canada. III. Energy transformations by sea urchins. *Mar. Biol.* **18,** 99–114.

Milne D.J. (1957) Recent British Columbia spring and coho salmon tagging experiments and a comparison with those conducted from 1925 to 1930. *Bull. Fish. Res. Bd Can.* **113,** 56 pp.

Moav R. & Wohlfarth G.W. (1967) Genetic improvement of yield in carp. *F.A.O. Fish. Rep.* **44** (4), 12–29.

MODE Scientific Council (1973) *MODE-I*: (Mid-Ocean Dynamics Experiment—One): *The program and the plan.* International Decade of Ocean Exploration—The National Science Foundation and the Office of Naval Research, Washington, D.C., 38 pp.

Moiseev P.A. (1969) *Living Resources of the World Ocean.* Pishchevaia promyshlennost, Moscow, 338 pp. (in Russian). English translation, 1971, Jerusalem, Israel Program for Scientific Translations, IPST Cat. No. 5954, 334 pp.

Monin A.S. & Yaglom A.M. (1971) *Statistical fluid mechanics: mechanics of turbulence, Vol.* 1. MIT Press, Cambridge, Mass., 769 pp.

Monod J. (1950) La technique de culture continue. Théorie et applications. *Ann. Inst. Pasteur.* **79,** 390–410.

Mooers C.N.K. & Allen J.S. (1973) *Final Report of the Coastal Upwelling Ecosystems Analysis Summer* 1973 *Theoretical Workshop.* School of Oceanography, Oregon State University, Corvallis, Oregon. 137 pp.

Moore H.B. (1949) The zooplankton of the upper waters of the Bermuda area of the N. Atlantic. *Bull. Bingham Oceangr. Coll.* **12,** 1–97.

Moore H.B. (1958) *Marine Ecology.* Wiley, New York, 493 pp.

Morton J. & Miller M. (1968) *The New Zealand Sea Shore.* Collins, London.

Mullin M.M. (1963) Some factors affecting the feeding of marine copepods of the genus *Calanus. Limnol. Oceanogr.* **8,** 239–50.

Mullin M.M. (1969) Production of zooplankton in the ocean: The present status and problems. *Oceanogr. Mar. Biol. Ann. Rev.* **7,** 293–314.

Mullin M.M. & Brooks E.R. (1970) Growth and metabolism of two planktonic, marine copepods as influenced by temperature and type of food. pp. 74–95. In *Marine Food Chains,* ed. Steele J.H. Oliver & Boyd, Edinburgh, 552 pp.

Mullin M.M. & Brooks E.R. (1973) Vertical distribution of juvenile *Calanus* and phytoplankton within the upper 50 m of water off La Jolla, California, pp. 347–54. In *Biological Oceanography of the northern North Pacific,* ed. Takenouti A.Y Idemitsu Shoten, Tokyo, 626 pp.

Munk W.H. (1950) On the wind-driven ocean circulation. *J. Meteor.* **7,** 79–93.

Munk W.H. (1966) Abyssal recipes. *Deep-Sea Res.* **13,** 707–30.

Munro A.L.S. & Brock T.D. (1968) Distinction between bacterial and algal utilization of soluble substances in the sea. *J. Gen. Microbiol.* **51,** 35–42.

Murphy G.I. (1966) Population biology of the Pacific sardine (*Sardinops caerulea*) *Proc. Calif. Acad. Sci.* 34 (1), 1–84.

Murray J. (1885) *Narrative report of the Challenger expedition.* London.

Murray J. & Hjort J. (1912) *The Depths of the Ocean.* MacMillan, London, 821 pp.

Namias J. (1972) Large-scale and long-term fluctuations in some atmospheric and oceanic variables. pp. 27–48. In *The changing chemistry of the oceans,* ed. Dyrssen D. & Jagner D., Wiley-Interscience, New York, 365 pp.

NAMIAS J. (1973) Response of the Equatorial Countercurrent to the subtropical atmosphere. *Science* **181,** 1244–5.

NEESS J. & DUGDALE R.C. (1959) Computation of production for populations of aquatic midge larvae. *Ecology* **40,** 425–30.

NEESS J., DUGDALE R.C., DUGDALE V.A. & GOERING J.J. (1962) Nitrogen metabolism in lakes. I. Measurement of nitrogen fixation with N-15. *Limnol. Oceanogr.* **7,** 163–9.

NEMOTO T. (1967) Feeding pattern of euphausiids and differentiations in their body characteristics. *Inf. Bull. Plankton. Japan (Comm. No. Dr. Y. Matsue)*, 157–71.

NEUMANN G. & PIERSON W.J. (1966) *Principles of physical oceanography*. Prentice-Hall, Englewood Cliffs, N.J. 545 pp.

NEVENZEL J.C. (1970) Occurrence, function and biosynthesis of wax esters in marine organisms. *Lipids* **5,** 308–19.

NICOL J.A.C. (1960) Spectral composition of the light of the lantern-fish, *Myctophum punctatum*. *J. Mar. biol. Assn. U.K.* **39,** 27–32.

NICOL J.A.C. (1967) In *The Biology of Marine Animals, 2nd edition*, pp. 532–77. Pitman, London, 699 pp.

NIKITINE B.N. (1929) Les migrations verticales saisonières des organismes plactoniques dans la Mer Noire. *Bull. Inst. Oceanogr. Monaco* **540,** 1–24.

NISHIZAWA S. (1969) Suspended material in the sea. II. Re-evaluation of the hypotheses. *Bull. Plankton Soc. Japan* **16,** 1–42.

O'BRIEN J.J. (1972) CUE-1 Meteorological Atlas. *Coastal Upwelling Ecosystems Analysis Atlas* **1,** 309 pp.

O'BRIEN J.J. (1974) Models of coastal upwelling. In *Symposium on numerical models of ocean circulation*. Nat'l. Acad. Sci. U.S.A., Washington, D.C., 204–215.

O'BRIEN J.J. & HURLBURT H.E. (1972) A numerical model of coastal upwelling. *J. Phys. Oceanogr.* **2,** 14–26.

O'BRIEN J.J. & REID R.O. (1967) The non-linear response of a two-layer, baroclinic ocean to a stationary, axially-symmetric hurricane: Part 1. *J. Atmos. Sci.* **24,** 197–207.

O'CONNELL C.P. & LEONG R.J.H. (1963) A towed pump and shipboard filtering system for sampling small zooplankters. *U.S. Fish. & Wildlife Serv. Spec. Sci. Rept. Fisheries No.* 452, 19 pp. U.S. Dept. Interior.

ODUM H.T. (1971) *Environment, Power, and Society*. Wiley-Interscience, New York, 331 pp.

OKUBO A. (1962) Horizontal diffusion from an instantaneous point source due to oceanic turbulence. *Tech. Rep. Chesapeake Bay Inst.* (32). Johns Hopkins University Press, Baltimore. 123 pp.

OKUBO A. (1971) Oceanic diffusion diagrams. *Deep-Sea Res.* **18,** 789–802.

OKUBO A. (1971) Horizontal and vertical mixing in the sea. pp. 89–168. In *Impingement of man on the oceans*, ed. Hood D.W. Wiley-Interscience, New York, 738 pp.

OMORI M. (1970) Variations of length, weight, respiratory rate, and chemical composition of *Calanus cristatus* in relation to its food and feeding, pp. 113–26. In *Marine Food Chains*, ed. Steele J.H. Oliver & Boyd, Edinburgh, 552 pp.

OSBORN T.R. (1974) Vertical profiling of velocity microstructure. *J. of Phys. Oceanogr.* **4,** 109–15.

OSTVEDT O.J. (1955) Zooplankton investigations from Weather Ship 'M' in the Norwegian Sea 1948–49 II: the annual vertical migration and its role in the life-history of copepods. *Hvalråd Skr.* **40,** 1–93.

OTHMER D.R. & ROELS O.A. (1973) Power, freshwater, and food from cold, deep sea water. *Science* **182,** 121–5.

OTSU T. (1960) Albacore migration and growth in the North Pacific ocean as estimated from tag recoveries. *Pacific Sci.* **14** (3), 257–66.

OTTESTAD P. (1942) On periodical variations in the yield of the great sea fisheries and the possibility of establishing yield prognoses. *Fisk Dir. Skr.* (*Ser. Havunders*) **7** (5), 11 pp.

PAASCHE E. (1973) Silicon and the ecology of marine plankton diatoms. I. *Thalassiosira pseudonana* (*Cyclotella nana*) grown in a chemostat with silicate as limiting nutrient. *Mar. Biol.* **19,** 117–26.

PAFFENHÖFER G.A. (1970) Cultivation of *Calanus helgolandicus* under controlled conditions. *Helgoländer Wiss. Meeresunters* **20,** 346–59.

PAINE R.T. (1965) Natural history, limiting factors and energetics of the opisthobranch *Navanax inermis. Ecology* **46,** 603–19.

PAINE R.T. (1966) Food web complexity and species diversity. *Amer. Natur.* **100,** 65–75.

PAINE R.T. (1969) A note on trophic complexity and community stability. *Amer. Natur.* **103,** 91–3.

PAINE R.T. (1971) A short-term experimental investigation of resource partitioning in a New Zealand rocky intertidal habitat. *Ecology* **52,** 1096–106.

PALOHEIMO J.E. & DICKIE L.M. (1966) Food and growth of fishes. III. Relation among food, body size and growth efficiency. *J. Fish. Res. Bd Can.* **23,** 1209–48.

PALOHEIMO J.E. & DICKIE L.M. (1970) Production and food supply. pp. 499–527. In *Marine Food Chains*, ed. Steele J.H. Oliver & Boyd, Edinburgh, 552 pp.

PAMATMAT M.M. (1971) Oxygen consumption by the seabed. 4. Shipboard and laboratory experiments. *Limnol. Oceanogr.* **16,** 536–50.

PANDIAN T.J. (1970) Intake and conversion of food in the fish *Limanda limanda* exposed to different temperatures. *Mar. Biol.* **5,** 1–17.

PARK O. (1963) Animal Ecology. *Encyclopaedia Britannica VII*, 912–23.

PARKER G.H. (1902) The reactions of copepods to various stimuli and the bearing of this on daily depth migration. *Bull. U.S. Fish. Comm.* **21,** 103.

PARKER R.R. & LARKIN P.A. (1959) A concept of growth in fishes. *J. Fish. Res. Bd Can.* **16** (5), 721–45.

PARRISH B.B. (ed.) (1973) Stock and recruitment. *Rapp. P.-V. Réun. Cons. int. Explor. Mer.* 164, 372 pp.

PARRISH B.B. & SAVILLE A. (1966) The biology of the north-east Atlantic herring population. *Oceanogr. Mar. Biol. Ann. Rev.* **3,** 323–73.

PARRISH B.B., SAVILLE A., CRAIG R.E., BAXTER I.G. & PRIESTLEY R. (1959) Observations on herring spawning and larval distribution in the Firth of Clyde in 1958. *J. Mar. Biol. Ass. UK. NS.* **38,** 445–53.

PARRISH J.D. & SAILA S.B. (1970) Interspecific competition, predation, and species diversity. *J. theor. Biol.* **27,** 207–20.

PARSONS T.R. (1969) The use of particle size spectra in determining structure of a plankton community. *J. oceanogr. Soc. Japan* **25,** 172–81.

PARSONS T.R. (1972) Size fractionation of primary producers in the subarctic Pacific Ocean. pp. 275–8. In *Biological Oceanography of the Northern North Pacific Ocean*, ed. Takenout, A.Y. Idemitsu Shoten, Tokyo, 626 pp.

PARSONS T.R. & LEBRASSEUR R.J. (1970) The availability of food to different trophic levels in the marine food chain. pp. 325–43. In *Marine Food Chains*, ed. Steele J.H. Oliver & Boyd, Edinburgh, 552 pp.

PARSONS, T.R., LEBRASSEUR R.J. & FULTON J.D. (1967) Some observations on the dependence of zooplankton grazing on the cell size and concentration of phytoplankton blooms. *J. oceanogr. Soc. Jap.* **23** (1), 10–17.

PARSONS T.R., LEBRASSEUR R.J., FULTON J.D. & KENNEDY O.D. (1969) Production studies in the Strait of Georgia. Part II. Secondary production under the Fraser River plume, February–May 1967. *J. exp. mar. Biol. Ecol.* **3,** 39–50.

PARSONS T.R., STEPHENS K. & TAKAHASHI M. (1972) The fertilization of Great Central Lake. I. Effect on primary production. *Fish. Bull. U.S.* **70** (1), 13–23.

PARSONS T.R. & TAKAHASHI M. (1973a) Environmental control of phytoplankton cell size. *Limnol. Oceanogr.* **18** (4), 511–15.

PARSONS T.R. & TAKAHASHI M. (1973b) *Biological Oceanographic Processes.* Pergamon, Oxford, 186 pp.

PATTEN B.C. (1968) Mathematical models of phytoplankton production. *Int. Rev. ges. Hydrobiol.* **53,** 357–408.

PATTEN B.C. (1971) A primer for ecological modeling and simulation with analog and digital computers. In *Systems Analysis and Simulation in Ecology, Vol. I,* ed. Patten B.C. Academic Press, New York, pp. 3–121.

PAULIK G.J. (1971) Anchovies, birds and fishermen in the Peru Current. In *Environment: resources, pollution and society,* ed. Murdoch W.W. Sinauer Associates, Inc. Stamford, Conn., pp. 158–85.

PAVLOU S.P., FRIEDERICHS G. & MACISAAC J.J. (1974) A modified tracer technique for the determination of nitrogen uptake by marine phytoplankton. *Analyt. Biochem.* **61,** 16–24.

PEARCY W.G. (1962) An estuarine population of Winter flounder *Pseudopleuronectes americanus* (Walbaum). *Bull. Bingham Oceanogr.* **18,** 1–78.

PEARL R. (1930) *The Biology of Population Growth.* Alfred A. Knopf, New York. 330 pp.

PEDLOSKY J. (1971) Geophysical fluid dynamics. *Lectures in Applied Mathematics* **13,** 1–60.

PEER D.L. (1970) Relation between biomass, productivity, and loss to predators in a population of a marine benthic polychaete *Pectinaria hyperborea. J. Fish Res. Bd Can.* **27,** 2143–53.

PELLA J.J. & TOMLINSON P.K. (1969) A generalized stock production model. *Bull. Inter-Am. Trop. Tuna Comm.* **13** (3), 421–96.

PENTELOW F.T.K. (1939) The relation between growth and food consumption in the brown trout (*Salmo trutta*). *J. exp. Biol.* **16,** 446–73.

PÉRÈS J.M. & DEVEZE L. (1963) *Océanographie biologique et biologie marine. II. La Vie pélagique.* Presse Univ. Paris, 514 pp.

PERKIN R.G. & WALKER E.R. (1972) Salinity calculations from *in situ* measurements. *J. Geophys. Res.* **77,** 6618–21.

PETERSEN C.G.J. (1893) On the biology of our flatfishes and on the decrease of our flatfish fisheries. *Rep. Dan. Biol. Stn* **4,** 1–146.

PETERSEN C.G.J. (1913) Valuation of the sea. II. The animal communities of the sea bottom and their importance for marine zoogeography. *Rep. Dan. Biol. Stn* **21,** 1–44.

PETERSEN C.G.J. (1918) The sea bottom and its production of fish food. *Rep. Dan. Biol. Stn* **25,** 1–62.

PETIPA T.S. (1964) Diurnal rhythms of the consumption and accumulation of fat in *Calanus helgolandicus* (Claus) in the Black Sea. *Dokl. Akad. Nauk. SSSR* **156,** 361–4.

PETIPA T.S. (1966) On the energy balance of *Calanus helgolandicus* in the Black Sea. *Physiol. Mar. Animals* (Akad. Nauk. SSSR Oceanogr. Comm. 60–87).

PETIPA T.S. & MAKAROVA N.P. (1969) Dependence of phytoplankton production on rhythm and rate of elimination. *Mar. Biol.* **3** (3), 191–5.

PETIPA T.S., PAVLOVA E.V. & MIRONOV G.N. (1970) The food web structure, utilization and transport of energy by trophic levels in the planktonic communities. pp. 142–167. In *Marine Food Chains,* ed. Steele J.H. Oliver & Boyd, Edinburgh, 552 pp.

PHILLIPS O.M. (1966) *The dynamics of the upper ocean.* Cambridge University Press. 261 pp.

PIELOU E.C. (1969) *An introduction to mathematical ecology.* Wiley-Interscience, New York, 286 pp.

PILLSBURY R.D., BOTTERO J.S. STILL R.E. & GILBERT W.E. (1974) A compilation of observations from moored current meters, v. VII. Oregon continental shelf, July–August 1973. *Data Report* 58, School of Oceanography, Oregon State University, ref. 74–7, 87 pp.

PLATT T. (1972) Local phytoplankton abundance and turbulence. *Deep-Sea Res.* **19,** 183–7.

PLATT T., DICKIE L.M. & TRITES R.W. (1970) Spatial heterogeneity of phytoplankton in a near-shore environment. *J. Fish. Res. Bd Can.* **27,** 1453–73.

PODUSHKO YU.N. (1970) The connection between the biological characteristics and the population dynamics of the smelt (*Osmerus eperlanus dentex* (Steindachner)) spawning in the Armur. *J. Ichthyol.* **10** (5), 602–9.

POMEROY L.R. (1959) Algal productivity in salt marshes of Georgia. *Limnol. Oceanogr.* **4,** 386–97.

POMEROY L.R. (1960) Residence time of dissolved phosphate in natural waters. *Science* **131,** 1731–2.

POMEROY L.R., MATHEWS H.M. & MIN H.S. (1963) Excretion of phosphate and soluble organic phosphorus compounds by zooplankton. *Limnol. Oceanogr.* **8** (1), 50–5.

POULET S.A. (1973) Grazing of *Pseudocalanus minutus* on naturally occurring particulate matter. *Limnol. Oceanogr.* **18,** 564–73.

POULSEN E.M. (1931) Biological investigations upon the cod in Danish waters. *Meddr. Komm Danm Fisk-og Havunders Serie Fiskeri.* **9** (1), 149 pp.

POURRIOT R. (1974) *Ann. Hydrobiol.* **5** (1) 43–55.

PRAKASH A. & TAYLOR F.J.R. (1966) A 'red water' bloom of *Gonyaulax acatenella* in the Strait of Georgia and its relation to paralytic shellfish toxicity. *J. Fish. Res. Bd Can.* **23,** 1265–70.

PRITCHARD A.L. (1948) A discussion of the mortality in pink salmon (*Oncorhynchus gorbuscha*) during their period of marine life. *Trans. Roy. Soc. Can.* **42** (V), 125–33.

PÜTTER A. (1909) *Die Ernährung der Wassertiere und der Stoffhaushalt der Gewässer* Gustav Fischer, Jena, 168 pp.

PÜTTER A. (1920) Studien über physiologische Ähnlichkeit. VI. Wachstums ähnlichkeiten. *Pflügers Arch. Ges. Physiol.* **180,** 298–340.

QUAYLE D.B. (1969) Paralytic shellfish poisoning in British Columbia. *Bull. Fish. Res. Bd Can.* **168,** 68 pp.

RABE F.W. (1970) Brook trout populations in Colorado beaver ponds. *Hydrobiologia* **35,** 431–48.

RABEN E. (1905) Über quantitative Bestimmung von Stickstoffverbindungen im Meerwasser nebst einem Anhang über die quantitative Bestimmung der im Meerwasser gelösten Kieselsaüre. *Wiss. Meeresunters. Abt. Kiel,* Bd VIII, 81–101.

RABINOWITCH E.I. (1945) *Photosynthesis and related processes. Vol. I.* New York, 599 pp.

RAKUSA-SUSZCZEWSKI S. (1969) The food and feeding habits of Chaetognatha in the seas around the British Isles. *Pol. Arch. Hydrobiol.* **16,** 213–32.

RAMSTER J.W., WYATT T. & HOUGHTON R. (1973) Towards a measure of the rate of drift of planktonic organisms in the vicinity of the Straits of Dover. *Int. Cons. Explor. Sea* C.M. 1973/L8, 10 pp. Mimeo.

RASMUSSEN E. (1973) Systematics and ecology of the Isefjord marine fauna (Denmark). *Ophelia* **11,** 1–507.

RAY D.L. (1966) IV. Seasonal aspects of food relationships and breeding. *Marine Biology, Vol.* 3, N.Y. Acad. Sci., New York, 313 pp.

RAYMONT J.E.G. (1963) *Plankton and Productivity in the Oceans.* Pergamon, Oxford. 660 pp.

RAYMONT J.E.G. & GAULD D.T. (1951) The respiration of some planktonic copepods. *J. mar. biol. Ass. U.K.* **29,** 681–93.

REDFIELD A.C., KETCHUM B.H. & RICHARDS F.A. (1963) The influence of organisms on the composition of seawater, pp. 26–77. In *The Sea, Vol.* 2, ed. Hill M.N. Wiley-Interscience, New York, 554 pp..

Reid J.L. (1965) *Intermediate Waters of the Pacific Ocean.* Johns Hopkins University Press, Baltimore, 85 pp.

Remsen C.C. (1971) The distribution of urea in coastal and oceanic waters. *Limnol. Oceanogr.* **16,** 732–40.

Rich W.H. (1937) Homing of Pacific salmon. *Science N.Y.* **85,** 477–8.

Richards F.A. (1968) Chemical and biological factors in the marine environment, pp. 259–303. In *Ocean Engineering*, ed. Brahtz J.F. Wiley, New York, 720 pp.

Richards S.W. & Riley G.A. (1967) The benthic epifauna of Long Island Sound. *Bull. Bingham Oceanogr. Coll.* **19,** 89–135.

Richman S. (1958) The transformation of energy by *Daphnia pulex. Ecol. Monogr.* **28,** 273–91.

Richman S. & Rogers J.N. (1969) The feeding of *Calanus helgolandicus* on synchronously growing populations of the marine diatom *Ditylum brightwelli. Limnol. Oceanogr.* **14,** 701–9.

Ricker W.E. (1937) The food and the food supply of sockeye salmon (*Oncorhynchus nerka* Walbaum) in Cultus Lake, British Columbia. *J. Biol. Bd Can.* **3** (6), 450–68.

Ricker W.E. (1948) Methods of estimating vital statistics of fish populations. *Indiana Univ. Publ. Sci. Ser.* **15,** 101 pp.

Ricker W.E. (1954) Stock and recruitment. *J. Fish. Res. Bd Can.* **11** (5), 559–623.

Ricker W.E. (1958) Handbook of computations for biological statistics of fish populations. *Bull. Fish. Bd Can.* **119,** 300 pp.

Ricker W.E. (1968) *Methods for Assessment of Fish Production in Fresh Waters. IBP Handbook No.* 3. Blackwell Scientific Publications, Oxford.

Ricklefs, R.E. (1968) Patterns of growth in birds. *Ibis* **110** (4), 419–51.

Riffenberg R.H. (1969) A stochastic model of interpopulation dynamics in marine ecology. *J. Fish. Res. Bd Can.* **26,** 2843–79.

Riley G.A. (1946) Factors controlling phytoplankton populations on Georges Bank. *J. Mar. Res.* **6,** 54–73

Riley G.A. (1947) A theoretical analysis of the zooplankton population of Georges Bank. *J. Mar. Res.* **6,** 104–13.

Riley G.A. (1956) Oceanography of Long Island Sound, 1952–54. IX. Production and utilization of organic matter. *Bull. Bingham Oceanogr. Coll.* **15,** 324–44.

Riley G.A. (1963) Theory of food-chain relations in the ocean. pp. 438–63. In: *The Sea Vol.* 2, ed. Hill M.N. 554 pp.

Riley G.A. (1963) Organic aggregates in sea water and the dynamics of their formation and utilization. *Limnol. Oceanogr.* **8,** 372–81.

Riley G.A. (1970) Particulate organic matter in sea water. *Adv. Mar. Biol.* **8,** 1–118.

Riley G.A., Stommel H. & Bumpus D.F. (1949) Quantitative ecology of the plankton of the Western North Atlantic. *Bull. Bingham Oceanogr. Coll.* **12,** 1–169.

Riley J.D. & Corlett J. (1966) The numbers of O group in Port Erin Bay. 1964–66 *Ann. Rep. Mar. biol Stn Port Erin* **78** (1965) 51–6.

Riley J.P. & Chester R. (1971) *Introduction to marine chemistry.* Academic Press, London, 365 pp.

Ringelberg, J. (1964) The positively phototactic reaction of *Daphnia magna* Strauss. *Neth. J. Sea Res.* **2,** 319–406.

Ritsraga S. (1971) On the changes of demersal fish catches taken from the otterboard trawling survey in the Gulf of Thailand from 1966 to 1970. In *Proceedings of the Second Symposium on Marine Fisheries*, Bangkok, Thailand, Marine Fisheries Research Laboratory.

ROBERTSON T.B. (1923) *The chemical basis of growth and senescence.* J.B. Lippincott, Philadelphia, 389 pp.

ROBSON D.S. (1966) Estimation of the relative fishing power of individual ships. *Bull. Int. Comm. NW. Atlantic Fish.* **3,** 5–14.

ROGER C. (1973) Recherches sur la situation trophique d'un groupe d'organismes pilagiques (Euphausiacea). I. Niveau trophiques des espèces. *Mar. Biol.* **18,** 312–16.

ROLLEFSEN G. (1953) Observations on the cod and the cod fisheries of Lofoten. *Rapp. P.-V. Réun Cons. int. Explor. Mer.* **136,** 40–7.

ROLLEFSEN G. (1955) The arctic cod. *Proc. UN International Technical Conf. on the Conserv. of the Living Resources of the Sea, Rome* 115–17.

ROSE M. (1924) *La biologie du plankton.* Thèse, Paris.

ROWE G.T. (1971) Benthic biomass and surface productivity. pp. 441–54. In *Fertility of the sea,* **2,** ed. Costlow J.D. Gordon & Breach Sci. Publ., New York, 622 pp.

RUDJAKOV J.A. (1970) The possible causes of diel vertical migration of planktonic animals. *Mar. Biol.* **6,** 98–105.

RUNNSTRØM S. (1936) A study on the life history and migrations of the Norwegian spring herring based on the analysis of the winter rings and summer zones of the scale. *Fiskdir. Skr. Ser. Hav.* **5** (2), 103 pp.

RUSBY J.S.M., SOMERS M.L., REVIE J., MCMARTNEY B.S. & STUBBS A.R. (1973) An experimental survey of a herring fishery by long-range Sonar. *Mar. Biol.* **22** (1), 271–91.

RUSSELL E.S. (1922) Report on market measurements in relation to the English cod fishery during the years 1912–14. *Fishery Invest.* (*Lond.*) *Ser.* 2 **5** (1), 76 pp.

RUSSELL E.S. (1937) Fish migrations. *Biol. Rev. Cambr. Philos. Soc. XII* (3), 320–37.

RUSSELL F.S. (1925) The vertical distribution of marine macro-plankton. An observation on diurnal changes. *J. mar. biol. Ass. U.K.* **13,** 769–809.

RUSSELL F.S. (1926) The vertical distribution of marine macroplankton IV. The apparent importance of light intensity as a controlling factor in the behaviour of certain species in the Plymouth area. *J. mar. biol. Ass. U.K.* **14,** 415–40.

RUSSELL F.S. (1927) The vertical distribution of plankton in the sea. *Biol. Rev.* **2** (3), 213–62.

RUSSELL F.S. (1939) Hydrographical and biological conditions in the North Sea as indicated by plankton organisms. *J. Cons. int. Explor. Mer.* **14,** 171–92.

RUSSELL F.S. & DEMIR N.S. (1971) On the seasonal abundance of young fish. XII. The years 1967, 1968, 1969 and 1970. *J. Mar. Biol. Ass. UK. NS.* **51** (1), 127–30.

RUSSELL F.S., SOUTHWARD A.J., BOALCH G.T. & BUTLER E.I. (1971) Changes in biological conditions in the English channel off Plymouth during the last half century. *Nature* (*Lond.*) **234,** 468–70.

RYLAND J.S. (1964) The feeding of plaice and sand-eel larvae in the Southern North Sea. *J. mar. biol. Ass. U.K.* **44,** 343–64.

RYLAND J.S. (1966) Observations on the development of larvae of the plaice, *Pleuronectes platessa* L. in Aquaria. *J. Cons. int. Explor. Mer.* **30** (2), 177–95.

RYTHER J.H. (1963) 17. Geographic variations in productivity. In *The Sea, Vol.* 2, ed. Hill M.N. Wiley-Interscience Publishers, New York, pp. 347–80.

RYTHER J.H. (1969) Photosynthesis and fish production in the sea. *Science* **166** 72–6.

RYTHER J.H. & DUNSTAN W.M. (1971) Nitrogen, phosphorus, and eutrophication in the coastal marine environment. *Science* **171,** 1008–13.

RYTHER J.H. & HULBERT E.M. (1960) On winter mixing and the vertical distribution of phytoplankton. *Limnol. Oceanogr.* **5,** 337–8.

RYTHER J.H., MENZEL D.W., HULBERT E.M., LORENZEN C.J. & CORWIN M. (1971) The production and utilization of organic matter in the Peru coastal current. *Inv. Pesq.* **35,** 43–59.

RYZHENKO M. (1961) 'Severyanka' in the schools of herring and cod. *Ryb. Rybolost.* **4,** 29–30.

SAETERSDAL G.S. & CADIMA E.L. (1960) A note on the growth of the Arctic cod and haddock. *I.C.E.S. C.M. Gadoid Fish Committee No. 90.*

SAILA S.B. (1961) A study of winter flounder movements. *Limnol. Oceanogr.* **6,** 292–8.

SAILA S.B. & FLOWERS M. (1969) Elementary applications of search theory to fishing tactics as related to some aspects of fish behaviour. *FAO Fish. Rep.* **62** (2), 343–55.

SAMEOTO D.D. (1971) Life history, ecological production, and an empirical mathematical model of the population of *Sagitta elegans* in St. Margaret's Bay, Nova Scotia. *J. Fish. Res. Bd Can.* **28,** 971–85.

SANDERS H.L. (1956) Oceanography of Long Island Sound, 1952–54. X. The biology of marine communities. *Bull. Bingham Oceanogr. Coll.* **15,** 345–414.

SAVAGE R.E. (1926) The plankton of a herring ground. *Fishery Invest.* (*Lond.*) **9** (1), 1–35.

SAVAGE R.E. (1937) The food of North Sea herring 1930–1934. *Fishery Invest. Ser. II* **15,** No. 5.

SCHAEFER M.B. (1954) Some aspects of the dynamics of populations important to the management of the commercial marine fish. *Bull. Inter-Am. Trop. Tuna Commn* **1,** 27–56.

SCHAEFER M.B. (1957) A study of the dynamics of the fishery for yellow tuna in the eastern tropical Pacific ocean. *Bull. Inter.-Am. Trop. Tuna Commn* **2,** 245–85.

SCHAEFER M.B. (1965) The potential harvest of the sea. *Trans. Am. Fish. Soc.* **94** (2), 123–8.

SCHELL D.M. (1974) Uptake and regeneration of free amino acids in marine waters of Southeast Alaska. *Limnol. Oceanogr.* **19,** 260–70.

SCHELTEMA R.S. (1971) Larval dispersal as a means of genetic exchange between geographically separated populations of shallow-water benthic marine gastropods. *Biol. Bull.* **140,** 284–322.

SCHMIDT J. (1922) The breeding places of the eel. *Phil. Trans. Roy. Soc. B.* **211,** 179–208.

SCHMIDT U. (1959) German investigations on commercial saithe landings in 1957. *Annls Biol., Copenh.* **14,** 136–43.

SEMINA H.J. (1972) The size of phytoplankton cells in the Pacific Ocean. *Int. Rev. Gesamten Hydrobiol.* **57,** 177–205.

SHAMARDINA I.P. (1968) Growth of the main species of Lake Glubokoye. *Probl. Ichthyol.* **8** (6), 828–33.

SHELBOURNE J.E. (1957) The feeding and condition of plaice larvae in good and bad plankton patches. *J. mar. biol. Ass. U.K.* **36,** 539–52.

SHELDON R.W. PRAKASH A. & SUTCLIFFE W.H. JR. (1972) The size distribution of particles in the Ocean. *Limnol. Oceanogr.* **17,** 327–40.

SHINDO S. (1973) General review of the trawl fishery and demersal fish stocks of the South China Sea. *FAO Fish. Tech. Pap.* **120,** 49 pp.

SHUMAM F.R. & LORENZEN C.J. (1975) Quantitative degradation of chlorophyll by a marine herbivore. *Limnol. Oceanogr.* **20** (4), 580–585.

SHUSHKINA E.A., ANISIMOV S.I. & KLEKOWSKI R.Z. (1968) Calculation of production efficiency in plankton copepods. *Pol. Arch. Hydrobiol.* **15,** 251–61.

SIMPSON R.H. (1973) Hurricane prediction: progress and problem areas. *Science* **181,** 899–907.

SINDERMANN C.J. & MAIRS D.F. (1959) A major blood group system in Atlantic Sea herring. *Copeia* **1959,** 228–32.

SKUD B.E. (1969) The effect of fishing on size composition and sex ratio of offshore lobster stocks. *Fisk. Dir. Skr. Ser. Hav.* **15,** 295–309.

SLOBODKIN L.B. (1963) *Growth and Regulation of Animal Populations.* Holt, Rinehart & Winston, New York, 184 pp.

SMAYDA T.J. (1966) A quantitative analysis of the phytoplankton of the Gulf of Panama.

III. General ecological conditions and the plankton dynamics at 8°46′N, 79°23′W from November 1954 to May 1957. *Inter. Amer. Trop. Tuna Comm. Bull.* **11** (5), 355–612.

SMAYDA T. (1970) The suspension and sinking of phytoplankton in the Sea. *Oceanogr. Mar. Biol. Ann. Rev.* 1970 **8,** 353–414.

SMITH F.E. (1969) Effects of enrichments in mathematical models. In *Eutrophication: Causes, Consequences, and Correctives.* Nat. Acad. Sci., Washington, D.C., pp. 631–45.

SMITH R.L. (1968) Upwelling. *Oceanogr. Mar. Biol. Ann. Rev.* **6,** 11–46.

SMITH R.L. (1974) A description of current, wind, and sea level variations during coastal upwelling off the Oregon coast, July–August, 1972. *J. Geophys. Res.* **79,** 435–43.

SOKAL R.R. & ROHLF F.J. (1969) *Biometry.* W.H. Freeman, 776 pp.

SOKOLOVA M.N. (1959) On the distribution of deep-water bottom animals in relation to their feeding habits and the character of sedimentation. *Deep-Sea Res.* **6,** 1–4.

SOKOLOVA M.N. (1972) Trophic structure of deep-sea macrobenthos. *Mar. Biol.* **16,** 1–12.

SØMME, J. (1934) Animal plankton of the Norwegian coast waters and the open sea, I. *Fisk. Dir. Skr. Ser. Hav.* **4** (9), 1–163.

SONINA M.A. (1969) Forecasting methods of autumn migrations of haddock in the Barents Sea. I.C.E.S. C.M. F:6.

SOROKIN Y.I. (1970) Abundance and production of bacteria in the water and bottom sediments of the central Pacific. *Dokl. Akad. Nauk SSSR Earth Sci. Sec.* **192,** 212–15.

SOROKIN Y.I. (1971) Bacterial populations as components of oceanic ecosystems. *Mar. Biol.* **11,** 101–5.

SOROKIN Y.I. (1973) Data on biological productivity of the Western tropical Pacific Ocean. *Mar. Biol.* **20,** 177–96.

SOROKIN Y.I. & WYSHKWARZEV D.I. (1933) Feeding on dissolved organic matter by some marine animals. *Aquaculture* **2,** 141–8.

SPRAGUE L. & VROOMAN A.M. (1962) A racial analysis of the Pacific sardine (*Sardinops caerulea*) based on studies of erythrocyte antigens. *Ann. N.Y. Acad. Sci.* **97,** 131–8.

SQUIRES H.J. (1970) Lobster (*Homarus americanus*) fishery and ecology in Port au Port Bay, Newfoundland, 1960–5. *Proc. Nat. Shellfish Ass.* **60,** 22–39.

STANEK E. (1964) The variability of the condition coefficient of the cod in the Bornholm Basin during the period 1955/1960. *Annls Biol., Copenh.* **19,** 100–1.

STEELE J.H. (1958) Plant production in the northern North Sea. *Scot. Home Dep. Mar. Res.* 1958 **7,** 1–36.

STEELE J.H. (1961) The environment of a herring fishery. *Mar. Res. Scot.* **6,** 19 pp.

STEELE J.H. (1962) Environmental control of photosynthesis in the sea. *Limnol. Oceanogr.* **7,** 137–50.

STEELE J.H. (1965) Some problems in the study of marine resources. *Spec. Publs. int. Comm. N.W. Atlant. Fish.* **6,** 463–76.

STEELE J.H. (1972) Factors controlling Marine Ecosystems. pp. 209–221. In *The Changing Chemistry of the Oceans.* eds. Dryssen D. & Jagner D. Wiley-Interscience, New York, 365 pp.

STEELE J.H. (1974a) *The Structure of Marine Ecosystems.* Harvard Univ. Press, Cambridge, Mass., 128 pp.

STEELE J.H. (1974b) Spatial heterogeneity and population stability. *Nature* (*Lond.*) **248,** 83.

STEELE J.H. (1974c) The stability of plankton ecosystems. pp. 179–194. In *Ecological Stability,* ed. Williamson M.H. & Usher M.B. Chapman & Hall, U.S.A. 120 pp.

STEELE J.H. & BAIRD I.E. (1972) Sedimentation of organic matter in a Scottish sea loch. *Mem. Ist. Ital. Idrobiol.* **29** (Suppl.), 74–88.

STEELE J.H., MCINTYRE A.D., JOHNSTON R., BAXTER I.G., TOPPING G. & DOOLEY H.D. (1973) Pollution studies in the Clyde Sea Area. *Mar. Pollut. Bull.* **4,** 153–7.

STEELE J.H. & MENZEL D.W. (1962) Conditions for maximum primary production in the mixed layer. *Deep-Sea Res.* **9,** 39–49.

STEEMANN-NIELSEN E. (1944) Havets Planteverden. *Skr. Komm. Danm. Fisk.-og. Havunders.* **13,** 1–108.

STEEMANN-NIELSEN E. (1952) The use of radioactive carbon (^{14}C) for measuring organic production in the sea. *J. Cons. int. Explor. Mer.* **18,** 117–40.

STEEMANN-NIELSEN E. (1964) On a complication in marine productivity work due to the influence of ultraviolet light. *J. Cons. int. Explor. Mer.* **29,** 130–5.

STEEMANN-NIELSEN E. & JØRGENSEN E.G. (1968) The adaptation of algae. I. General part. *Physiol. Plant.* **21,** 401–13.

STEPHENS K., SHELDON R.W. & PARSONS T.R. (1967) Seasonal variations in the availability of food for benthos in a coastal environment. *Ecology* **48,** 852–5.

STEPHENSON T.A. & STEPHENSON A. (1949) The universal features of zonation between tide-marks on rocky coasts. *Jour. Ecol.* **37,** 289–305.

STEWART R.W. (1967) Getting data from the ocean. *The Collection and Processing of Field Data, a CSIRO Symposium,* eds, Bradley F.F. & Denmead O.T. Wiley-Interscience, New York, 597 pp.

STOMMEL H. (1948) The westward intensification of wind-driven ocean currents. *Trans. Am. Geophys. Un.* **29,** 202–6.

STOMMEL H. (1960) *The Gulf Stream.* Univ. California Press, Berkeley, 202 pp.

STOMMEL H. & ARONS A.B. (1960) On the abyssal circulation of the world ocean—II. An idealized model of circulation pattern and amplitude in oceanic basins. *Deep-Sea Res.* **6,** 217–33.

STRICKLAND J.D.H. (1958) Solar radiation penetrating the ocean. A review of requirements, data and methods of measurement, with particular reference to photosynthetic productivity. *J. Fish. Res. Bd Can.* **15,** 453–93.

STRICKLAND J.D.H. (1960) Measuring the production of marine phytoplankton. *Fish. Res. Bd. Can. Bull.* **122,** 172.

STRICKLAND J.D.H. (1965) Production of organic matter in the primary stages of the marine food chain. pp. 477–610. In *Chemical Oceanography Vol.* 1, eds Riley J.P. & Skirrow G. Academic Press, London and New York, 712 pp.

STRICKLAND J.D.H. (1968) A comparison of profiles of nutrient and chlorophyll concentrations taken from discrete depths and by continuous recording. *Limnol. Oceanogr.* **13,** 388–91.

STRICKLAND J.D.H. (1970) The ecology of the plankton off La Jolla, California, in the period April through September, 1967. *Bull. Scripps Inst. Oceanogr.* **17,** 1–22.

STRICKLAND J.D.H. (1972) Research on the marine planktonic food web at the Institute of Marine Resources: A review of the past seven years of work. *Oceanogr. Mar Biol. Ann. Rev.,* 1972 **10,** 349–414.

STRICKLAND J.D.H. & PARSONS T.R. (1968) A practical handbook of seawater analysis. *Fish. Res. Bd Can. Bull.* **167,** 311.

STUBBS A.R. & LAWRIE R.G.G. (1962) Asdic as an aid to spawning ground investigation. *J. Cons. int. Explor. Mer.* **27** (3), 248–60.

SUDA A. (1973) Development of fisheries for non-conventional species. *J. Fish. Res. Bd Can.* **30** (12), Pt. 2, 2121–58.

SUSHCHENYA L.M. (1970) Food rations, metabolism and growth of crustaceans. pp. 127–41. In *Marine Food Chains,* ed. Steele J.H. Oliver & Boyd, Edinburgh, 552 pp,

SUTCLIFFE W.H., BAYLOR E.R. & MENZEL D.W. (1963) Sea surface chemistry and Langmuir circulation. *Deep-Sea Res.* **10,** 233–43.

SVÄRDSON G. (1949) The Coregonid problem. I. Some general aspects of the problem. Rep. Inst. Freshwat. Res. Drottningholm, *Fish. Bd of Sweden Ann. Rep.* 29, 1948, 89–101.

SVERDRUP H.U. (1947) Wind driven currents in a baroclinic ocean; with application to the equatorial currents of the Eastern Pacific. *Proc. natn. Acad. Sci.* (U.S.A.) **33,** 318–26.

SVERDRUP H.U. (1953) On conditions for the vernal blooming of phytoplankton. *J. Cons. int. Explor. Mer.* **18,** 287–95.

SVERDRUP H.U. (1957) Oceanography. *Handb. Phys.* **48,** 608–70.

SVERDRUP H.U., JOHNSON M.W. & FLEMING R.H. (1942) *The Oceans.* Prentice-Hall, Englewood Cliffs, N.J. 1087 pp.

SWINGLE H.S. & SMITH E.V. (1943) The management of ponds with stunted fish populations. *Trans. Am. Fish. Soc. 71st Ann. Meeting,* 102–5.

SYRETT P.J. (1962) Nitrogen assimilation, pp. 171–88. In *Physiology and Biochemistry of Algae,* ed. Lewin R.A. Academic Press, New York, 929 pp.

SZEKIELDA K.H. (1972) Ozeanische strukturen in satellitenbildern. *UMSCHAU* 72 *Helt* **3,** 95–7.

SZEKIELDA K.H. (1973) Biomass in the upwelling areas along the north-west coast of Africa as viewed with ERTS-1. *Symp. on Significant Results Obtained from ERTS*-1 **1,** 1385–401, NASA SP-327.

SZEKIELDA K.H. (1973b) Validity of ocean temperatures monitored from satellites. *J. Cons. int. Explor. Mer.* **35,** 78–86.

TAFT A.C. & SHAPOVALOV L. (1938) Homing instinct and straying amongst steelhead trout (*Salmo gairdnerii*) and silver salmon (*Oncorhynchus kisutch*). *Calif. Fish and Game* **24,** 118–25.

TAKAHASHI M., FUJII K. & PARSONS T.R. (1973) Simulation study of phytoplankton photosynthesis and growth in the Fraser River estuary. *Mar. Biol.* **19,** 102–16.

TAKAHASHI M. & PARSONS T.R. (1972) Maximization of the standing stock and primary productivity of marine zooplankton under natural conditions. *Indian J. Mar. Sci.* **1,** 61–2.

TALLING J.F. (1966) Photosynthetic behavior in stratified and unstratified lake populations of a planktonic diatom. *J. Ecol.* **54,** 99–127.

TANIGUCHI A. (1973) Phytoplankton–Zooplankton relationships in the Western Pacific Ocean and adjacent seas. *Mar. Biol.* **21,** 115–21.

TAYLOR C.C. (1958) Cod growth and temperature. *J. Cons. int. Explor. Mer.* **23,** 366–70.

TAYLOR C.C. (1962) Growth equations with metabolic parameters. *J. Cons. int. Explor. Mer.* **27,** 270–86.

TEMPLEMAN W. (1965) Relation of periods of successful year classes of haddock on the Grand Banks to Periods of success of year-classes for cod, haddock and herring in areas to the North and East. *Spec. Publ. Int. Comm. NW. Atlant. Fish.* **6,** 523–34.

TEMPLEMAN W. (1971) Year class success in some North Atlantic stocks of cod and haddock. *ICNAF Environmental Symposium Bedford Inst. Dartmouth Nova Scotia,* 18–19 *May* 1971, Contrib. 7, 32 pp. Mimeo.

TENNEKES H. & LUMLEY J.L. (1972) *A first course in turbulence.* MIT Press, Cambridge, Mass. 300 pp.

THOMPSON C.W. (1878) *The Voyage of the Challenger. Vol. I. The Atlantic.* Harper & Bros., New York, 391 pp.

THOMPSON J.D. & O'BRIEN J.J. (1973) Time-dependent coastal upwelling. *J. Phys. Oceanogr.* **3,** 33–46.

THOMPSON W. (1874) *The Depth of the Sea.* Macmillan, London, 527 pp.

THORSON G. (1946) Reproduction and larval development of Danish marine invertebrates. *Medd. Komm. Danm. Fiskeri. Hav. Ser. Plankton* **4,** 1–523.

THORSON G. (1950) Reproductive and larval ecology of marine bottom invertebrates. *Biol. Rev. Cambridge Philos. Soc.* **25,** 1–45.

THORSON G. (1957) Bottom communities. *Mem. geol. Soc. Am.* **67,** 461–534.

TIEWS K., SUCONDHAMARN P. & ISARANKURA A.P. (1967) On the changes in the abundance of demersal fish stocks in the Gulf of Thailand from 1963–64 to 1966 as consequences of trawl fisheries development. *Contrib. Mar. Fish. Lab. Bangkok* (**8**), 39 pp.

DI TORO D., O'CONNOR D.J., MANCINI J.L. & THOMANN R.V. (1973) Preliminary phytoplankton-zooplankton-nutrient model of Western Lake Erie. pp. 423–474. In *Systems Analysis and Simulation in Ecology*, 3, ed. Patten B.C. Academic Press, New York, 601 pp.

TOWNSEND C.H. (1935) The distribution of certain whales as shown by logbook records of American whaleships. *Zoologica, N.Y.* **19,** 1–50.

TRANTER D.J. (1973) Seasonal studies of a pelagic ecosystem (Meridian 110°E). In *Ecological Studies. Analysis and Synthesis, Vol.* 3, ed. Zeitschel B. Springer-Verlag, Berlin.

TROUT, G.C. (1957) The Bear Island cod; migrations and movements. *Fishery Invest.* (*Lond.*) *Ser.* 2 **21** (6), 51 pp.

TUCKER D.G. & GAZEY B.K. (1966) *Applied Underwater Acoustics*. Pergamon, Oxford, 244 pp.

TUREKIAN K.K. (1968) *Oceans*. Prentice-Hall, Englewood Cliffs, N.J. 120 pp.

TYLER J.E. (1968) The Secchi disk. *Limnol. Oceanogr.* **13,** 1–6.

TYLER J.E. (1973) Applied radiometry. *Oceanogr. Mar. Biol. Ann. Rev.* **11,** 11–15.

TYTLER P. (1969) Relationship between oxygen consumption and swimming speed in the haddock, *Melanogrammus aeglefinus. Nature* (*Lond.*) **221,** 274–5.

UDA M. (1937) Researches on 'Siome' or current rip in the seas and oceans. *Geophys. Mag. XI.* **4,** 307–72.

URICK R.J. (1967) *Principles of underwater sound for engineers*. McGraw-Hill, New York, 342 pp.

URSIN E. (1967) A mathematical model of some aspects of fish growth, respiration and mortality. *J. Fish. Res. Bd Can.* **24,** 2355–453.

URSIN E. (1973) On the prey size preferences of cod and dab. *Medd. Danm. Fisk.-og Havunders. N.S.* **7,** 85–98.

VACCARO R.F. (1965) Inorganic nitrogen in seawater, pp. 365–408. In *Chemical Oceanography*, Vol. I, eds Riley J.P. & Skirrow G. Academic Press, New York, 712 pp.

VAN DYNE G.M. (1966) Ecosystems, systems ecology, and systems ecologists. *Oak Ridge Natl. Lab. Tech. Memo*. 3957, Oak Ridge, Tennessee, 40 pp.

DE VEEN J.F. (1962) On the subpopulations of plaice in the southern North Sea. *Int. Cons. Explor. Sea. CM* 1962 **94,** 6 pp.

DE VEEN J.F. (1970) On the orientation of the plaice (*Pleuronectes platessa* L.) 1. Evidence for orientating factors derived from the ICES transplantation experiments in the years 1904–09. *J. Cons. int. Explor. Mer.* **33** (2), 192–227.

VENRICK E.L., MCGOWAN J.A. & MANTYLA A.W. (1973) Deep maximum of photosynthetic chlorophyll in the Pacific Ocean. *Fishery Bull. Fish. Widl. Serv. U.S.* **71,** 41–52.

VERNON L.P. & SEELY G.R. (1966) *The Chlorophylls*. Academic Press, New York and London, 697 pp.

VERVOORT V. (1963) Pelagic copepoda I. *Atlantide Rep.* **7,** 77–194.

VINOGRADOV M.E. (1953) The role of vertical migration of the zooplankton in the feeding of deep sea animals. *Priroda, Mosk.* **6,** 95–6.

VINOGRADOV M.E. (1955) Vertical migrations of zooplankton and their importance for the nutrition of abyssal pelagic fauns. *Trudy Inst. Okeanol. USSR* **13,** 71–6.

VINOGRADOV M.E. (1962) Feeding of the deep sea zooplankton. *Rapp. P.-V. Réun. Cons. int. Explor. Mer.* **153,** 114–20.

VINOGRADOV M.E. (1968) *Vertical distribution of the oceanic zooplankton*. Moscow. (Transl. U.S. Dept. Comm. 1970), 339 pp.

VINOGRADOV M.E., MENSHUTKIN V.V. & SHUSHKINA E.A. (1972) On Mathematical simulation of a pelagic ecosystem in tropical waters of the ocean. *Mar. Biol.* **16,** 261–8.

VINOGRADOVA N.G. (1962) Some problems of the study of deepsea bottom fauna. *J. oceanogr. Soc. Japan*, 20th Anniversary Volume, 724–41.

VLYMEN W.J. (1970) Energy expenditure of swimming copepods. *Limnol. Oceanogr.* **15** (3), 348–56.

VOLLENWEIDER R.A. (ed.) (1969) *A manual on methods for measuring primary production in aquatic environments. IBP Handbook No.* 12. Blackwell Scientific Publications, Oxford, 213 pp.

VOLTERRA V. (1928) Variations and fluctuations of the number of individuals in animal species living together. *J. Cons. int. Explor. Mer.* **3,** 1–51.

WALFORD L.A. (1946) A new graphic method of describing the growth of animals. *Biol. Bull. mar. biol. Lab., Woods Hole* **90,** 41–7.

WALSH J.J. (1971) Relative importance of habitat variables in predicting the distribution of phytoplankton at the ecotone of the Antarctic Upwelling Ecosystem. *Ecol. Monogr.* **41,** 291–309.

WALSH J.J. (1972) Implications of a systems approach to oceanography. *Science* **176,** 969–75.

WALSH J.J. (1975) A spatial simulation model of the Peru upwelling ecosystem. *Deep-Sea Res.* 22:201–236.

WALSH J.J. (1974) Primary production in the sea. *Proc. 1st. Int. Congr. Ecology. Pudoc, Netherlands.* 150–154.

WALSH J.J. & DUGDALE R.C. (1971) A simulation model of the nitrogen flow in the Peruvian upwelling system. *Inv. Pesq.* **35** (1), 309–30.

WALSH J.J. & HOWE S.O. (in press) Protein from the sea: a comparison of the simulated nitrogen and carbon productivity in the Peru upwelling ecosystem. In *Systems Analysis and Simulation in Ecology, Vol. IV*, ed. Patten B.C. Academic Press, New York.

WALSH J.J., KELLEY J.C., DUGDALE R.C. & FROST B.W. (1971) Gross features of the Peruvian upwelling system with special reference to possible diel variation. *Inv. Pesq.* **35** (1), 25–42.

WALSH J.J., KELLEY J.C., WHITLEDGE T.E., MACISAAC J.J. & HUNTSMAN S.A. (1974) Spin-up of the Baja California upwelling ecosystem. *Limnol. Oceanogr.* **19** (4):553–572.

WARREN B.A. (1971) Evidence for a deep western boundary current in the South Indian Ocean. *Nature (Lond.)* **229,** 18–19.

WARREN B.A. & VOORHIS A.D. (1970) Velocity measurements in the deep western boundary current of the South Pacific. *Nature (Lond.)* **228,** 849–50.

WARREN C.E. & DAVIS G.E. (1966) Laboratory studies on the feeding, bioenergetics, and growth of fish, pp. 175–214. In *The Biological Basis of Freshwater Fish Production*, ed. Gerking S.D. Blackwell Scientific Publications, Oxford.

WATERMAN T.H. & BERRY D.A. (1971) Evidence for diurnal vertical plankton migration below the photic zone. (NSF grant report).

WATERS T.F. (1969) The turnover ratio in production ecology of freshwater invertebrates. *Am. Nat.* **103,** 173–85.

WATT K.E.F. (1968) *Ecology and Resource Management*. McGraw-Hill, New York, 450 pp.

WATT W.D. & HAYES F.R. (1963) Tracer study of the phosphorus cycle in seawater. *Limnol. Oceanogr.* **8,** 276–85.

WEBSTER F. (1961) The effect of meanders on the kinetic energy balance of the Gulf Stream. *Tellus* **13,** 392–401.

WEIDEMANN H. (ed.) (1973) The ICES diffusion experiment RHENO 1965. *Rapp. P.-V. Réun. Cons. int. Explor. Mer.* **163,** 111 pp.

WEIHS D. (1973) Hydromechanics of fish schooling. *Nature* (*Lond.*) **241,** 290–1.

WELLS J.W. (1957) Coral reefs. *Mem. Geol. Soc. Am.* **67** (1), 609–31.

WESTON D.E., HORRIGAN A.A., THOMAS S.J.L. & REVIE J. (1969) Studies of sound transmission fluctuations in shallow coastal waters. *Phil. Trans. R. Soc. A.* **265** (1169), 567–608.

WHEELER P.A., NORTH B.B. & STEPHENS G.S. (1974) Amino acid uptake by marine phytoplankers. *Limnol. Oceanogr.* **19,** 249–59.

WHITLEDGE T.E. (1972) The regeneration of nutrients by nekton in the Peru upwelling system. Ph.D. dissertation, University of Washington, 114 pp.

WHITLEDGE T.E. & PACKARD T.T. (1971) Nutrient excretion by anchovies and zooplankton in Pacific upwelling regions. *Inv. Pesq.* **35,** 243–50.

WIBORG K.F. (1954) Investigations on zooplankton in coastal and offshore waters of western and northwestern Norway. *Rep. Norw. Fishery Mar. Invest.* **II** (1), 1–246.

WICKSTEAD J.H. (1962) Food and feeding in pelagic copepods. *Proc. Zool. Soc. London* **139,** 545–56.

WIEBE P.H. (1970) Small-scale spatial distribution in oceanic zooplankton. *Limnol. Oceanogr.* **15,** 205–17.

WILDER D.G. (1965) Lobster conservation in Canada. *Rapp. P.-V. Réun. Cons. int. Explor. Mer.* **156,** 21–9.

WILLIAMS R. (1973) In press.

WILLIAMSON D.I. (1969) Names of larvae in the Decapoda and Euphausiacea. *Crustaceana* **16** (2), 210–13.

WILSON D.S. (1973) Food size selection among copepods. *Ecology* **54** (4), 909–14.

WIMPENNY R.S. (1938) Diurnal variation in the feeding and breeding of zooplankton related to the numerical balance of the zoo-phytoplankton community. *J. Cons. int. Explor. Mer.* **13,** 323–36.

WINBERG G.G. (1956) *Rate of metabolism and food requirements of fishes.* Nauchnye Trudy Belorusskovo Gosudarstvennovo Universiteta Imeni V.I. Lenina, Minsck, 253 pp; and *Transl. Fish. Res. Bd Can.*, Ser. 194, 202 pp.

WINBERG G.G. (1971) *Methods for the Estimation of Production of Aquatic Animals*, Transl. Duncan A. Academic Press, New York.

WOOSTER W.S. & JONES J.H. (1970) California undercurrent off northern Baja California. *J. Mar. Res.* **28** (2), 235–50.

WOOSTER W.S. & REID J.L. (1963) Eastern boundary currents, pp. 253–80. In *The Sea, Vol.* 2, ed. Hill M.N. Wiley-Interscience, New York, 554 pp.

WOOSTER W.S. & SIEVERS H. (1970) Seasonal variations of temperature, drift, and heat exchange in surface waters off the west coast of South America. *Limnol. Oceanogr.* **15** (4), 595–605.

WORTHINGTON E.B. (1931) Vertical movements of fresh water macroplankton. *Int. Rev. hydrobiol.* **25,** 394–436.

WORTHINGTON L.V. (1970) The Norwegian Sea as a mediterranean basin. *Deep-Sea Res.* **17,** 77–84.

WRIGHT R. (1969) Deep water movement in the western Atlantic as determined by use of a box model. *Deep-Sea Res.* **16** (Suppl.), 433–46.

WYATT T. (1971) Production dynamics of *Oikopleura dioica* in the Southern North Sea, and the role of fish larvae which prey on them. *Thalassia Jugosl.* **7,** 435–44.

WYATT T. (1973) The biology of *Oikopleura dioica* and *Fritillaria borealis* in the Southern Bight. *Mar. Biol.* **22,** 137–58.

WYATT T. (1974) The feeding of plaice and sandeel larvae in the Southern Bight in relation to the distribution of their food organisms. pp. 245–51. In *The Early Life History of Fish*, ed. Blaxter J.H.S. Springer-Verlag, Berlin, Heidelberg, New York, 765 pp.

WYRTKI K. (1962) The oxygen minima in relation to ocean circulation. *Deep-Sea Res.* **9,** 11–24.

WYRTKI K. (1973) Teleconnections in the Equatorial Pacific Ocean. *Science* **180,** 66–8.

YABLONSKAYA E.A. (1962) A study of the seasonal population dynamics of the plankton copepods as a method of determination of their production. Rapp. *P–v. Réun. Cons. int. Explor. Mer.* **153,** 224–6.

YENTSCH C.S. & LEE R.W. (1966) A study of photosynthetic light reactions, and a new interpretation of sun and shade phytoplankton. *J. Mar. Res.* **24,** 319–37.

YONGE C.M. (1957) Symbiosis. In *Treatise on marine ecology and palaeoecology Vol. I,* ed. Hedgpeth J.W. *Mem. Geol. Soc. Amer.* **67,** 429–42.

YOSHIDA K. (1967) Circulation in the eastern tropical oceans with special reference to upwelling and undercurrents. *Jap. J. Geophys.* **4,** 1–75.

ZAIKA V.E. (1970) Relationship between the productivity of marine molluscs and their life span. *Oceanology* **10,** 547–52.

ZAIKA V.E. (1972) *Specific production of aquatic invertebrates.* Naukova Dumka Publishers, Kiev. 147 pp. (In Russian).

ZAITSEV Y.P. (1961) Surface pelagic biocoenosis of the Black Sea. *Zool. Zhurnal* **40,** 818–25.

ZEITZSCHEL B. (1965) Zur Sedimentation von Seston, eine produktionsbiologische Untersuchungen von Sinkstoffen und Sedimenten der westlichen und mittleren Ostsee. *Kieler Meeresforsch.* **21,** 55–80.

ZENKEVITCH L.A. (1963) *Biology of the seas of the USSR.* Allen & Unwin, London, 955 pp.

ZENKEVITCH L.A., BARSANOVA N.G. & BELYAEV G.M. (1960) Quantitative distribution of bottom fauna in the Northern part of the Pacific at a depth below 200 m. *Dokl. Akad. Nauk.* **133** (In Russian).

ZENKEVITCH L.A., BIRSHTEIN J.A. & BELYAEV G.M. (1954) Exploration of the fauna of the Kurile-Kamchatka trench (from the material of the Pacific Ocean expedition of the Institute of Oceanology of the Academy of Sciences, USSR). *Priroda,* Feb. 1954, 61–73.

ZENKEVITCH, L.A., BIRSHTEIN J.A. & BELYAEV G.M. (1955) Quantitative investigation of the fauna of the Kurile-Kamchatka trench. *Trans. Inst. Oceanol.* **12,** 345–71.

ZENKEVICH L.A. & FILATOVA Z.A. (1960) Quantitative distribution of the bottom fauna in the North Pacific at depths of more than 2000 m. *Dokl. Acad. Nauk SSSR* **133,** 451–3. (In Russian).

ZEUTHEN E. (1953) Oxygen uptake as related to body size in organisms. *Q. Rev. Biol.* **28** (1), 1–12.

ZIJLSTRA J.J. (1972) On the importance of the Wadden Sea as a nursery area in relation to the conservation of the southern North Sea fishery resources. *Symp. Zool. Soc.* (*Lond.*) (1972) **29,** 233–58.

Author Index

Acton F.S. 391
Adachi R. 109
Adams J.A. 16
Agassiz A. 317
Albers V.M. 3
Alexander R.McN. 272
Allen J.S. 383
Allen K.R. 196, 198, 204, 240
Alm G. 263, 264
Alvarino A. 123, 124
Alverson D.L. 323
Ancellin J. 328
Andrews K.J. 123
Andrewartha H.G. 197
Anderson G.C. 83
Anderson V. 131
Angel M.V. 124, 130
Anraku M. 20
Ansell A.D. 67
Antia N.K. 179
Anraku M. 352
Apstein C. 197
Aristotle 9
Armstrong F.A.J. 83, 149, 391
Arons A.B. 54, 55, 56
Arvesen J.C. 381
Atkins W.R.G. 12

Backiel T. 274
Backus R.H. 32, 126, 134
Bagenal T.B. 264
Bainbridge R. 83, 100, 105, 120, 122, 334
Baird R.C. 131, 133
Bakun A. 45
Ballester A. 386
Baldina E.P. 214, 216
Banse K. 120, 121
Bary McK.B. 129
Barraclough W.E. 84, 122, 403
Barham E.G. 119
Barnes H. 2, 101
Barnett A.M. 137
Bartlett M.S. 5
Battle H.I. 336
Baxter G.C. 324
Baylor E.R. 127
Bé A.W.H. 117
Beamish F.W.H. 271, 272
Beckman W.C. 265
Becker P. 381
Beers, J.R. 84, 85, 86, 87, 118, 348, 386
Beklemishev C.W. 134
Belyaev G.M. 232
Bertalanffy L. von 254, 256, 276
Bernard C. 4
Berner A. 209
Berry D.A. 121
Beukema J.J. 278
Beverton R.J.H. 18, 257, 276, 293, 298, 299, 300, 301, 310, 322, 323, 325
Bien G.S. 53
Bierstein J.A. 67
Birch L.C. 197
Birge E.A. 1, 7
Birkett L. 239, 240
Blackman R.B. 366
Blackburn M. 391, 403
Blaxter J.H.S. 126, 127, 130, 320
Blegvad H. 261
Bledsoe L.J. 391
Boden B.P. 127, 128
Bogorov V.G. 123, 124
Bogdanov Yu.A. 231
Bolin B. 57, 389
Bolster G.C. 324
Bonham Carter G. 389
Bostrøm O. 324
Bowden K.F. 38, 104
Boyd C.M. 386, 390, 402
Braarud T. 174

Brandt K. 12
Breen P.A. 74, 247, 274, 319, 350
Brett J.R. 272, 273
Bridger J.P. 323, 324
Brinton E. 123, 124
Brock V. 337
Brock T.D. 228
Brocksen R.W. 189, 190
Brody S. 265
Brooks E.R. 91, 198, 205, 206, 210, 214, 217
Brooks J.L. 93, 94, 132
Brown M.E. 264, 268, 269
Bryan K. 41, 388, 389
Buchanan J.B. 98
Buchanan-Wollaston H.J. 197, 198
Bückmann A. 328
Bumpus D.F. 174
Burd A.C. 260, 264
Burke M.C. 240, 242
Burnet A.M.R. 261
Burns C.W. 89
Burukowski R.N. 313

Cadima E.L. 257
Cadwalader G. 404
Caperon J. 164, 165
Carpenter E.J. 160, 170
Carlander K.D. 261, 265, 278
Carroz J.E. 312
Caspers H. 131
Cassie R.M. 83, 100
Caswell H. 197
Chapman D.G. 296
Chester R. 25, 26, 28
Chindonova Yu.G. 121, 122
Clarke G.L. 2, 34, 121, 125, 133
Clayden A.D. 19
Clemens W.A. 323
Cleve P.T. 10
Cochran W.G. 370, 372, 375
Connell J.H. 342
Conover R.J. 89, 191, 192, 202, 208, 209, 212, 214, 217, 221, 222
Conway H.L. 143, 156, 164, 165
Cooper L.H.N. 132
Corkett C.J. 93
Corlett J. 320
Corner E.D.S. 15, 202, 209, 211
Coull B.C. 229
Cox M.D. 41, 389
Cox R.A. 26
Craig H. 53, 54
Crisp D.J. 2
Currie R.I. 109, 110, 126
Curl H. 381
Cushing D.H. 2, 10, 11, 14, 15, 16, 17, 82, 86, 95, 101, 102, 109, 117, 120, 125, 180, 206, 216, 260, 264, 292, 302, 308, 309, 320, 322, 323/324, 325, 330, 331, 336, 337 356, 389
Cuvier le Baron ix, 117

Darwin C. 2
Davies A.G. 202
Davis C.O. 161, 164
Davis G.E. 267, 274, 275
Dawes B. 268
Day F. 2
Dayton P.K. 67, 75
De Ligny W. 328
Demir N.S. 3
Denton E. 2, 121
Dévèze L. 61
Devold F. 323
Dice L. 7
Dickie L.M. 95, 223, 247, 249, 265, 266
Dickson R.R. 3, 324
Dietrich G. 45, 323
Digby P.S.B. 123
Dittmar W. 26
Dodson S.I. 93, 94
Doty M.S. 318
Dow R.L. 245
Droop M.R. 13, 164
Dugdale R.C. 12, 13, 15, 144, 151, 152, 153, 154, 155, 156, 157, 158, 159, 160, 166, 167, 168, 169, 170, 171, 394
Dunbar M.J. 130
Dunstan W.M. 88, 165
Duntley S.Q. 32, 33, 34
Dussart B.H. 61
Duursma E.K. 356
Dyrssen D. 25, 58

Ebling F.J. 74
Eddy S. 261, 265, 278
Edmondson W.T. 196, 197, 199, 203, 204, 207
Edwards R.R.C. 266, 275
Eibl-Eibesfeldt I. 68
Ekman V.W. 43, 44, 45
Elliott H.W. 2
Elton C. 62, 343
Emery K.D. 141, 142, 318
English T.S. 292
Enright J.T. 126
Eppley R.W. 16, 153, 154, 162, 168, 169, 170, 171
Esterly C.O. 117, 120, 125, 126
Ewart T. 386

Fager E.W. 198, 318
Farmer G.J. 272
Feldmann J. 68
Fenchel T. 228
Filatova Z.A. 232
Finlayson D.M. 266
Fisher L.R. ix, 208
Flowers M. 337
Foerster R.E. 325
Fofonoff N.P. 24, 28
Forbes S.T. 9, 292
Ford E. 256
Forel F. 1
Fournier R.O. 122
Francis V. 221, 222
Frasseto R. 134
Frazer F.C. 117, 123
Freund J.E. 371
Frosch R.A. 32, 33
Frost B.W. 16, 131
Fryer G.M. 2
Fuchs T. 137
Fugino K. 328
Fuglister F.C. 381
Fuhs G.W. 162, 163, 164

Gaarder T. 174
Gambell R. 312
Gardiner A.C. 13
Garstang W. 18
Gaten R.R. 135
Gauld D.T. 201, 209
Gause G.F. 343
Gazey B.K. 33
Gerking S.O. 261, 271
Gilbert C.H. 325
Goering J.J. 153, 159, 161, 166, 167, 169, 170, 171
Gold A 81, 95
Goldman J.C. 405
Goldman C.R. 174
Goodwin T.H. 174
Gosse P. 2
Graham M. 18, 253
Gran H.H. 174
Gray J. 257
Graybill F.A. 370, 372
Gregg M. 27
Grenney W.J. 389
Greze B.S. 214, 215, 216
Grill E.V. 389
Grindley J.R. 134
Guillard R.L. 161
Gulland J.A. 8, 293, 297, 300, 303, 305, 306, 309, 311, 312, 314, 402
Gunther E.R. 117, 123, 134

Haeckel E. 1, 2
Hall W.B. 344
Halpern D. 58, 375, 381
Hammer W.M. 126
Hansen W.J. 130
Harbough J.W. 389
Harder W. 132
Harden Jones F.R. 17, 98, 321, 322, 323, 325
Hardin G. 346
Harding D. 198, 320, 322
Hardy A.C. 11, 117, 123, 133, 134, 135, 319, 344
Hargrave B.T. 231, 232, 233, 234
Harris J.G.K. 302, 330
Harris J.E. 126
Harrison P.J. 161, 162, 164, 165, 166, 171
Hart T.J. 109, 110
Harvey H.W. 9, 14, 15, 162, 174, 216
Haury L.B. 101, 386
Haxo F.T. 175
Hayes F.R. 162
Hays E. 134
Hedgpeth J.W. 60
Heincke F. 18, 253, 320
Heinle D.R. 196, 198, 205, 206, 217, 219
Heinrich A.K. 76, 180
Hellerman S. 43
Hempel G. 320
Hensen V. 9, 11, 196, 197
Hentschel E. 122
Herbert D. 164
Herbland A. 392
Herman S.S. 126
Heron A.C. 205, 207
Hersey J.B. 32, 119
Hesse H. 23, 230
Hessler R.R. 236
Hewitt E.J. 177
Heyerdahl T. 10
Hickel W. 109
Hida T.S. 124
Hiemstra W.H. 319
Hile R. 261, 263
Himmelman J.H. 247
Hirota J. 137
Hjort J. 2, 317, 356
Hobson L.A. 131
Hodgson W.C. 320
Hoglund H. 323
Holmes R.W. 174
Holland W.R. 57
Holling C. 330
Holt S.J. 18, 257, 262, 276, 298, 299, 300, 310, 323
Hood D.W. 57

Hopkins T.S. 398
Horne R.A. 25
Hulbert E.M. 83, 88
Hure J. 120
Hurlburt H.E. 45, 402
Hutchinson G.E. 135, 343
Huyer A. 58

Idyll C.P. 58, 405
Ikeda T. 90, 91, 92, 202, 212
Iles T.D. 2, 261
Innis G. 391
Isaacs J.D. 129, 231
Ivanenkov U.H. 131
Ivanov A.V. 10
Ivlev V.S. 80, 84, 201, 343

Jagner D. 25, 58
Jenkins G.M. 368
Jerlov N.G. 33, 126, 174, 175, 176
Johannes R.E. 13, 94
Johnston R. 26, 386
Johnstone J. 253
Johnson P.O. 320, 323
Jones R. 115
Jones J.H. 398
Jones R. 310, 320, 325, 331, 344
Jorgensen C.B. 90, 177, 191, 230
Joseph J. 106
Juday C. ix, 1, 7, 118, 204
Junars P.A. 236

Kalmijn A.J. 3
Kampa E.M. 126, 127, 128
Kamshilov M.M. 216
Kamykowski D. 364
Kang T. 328
Kashkin N.I. 134
Kawajiri M. 265
Keeling D.C. 57, 389
Kelly J.C. 362, 364, 370, 372, 374, 375
Kennedy O.D. 403
Kent W. Saville 2
Kerner E.H. 343
Kerr S.R. 266
Ketchum B.H. 12, 143, 162, 389
Khmeleva N.N. 92
Kierstead H. 107, 109, 113
King J.E. 124
Kinne O. 262
Kinzer J. 118
Kipling R. 2
Kirpichnikov V.S. 264
Kishimoto R. 401
Kitching J.A. 74
Kleiber M. 268
Koblentz-Mishke O.J. 147, 148, 182
Koczy F.F. 121
Kowal N.E. 391
Kraus E.B. 41, 44, 57
Krefft G. 10
Kriss S. 139
Krogh A. 14
Krumbein W.C. 370, 372
Kullenberg G. 104
Kumlov S.K. 84
Kuo H.H. 56, 57

Lack D. 8
La Fond E.C. 83, 93
Lamb H.H. 3
Langmuir I. 105
Laplace P.S. 5
Larkin P.A. 257
Lasker R. 96, 208, 209, 352
Lauff G.H. 57
Lawrie R.G.G. 324
Lawry J.V. 69
Lea E. 320
Leach J.H. 228
Lebour M.V. 319
Le Brasseur R.J. 189, 352, 403
Le Cren E.D. 253
Lee A.J. 4
Lee R.F. 135, 137
Lee R.W. 177
Legand M. 124
Lendenvald R.U. 126
Leong R.J.H. 386
Levenstein R.J. 232
Lewis J.R. 73
Li C.C. 329
Liebermann L.N. 32
Lindeman R.L. 1, 204, 341, 346, 390
Lissmann H.W. 3
Lloyd I.J. 403
Lohmann H. ix, 2, 9, 13, 15, 191
Longhurst A.R. 83, 101, 118, 122, 131, 134, 318, 399, 402
Longuet-Higgins M.S. 46, 58
Lorenzen C.J. 100, 101, 131, 181, 184, 226, 391
Lotka A.J. 341
Love C.M. 132
Luckinbill L.S. 343
Lumley J.L. 38
Lund J.W.G. 13, 16
Lyubimova T.G. 313

Ma Y.H. 391
McAllister C.D. 88, 134

MacArthur R.H. 61, 71, 84, 342
McCarthy J.J. 160, 169
MacFadyen A. 213, 214
MacGinitie G.E. 194
MacGinitie N. 194
McGowan J.A. 132
McIntyre A.D. 229, 349
McIntyre D.B. 370
MacIsaac J.J. 151, 152, 153, 154, 155, 156, 157, 158, 162, 168
Mackintosh N.A. 123, 334
McLaren I.A. 133, 134, 135, 136, 195
McManus D.A. 370, 372, 375
McNae W. 318
McNeil W. 330
McNider R.T. 45, 390
McQuillen K. 165
Maeda S. 401
Mague T.H. 170
Mairs D.F. 328
Makarova N.P. 92, 134
Malone T.C. 87, 89
Manabe S. 388, 389
Mandelstam J. 165
Mann K.H. 74, 196, 198, 228, 231, 232, 240, 247, 274, 319, 350, 390
Mantyla A.W. 132
Margalef R. 61, 75, 76, 384
Marr J.W.S. 88, 352
Marshall S.M. 15, 101, 113, 122, 123, 124, 137, 191, 196, 200, 201, 202, 204, 205, 206, 209, 210
Martin N.V. 263
Maslov N.A. 323
Matsumoto W. 320, 325
Mauchline J. 123, 136, 208
May R.M. 8, 114, 391
Meek A. 16, 17
Menasveta D. 303
Menzel D.W. 151
Merriman D. 3
Meyer J. 164, 165
Michael E.L. 117, 125
Mileikovsky S.A. 237
Miller M. 73
Miller R.J. 240
Milliken D.M. 131
Milne D.J. 323
Mironov G.N. 353
Moav R. 264
Moiseev P.A. 305
Monin A.S. 38
Monod J. 150, 151
Mooers C.N.K. 383
Mookherjii P.S. 271
Moore H.B. 119, 120, 133
Morton J. 73
Mullin M.M. 15, 89, 91, 93, 120, 122, 131, 132, 196, 198, 205, 206, 214, 216, 217, 218, 220
Munk W.H. 49, 50, 53, 54
Munro A.L.S. 228
Murphy G.I. 22, 311, 313
Murray J. 2, 117, 317

Nakken O. 292
Namias J. 58, 405
Nansen F. 43
Nédelèc C. 328
Neess J. 150, 166
Nemoto T. 96
Neumann G. 25, 28, 30, 36, 37, 42, 57, 58
Nevenzel J.C. 135
Nicol J.A.C. 2, 69
Nikitine B.N. 130
Nishizawa S. 97

O'Brien J.J. 41, 45, 381, 390
O'Connell C.P. 386
Odum H.T. 389
Okubo A. 57, 106, 108
Omori M. 99, 352
Orr A.P. 113, 122, 123, 124, 137, 191, 200, 202, 204, 205, 209, 321, 346
Osborn R.R. 93
Ostvedt O.J. 122
Othmer D.R. 405
Otsu T. 323
Ottestad P. 3, 329

Paasche E. 161, 164
Packard T.T. 169, 170
Paffenhöfer G.A. 93, 210
Paine R.T. 67, 74, 353
Paloheimo J.E. 95, 223, 265, 266
Park O. 1, 2
Parker G.H. 125
Parker R.R. 257
Parrish B.B. 130, 243, 302, 320, 324
Parsons T.R. 14, 86, 89, 96, 97, 143, 174, 189, 196, 200, 202, 206, 209, 211, 212, 214, 352, 390
Pascal B. 5
Patmatmat M. 231
Patten B.C. 389
Paulik G.J. 309
Pavlov S.P. 153, 168
Pavlova E.V. 353
Pearcy W.G. 330
Pearl R. 5
Pedlosky J. 23, 39, 41

Peer D.L. 240
Pella J.J. 296
Pentelow F.T.K. 268
Pérès J.M. 61
Perkin R.G. 26
Peterson C.G.J. 18, 131, 309, 317
Petipa T.S. 92, 134, 135, 206, 210, 211, 217, 353
Phillips O.M. 38, 40, 57, 58
Pielou E.C. 113
Pierson W.J. 25, 28, 30, 36, 37, 42, 57, 58
Pillsbury R.D. 382
Platt T. 4, 100, 109
Podushko Yu.N. 265
Pomeroy L.R. 13, 162, 228
Poulet S.A. 89, 90, 96, 201
Pourriot R. 65
Prakash A. 109
Pritchard A.C. 325
Pütter A. 14, 15, 276, 356

Quale D.B. 109

Raben E. 12
Rabinowitch E.I. 68
Rakusa-Suszczewski S. 346
Ramster J.W. 323
Rasmussen E. 349
Ray D.L. 195
Raymont J.E. 174, 208, 209
Redfield A.C. 142, 146, 174
Reid J.L. 41, 51, 146, 147
Remsen C.C. 160
Renger E.M. 164
Rich W.H. 325
Richards F.A. 143
Richards S.W. 240
Richman S. 15, 95, 201
Ricker W.E. 8, 18, 19, 198, 265, 298, 299, 300, 301, 302
Riffenburgh R. 337, 389
Riley G.A. 1, 11, 15, 16, 82, 97, 151, 174, 180, 206, 214, 216, 218, 231, 232, 240, 356
Riley J.D. 320, 357, 358
Riley J.P. 25, 26, 28
Ringelberg J. 127, 129
Ristraga S. 304
Robertson T.B. 256
Robinson D. 403
Robinson M.A. 312
Robson D.S. 293
Roels O.A. 405
Roger C. 347
Rogers J.N. 15, 201
Rohlf F.J. 372
Rollefsen G. 260, 320, 324
Rose M. 125
Rowe G.T. 233
Rozanov A.G. 131
Rudjakov J.A. 132
Rusby J.S.M. 11
Russell E.S. 253
Russell F.S. 3, 16, 17, 82, 83, 117, 120, 125
Ryland J.S. 320
Ryther J.H. 45, 83, 85, 86, 87, 88, 95, 146, 165, 174, 179, 228, 309, 354
Ryzhenko M. 130

Saetersdal G.S. 257
Saila S.B. 323, 337, 343
Sameoto D.D. 389
Sanders H.L. 240, 242
Sargent J.R. 135
Savage R.E. 130, 338, 344
Saville A. 320
Schaefer M.B. 18, 295, 305, 306
Schell D.M. 161
Scheltema R.S. 237
Schmidt J. 328
Schmidt U. 256
Schwartzlose R. 129
Scotto di Carlo B. 120
Semina H.J. 87
Sendner H. 106
Shapovalov L. 325
Shelbourne J.E. 320, 351
Sheldon R.W. 90
Shindo S. 303
Shushkina E.A. 92, 95
Sieburth H. 139
Sievers H. 401
Sindermann C.J. 328
Skud B.E. 246
Slobodkin L.B. 107, 109, 113
Smayda T. 14, 180, 217
Smith E.V. 265
Smith F.E. 127, 402
Smith R.L. 45, 47, 146, 381
Sokal R.R. 372
Sokolova M.N. 234, 236
Sømme J. 124
Sorokin Y.I. 192, 231, 357
Sprague L. 328
Squires H.J. 244
Stanck E. 253
Steele J.H. 7, 14, 16, 82, 84, 101, 102, 106, 109, 111, 112, 113, 114, 151, 174, 190, 201, 214, 216, 222, 223, 232, 309, 310, 357, 389
Steemann-Nielsen E. 175, 176, 177, 181, 349

Stephens K. 232
Stephenson A. 74
Stephenson T.A. 74
Stewart B. 58
Stewart G.L. 84, 118, 348
Stewart R.W. 364
Stommel H. 29, 48, 49, 54, 55, 56, 57, 103, 174
Strickland J.D.H. 9, 33, 35, 93, 174, 177, 356
Stubbs A.R. 324
Sushchenya L.M. 95, 200, 201, 202, 211
Sutcliffe W.H. 356
Sutherland D.R. 272, 273
Svärdson G. 261
Sverdrup H.U. 25, 30, 35, 42, 49, 55, 83, 180
Swingle H.S. 265
Syrett P.J. 155
Szekielda K.H. 380

Taft A.C. 325
Takahashi M. 86, 88, 89, 143, 174, 196, 200, 202, 206, 211, 212, 214, 289, 390
Talbot J.W. 198, 320, 322
Talling J.F. 183
Taniguchi A. 94
Taylor C.C. 262, 276
Taylor F.J.R. 109
Templeman W. 329
Tennekes H. 38
Thompson C.W. 388
Thompson J.D. 45
Thompson J.Vaughan 11
Thompson W. 317
Thorne R.E. 391
Thorson G. 72, 73, 98, 193, 195, 196, 237, 238, 317
Tiews K. 304
Tigeguchi K. 15
Tomlinson P.K. 296
di Toro D. 14
Townsend C.H. 308
Tranter D.J. 188
Trevallion A. 266
Trout G.C. 324
Tucker D.G. 33
Tukey J.W. 366
Tungate D.S. 101
Turekian K.K. 27
Tyler J.E. 34, 35
Tytler P. 272

Uda M. 4
Urick R.J. 33
Ursin E. 257, 276, 335, 354

Vaccaro R.F. 149
Van Dyne G.M. 389
de Veen J.F. 323, 325
Venrick E.L. 83, 132
Veronis G. 56, 57
Vervoort V. 124
Vinogradov M.E. 118, 120, 121, 122, 123, 124, 130, 133, 389
Vinogradova N.G. 232
Vlymen W.J. 81, 95, 135
Vollenweider R.A. 174
Volterra V. 341
Von Liebig J. 12
Voorhis A.D. 56
Vrooman A.M. 328
Vúcetic T. 15, 16

Walford L.A. 256
Walker E.R. 26
Walker M.G. 11
Walsh J.J. 14, 85, 109, 110, 144, 231, 388, 389, 390, 392, 393, 394, 395, 397, 400, 401
Warren B.A. 56
Warren C.E. 267, 274, 275
Waterman T.H. 121
Waters T.F. 243
Watt K.E.F. 389
Watt W.D. 162
Watts D.C. 368
Webster F. 113, 232
Weichs D. 337
Weidemann H. 106
Wells J.W. 318
Weston D.E. 11
Wheeler P.A. 161
Whitledge R.F. 169, 170
Wiborg K.F. 123
Wickstead J.H. 346
Wiebe P.H. 83, 101
Wilder D.G. 244
Williams R. 118, 119
Williamson D.I. 195
Wilson D.F. 131
Wilson D.S. 90, 201, 390
Wimpenny R.S. 134
Winberg G.G. 195, 196, 199, 203, 204, 268, 273, 274
Wohlfarth G.W. 264
Wooster W.S. 51, 146, 398, 401
Worthington E.B. 126
Worthington L.V. 53, 381

Wright R. 57
Wyatt T. 350, 351, 352, 353
Wyrtki K. 53, 54, 131, 405
Wyshkwarzev D.I. 192

Yablonskaya E.A. 196, 216
Yaglom A.M. 38
Yentsch C.S. 177
Yonge C.M. 355
Yoshida K. 110

Zaika V.Ye. 92, 242, 243
Zaitsev Y.P. 69
Zeitschel B. 232
Zenkevitch L.A. 10, 232, 233, 317
Zeuthen E. 268
Zilstra J.J. 323
Zobell C. 139

Subject Index

Abyssal circulation 54–5
Abyssal water 52, 55
Accessory pigments 175
Acoustic techniques 11, 119
Adiabatic effect 28, 29
Advective flux 51
Aggregation on phytoplankton 100
Aggregation
  feeding 129
  fish 130
  vertical migrations 131–2
Algae 62
Algal mortality 16
Algal reproduction 12, 14, 16
Amino acids, uptake by phytoplankton 161
Ammonia 143, 150, 170
Anchovy
  Peru fishery 285, 306
  production 309
Animal colour 68, 69
Antarctic Bottom waters 53
Antarctic Circumpolar Current 50
Antarctic Intermediate Water 52
Appendicularians 62
Archimedes' principle 31
Assimilation efficiency 92, 95, 200, 202, 209, 211, 218
Attenuation lengths
  of electromagnetic radiation 32
  of light 32
  of sound 32

Bacteria 62, 139, 170
Baltic outflow 106
Bathypelagic zone 60
Beam attenuation coefficient 33–4
Benthic flora 74
Benthic distribution
  biomass in world ocean 232
  by feeding habit 234
  Indian Antarctic 232
  Pacific 232
  N. Pacific 232
Benthos 62, 226
  animals 228
  animal input to the benthos 231
  biomass distribution 232
  collecting methods in deep sea 236
  eutrophic 236
  epibenthic algae 228
  fish in the deep sea 231
  infauna 228
  life histories 237
  macrobenthos 228
  meiobenthos 228
  microbenthos 228
  oligotrophic 236
  oxygen uptake 231
  plants 226
Biology of fishes 319
  fixed spawning ground 325
  fixed spawning season in temperate seas 325
  genetic isolation 328
  haemoglobins 329
  Heincke's Law 320
  larval drift 321, 328
  larval drift of plaice 322, 328
  larval life 320
  Life history 319
  life span 320
  migration range 323
  migratory circuit 321
  Nursery ground life 320, 323
  population processes 329
  predation 330
  reproductive isolation 324, 328
  restricted spawning grounds 324, 325
  stocklets 329
  transferrins 329

Bioluminescence 69
Birds 62
Black Sea plankton 353
  change of trophic level in life cycle 354
  five trophic levels 354
  mixed food consumers 353
Body size 61, 65, 79, 80, 82, 86, 87, 88, 89, 189, 200, 207–8, 221–2, 90–2, 93, 94
Brunt-Väisälä 31, 40, 55

Cabbelling 30
California current 48, 134, 144
Capillary waves 24
Carbon budget 198–202
Carbon-14 99, 150, 175, 178
Carrying capacity 6, 8
Cell quota of nutrients 164–5
Chaetognaths 62
Challenger expedition 9, 26
Chlorophyll 11, 86, 87, 93
  microlayers 93
Circadian rhythms 126
Climatic change 6
Cobalt 26
Coelacanth 10
Colour of algae 67
Communities
  Danish benthic 317, 318
  demersal fish 318
  eel grass community 349
  Eltonian pyramid of numbers 318
  fishes in the pelagic community 319
  pelagic 318, 334
Compensation point 176–7
Competition 8, 6, 9, 19
Compressibility of sea water 31
Concentration equation 51
Condition factor 253
Conservation
  of mass 35
  of momentum 35
Convergence 44
Copepods 8, 14, 15
Copper 26
Coprophagy 121
Coral reefs 67, 74
Corioli's Force 36, 41, 42, 43, 45, 46
Costa Rica dome 168
Crustaceans 62
Cryptic coloration 68
Ctenophores 62
Currents
  Antarctic circumpolar 50
  California 48, 144, 134
  Canary 144
  Eastern boundary 48, 51, 146
  Equatorial 48, 94, 146
  Equatorial counter 48, 49, 50, 51
  Gulf Stream 24, 48, 50
  Kuroshio 4, 50
  Peru 144
  Somali 50

Death rates 4
Deep scattering layer 11, 32, 119, 128, 129, 131, 134
Denitrification 143
Density of sea water 24, 25, 28
Density dependent processes 6, 8, 18, 62
Detritus 226
  plant 226
  role in eel grass community 349
  role of particle size 350
Diatoms 65
Diffuse attenuation coefficient 33–4
Discovery Investigations 9
Dissolved organic matter 228
Divergences 10, 44, 45
Diversity Index 8
Doldrums 48, 49, 50
Dynamic topography 43

Eastern boundary current 48, 51, 146
Eastropac 10
Echinoids 67
Echo-sounding 11
Eddy transport 39, 40
Eddy viscosity 39, 40
Ekman layer 44
Ekman spiral 44
Ekman transport 44
El Niño 24, 55
Eltonian pyramid 9
Emigration 5
Encounter theory 16
Energy flow 7
Epifauna 10
Epipelagic zone 60
Equation of continuity 40–1
Equations of motion 35–6, 41
Equations of state 35
Equatorial Counter Current 49–51
Equatorial Current 48, 146
Eulerian motion 36
Euphotic zone 75, 79, 104
Euphausiids 11
Eutrophication 14, 144, 147, 157, 223
Excretion 13
  by zooplankton 169–71

Extinction coefficient of light 34, 35, 88, 176

Feeding 93, 94, 96
benthic predators 236
by burrowing 230
by scooping detritus 234
carnivores 226
deposit feeders 226, 228, 229, 236
filter feeders 226, 228, 234
Filter feeding 69
Fish 62
Fish larvae 11, 17, 62
Fish larvae in Southern Bight 351
behavioural changes 353
competition 351
food of fish larvae 351
food switching 352
plaice and sandeels 351
sizes of food 352
Fish population dynamics 287
acoustic surveys 292
analytical models 298
catch statistics 287
dependence of catch per effort on effort 297
dependence of recruitment on parent stock 300, 301
ecological efficiency 309, 311
egg surveys 292
fishing effort 293, 294
herring and cod 311
increase in North Sea gadoids in the sixties 310
interspecific relationships 303
management 312, 314
lobster 244
scallop 247
Schaefer model 295, 296
stock density 287, 293, 294
tagging 292
yield per recruit model 299, 300
Fish populations 329
Baltic cod 331
Boothbay harbour herring 331
Southern North Sea plaice 331
Fisheries
Atlanto Scandian herring 311
Antarctic blue whale 296
Californian sardine 303, 311, 313
catches in upwelling areas 308
Development 313
Eastern tropical yellowfin 306
expansion 285
Gulf of Thailand 303, 304
Iceland plaice 298
Industrial scale 283
North Sea 309
North Sea haddock 302
Pacific halibut 312
Peruvian anchovy 285, 306, 313
South African pilchard 313
territorial limits 283
world distribution 283
Fisheries Commissions 300
FAO Commissions 305
IATTC 305
ICES 300, 305
ICNAF 300, 305, 312
International Whaling Commission 312
Northeast Atlantic Fisheries Commission 312
North Pacific Fur Seal Commission 312
UN Conference on the Law of the Sea 313
Fisheries management 82
Fishing mortality 18
Food chains 341
bacterial and autotrophic production 307
Black Sea plankton 353
blurring of 'trophic level' 347
competition between plaice and sandeel 346, 350
feeding by Sagitta 346
food chain models 358
food webs 341
Gause's axiom 346
Gause's experiments 343
lengths of food chains 354
Limfjord 349
North Sea herring 344
number of links 343
organic carbon 356
predator/prey relation 341
Pütter's theory 356
pyramid of numbers 343
role of bacteria 356
Southern Bight fish larvae 350
top predators 355
trophic dynamic approach 341
trophic links 342
Food selection 191
Food webs 7, 9, 84, 94, 95, 96, 97
Fram 43
Friction 37
Fungi 62

Geostrophic flow 55
velocity 43

Geostrophy 41–3
Gigantism 67
Grazing 12, 13, 14–16, 76, 87, 88, 101, 103, 107, 109–12, 179–81
  encounter theory 14
  selectivity 90
Growth 6
  allometric 253
  changes following transplantation 261
  density effects 265
  effect of length of growing season 261
  effect of maturity 263
  Fishes 251
  geographical variation (cod, haddock) 258
  genetic factors 264
  hierarchical effects 264
  length 251
  predetermined level 278
  rate 251
  temperature effects 261
  temporal changes in rate 260
  variation between species 257
  weight 252
Growth efficiency 91, 92, 94, 95, 202, 209–11, 213, 214
Growth equations 254
  dependence K and L_∞ on temperature 262
  Ford–Walford plot 256
  Gompertz 257
  L_∞ and K 256
  Parker and Larkin 257
  relationship between K and L_∞ 258
  Robertson 256
  von Bertalanffy 254
Growth physiology 265
  basic metabolism 268, 278
  conversion efficiency 269, 271, 272
  dependence on food intake 265
  dependence of digestion rate on temperature 272
  dependence of maintenance ration on body weight 268, 269, 278
  energy needed for swimming 272
  food intake and growth efficiency 266
  gross and net efficiency 269
  maintenance ration 267
  metabolic growth models 273
  metabolic cost 278
  temperature on metabolism 271
  Van't Hoff rule 271
Gulf Stream 24, 42, 48, 50

Herring food chain 344
  Gause's axiom 346
  number of links 344
  private food supply 344
Hurricanes 41
Hydrostatic equation 36
Hypopneuston 68

Immigration 5
Inertial oscillations 45
  period 45, 46
Infauna 62
Internal Waves 83
International Indian Ocean Expedition 10
Ivlev curve 14
Ivlev equation 201
Ivlev's principle 80

Kinematic viscosity 37, 39
Kuroshio current 4, 48, 50

Ladder of migrations 122
Lagrangian motion 36
Langmuir circulation 105
Laminar flow 37
Larval life 72, 73
Law of the Minimum 12
Lengths of food chains 354
  coastal and oceanic chains 354
  commensalism 355
  filter feeders and deep sea fishes 355
  parasitism 355
  predator/prey ratio 354
  shortened food chains 355
  symbiosis 355
Level of no motion 42
Life cycles 71
Light 33–5
  absorption 33
  scattering 33
Light and photosynthesis 174–7
Littoral environment 73–74
Lobsters 244
  capture 244
  dependence of catches on sea temperature 246
  development of Canadian fishery 245
  eat sea urchins which eat kelp 247
  fishing pressure off eastern Canada 245
  population dynamics 244
  size limits 245
Logistic curve 3, 5, 6, 18
Lotka-Volterra equations 7, 113
Luminescence 2

Macroplankton 4, 62
Mammals 62
Mangrove swamps 228, 350
Mediterranean Water 52
Medusae 65
Meroplankton 237, 350
  immigration 324
  loss of larvae from population 237
  production 240
  pseudopopulations 237
  with little food reserves 237, 238
  with large food reserves 237, 238
Mesopelagic zone 60
Mesoplankton 62
Metabolic growth models 273
  Pütter 276
  Von Bertalanffy 276
  Winberg 273
Metamorphosis 73
Michaelis constant 151–3, 161
Michaelis–Menten equation 13, 15, 151
Microlayers of chlorophyll 93
Microplankton 62
Microneuston 129
Migration 18, 19
  seasonal 5, 79
Migration of fish 16, 17, 321
  Albacore 323
  dependence on current systems 324
  hake 323
  herring 323
  larval drift 321
  migratory circuit 321
  Pacific salmon 323
  plaice 323
  salmon 98
  sprats 323
  tuna 98
  whales 98
Migrations
  daily vertical 119, 122, 123
Mixing 103–4
Models 5, 12, 19, 83, 388
  agricultural 12
  Bahia California 392
  Baja California model 397
  Challenger 338
  characteristics of the three upwelling areas 393
  data lack 389
  fertilization 403
  lateral mixing 106
  length of food chain 402
  North West Africa 392, 400
  Peru model 392, 396
  philosophy 390
  phytoplankton 88, 112
  size fractionation 394
  specification 394
  stochastic 5
  upwelling systems 391
  use of partial difference equations 395
  variety 389
  zooplankton 112
Molecular dynamic viscosity 37
Momentum equations 45
Monod equation 150
Monsoon 50

Nanoplankton 62
Navier Stokes equations 37, 38
Nekton 61, 82
Newtons Second Law 35–6
Nitrate 11, 12, 26, 87, 88, 142, 156
Nitrite 11, 156
Nitrogen uptake 154–7
Nitrogen-15 technique 150, 151, 153–4, 156, 167
Nitrogen fixation 143, 170
Norpac 10
North Equatorial Current 94
North Atlantic Deep Water 53
North Pacific Central Gyre 169
  Intermediate Water 52
NORWESTLANT 10
Nursery grounds 17
Nutrient exhaustion 12
  limitation 164
  regeneration 12, 13, 94, 97, 169–72, 178–9, 221
Nutrients 16, 26, 27, 79, 83, 88, 141–72
  excretion 13
  and growth 163
  and light 154, 157, 161
  limitation 15
  luxury consumption 13
  and upwelling 45, 145

Oceanic circulation 47
  gyres 103
Ontogenetic migrations 122, 124
Oozes globigerine 234
  radiolarian 234
Organic matter detritus 97, 228
  in coastal water 232
  in Limfjord 349
  in Pacific 236
  in suspension 229
  reaching sea floor 231
  sinking 231
Overdispersion of phytoplankton 99
  of zooplankton 101

Overfishing 18
Oxygen 26
  minimum layer 54, 83, 131

Parental care 6
Patchiness 80, 83, 84, 97, 98–115, 134
  of chlorophyll 100–2, 111
  critical size 107, 108, 114
  and sampling 119
  of zooplankton 112
Pelagic community 334
  aggregation of filterers as they feed 337
  cohort grows through a series of predatory fields 338
  cruising speed and shoal size 336
  extent in world ocean 334
  filterers 336
  filterers and shoaling 336
  food chain 339
  lack of shoaling at night 337
  network of growth and death 338
  pelagic territory 335
  predator/prey ratio in length and weight 335
  predatory food chain 335, 343
  search theory and shoaling 337
  shoaling to avoid predation 337
  top predators 335
  visual volume of a predator 335
Pelagic environment 75
Pennatularians 67
Petersen's Limfjord food chain 348
  destruction of the eel grass community 399
  eel grass meadows 349
  role of microbenthos and meiobenthos 349
  role of organic detritus 349
Primary production 45
  dependence of macrofauna on 234
  dependence of world fish production on 305
  epibenthic algae 228
  proportion lost to benthos 231, 232, 233
  proportion lost by zooplankton faeces 232
  sea grasses 228
  seaweeds 228
pH and photosynthesis 101
Phaeopigments 87, 181
Pheromones 132
Phosphate 11, 12, 26, 76, 87, 88, 142, 162–3
Phosphorus 4, 14, 16, 82
  cell quota 13
  storage 12, 13
Photic zone 60
Photosynthesis 27, 79, 101, 174–5
Phycoerythrin 67
Phytoplankton blooms 109–10
Phytoplankton/zooplankton interrelations 93, 94, 95, 101, 112, 188–90
Plankton 61, 65, 82
  distributions and ocean currents 103
  Expedition 9
Pneuston 68
Pogonophora 10
Poisson distribution 99
Pollution 97
Population dynamics 18–19
Population processes 329
  competition 329, 331
  density dependent processes 329, 331, 332
  dependence of density processes on food 331
  dependence of predation on age 330
  magnitude of recruitment 329
  maximization biomass within a cohort 334
  mortality as function of age 330
  natural mortality 332, 338
  predatory fields 332, 336, 344
  private food supply 331, 340, 344
  stabilization mechanisms 329, 340
  use of predation 330
Potential density 28–9, 31, 52
  temperature 28, 29, 30, 52, 53
Predation
  avoidance of 132
Predator–prey relations 4, 7, 14, 93, 94, 101
Pressure of sea water 24
Production 85, 95, 196–8, 214
  agricultural model 12
  Allen curve 240, 242
  Anchovy 309
  Benthic 225
    in different regions 231
  Birkett's method of estimation 239
  elimination 239
  methods of estimating benthic 238
  North Sea 15–16
  P/B ratio 238, 242
  primary 27, 33, 75, 86, 95, 147–8, 173–85
  production cycle 13, 14, 72, 76
  seagrasses 228
  seaweeds 228
  secondary 95, 186–224, 216–20

Productivity index 184–5
Protozoans 62
Pycnocline and vertical migration 131

Q_{10} 92

Red tides 75, 107, 109
Recruitment 3, 6, 8, 17, 18
Reproduction 4, 6
Reproductive strategy 72, 82
Respiration rate 202, 209, 212
Reynolds stress 39, 43
Richardson number 40, 104
Rossby number 41
  waves 55
Rotifers 65
Russell cycle 3

Salinity 24, 25, 51
Sampling 367
  adaptive sampling 374
  aliasing 370
  areal sampling 374
  Autoanalyzer 386
  biases 365
  choices of sampling 362
  cost benefit 370
  cost of sampling and precision 372
  distribution of discrete samples 372
  Fourier series in sampling 367
  Fluorometry 384
  Moored arrays 381
  navigational error 365
  Nyquist sampling theorem 366
  power spectrum 370
  random sampling techniques 373
  remote sensing 380
  sampling population 363
  sampling problem 362
  shipboard sampling 381, 383
  stratified samples 373
  systematic samples 373
Scallops 247
  dependence of recruitment 247
  dependence of year classes on temperature 247
  mesh size control in Bay of Fundy 249
Sea water 24, 26
  compressibility 31
Secchi disc 34, 35
Secondary production 216–20
Sediments organic content 234
Sharks 6, 62
Shellfish 243
  lobster population dynamics 244
  scallop population dynamics 244
  value of catches in eastern Canada 244
Sigma – t 28
Silica limitation 13
Silicate 11, 142, 161–2
Sinking 179–80
Siphonophores 65
Size fractionation 220
  selection of food 210
SOFAR 32
Solar radiation 33
Somali Current 50
Sonar 11, 119
Sound
  attenuation 32
  propagation 31, 32
  refraction 31
  scattering 32
Snell's Law 31
Spawning grounds 17
Sperm whale 2
Squids 2, 62
Stability 5, 8, 77
  of ecosystems 19, 80, 84, 97
  of numbers 7
  of populations 114, 140
  stabilizing mechanisms 7
  of water 31, 86, 180
Submersibles 119
Succession 75, 76
Suctorial feeding 70
Superfluous feeding 202

Tagging experiments 18
Temperature 24, 25, 27, 31, 51
  and adiabatic pressure 28
  and metabolism 94
  microstructure 24, 27
  potential 28, 29, 30, 52, 53
Temperature/Salinity diagram 28, 29, 30, 52
Thaliaceans 62, 65
Thermal wind equations 42
Thermocline 27, 53, 54, 80
Thermohaline circulation 47
  processes 52
Tides 24, 36, 42
Top predators 355
  control of numbers 356
  fecundity 356
  metabolic efficiency 356
Total scattering coefficient 34
Trace elements 26
Trade winds 47

Transfer coefficients 8, 94, 95, 97
Transparency 34
Tree rings 3
Tunicates 65
Turbulence 4, 37–40, 51, 54, 55, 61, 62, 71, 76, 179, 180
  eddy diffusion coefficient 103–4, 108
  and nutrients 142, 144
  and plankton 103

Ultraplankton 62
Underwater acoustics 33
Upwelling 10, 15, 45, 47, 55, 79, 88, 105, 178, 179
  of abyssal water 54
  and nutrients 45, 145, 146
  Peruvian 85, 100, 101, 157, 167–9
  and phytoplankton 109, 110
  and seasonal migration 122

Vertical distribution 116
  life zones 116
Vertical migration 116–37
  and bioenergetics 135
  and biological significance 132–7
  control of 125
  and polarised light 127
  and pycnocline 131
  suppression of 137
Viscosity 103
Vorticity 46, 47, 49
  equation 46, 48, 55

Water masses 29, 30
Water types 29
Western boundary currents 48, 50, 51
Westerly winds 47
Wind driven circulation 44, 47
Wind stress 43, 51
World potential fish production 304
  demersal fishes 308
  herring-like fishes 308
  management 312
  potential harvest 306, 307
  from primary production 305, 309, 310

Yield from the sea 281
  management 312, 314
  small fish of the deep ocean 281

Zooplankton 83, 94

Index of Plants and Animals

Acanthaster 74
Acanthephyra 121
Acartia 76, 126, 210, 353
Acartia tonsa 90, 198, 205, 206, 217, 218, 219
Acipenser 257
Acmaea 74
Ammodytes 344
Ammodytes marinus 71, 319, 351
Amphitrite 230
Amphisolenia spinulosa 65
Amphora 156
Ampelisca 242
Anchovy 85
Archidoris pseudoargus 191
Arenicola 230
Aspidosiphon 70
Asterionella formosa 12, 13, 16
Ascophyllum 227
Asteroidea 193

Balaenoptera musculus 287
Bacillariophyceae 75
Balaeonophilus 71
Bentheuphausia 121
Biddulphia 351
Biddulphia sinensis 15
Boreomysis 121
Bosmina 89
Bregmaceros nectabanus 131

Calanoides acutus 123
Calanoides carinatus 122, 124
Calanus 15, 91, 101, 102, 113, 123, 135, 137, 200, 201, 206, 210, 338, 353
Calanus cristatus 91, 123, 124, 214, 216
Calanus finmarchicus 76, 99, 118, 122, 124, 130, 134, 135, 191, 204, 205, 216
Calanus helgolandicus (pacificus) (Brodskii) 91, 122, 126, 132, 135, 205, 206, 209, 214, 215, 217
Calanus hyperboreus 124, 192, 202, 207, 209, 211, 214
Calanus helgolandicus 399
Calanus finmarchicus 346
Calanus plumchrus 122, 123, 124, 206, 209, 214, 216
Calyptraeidae 69
Candacia 70, 124
Capitellid 70
Cardium 72
Centropages ponticus 215
Cerataulina 65
Ceratium belone 65
Cerataulina bergonii 66
Centengraulis mysticetus 402
Centropages 352, 353
Cetorhinus maximus 336
Chaetoceros contortum 63, 65
Chaetoceros debilis 85
Chaetoceros didymus 10
Chaetoceros lorenzianus 85
Chlamydomonas 205
Chlorella 174
Chrysophyceae 75
Chrysophytes 85
Ciliates 85
Clupea harengus 71, 130, 259, 310, 320, 344
Clupea sprattus 320
Coccolithus huxleyi 88, 89
Colossendeis 67
Collozoum 76
Conacon foliaceus 63
Coscinodiscus 65, 352, 351
Cryptomonads 85
Cyprinodon macularis 262

Daphnia 89

Daphnia magna 126, 127
Daphnia pulex 95
Diatoms 85
Diaptomus salinus 216
Dicerobatis eregoodoo 336
Dinophyceae 75
Ditylum brightwellii 88, 89, 207, 211
Dunaliella tertiolecta 164, 208

Echinoidea 193
Ecklonia 227
Engraulis mordax 169, 402
Engraulis ringens 169, 402
Enteromorpha 74
Ethmodiscus rex 65
Eucalanus bungii 123, 216
Euchaeta 120, 124
Eucopia 121
Eukrohnia 121
Euphausia eximia 136
Euphausia pacifica 127, 206, 208, 352, 354
Euphausia superba 79, 123, 313, 334, 352
Eurycope 70, 121
Evadne 215

Flagellates 85
Fragilariopsis antarctica 352
Fucaceae 74
Fucus 227

Gadus morhua 258, 287, 320, 351
Gaetanus 121
Gambusia holbrookii 25
Gaussia princeps 137
Gennadas 121
Geryon 67
Glaucas 68
Gonyaulax polyedra 208, 398, 399

Homarus Americanus 244
Holothuroidea 193
Hymenodora 123
Hyperamblyops 121

Ianthina 68

Labidesthes 258
Labidocera 135
Laminaria 227, 247
Laminariales 74
Labistes 251

Lepomis gibbosus 273
Lepomis macrochirus 271
Leucichthys artedi 261
Leurestes tenuis 72
Limanda limanda 309, 320, 351
Littorina 240
Littorina saxatilis 242
Lophius piscatorius 63

Macoma 242
Macrocylops albidus 92
Macrocystis 227
Macrostella gracilis 65
Mactra 239
Mactra stultorum 240
Maldanid 70
Meganyctiphanes norvegica 208
Melanogrammus aeglefinus 258, 287
Merlangius merlangus 310, 335
Merluccius productus 323
Metridia lucens 118
Metridia ochotensa 124
Microcanthodium setiferum 63
Modiolaria 72
Mola 355
Mugil auratus 69
Mya 242

Natica aldai 67
Navanax inermis 67
Nemertini 193
Neomysis americana 126
Nitzschia 65, 351
  closterium 65
Noctiluca miliaris 215

Oikopleura 2, 214, 320, 321, 351, 352, 353, 354
Oikopleura albicans 191
Oikopleura dioica 215, 351
Oithona 124, 209
Oithona minuta 215, 353, 354
Oithona similis 76, 215
Oncaea 120
Oncorhynchus gorbuscha 325
Oncorhynchus tshawytscha 257
Ophiothrix fragilis 65

Paracalanus parvus 215, 353
Parapeneus longipes 69
Pectinaria 240
Penilia 215
Phaeocystis 65, 66, 351

Phaeodactylum tricornutum 143, 162
Physalia 68, 70
Physeter macrocephalus 335
Pisaster ochraceus 74
Placopecten magellanicus 247
Planktoniella sol 65, 66
Platichthys flesus 351
Platymonas subcordiformis 208
Pleurobrachia pileus 353
Pleuromamma 124
Pleuromamma robusta 118, 120
Pleuroncodes planipes 134, 399, 402
Pleuronectas platessa 261, 287, 320, 350
Pogonophora 10
Pollachius virens 256
Polychaeta 193
Pontella 120
Porcellana platycheles 65
Porpita 68
Pseudocalanus 338, 344, 352, 353
Pseudocalanus elongatus 124, 215
Pseudocalanus minutus 90, 91, 96, 195, 196, 209
Pseudopleuronectes americanus 323
Pterois volitans 68

Rhincalanus 91, 124, 210
Rhincalanus gigas 123
Rhizosolenia 10, 65
Rhizosolenia hebetata 65
Rhizosolenia setigera 211
Rostraria 63

Sabellidae 69
Sagitta elegans 82, 346
Sagitta setosa 82, 215
Sagitta spp. 124
Salmo gairdneri 257, 325
Salmo salar 263
Salmo trutta 261, 263
Salpingella glockentogeri 65
Sapphirina 69
Sardinella eba 402
Sardinops caerulea 303
Sardinops ocellata 402
Sargassum 68, 70
Scolecithrix 124
Scrobicularia plana 230
Sergestes 63
Serpulidae 69
Siddharta 23
Skeletonema costatum 143, 156, 162, 165, 166, 208
Solea solea 287
Spinocalanus 120
Spiratella helicina 130
Storthyngura 67
Strongylocentrotus 240, 247

Tarletonbeania crenularis 69
Temora 351
Temoropia 120
Thalassiosira fluviatilis 163, 208, 211
Thalassiosira gravida 65, 66, 102
Thalassiosira pseudonana 161, 163
Thalia democratica 205, 207
Thynnus thynnus 257
Thysanoëssa sp. 208
Tilapia 261, 264
Tintinnids 85
Trichodesmium 65, 75, 170
Trichodesmium erythaeum 66
Tridacna 355
Turritellidae 69

Undinula 124

Velella 68
Venus striatula 62

Zostera 349

Geographical Index

Adriatic 120
Aegean Sea 156
Africa 45, 109, 123, 124
Alaska 147, 161
Amur River 265
Antarctic 27, 52, 53, 55, 57, 88, 117, 123, 134
Antarctica 147
Arabian Sea 131
Aral Sea 216
Arctic 124
Athens 155
Atlantic 32, 48, 49, 52, 53, 56, 60, 106, 122, 124, 125, 131, 146, 170, 215
Australia 50, 146

Baltic ix, 4, 106
Ballantrae Bank 324
Barents Sea 214, 216
Bay of Fundy 248, 249
Bear Island 4
Bedford Basin 234
Benguela 50
Bering Sea 94
Bermuda 169, 170
Black Sea 130, 131, 135, 206, 211, 214, 215, 216, 217
Brazil 50
British Columbian waters 209
Britain 73

California 45, 117, 118, 122, 123, 131, 132, 160, 205
Canary Islands 50, 128
Cape Roxo 318
Caribbean 32, 131
Carioca Trench 131
Chesapeake Bay 217, 218
Clyde Sea Area 106, 107, 110, 111
Connecticut 246
Costa Rica Dome 168, 170

Denmark Strait 53

Ellerslie Island 234
English Channel 4, 117, 216
Esbjerg 323

Faroe Islands 16
Firth of Clyde 137
Fladen Ground 234

Galapagos Islands 32
Gaspé Passage 234
Georges Bank 206, 214, 216, 218, 228, 249
Georgia 228
Germany 18
Ghana 122
Great Barrier Reef 2
Greece 155
Gulf of Mexico 233
Gulf of Maine 180
Gulf of Panama 214, 217
Gulf of St Lawrence 109
Gulf of Thailand 303, 304

Holland 18

Iceland 16
Iceland–Scotland Ridge 53
'India' Weather Ship 119
Indian Ocean 50, 60, 188
Ionian Sea 215

Kiel Bay 13
Kuroshio 4, 48, 50

La Jolla 120, 217
Lake Erie 14
Lake Mendota ix, 1, 7
Lake Tanganyika 2
Lake Windermere 12, 13, 16
Loch Nevis 234
Loch Striven 195
Long Island Sound 180, 217, 234, 240, 242

Magdalen Shallows 234
Maine 247
Maas 323
Massachusetts 246
Mediterranean 52, 106, 146, 166, 168, 171
Monona (Wisconsin, U.S.A.) 1
Morocco 130

New Brunswick 246
New England 246
New York 246
New Zealand 73
North Atlantic 17, 19, 53, 54, 55, 114, 124, 134
North Sea 14, 15, 16, 79, 98, 99, 101, 102, 112, 113, 114, 130, 180, 197, 216
North Equatorial Current 94
Norway 3
Norwegian Deeps 105
Norwegian Sea 27, 53, 98
Nova Scotia 100, 247

Oregon 45, 131

Pacific 17, 32, 48, 60, 83, 84, 87, 88, 97, 100, 121, 122, 123, 124, 134, 146, 147, 168, 169, 170, 171, 180, 216, 312
Patuxent River 198, 205, 218
Peru 45, 85, 97, 101, 109, 122, 131, 144, 146, 155, 157, 167, 168, 169, 170, 171, 172
Plymouth 14, 82
Pribilov Islands 2
Prince Edward Island 246

Quebec 246

Rhine 323
River Blackwater 324
Ross Sea 53

Sandettié Bank 324
San Diego 126
San Francisco Bay 3
Sargasso Sea 65, 88, 97, 98, 166, 169, 170, 231
Saronikos Gulf 156
St Margaret's Bay 232, 234, 240
Sevastopol Bay 214, 215
South America 146
Southern Ocean 79
South Georgia 123
Spitzbergen 123
Straits of Dover 310
Straits of Gibraltar 52, 134

Tenerife 126
Texel Island 323
Tromsø Sound 123

Vestfjord 3

Waddensea 323
Wash 323
Washington 74
Weddell Sea 52, 53